VOLUME FIVE HUNDRED AND SEVENTY NINE

METHODS IN ENZYMOLOGY

The Resolution Revolution: Recent Advances In cryoEM

METHODS IN ENZYMOLOGY

VOLUME FIVE HUNDRED AND SEVENTY NINE

METHODS IN ENZYMOLOGY

The Resolution Revolution: Recent Advances In cryoEM

Edited by

R.A. CROWTHER
MRC Laboratory of Molecular Biology
Cambridge, United Kingdom

AMSTERDAM • BOSTON • HEIDELBERG • LONDON
NEW YORK • OXFORD • PARIS • SAN DIEGO
SAN FRANCISCO • SINGAPORE • SYDNEY • TOKYO
Academic Press is an imprint of Elsevier

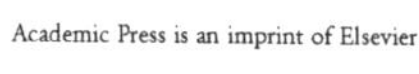

Academic Press is an imprint of Elsevier
50 Hampshire Street, 5th Floor, Cambridge, MA 02139, United States
525 B Street, Suite 1800, San Diego, CA 92101–4495, United States
The Boulevard, Langford Lane, Kidlington, Oxford OX5 1GB, United Kingdom
125 London Wall, London, EC2Y 5AS, United Kingdom

First edition 2016

ISBN: 978-0-12-805382-9
ISSN: 0076-6879

For information on all Academic Press publications
visit our website at https://www.elsevier.com/

Publisher: Zoe Kruze
Acquisition Editor: Zoe Kruze
Editorial Project Manager: Helene Kabes
Production Project Manager: Magesh Kumar Mahalingam
Cover Designer: Maria Ines Cruz

Typeset by SPi Global, India

CONTENTS

CONTRIBUTORS

J.A.G. Briggs
Structural and Computational Biology Unit, European Molecular Biology Laboratory, Heidelberg, Germany

B. Carragher
Simons Electron Microscopy Center, New York Structural Biology Center, The National Resource for Automated Molecular Microscopy; Columbia University, New York, NY, United States

A. Cheng
Simons Electron Microscopy Center, New York Structural Biology Center, The National Resource for Automated Molecular Microscopy, New York, NY, United States

W. Chiu
National Center for Macromolecular Imaging, Baylor College of Medicine, Houston, TX, United States

V.P. Dandey
Simons Electron Microscopy Center, New York Structural Biology Center, The National Resource for Automated Molecular Microscopy, New York, NY, United States

F. DiMaio
University of Washington; Institute for Protein Design, University of Washington, Seattle, WA, United States

A.R. Faruqi
MRC Laboratory of Molecular Biology, Cambridge, United Kingdom

S.A. Fromm
EMBL—European Molecular Biology Laboratory, Structural and Computational Biology Unit, Heidelberg, Germany

R.M. Glaeser
Lawrence Berkeley National Laboratory, University of California, Berkeley, CA, United States

T. Gonen
Janelia Research Campus, Howard Hughes Medical Institute, Ashburn, VA, United States

N. Grigorieff
Janelia Research Campus, Howard Hughes Medical Institute, Ashburn, VA, United States

R. Henderson
MRC Laboratory of Molecular Biology, Cambridge, United Kingdom

C.L. Lawson
Research Collaboratory for Structural Bioinformatics, Center for Integrative Proteomics Research, Rutgers, The State University of New Jersey, Piscataway, NJ, United States

S.J. Ludtke
National Center for Macromolecular Imaging, Baylor College of Medicine, Houston, TX, United States

G. McMullan
MRC Laboratory of Molecular Biology, Cambridge, United Kingdom

G.N. Murshudov
MRC Laboratory of Molecular Biology, Cambridge, United Kingdom

L.A. Passmore
MRC Laboratory of Molecular Biology, Cambridge, United Kingdom

A. Patwardhan
Protein Data Bank in Europe, European Molecular Biology Laboratory, European Bioinformatics Institute, Wellcome Genome Campus, Hinxton, United Kingdom

C.S. Potter
Simons Electron Microscopy Center, New York Structural Biology Center, The National Resource for Automated Molecular Microscopy; Columbia University, New York, NY, United States

Z.A. Ripstein
Molecular Structure and Function Program, The Hospital for Sick Children; University of Toronto, Toronto, ON, Canada

J.A. Rodriguez
UCLA-DOE Institute, University of California, Los Angeles, CA, United States

P.B. Rosenthal
Francis Crick Institute, Mill Hill Laboratory, London, United Kingdom

J.L. Rubinstein
Molecular Structure and Function Program, The Hospital for Sick Children; University of Toronto, Toronto, ON, Canada

C.J. Russo
MRC Laboratory of Molecular Biology, Cambridge, United Kingdom

C. Sachse
EMBL—European Molecular Biology Laboratory, Structural and Computational Biology Unit, Heidelberg, Germany

S.H.W. Scheres
MRC Laboratory of Molecular Biology, Francis Crick Avenue, Cambridge Biomedical Campus, Cambridge, United Kingdom

Y.Z. Tan
Simons Electron Microscopy Center, New York Structural Biology Center, The National Resource for Automated Molecular Microscopy; Columbia University, New York, NY, United States

W. Wan
Structural and Computational Biology Unit, European Molecular Biology Laboratory, Heidelberg, Germany

PREFACE

Every so often a field will make huge progress in a relatively short space of time, generating much excitement and insight as a consequence. Such is now the case in electron microscopy of biological molecules, where a "resolution revolution" has been declared (Kühlbrandt, 2014) and recent developments have led to the determination of the atomic structures of many previously intractable complexes. This volume describes the technical advances in instrumentation and software that have come together to create a situation where spectacular progress has been possible.

For a full understanding of the function and interactions of macromolecules and subcellular complexes, it is essential to visualize their structure. Hitherto, atomic structures have been accessible only by X-ray crystallography, provided suitable crystals can be made, or by NMR, which can provide valuable information about dynamics and atomic structure for smaller complexes. Both approaches require considerable amounts of material, which must be pure and homogeneous. Until recently, electron microscopy was limited to lower resolution, except in the special cases of regular two-dimensional or helical arrays. However, new technical advances in specimen preparation and electron detectors and improvements in computer image processing have come together in the last 3 to 4 years to create a situation where atomic structures can now be determined for particles of molecular mass greater than about 150 kDa. Basic aspects of electron cryo-microscopy (cryo-EM) were covered thoroughly and in great detail by *Methods in Enzymology* in 2010, in Volumes 481–483 edited by Grant Jensen. This volume therefore concentrates on developments since then, particularly those dealing with dispersed suspensions of molecules, now known as single-particle cryo-EM. The focus is on methods rather than specific biological results, as the latter are arriving at a rate that would make a review out of date by the time it was published. Such is the current excitement and activity in the field that new structures are published every week.

Before introducing this volume, it is worth giving a brief summary of some of the key historical developments in the field prior to the recent advances. The start of computing three-dimensional maps from electron micrographs can be clearly dated to 1968 with publication by DeRosier and Klug of a three-dimensional reconstruction of the tail of bacteriophage T4 (DeRosier & Klug, 1968). Although applied to a special case with helical symmetry, where a single view gives sufficient information to make a map at least to limited resolution, they laid out general principles for any kind of specimen, based on the use of Fourier transforms. The next development was to tackle icosahedral viruses, where data had to be combined from several particles in different orientations (Crowther, Amos, Finch, DeRosier, & Klug, 1970). This involved finding view parameters by so-called common lines procedures, which also served to fix the relative magnification, contrast, and handedness for combining the different images. These maps represented the first single-particle reconstructions. The particle symmetry, helical or icosahedral, was used both to aid the computer processing procedures on the limited computing power then available and to average over multiple copies of the asymmetric unit to generate a map with improved signal-to-noise ratio.

Analysis of images of isolated nonsymmetric individual particles is more difficult, because of the small signal and large noise in each image of a single particle. An important first step was the introduction by van Heel and Frank of statistical methods of image classification, which allowed noisy images of similar views of a particle to be grouped together in classes and then averaged (Frank & van Heel, 1982; van Heel & Frank, 1981). They then developed methods to determine the relative orientation of the views in the different classes, using random conical tilt (Radermacher, Wagenknecht, Verschoor, & Frank, 1986) or angular reconstitution (van Heel, 1987), to allow a three-dimensional map to be calculated. Once an initial model had been made, the view parameters could be refined iteratively and a series of improved maps calculated. These approaches for processing images of particles with low or no symmetry were encoded in the comprehensive software packages SPIDER (Frank, Shimkin, & Dowse, 1981) and IMAGIC (van Heel & Keegstra, 1981), that were developed further over the years and were used by many investigators.

The early work with negatively stained specimens gave only molecular envelopes, so to obtain more detailed information about the biological material rather than the stain, it was necessary to find alternative ways of preparing grids. One approach was taken by Unwin and Henderson, who used glucose to embed thin crystals (Unwin & Henderson, 1975). However, such specimens without heavy metal stain are extremely radiation sensitive and so have to be imaged with low electron dose. The resulting images are therefore very noisy, being limited by the number of electrons recorded in each image element. To extract the biological signal from the noise, it was therefore necessary to average a large number of images of the molecule, a procedure greatly expedited by having a two-dimensional crystal. This approach led to the 7 Å structure of bacteriorhodopsin (Henderson & Unwin, 1975), giving the first direct picture of a membrane protein. Later using improved images and computational methods, the atomic structure of bacteriorhodopsin was solved (Henderson et al., 1990) and refined (Grigorieff, Ceska, Downing, Baldwin, & Henderson, 1996) to produce a model of the quality previously produced only by X-ray crystallography. The power of this approach for solving well-ordered two-dimensional crystals was emphasized by the production of a 3.4 Å resolution map of the light-harvesting complex (Kühlbrandt, Wang, & Fujiyoshi, 1994), central to photosynthesis in green plants; of a 3.7 Å map of tubulin (Nogales, Wolf, & Downing, 1998), which gave the first atomic model of this key cellular component; and of a 1.9 Å map of aquaporin (Gonen et al., 2005), which showed details of the pore and of interactions between protein and lipids.

However, the majority of specimens cannot be persuaded to form suitable two-dimensional crystals and so a method of preserving dispersed particles was needed. Glucose embedding does not give satisfactory contrast for single molecules, as the sugar and protein are too closely matched in density. Taylor and Glaeser had earlier shown that frozen thin crystals, such as catalase, could give electron diffraction patterns out to spacings of 3.4 Å, demonstrating that the molecules were well preserved under these conditions (Taylor & Glaeser, 1974). Dubochet and colleagues then developed plunge freezing as a way of preserving particulate specimens (Adrian, Dubochet, Lepault, & McDowell, 1984; Dubochet et al., 1988; Dubochet, Lepault, Freeman, Berriman, & Homo, 1982). The freezing happens so rapidly that the dispersed particles are trapped in a thin layer of amorphous or vitreous ice, which then has to be imaged at liquid nitrogen temperatures to prevent crystallization of the water. As with glucose-embedded material, the frozen samples are very radiation sensitive and the images therefore very noisy. Image averaging is essential to extract detailed information.

For helical tubes and icosahedral viruses, this could be readily achieved by exploiting the inbuilt symmetry, as had been done earlier for stained specimens. The structure of the acetylcholine receptor was determined to 4 Å from helical tubes (Miyazawa, Fujiyoshi, & Unwin, 2003) and an atomic model refined to show the receptor in its closed configuration (Unwin, 2005). An atomic model of the bacterial flagellar filament revealed details of the intricate molecular packing and hydrophobic interactions that stabilize the structure (Yonekura, Maki-Yonekura, & Namba, 2003). For icosahedral viruses, a 7.4 Å map of hepatitis B virus core protein gave a numbered chain trace for the first time from single-particle images (Böttcher, Wynne, & Crowther, 1997). Further progress led to a 4.5 Å map of epsilon15 phage, which showed the backbone structure of the capsid protein (Jiang et al., 2008); ~4 Å maps of the rotavirus double-layer particle (Zhang et al., 2008) and of the triple-layer particle (Chen et al., 2009) which showed interactions important for assembly and uncoating; and a 3.3 Å map of a subviral particle of aquareovirus (Zhang, Jin, Fang, Hui, & Zhou, 2010). The high molecular mass and symmetry of these particles aided the image processing procedures and substantially increased the number of available views of the asymmetric unit. The results established the viability of cryo-EM to produce maps from images of symmetrical single particles of a quality equivalent to those obtained by X-ray crystallography. The map of the nonsymmetrical 70S ribosome at 11.5 Å resolution (Gabashvili et al., 2000) set the stage for the advances that led to the atomic structures of much smaller particles, such as TRPV1 (Liao, Cao, Julius, & Cheng, 2013), gamma-secretase (Bai et al., 2015), and beta-galactosidase (Bartesaghi et al., 2015). Some of the technical developments underlying these advances and causing the current excitement in the field are described in this volume.

Probably the most significant development underpinning the advances has been the production of efficient direct electron detectors. The key to high-quality images is to add as little noise as possible, particularly at high resolution, to the biological signal while recording the image, a property measured by the detective quantum efficiency (DQE) of the detector. Although photographic film is an adequate recording medium, its DQE at image spacings corresponding to atomic detail is relatively poor. Presently available direct electron detectors have a much superior DQE and should eventually reach 100%. The development and properties of direct detectors are described by Henderson and colleagues in Chapter 1.

Besides an improved DQE, electronic detectors give the possibility of time-slicing the recording of an image to allow particle movements to be analyzed, tracked, and corrected. This has allowed a more insightful analysis of the behavior of the frozen specimen when exposed to the electron beam, in terms of physical movement, charging, and radiation damage, as described in Chapter 2 by Glaeser. New kinds of specimen support are reducing beam-induced specimen movement and give the promise of more sophisticated ways of making particulate specimens adhere to the grid material, as shown in Chapter 3 by Passmore and Russo. The new detectors have also opened new opportunities for automated data collection and Carragher and colleagues give an overview what is now possible in Chapter 4. The time-sliced recordings are called movies and their processing to correct for movement of the particles within the image during recording is described by Ripstein and Rubinstein in Chapter 5.

The other major area where there has been a significant advance since 2010 has been in the making of three-dimensional maps from the recorded images. In particular the use of appropriate statistical methods for dealing with signal and noise, involving the use of maximum likelihood approaches as originally suggested by Sigworth (1998), has proved very fruitful. The three main programming packages that have been used for the majority of the recent high-resolution studies, RELION, EMAN2.1, and FREALIGN, are described in Chapters 6–8 by Scheres, Ludtke, and Grigorieff, respectively. It is notable that these packages are increasingly able to sort and analyze heterogeneous populations of particles, whether arising from partial occupancy of components or flexibility within a complex (Scheres et al., 2007). Such analysis gives the promise in appropriate circumstances of determining the structures of different functional states of a complex from images of a single biological preparation. This is an added strength of cryo-EM compared with crystallography, although achieving good biological preparations should always be a priority.

It is important that a reconstructed map can be trusted to be a true representation of the underlying biological structure. Problems potentially arise from the over-fitting of noise in the images, leading to erroneous features in the map. Methods for avoiding such over-fitting and for assessing the degree of reliable detail in the map are discussed by Rosenthal in Chapter 9. Once a map has been validated and the resolution is sufficiently high, an atomic model can be built as described by DiMaio and Chiu in Chapter 10 and refined as described by Murshudov in Chapter 11. There is a long history of atomic model building and refinement in X-ray protein crystallography, so cryo-EM has much to learn from that discipline but there is an important difference, as experimentally measured phases are available in cryo-EM.

Many important biological specimens cannot be prepared as single particles, as, for example, long helical arrays or subcellular structures studied in situ by tomography. Although not strictly "single particles," such specimens are often best analyzed by methods closely related to single-particle methods. We have therefore included Chapter 12 by Fromm and Sachse on segmented processing of helical arrays (Beroukhim & Unwin, 1997; Egelman, 2000) and Chapter 13 by Wan and Briggs on subtomogram averaging, both of which are leading to the extraction of higher resolution features than can be obtained from simple helical processing or direct inspection of the tomograms. A new technique using electron diffraction for studying three-dimensional microcrystals too small for X-ray diffraction is presented in Chapter 14 by Rodriguez and Gonen.

Finally, it is crucially important that all the data and maps generated by the approaches described earlier are adequately stored and curated, so as to be readily available to the widest possible audience. In Chapter 15, Patwardhan and Lawson describe the current state of the relevant databases.

I am grateful to all my colleagues who have taken the time and trouble to contribute to what I hope will prove a useful reference volume for the rapidly expanding field of electron cryo-microscopy and I look forward to seeing in the literature ever more impressive examples of the biological understanding that ensues.

R.A. Crowther
MRC Laboratory of Molecular Biology, Cambridge, United Kingdom

REFERENCES

Adrian, M., Dubochet, J., Lepault, J., & McDowell, A. W. (1984). Cryo-electron microscopy of viruses. *Nature*, *308*, 32–36.

Bai, X. C., Yan, C. Y., Yang, G. H., Lu, P. L., Ma, D., Sun, L. F., et al. (2015). An atomic structure of human gamma-secretase. *Nature*, *525*, 212–217.

Bartesaghi, A., Merk, A., Banerjee, S., Matthies, D., Wu, X., Milne, J. L. S., et al. (2015). 2.2 Å resolution cryo-EM structure of beta-galactosidase in complex with a cell-permeant inhibitor. *Science*, *348*, 1147–1151.

Beroukhim, R., & Unwin, N. (1997). Distortion correction of tubular crystals: Improvements in acetylcholine receptor structure. *Ultramicroscopy*, *70*, 57–81.

Böttcher, B., Wynne, S. A., & Crowther, R. A. (1997). Determination of the fold of the core protein of hepatitis B virus by electron cryomicroscopy. *Nature*, *386*, 88–91.

Chen, J. Z., Settembre, E. C., Aoki, S. T., Zhang, X., Bellamy, A. R., Dormitzer, P. R., et al. (2009). Molecular interactions in rotavirus assembly and uncoating seen by high resolution cryo-EM. *Proceedings of the National Academy of Sciences of the United States of America*, *106*, 10644–10648.

Crowther, R. A., Amos, L. A., Finch, J. T., DeRosier, D. J., & Klug, A. (1970). Three dimensional reconstructions of spherical viruses by Fourier synthesis from electron micrographs. *Nature*, *226*, 421–425.

DeRosier, D. J., & Klug, A. (1968). Reconstruction of three dimensional structures from electron micrographs. *Nature*, *217*, 130–134.

Dubochet, J., Adrian, M., Chang, J.-J., Homo, J.-C., Lepault, J., McDowell, A. W., et al. (1988). Cryo-electron microscopy of vitrified specimens. *Quarterly Reviews of Biophysics*, *21*, 129–228.

Dubochet, J., Lepault, J., Freeman, R., Berriman, J. A., & Homo, H. C. (1982). Electron microscopy of frozen water and aqueous suspensions. *Journal of Microscopy*, *128*, 219–237.

Egelman, E. H. (2000). A robust algorithm for the reconstruction of helical filaments using single particle methods. *Ultramicroscopy*, *85*, 225–234.

Frank, J., Shimkin, B., & Dowse, H. (1981). SPIDER—A modular software system for electron image processing. *Ultramicroscopy, 6*, 343–358.
Frank, J., & van Heel, M. (1982). Correspondence-analysis of aligned images of biological particles. *Journal of Molecular Biology, 161*, 134–137.
Gabashvili, I. S., Agrawal, R. K., Spahn, C. M. T., Grassucci, R. A., Svergun, D. I., Frank, J., et al. (2000). Solution structure of the E. coli 70S ribosome at 11 Å resolution. *Cell, 100*, 537–549.
Gonen, T., Cheng, Y., Sliz, P., Hioaki, Y., Fujiyoshi, Y., Harrison, S. C., et al. (2005). Lipid-protein interactions in double-layered two-dimensional AQP0 crystals. *Nature, 438*, 633–638.
Grigorieff, N., Ceska, T. A., Downing, K. H., Baldwin, J. M., & Henderson, R. (1996). Electron-crystallographic refinement of the structure of bacteriorhodopsin. *Journal of Molecular Biology, 259*, 393–421.
Henderson, R., Baldwin, J. M., Ceska, T. A., Zemlin, F., Beckmann, E., & Downing, K. H. (1990). Model for the structure of bacteriorhodopsin based on high resolution electron microscopy. *Journal of Molecular Biology, 213*, 899–929.
Henderson, R., & Unwin, P. N. T. (1975). Three-dimensional model of purple membrane obtained by electron microscopy. *Nature, 257*, 28–32.
Jiang, W., Baker, M. L., Jakana, J., Weigele, P. R., King, J., & Chiu, W. (2008). Backbone structure of the infectious epsilon15 virus revealed by electron microscopy. *Nature, 451*, 1130–1134.
Kühlbrandt, W. (2014). The resolution revolution. *Science, 343*, 1443–1444.
Kühlbrandt, W., Wang, D. N., & Fujiyoshi, Y. (1994). Atomic model of plant light-harvesting complex by electron crystallography. *Nature, 367*, 614–621.
Liao, M., Cao, E., Julius, D., & Cheng, Y. (2013). Structure of the TRPV1 channel determined by electron cryo-microscopy. *Nature, 504*, 107–112.
Miyazawa, A., Fujiyoshi, Y., & Unwin, N. (2003). Structure and the gating mechanism of the nicotinic acetylcholine receptor at 4 Å resolution. *Nature, 423*, 949–955.
Nogales, E., Wolf, S. G., & Downing, K. H. (1998). Structure of the αβ tubulin dimer by electron crystallography. *Nature, 391*, 199–203.
Radermacher, M., Wagenknecht, T., Verschoor, A., & Frank, J. (1986). Three-dimensional reconstruction from a single exposure, random conical tilt series applied to the 50S ribosomal subunit of Escherichia coli. *Journal of Microscopy, 146*, 113–136.
Scheres, S. H. W., Gao, H., Valle, M., Herman, G. T., Eggermont, P. P. B., Frank, J., et al. (2007). Disentangling conformational states of macromolecules in 3D-EM through likelihood optimization. *Nature Methods, 4*, 27–29.
Sigworth, F. J. (1998). A maximum-likelihood approach to single-particle image refinement. *Journal of Structural Biology, 122*, 328–339.
Taylor, K. A., & Glaeser, R. M. (1974). Electron diffraction of frozen, hydrated protein crystals. *Science, 186*, 1036–1037.
Unwin, N. (2005). Refined structure of the nicotinic acetylcholine receptor at 4 Å resolution. *Journal of Molecular Biology, 346*, 967–989.
Unwin, P. N. T., & Henderson, R. (1975). Molecular structure determination by electron microscopy of unstained crystalline specimens. *Journal of Molecular Biology, 94*, 425–440.
van Heel, M. (1987). Angular reconstitution: A posteriori assignment of projection directions for 3D reconstruction. *Ultramicroscopy, 21*, 111–123.
van Heel, M., & Frank, J. (1981). Use of multivariate statistics in analysing the images of biological macromolecules. *Ultramicroscopy, 6*, 187–194.
van Heel, M., & Keegstra, W. (1981). IMAGIC: A fast, flexible and friendly image analysis software system. *Ultramicroscopy, 7*, 113–130.

Yonekura, K., Maki-Yonekura, S., & Namba, K. (2003). Complete atomic model of the bacterial flagellar filament by electron cryomicroscopy. *Nature*, *424*, 643–650.

Zhang, X., Jin, L., Fang, Q., Hui, W. H., & Zhou, Z. H. (2010). 3.3 Å cryo-EM structure of a non-enveloped virus reveals a priming mechanism for cell entry. *Cell*, *141*, 472–482.

Zhang, X., Settembre, E., Xu, C., Dormitzer, P. R., Bellamy, R., Harrison, S. C., et al. (2008). Near-atomic resolution using electron cryomicroscopy and single-particle reconstruction. *Proceedings of the National Academy of Sciences of the United States of America*, *105*, 1867–1872.

CHAPTER ONE

Direct Electron Detectors

G. McMullan, A.R. Faruqi, R. Henderson[1]
MRC Laboratory of Molecular Biology, Cambridge, United Kingdom
[1]Corresponding author: e-mail address: rh15@mrc-lmb.cam.ac.uk

Contents

Abstract

Direct electron detectors have played a key role in the recent increase in the power of single-particle electron cryomicroscopy (cryoEM). In this chapter, we summarize the background to these recent developments, give a practical guide to their optimal use, and discuss future directions.

1. INTRODUCTION

Electron cryomicroscopy (cryoEM) has experienced a surge in its capability in recent years, due to improved microscopes, better detectors, and better software. The role of detectors has arguably been the central factor because it allowed the full benefit of many electron-optical improvements over the last decade to be exploited and has driven the development of new software to deal with the increased information content of the images.

Three companies produce the currently available direct electron detectors, Gatan, FEI, and Direct Electron. These are sometimes called Direct Detection Devices (DDD). In each case, the detectors produce images from 300 keV electrons that are significantly better than obtained with film (McMullan, Faruqi, Clare, & Henderson, 2014). All three products are based on similar sensor technology in which electrons directly impinge on a lightly doped silicon epilayer supported on a more highly doped silicon substrate,

Methods in Enzymology, Volume 579
ISSN 0076-6879
http://dx.doi.org/10.1016/bs.mie.2016.05.056

with each frame of the exposure being read out continuously in rolling-shutter mode as a "movie."

We present a brief history of how we reached this point, a methods section on the comparative merits of the new detectors and strategies for data collection, and conclude with a glimpse into the future to anticipate further developments.

2. PAST

Electronic image sensors developed over many years. The two technologies that have been most important in electron microscopy (EM) are charge-coupled devices (CCDs) and monolithic active pixel sensors (MAPSs) fabricated with standard complementary metal oxide semiconductor (CMOS) technology. Boyle and Smith from Bell Laboratories received the 2009 Nobel Prize in Physics for inventing the CCD sensor in 1969, which has had enormous success in light imaging. The name active pixel sensor (APS) was coined in 1985 by Nakamura and generalized by Fossum (1993) to describe any sensor with at least one active transistor within each pixel. In practice, at least two more transistors are required for row selection and reset. Fossum also pointed out that CMOS APSs allowed bigger arrays, faster readout, reduced noise, and reduced sensitivity to radiation damage compared with CCDs, and indeed CMOS sensors have eclipsed CCDs in most current applications where these advantages were critical. The onward march of Moore's law has reduced the size of lithography features in sensor manufacture. This has underpinned the evolution of better electronic image sensors and will continue to drive future developments.

Two parameters important in specifying the performance of detectors are Detective Quantum Efficiency (DQE) and Modulation Transfer Function (MTF). DQE, defined by

$$\mathrm{DQE} = (S/N)^2{}_{\mathrm{OUT}} / (S/N)^2{}_{\mathrm{IN}}$$

shows how much the noise or the physics of the mechanism of signal conversion in the detector degrades the original signal in the image, in terms of the signal-to-noise (S/N) ratio in the output compared with that in the input. DQE can be plotted as a function of spatial frequency and a DQE of unity implies a perfect detector adding no noise. For a pixelated detector, the pixel spacing fixes the maximum spatial frequency in an image that can be recorded by the detector, since the shortest wavelength must be sampled at least twice. This sets the so-called Nyquist cut-off frequency as 1/(2*pixel_spacing). A representative value of DQE is often specified at half the maximum frequency, referred to as half Nyquist. The MTF specifies how strongly the various spatial frequencies in the image out to Nyquist are recorded and is set by the pixel size and other factors, such as the spreading of the electrons in the active layer of the detector. MTF is plotted as a function of spatial frequency and a value of unity implies perfect retention of the relative amplitude of that spatial frequency.

In EM, the introduction of electronic detectors in the 1990s as a replacement for photographic film (Krivanek & Mooney, 1993) was based on CCD sensors with a phosphor and fiber-optic coupling. The phosphor/fiber-optic, which converts the signal from a high-energy incident electron into a pulse of light photons, was essential because CCDs are easily destroyed by direct illumination with electrons. Their use for recording images or electron diffraction patterns (Downing & Hendrickson, 1999; Faruqi, Henderson, & Subramaniam, 1999) had both advantages and disadvantages. The advantage was that the microscopist could immediately see the quality of their specimen and thus avoid the long delay between exposure and observation of the image that occurred when using film. The disadvantage was that these early electronic cameras for EM were based on phosphor/fiber-optic/CCD technology, in which the DQE was only slightly better than that of film when used with 80 or 100 keV electrons. At higher voltages, although the microscope resolution, depth-of-focus, aberrations, and beam penetration improve, and the undesirable effects of beam-induced specimen charging are reduced, the DQE of these detectors was actually a lot worse than film. This was especially clear once techniques for measuring the frequency dependence of the DQE (De Ruijter, 1995; Meyer & Kirkland, 2000) were developed. Typically, the DQE of a phosphor/fiber-optic/CCD electron camera operated at 300 keV was 7–10% at half Nyquist resolution, much lower than film, which has a DQE at half Nyquist of 30–35% when using a 7 μm pixel size (McMullan, Chen, Henderson, & Faruqi, 2009). As a result, the highest-resolution single-particle structures determined prior to 2012 were largely obtained with images recorded on film (Grigorieff & Harrison, 2011; Zhang, Jin, Fang, Hui, & Zhou, 2010).

The reduced sensitivity to radiation damage of CMOS/MAPs compared with CCDs attracted the attention of the charged particle detection community beginning around 1998 (Caccia et al., 1999; Kleinfelder et al., 2002; Turchetta et al., 2001). The electron microscope community began to take note around 2003 (Evans et al., 2005; Faruqi et al., 2005; Milazzo et al., 2005; Turchetta, 2003; Xuong et al., 2004), and at this point in 2016, it is fair to say that CMOS/MAPS technology is expected to dominate in TEM and probably in high-energy physics charged particle detection, such as at the Large Hadron Collider. Technical development of these CMOS/MAPS detectors was carried out between 2005 and 2010, with product announcements in 2009 and eventual delivery of the first commercial products in 2012.

3. PRESENT

CMOS/MAPS detectors have a rolling readout mechanism and mostly work with a rolling-shutter. They can read out images continuously with a frame rate that can range from 1 to 1000 Hz or more. The original designs for digital consumer cameras (Sunetra, Kemeny, & Fossum, 1993) were adapted for the detection of charged particles through some changes to the basic design of the pixels. The primary difference for the revised design was the inclusion of a thin p-epitaxial layer (Turchetta et al., 2001) above the substrate, shown schematically in Fig. 1A. The top surface consists of a passivation layer a few μm thick containing the electronics and interconnects for the readout. Fig. 1B shows an example of one type of pixel design with three transistors (3T) in the readout electronics. This design is the simplest type of active pixel and is still popular; however, there are a number of more complex designs with additional transistors, which can provide some advantages, such as reducing the readout noise by correlated double sampling of the pixel charge.

The basic operation of this 3T pixel geometry is as follows (Mendis et al., 1997). Part of the energy of the incoming primary electron is converted into electron–hole pairs in the epilayer, which has a typical thickness of 5–20 μm. Due to doping differences in p− and p+ silicon there is a potential barrier at the boundary between the epilayer and the substrate on one side and the P-wells on the other side. This means that radiation-generated electrons are kept within the epilayer and eventually will be collected by the N-well diode through a mixture of drift and diffusion processes. Prior to starting an exposure in each frame, transistor T1 resets the diode capacitance by charging it up to the reset voltage, which is usually between 1 and 3 V. The collected charge from the electron/hole pairs that are created by the electric field of the high-energy incident electrons then discharges the capacitance on the input of T1 (Fig. 1B) and constitutes the signal (Prydderch et al., 2003). The T1/diode reset also introduces unwanted kTC noise to the collection node voltage. kTC noise refers to the variation

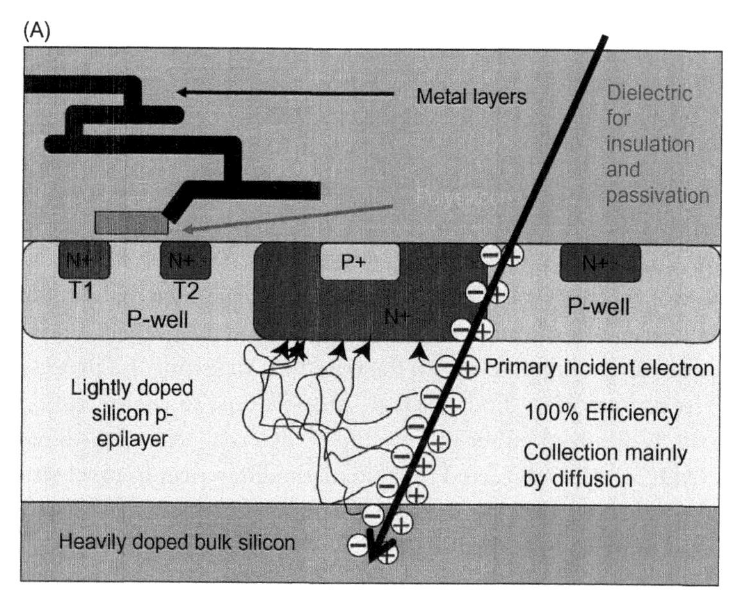

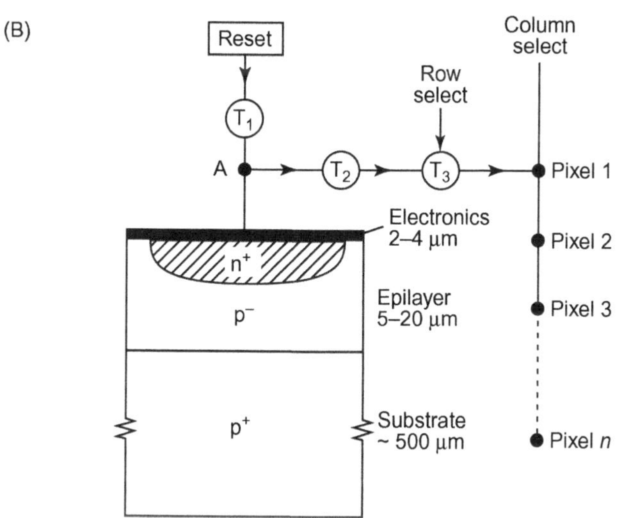

Fig. 1 (A) Schematic diagram viewed parallel to the sensor surface, of a single pixel in a typical CMOS sensor. The incident electron is represented by the *black arrow* from *top right* to *bottom left*. The + and − symbols indicate the electron–hole pairs that are created by the transient electric field as the high-energy electron passes. The mini-electrons are collected by the N-well whose potential drops during the exposure. (B) Schematic of readout for a single pixel, showing the 3T logic and its relationship to the N+ diode/capacitor. *Panel (A): Adapted from Turchetta, R. (2003). CMOS monolithic active pixel sensors (MAPS) for scientific applications. In:* 9th Workshop on electronics for LHC experiments Amsterdam, 2003 *(pp. 1–6). http://lhc-electronics-workshop.web.cern.ch/lhc-electronics-workshop/2003/PLENARYT/TURCHETT.PDF. Panel (B): Reproduced from Faruqi, A. R. (2007). Direct electron detectors for electron microscopy.* Advances in Imaging and Electron Physics, 145, *55–93.* (See the color plate.)

in the reset voltage due to thermal fluctuations in the charge on the capacitance of the diode. It is possible to reduce the effects of kTC noise by reading out twice: once immediately after reset and once after arrival of the signal followed by subtraction of the first value from the second; this is called correlated double sampling (CDS). On readout after the exposure, transistor T2 acts as source follower transferring the voltage signal to the readout transistor T3. The charge is read out by selecting rows (by activating T3) and reading out the corresponding pixels in all the columns, repeating the process for all rows. The reduction in diode voltage, proportional to charge collected on each pixel, is digitized either by on-chip or external analog-to-digital converters (ADCs), then corrected for systematic differences in pixel sensitivity by bright-field and dark-field calibrations, before being formed into an image that is used for further analysis.

Fig. 2A shows a typical layout of the pixels in a 4k × 4k sensor. Fig. 2B shows how several of these sensors are manufactured on a typical 200 mm wafer, before being diced, backthinned, mounted, and bonded. Finally, the fully mounted sensor is housed in an evacuated camera chamber, which requires electrical feedthroughs, cooling to reduce electronic noise, and possibly a mechanism for retraction if the microscope is used in other modes or with other detectors.

The frame readout rate depends on how many ADCs are used simultaneously to digitize the voltages from the diodes. The effect of radiation damage on the detector can be minimized both by using special radiation-hard design features, but also by increasing the frame rate, since the main effect of radiation damage is to increase the leakage current, which discharges the diode capacitance even in the absence of illumination, and this is less important at fast frame rates. Backthinning can be used to make the sensors very thin (~30 μm) and this reduces the noise contribution from backscattered electrons (Fig. 3). The MTF and DQE of a direct electron detector can be improved by a factor of 2 or more by backthinning (McMullan, Faruqi, et al., 2009). The reduction of backscattering also results in a two- to threefold improvement in detector lifetime due to reduced overall energy deposition in the sensitive layer.

The new detectors can operate in either integrating mode or counting mode. In integrating mode, the amount of energy deposited in each pixel is read out directly as an analog voltage that is digitized and represents the image after dark-field and bright-field corrections. Because the interaction of each high-energy electron with the sensor is stochastic, as illustrated in Fig. 3, the amount of energy deposited by each incident electron can vary

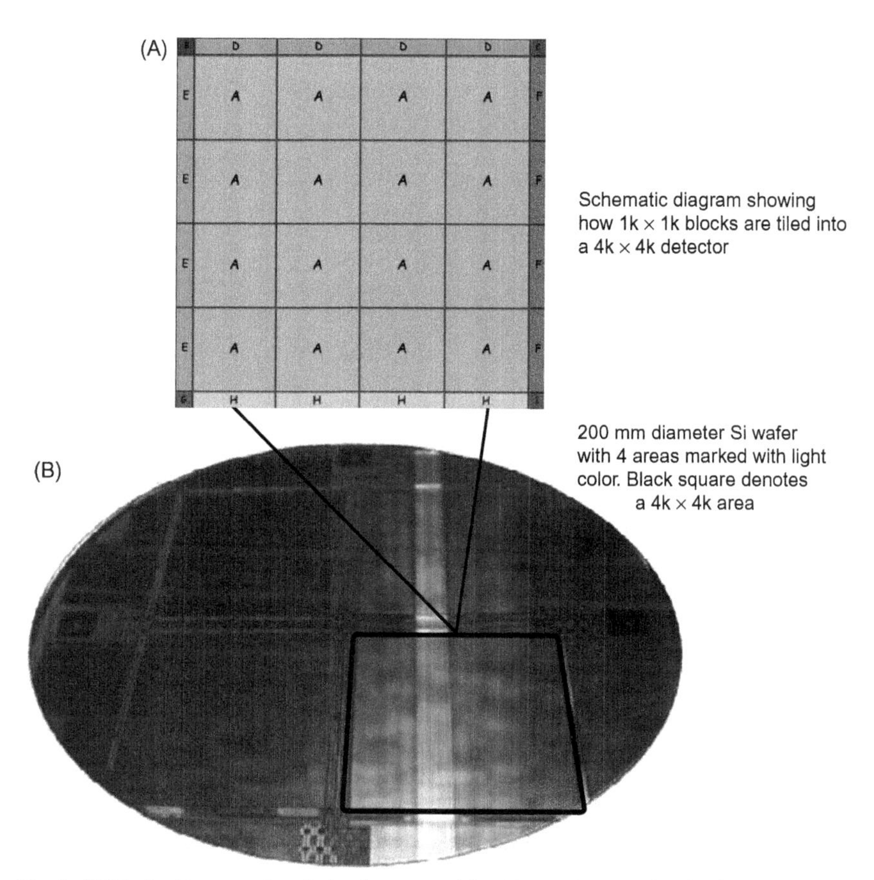

Fig. 2 (A) Typical layout of a stitched sensor with 4096 × 4096 pixels. Block A consists of a 1024 × 1024 pixel array that is tiled four times in each direction, but the number of blocks can be increased to produce larger area detectors. The edge blocks B through I control the addressing, reset and readout. (B) 200 mm silicon wafer showing the arrangement of four detectors. *Adapted from Guerrini, N., Turchetta, R., Van Hoften, G., Henderson, R., McMullan, G., & Faruqi, A. R. (2011). A high frame rate, 16 million pixels, radiation hard CMOS sensor.* Journal of Instrumentation, 6 C03003.

by a factor of 40 or more. This limits the DQE to about 60%, though parameter choice during design and manufacture, such as use of a thin epilayer and small pixels, can make this much lower. In counting mode, the intensity of the illumination is reduced by a factor of several hundred to a dose rate of perhaps only one electron per 100 pixels per frame. Individual electron events are then identified and replaced computationally by a delta function or a more sophisticated function (Li, Mooney, et al., 2013; McMullan, Clark, Turchetta, & Faruqi, 2009; Turchetta, 1993), so that each event gets

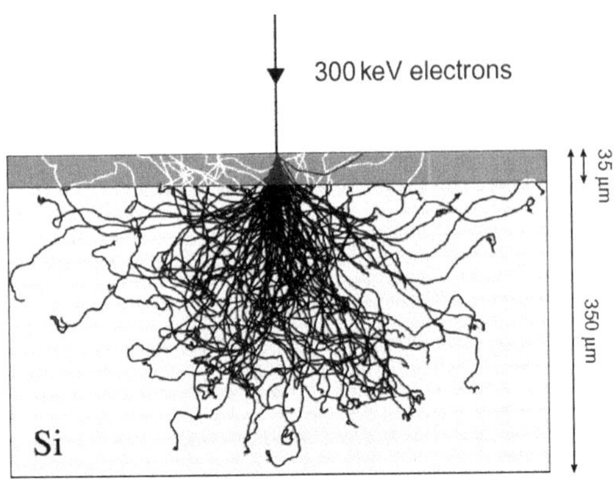

Fig. 3 Schematic of 300 keV electron trajectories, showing a Monte Carlo simulation of 300 keV electron tracks in silicon. After backthinning to 35 µm, only those parts of the electron tracks highlighted in *red* would contribute to the recorded signal. Before backthinning, the additional *white tracks* would contribute a low-resolution component to the signal together with contributions to the noise at all spatial frequencies. The overall thickness of the silicon in the figure is 350 µm with the 35 µm layer that remains after backthinning shown in *gray*. *Reproduced from McMullan, G., Faruqi, A. R., Henderson, R., Guerrini, N., Turchetta, R., Jacobs, A., & van Hoften, G. (2009). Experimental observation of the improvement in MTF from backthinning a CMOS direct electron detector.* Ultramicroscopy, 109*(9), 1144–1147.* (See the color plate.)

the same weight. In principle, if the signal-to-noise ratio for each event is high enough and its location can be determined accurately enough, this can give much higher DQE(0) and DQE(Nyquist), respectively.

The currently available direct electron detectors are the K2 from Gatan, recently upgraded to K2-XP, the Falcon II from FEI, recently upgraded to Falcon III, and DE-20 from Direct Electron, with a recent addition of DE-64. In each case, their detectors produce images using 300 keV electrons that are significantly better than obtained with film (Fig. 4 and McMullan et al., 2014). The DQE at half the Nyquist frequency is in the range 40–60% for the detectors when used in integrating mode, but typically drops to ~25% at the Nyquist frequency. Higher DQE values than these can only be obtained by operating in counting mode where the image of each incident electron is substituted by an idealized single count, so all future improved detectors will need to operate in counting mode. Higher frame rates are also required for counting, to avoid either double hits on individual pixels or very long exposures times, and at present only the Gatan K2, when operated in counting mode, can produce a DQE(0) as high as 80% in conjunction with reasonably small exposure times (Li, Mooney, et al., 2013;

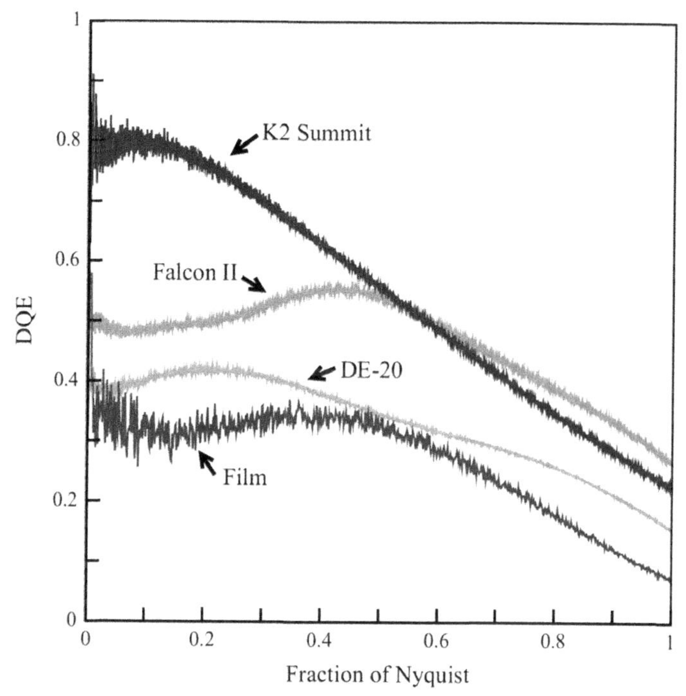

Fig. 4 Comparison of DQE at 300 keV as a function of spatial frequency for the DE-20 (*green* (*light gray* in the print version)), Falcon II (*red* (*gray* in the print version)), and K2 Summit (*blue* (*dark gray* in the print version)). The corresponding DQE of photographic film is shown in *black*. *Reproduced from McMullan, G., Faruqi, A. R., Clare, D., & Henderson, R. (2014). Comparison of optimal performance at 300 keV of three direct electron detectors for use in low dose electron microscopy.* Ultramicroscopy, 147, *156–163.*

Li, Zheng, Egami, Agard, & Cheng, 2013). The K2 detector frame rate is about 10 times higher than that available with the two other detector brands. If the arrival point of each incident electron can be determined to subpixel accuracy with sufficient precision, there is no reason why the DQE(ω) should not approach 100% without much drop at Nyquist, and the detectors might then be usable beyond Nyquist in super-resolution mode. The K2 detector already allows operation in super-resolution mode, but has relatively low DQE(Nyquist). An increased DQE at and beyond Nyquist would make it and other detectors that can operate in the counting mode much more powerful. FEI and Direct Electron have both announced that the Falcon and DE can be run in counting mode, producing higher DQE but at the expense of longer exposure times. It is too early to see whether practical data collection is viable.

The improved DQE of the new direct electron detectors, together with the ability to record high frame rate movies, has brought synergistic advantages. The improved DQE alone provides a big gain in the signal-to-noise ratio in this recorded image. The availability of movie recording allows the effect of beam-induced specimen movement and image blurring to be reduced by subsequent alignment and appropriate weighting of the individual frames, thus increasing the sharpness of the particle images (Bai, Fernandez, McMullan, & Scheres, 2013; Brilot et al., 2012; Campbell et al., 2012; Grant & Grigorieff, 2015; Li, Mooney, et al., 2013; Ripstein & Rubinstein, 2016; Rubinstein & Brubaker, 2015; Scheres, 2014; Vinothkumar, McMullan, & Henderson, 2014). An example of this is shown in Fig. 5, and another example can be seen in Brilot et al. (2012). These improved motion-corrected images then allow the more accurate determination of the position and orientation of each particle, which produces a 3D map of the structure at higher resolution than it would be without the corrections applied. This higher resolution map then provides a better target for the orientation determination in an iterative manner. Thus the new generation of detectors has produced a bigger impact on the attainable resolution than might have been expected from the increased DQE alone.

The improvement in DQE that is possible by electron counting is most apparent presently at low resolution on the K2 detector (Fig. 4). This improvement can be exploited by recording images at higher magnification than might normally be chosen, for example, at 0.7 or 1.0 Å per pixel, and by carrying out subsequent 2×2 or 3×3 binning. We now discuss briefly the practical choice of parameters for cryoEM illumination conditions to optimize image quality for each of the new detectors.

3.1 Practical Advice for the User

This paragraph may be the most useful for the user who simply wants to know what settings to use to get the most out of the new detectors. Typical microscope parameters used for acquisition of high-resolution images and movies with the three detector types are given in Table 1, and are discussed in more detail by Passmore and Russo (2016). Briefly, each detector has an optimum dose rate at which the DQE is maximal. At higher doses, the detectors all saturate so that each frame of the movie would then effectively contain no information. Likewise, at very low dose rates, many frames have to be added together to give an image with adequate total exposure and these

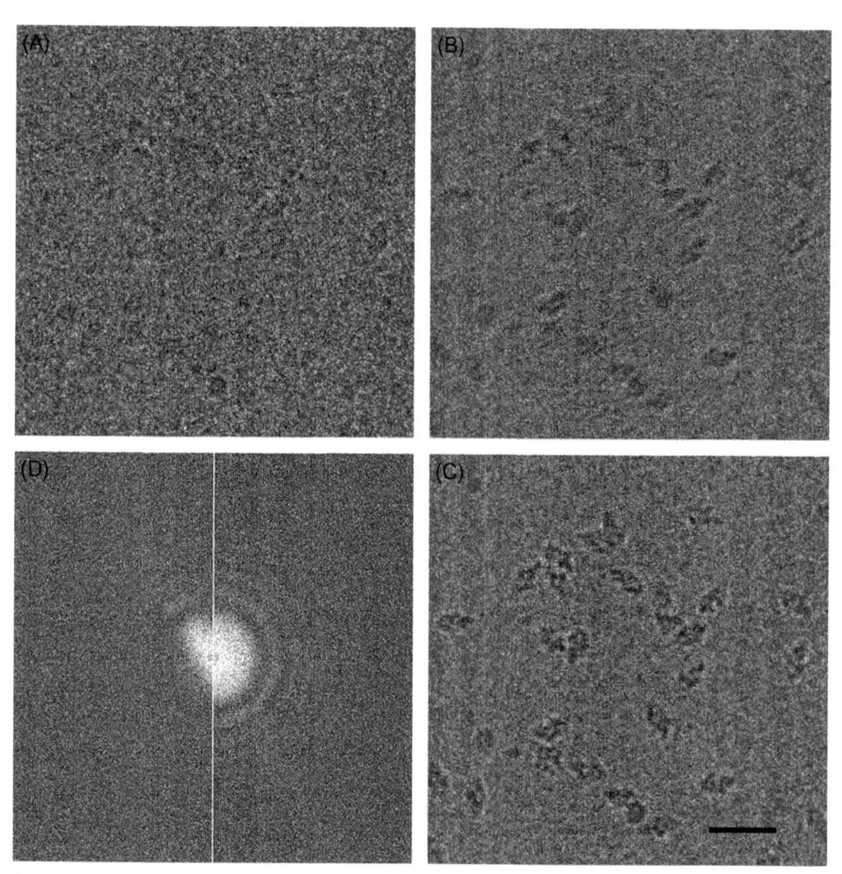

Fig. 5 Image of a 1024 × 1024 pixel area of a cryoEM image (20.21.43) of a typical structural biology specimen of the *E. coli* enzyme β-galactosidase, showing (A) single frame with an electron dose of ~6 electrons/Å^2, the sum of movie frames with a total electron dose of ~50 electrons/Å^2 without (B) and with (C) alignment, and (D) the power spectra (Fourier transforms) of the unaligned (*left*) and aligned (*right*) images out to half Nyquist (1/7 Å^{-1}). The images were recorded at 300 keV on an FEI Krios microscope with a Falcon II detector using 1.75 Å pixel size and ~3.5 μm defocus. Scale bar 300 Å.

summed images would then be dominated by electronic readout noise. In practice, the DQE versus dose rate curve is fairly broad, so a range of exposures is acceptable. For the Falcon II detector the range from 0.2 to 4.5 electrons/pixel/frame is usable (Kuijper et al., 2015). For DE, the range from 0.15 to 3.0 is viable. For K2 Summit, the range from 0.002 to 0.025 electrons/pixel/400 Hz frame is reasonable (Li, Mooney, et al., 2013; Li, Zheng, et al., 2013). For typical magnification choices, this translates into 15–40 electrons/Å^2/s for Falcon and DE, or 1–8 electrons/Å^2/s for K2.

Table 1 Typical Microscope Parameters Used for Acquisition of High-Resolution Images and Movies with the Three Detector Types

Camera	Direct Electron DE-20 (Integrating Mode)	FEI Falcon II (Integrating Mode)	Gatan K2 Summit (Counting Mode)
Pixel size at detector (μm)	6.4	14	5
Pixel size at specimen (Å)	1.0–1.3	1.7	1.0–1.4
Magnification (detector/specimen)	50–60 k ×	80 k ×	35–50 k ×
Movie frames/s (Hz)	20–32	17	400
Electron dose at detector (electrons/pixel/frame)	1.4–2	3	0.012–0.02
Electron dose at specimen (electrons/Å^2/s)	24–38	17	2.5–8
Total exposure time (seconds)	1.5–3.0	3	6–16
Total electron exposure (electrons/Å^2)	52–75	51	40–48

The saturating dose rate is several hundred times higher when the detector is used in integrating mode compared with counting mode, so both the Falcon and DE detectors can be operated as integrating detectors at dose rates up to 2 or 3 electrons/pixel/frame, whereas the dose rate in counting mode should be kept below 0.01–0.025 electrons/pixel/frame. In practice, this means the total exposure time for Falcon and DE is in the range 1–3 s, whereas with the K2, exposures are normally in the range of 6–16 s, or even longer if a bigger pixel size is used, as shown in Table 1. This means that the Falcon and DE detectors can easily be used on microscopes that have less stable cold stages with significant stage drift. In comparison, the K2 detector with its thin epilayer has a relatively poor DQE when operated in integrating mode, so in practice the K2 should always be used in counting mode (either with or without super-resolution, subpixel interpolation). Conversely, the exposure times for Falcon and DE detectors in counting mode would be very long, possibly more than 60 s, so at this moment there is no published structural determination using these detectors in counting mode.

4. FUTURE

Further improvements in detector performance require solving one or two remaining problems. First it is important that detectors should be developed so that their DQE is closer to 100%. This will require universal adoption of electron counting since the distribution of energy deposition when individual electrons pass through any detector is stochastic (Fig. 6) and the so-called Landau distribution (Fig. 7) of energy deposited by individual electron events has a very wide spread. Electron counting can be done on any of the CMOS detectors but is too slow for normal use unless the frame rate is high. A reasonably high frame rate is already available commercially with the Gatan K2/Summit detector (Li, Mooney, et al., 2013; Li, Zheng, et al., 2013), and a paper describing the counting mode of the FEI Falcon III camera was recently published (Kuijper et al., 2015). Any implementation of electron counting will benefit from minimizing missed or overlapping events, which requires a high signal-to-noise ratio and a high frame rate.

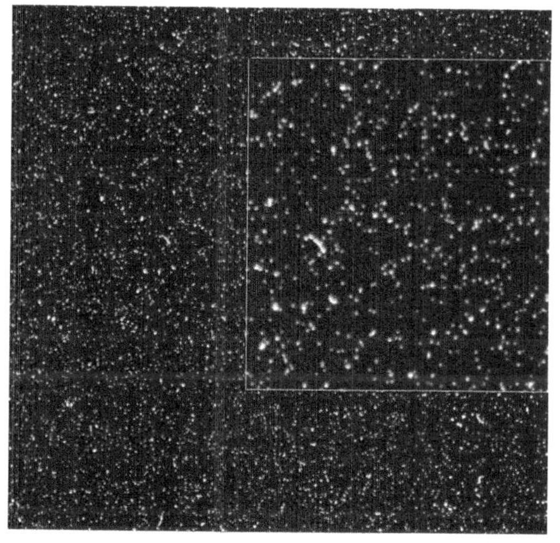

Fig. 6 Single frame image recorded using 300 keV electrons, showing single electron events on an FEI Falcon III detector, with excellent signal-to-noise ratio. A magnified portion is inset. Falcon III is a highly backthinned sensor with very low noise level. The single frame shows about 100,000 electron events (ie, one electron per 150 pixels). *This figure is similar to that for Falcon III in Kuijper, M., van Hoften, G., Janssen, B., Geurink, R., De Carlo, S., Vos, M., … Storms, M. (2015). FEI's direct electron detector developments: Embarking on a revolution in cryo-TEM.* Journal of Structural Biology, 192*(2), 179–187.*

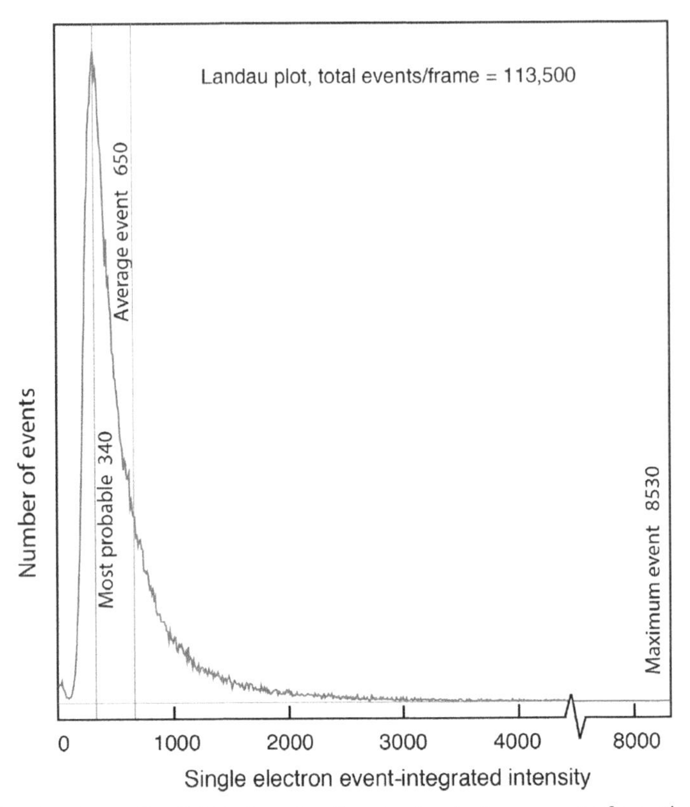

Fig. 7 Landau intensity distribution for the single electron events from the image shown in Fig. 6. Single electron events are identified by an initial threshold criterion and then all pixels contributing to each event are added together to determine the total signal from each electron. The Landau plot is the histogram of the single electron event distribution.

Improving the signal-to-noise ratio of the readout helps to minimize missed events. A larger pixel size would allow the use of a thicker epilayer with a resulting increased signal; a greater degree of backthinning reduces the incidence of extended electron tracks. Accurate subpixel localisation is incompatible with events that appear as tracks. However, although it may be possible to get DQE(0) to approach 100%, it is unrealistic to expect DQE(Nyquist) to exceed 80–90%, since this would require the accuracy of subpixel localization of electron events to be better than ±0.15 pixels. A very small number of electrons will always be backscattered elastically from the detector surface without depositing any energy. Others may pass through and deposit very little energy so they may remain undetected, unless the epilayer is thick enough to improve signal-to-noise ratio. Others will be scattered sideways (Fig. 3) to leave tracks of deposited energy that contain no high-resolution information. Finally, unless the frame rate is very high there will always be the temptation to increase the beam intensity, thus increasing the number of overlapping events and reducing the overall DQE.

The technology used for direct electron detectors is very similar to that used in digital cameras and phones. The feature sizes used in sensor lithography have dropped from around 1 μm in the early 1990s to 65 nm in some of the latest mobile phone cameras in 2016. In corresponding state-of-the-art central processor units, feature sizes as small as 14 nm are being used. The current EM sensors use 350 nm or 180 nm technology that is years behind the industry leaders. With reasonable further investment in direct electron detector development, it is possible to envisage bigger, faster, electron counting sensors that would further revolutionize the speed and quality of image recording. We can certainly look forward to being able to tackle even more challenging biological structures, whether using single particle or tomography approaches, to determine smaller (and larger) structures using fewer images to higher resolution and with the ability to distinguish more three-dimensional states than at present. The future holds great promise.

REFERENCES

Bai, X. C., Fernandez, I. S., McMullan, G., & Scheres, S. H. W. (2013). Ribosome structures to near-atomic resolution from thirty thousand cryo-EM particles. *eLife, 2,* e00461.

Brilot, A. F., Chen, J. Z., Cheng, A. C., Pan, J. H., Harrison, S. C., Potter, C. S., … Grigorieff, N. (2012). Beam-induced motion of vitrified specimen on holey carbon film. *Journal of Structural Biology, 177*(3), 630–637.

Caccia, M., Campagnolo, R., Meroni, C., Kucewicz, W., Deptuch, G., Zalewska, A., … Turchetta, R. (1999). *High resolution pixel detectors for e+e- linear colliders.* arXiv:hep-ex/9910019v1.

Campbell, M. G., Cheng, A. C., Brilot, A. F., Moeller, A., Lyumkis, D., Veesler, D., … Grigorieff, N. (2012). Movies of ice-embedded particles enhance resolution in electron cryo-microscopy. *Structure, 20*(11), 1823–1828.

De Ruijter, W. J. (1995). Imaging properties and applications of slow-scan charge-coupled-device cameras suitable for electron-microscopy. *Micron, 26*(3), 247–275.

Downing, K. H., & Hendrickson, F. M. (1999). Performance of a 2 k CCD camera designed for electron crystallography at 400 kV. *Ultramicroscopy, 75*(4), 215–233.

Evans, D. A., Allport, P. P., Casse, G., Faruqi, A. R., Gallop, B., Henderson, R., … Waltham, N. (2005). CMOS active pixel sensors for ionising radiation. *Nuclear Instruments & Methods in Physics Research Section A—Accelerators, Spectrometers, Detectors and Associated Equipment, 546*(1–2), 281–285.

Faruqi, A. R., Henderson, R., Pryddetch, M., Allport, P., Evans, A., & Turchetta, R. (2005). Direct single electron detection with a CMOS detector for electron microscopy. *Nuclear Instruments & Methods in Physics Research Section A—Accelerators, Spectrometers, Detectors and Associated Equipment, 546*(1–2), 170–175.

Faruqi, A. R., Henderson, R., & Subramaniam, S. (1999). Cooled CCD detector with tapered fibre optics for recording electron diffraction patterns. *Ultramicroscopy*, *75*(4), 235–250.

Fossum, E. R. (1993). Active pixel sensors—Are CCDs dinosaurs. In *Charge-coupled devices and solid state optical sensors III: Vol. 1900*. (pp. 2–14). http://dx.doi.org/10.1117/12.148585.

Grant, T., & Grigorieff, N. (2015). Measuring the optimal exposure for single particle cryo-EM using a 2.6 Å reconstruction of rotavirus VP6. *eLife*, *4*, e06980. http://dx.doi.org/10.7554/eLife.06980.

Grigorieff, N., & Harrison, S. C. (2011). Near-atomic resolution reconstructions of icosahedral viruses from electron cryo-microscopy. *Current Opinion in Structural Biology*, *21*(2), 265–273.

Kleinfelder, S., Bichsel, H., Bieser, F., Matis, H. S., Rai, G., Retiere, F., … Yamamoto, E. (2002). Integrated X-ray and charged particle active pixel CMOS sensor arrays using an epitaxial silicon sensitive region. *Proceedings of SPIE*, *4784*, 208–217. http://dx.doi.org/10.1117/12.450826.

Krivanek, O. L., & Mooney, P. E. (1993). Applications of slow-scan CCD cameras in transmission electron-microscopy. *Ultramicroscopy*, *49*(1–4), 95–108.

Kuijper, M., van Hoften, G., Janssen, B., Geurink, R., De Carlo, S., Vos, M., … Storms, M. (2015). FEI's direct electron detector developments: Embarking on a revolution in cryo-TEM. *Journal of Structural Biology*, *192*(2), 179–187.

Li, X. M., Mooney, P., Zheng, S., Booth, C. R., Braunfeld, M. B., Gubbens, S., … Cheng, Y. F. (2013). Electron counting and beam-induced motion correction enable near-atomic-resolution single-particle cryo-EM. *Nature Methods*, *10*(6), 584–590.

Li, X. M., Zheng, S. Q., Egami, K., Agard, D. A., & Cheng, Y. F. (2013). Influence of electron dose rate on electron counting images recorded with the K2 camera. *Journal of Structural Biology*, *184*(2), 251–260.

McMullan, G., Chen, S., Henderson, R., & Faruqi, A. R. (2009). Detective quantum efficiency of electron area detectors in electron microscopy. *Ultramicroscopy*, *109*(9), 1126–1143.

McMullan, G., Clark, A. T., Turchetta, R., & Faruqi, A. R. (2009). Enhanced imaging in low dose electron microscopy using electron counting. *Ultramicroscopy*, *109*(12), 1411–1416.

McMullan, G., Faruqi, A. R., Clare, D., & Henderson, R. (2014). Comparison of optimal performance at 300 keV of three direct electron detectors for use in low dose electron microscopy. *Ultramicroscopy*, *147*, 156 163.

McMullan, G., Faruqi, A. R., Henderson, R., Guerrini, N., Turchetta, R., Jacobs, A., & van Hoften, G. (2009). Experimental observation of the improvement in MTF from back-thinning a CMOS direct electron detector. *Ultramicroscopy*, *109*(9), 1144–1147.

Mendis, S. K., Kemeny, S. E., Gee, R. C., Pain, B., Staller, C. O., Kim, Q. S., & Fossum, E. R. (1997). CMOS active pixel image sensors for highly integrated imaging systems. *IEEE Journal of Solid-State Circuits*, *32*(2), 187–197. http://dx.doi.org/10.1109/4.551910.

Meyer, R. R., & Kirkland, A. I. (2000). Characterisation of the signal and noise transfer of CCD cameras for electron detection. *Microscopy Research and Technique*, *49*(3), 269–280.

Milazzo, A. C., Leblanc, P., Duttweiler, F., Jin, L., Bouwer, J. C., Peltier, S., … Xuong, N. H. (2005). Active pixel sensor array as a detector for electron microscopy. *Ultramicroscopy*, *104*(2), 152–159.

Passmore, L. A., & Russo, C. J. (2016). Specimen preparation for high-resolution cryo-EM. *Methods in Enzymology*, *579*, 51–86.

Prydderch, M. L., Waltham, N. J., Turchetta, R., French, M. J., Holt, R., Marshall, A., … Mapson-Menard, H. (2003). A 512 x 512 CMOS monolithic active pixel sensor with integrated ADCs for space science. *Nuclear Instruments & Methods in Physics Research Section A—Accelerators, Spectrometers, Detectors and Associated Equipment*, *512*(1–2), 358–367. http://dx.doi.org/10.1016/S0168-9002(03)01914-4.

Ripstein, Z. A., & Rubinstein, J. L. (2016). Processing of cryo-EM movie data. *Methods in Enzymology*, *579*, 103–124.

Rubinstein, J. L., & Brubaker, M. A. (2015). Alignment of cryo-EM movies of individual particles by optimization of image translations. *Journal of Structural Biology*, *192*(2), 188–195.

Scheres, S. H. W. (2014). Beam-induced motion correction for sub-megadalton cryo-EM particles. *eLife*, *3*, e03665.

Sunetra, K. M., Kemeny, S. E., & Fossum, E. R. (1993). A 128x128 CMOS active pixel image sensor for highly integrated imaging systems. In *IEEE, IEDM'93* (pp. 583–586). http://dx.doi.org/10.1109/IEDM.1993.347235.

Turchetta, R. (1993). Spatial-resolution of silicon microstrip detectors. *Nuclear Instruments & Methods in Physics Research Section A—Accelerators, Spectrometers, Detectors and Associated Equipment*, *335*(1–2), 44–58.

Turchetta, R. (2003). CMOS monolithic active pixel sensors (MAPS) for scientific applications. In *9th Workshop on electronics for LHC experiments, Amsterdam, 2003* (pp. 1–6). http://lhc-electronics-workshop.web.cern.ch/lhc-electronics-workshop/2003/PLENARYT/TURCHETT.PDF.

Turchetta, R., Berst, J. D., Casadei, B., Claus, G., Colledani, C., Dulinski, W.,Winter, M. … (2001). A monolithic active pixel sensor for charged particle tracking and imaging using standard VLSI CMOS technology. *Nuclear Instruments & Methods in Physics Research Section A—Accelerators, Spectrometers, Detectors and Associated Equipment*, *458*(3), 677–689. http://dx.doi.org/10.1016/S0168-9002(00)00893-7.

Vinothkumar, K. R., McMullan, G., & Henderson, R. (2014). Molecular mechanism of antibody-mediated activation of beta-galactosidase. *Structure*, *22*(4), 621–627.

Xuong, N. H., Milazzo, A. C., LeBlanc, P., Duttweiler, F., Bouwer, J., Peltier, S., … Kleinfelder, S. (2004). First use of a high sensitivity active pixel sensor array as a detector for electron microscopy. *Proceedings of SPIE*, *5301*, 242–249. http://dx.doi.org/10.1117/12.526021.

Zhang, X., Jin, L., Fang, Q., Hui, W. H., & Zhou, Z. H. (2010). 3.3 angstrom cryo-EM structure of a nonenveloped virus reveals a priming mechanism for cell entry. *Cell*, *141*(3), 472–482.

CHAPTER TWO

Specimen Behavior in the Electron Beam

R.M. Glaeser[1]
Lawrence Berkeley National Laboratory, University of California, Berkeley, CA, United States
[1]Corresponding author: e-mail address: rmglaeser@lbl.gov

Contents

Methods in Enzymology, Volume 579
ISSN 0076-6879
http://dx.doi.org/10.1016/bs.mie.2016.04.010

Abstract

It has long been known that cryo-EM specimens are severely damaged by a level of electron exposure that is much lower than what is needed to obtain high-resolution images from single macromolecules. Perhaps less well appreciated in the cryo-EM literature, the vitreous ice in which samples are suspended is equally sensitivity to radiation damage. This chapter provides a review of several fundamental topics such as inelastic scattering of electrons, radiation chemistry, and radiation biology, which—together—can help one to understand why radiation damage occurs so "easily." This chapter also addresses the issue of beam-induced motion that occurs at even lower levels of electron exposure. While specimen charging may be a contributor to this motion, it is argued that both radiation-induced relief of preexisting stress and damage-induced generation of additional stress may be the dominant causes of radiation-induced movement.

1. INTRODUCTION

This chapter describes how and why cryo-EM specimens—and their images—progressively change with time as images are being recorded. The fact that electrons are ionizing radiation is the underlying, fundamental cause of such changes.

Biological specimens are easily damaged by ionizing radiation. As a result, changes in molecular structure that accumulate during the exposure are a primary concern. Ultimately, the molecular damage becomes so extensive that it is futile to extend the exposure any further.

Even at the very beginning of an exposure, it is possible that irradiating a specimen may relieve preexisting stress. At the same time, the accumulated structural damage may also generate new stress as the exposure continues. Either way, dynamic changes in stress may result in collective (beam-induced) movement of the specimen as a whole. This specimen movement, possibly along with movement of the image caused by electrostatic charging of the specimen, causes blurring of high-resolution features of the image. It thus is important to be aware of these effects when collecting images.

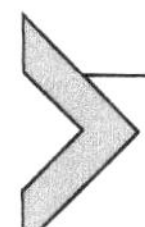

2. HIGH-ENERGY ELECTRONS ARE A FORM OF IONIZING RADIATION AS WELL AS BEING A FORM OF SHORT-WAVELENGTH RADIATION THAT CAN BE FOCUSED

2.1 Electron-Scattering Events Can Be Either Elastic or Inelastic

When an incident electron is scattered by the specimen, the outcome is described as being either an elastic or an inelastic event, depending upon whether the electrons within the specimen remain in their ground state or not. The relative probabilities for elastic vs inelastic-scattering events are proportional to their respective "total scattering cross sections." The values of these cross sections depend primarily upon the atomic numbers of the atoms making up the specimen, and one can ignore the extent to which they depend upon the chemical bonding between the atoms. Section 5.2.4 in Reimer and Kohl (2008) estimates that the ratio of inelastic scattering to elastic-scattering scales theoretically as $\sim 26/Z$, where Z is the atomic number, whereas experimentally is seems to scale as $\sim 20/Z$. This difference is of minor importance. What is important is that the relative amounts of elastic and inelastic scattering are similar for different types of biological materials, differing mainly to the extent that their chemical compositions may differ.

Both the elastic and inelastic-scattering cross sections decrease strongly (roughly as $1/v^2$, where v is the electron speed) as the energy of the incident electrons increases. The decrease in the elastic-scattering cross section, at higher voltage, leads to a decrease in the contrast of weak phase objects, of course. At the same time, however, the decrease in the inelastic-scattering cross section makes it possible to use a correspondingly larger electron exposure, while still staying within the same limit of what is a "safe" radiation dose.

The ratio of elastic to inelastic-scattering cross sections for carbon decreases by only $\sim 13\%$ as the energy of the incident electrons is increased from 100 to 300 keV, as can be calculated from equation 5.65 in Reimer and

Kohl (2008). As a result, the signal-to-noise ratio in the image does not change significantly as the voltage is increased, provided that one takes advantage of the increased electron exposures that are allowed at higher voltages, as is mentioned in the previous paragraph.

On the other hand, there are incremental benefits that come with using higher electron energies, such as the fact that the depth of focus increases in inverse proportion to the electron wavelength. The depth of focus relative to the thickness of the sample becomes increasingly important as the image resolution improves (Agard, Cheng, Glaeser, & Subramaniam, 2014). Another consideration, important at all resolutions, is that plural scattering decreases as the voltage increases. More specifically, there is a smaller loss of signal due to some of the (single-scattering) elastic events being either preceded by, or followed by, one or more inelastic events. The extent to which this type of plural scattering is important depends upon the ratio of the sample thickness relative to the mean-free path for inelastic scattering. For samples much thinner than the mean-free path, a fraction t/Λ of the elastically scattered electrons are lost to a second, inelastic event, where t is the sample thickness, and Λ is the mean-free path. Perhaps the best estimate for the inelastic mean-free path in ice is ~200 nm at an energy of 120 keV (Grimm, Typke, Barmann, & Baumeister, 1996). This value, in turn, suggests that the mean-free path for inelastic scattering of 300 keV electrons is ~350 nm.

2.2 Energy Is Deposited in the Specimen as a Result of Inelastic Scattering

Variable amounts of energy are lost by incident, high-energy electrons as a result of individual, inelastic-scattering events. Most of this energy is deposited within the specimen, and only a small fraction is carried away in the form of the kinetic energy of secondary electrons.

When the amount of energy lost in individual scattering events is measured for a large number of electrons, the probabilities for various energy losses can be presented as an energy-loss spectrum. Fig. 1, reproduced from Aronova and Leapman (2012), shows examples of such spectra for a few, representative biological materials.

If the specimen is thin enough, most of the electrons remain unscattered. These unscattered electrons make up the majority of the large peak that is seen at zero energy loss in Fig. 1B. The elastically scattered electrons also show up in the zero-loss peak, of course.

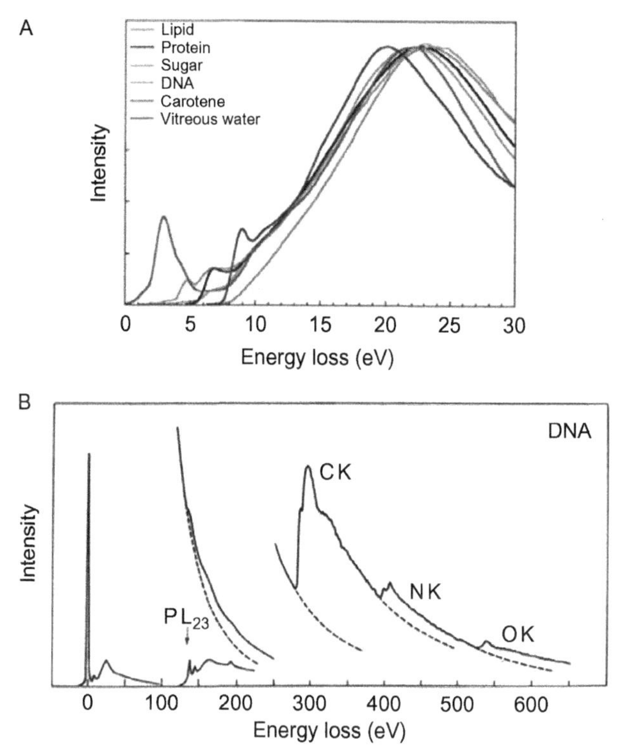

Fig. 1 Electron energy-loss spectra of various biological materials. Panel (A) shows just the region of low energy losses, from 0 to 30 eV. The various energy-loss peaks produced by different materials in the region 0–10 eV correspond to peaks in the UV–vis spectra of the same materials. The broad, intense peak extending from about 10 eV to almost 100 eV is due to the "plasmon" loss, or collective dielectric loss of the materials. Plasmon losses have limited usefulness to distinguish different types of biological materials. They are, on the other hand, the main events that lead to ionization and radiation damage of biological materials. Panel (B) shows a broader range of the energy-loss spectrum for DNA. Various peaks are shown in the range of ~150–600 eV energy loss, corresponding to inner-shell ionization energies of various elements. *Reproduced with permission from Aronova, M. A., & Leapman, R. D. (2012). Development of electron energy loss spectroscopy in the biological sciences.* MRS Bulletin/Materials Research Society, 37*(1), 53–62. doi: 10.1557/mrs.2011.329.*

Different regions of the energy-loss spectrum correspond to quite different physical events. Peaks in the spectrum at energies below about 10 eV, seen in some of the curves in Fig. 1A, correspond to peaks in the optical absorption (UV–vis) spectrum. This part of the (energy-loss) spectrum is mainly associated with single-electron excitations of conjugated bond systems, such as aromatic groups or linear polyenes.

The large, broad peak between 10 and 100 eV, seen for all materials, corresponds to the simultaneous (collective) excitation of many electrons in the specimen. In metals, the process is referred to as plasmon excitation. The same terminology is often used even when the specimen is not a metallic conductor. Alternatively, this collective excitation can be called a "dielectric loss." In an insulator, excitation of a mode in which many electrons move in synchrony soon decays into lower energy, single-electron excitations and ionizations, plus heat. Such single-electron ionization events are very likely to result in radiolysis of aliphatic organic molecules, while aromatic molecules—as we know empirically—are much less likely to be damaged. The distinction, of course, is that the loss of a single valence electron is localized to a single bond in aliphatic molecules, while in an aromatic molecule the loss is delocalized over multiple covalent bonds.

Finally, in the energy-loss range of a few hundred eV (for organic molecules), there are small peaks—like those shown in Fig. 1B—that correspond to ionization of K-shell (1S-state) electrons. K-shell ionization of low-Z atoms usually decays by the Auger process (Chattarji, 1976). Such events often cause ejection of two or more valence electrons from a single atom. As a result, K-shell ionization is very likely to result in bond rupture, even for aromatic molecules. The relative frequency of such events is low, however, since the cross section for K-shell ionization is about 1000 times smaller than for plasmon excitation. The net result is that K-shell ionization makes a negligible contribution to radiation damage at the electron exposures used in cryo-EM.

2.3 Values of the Linear Energy Transfer (LET) Can Be Used to Estimate the Energy Deposited

The term "linear energy transfer (LET)" is used to indicate the average amount of energy that is lost per unit path-length as a charged particle travels through a given material. The LET for electrons is traditionally expressed in units of MeV/cm, or, when divided by the mass density, in units of MeV-cm^2/g. Values of the LET for electrons have been tabulated for many materials and for a wide range of energies of the incident electrons (Berger & Selzer, 1964).

The average amount of energy deposited in a thin sample, per electron, can be estimated by multiplying the LET by the sample thickness, t. Similarly, the total energy deposited per gram of a specimen, following an exposure of N electrons/area, is

$$E = \frac{\mathrm{LET} \cdot N}{\rho} \tag{1}$$

where ρ is the mass density of the specimen material.

The energy deposited per gram is referred to as the radiation dose. Radiation doses are usually expressed in rads in the older literature, where 1 rad is equal to 100 erg/g. Alternatively the dose is expressed in the Standard International (SI) units of gray (Gy), where 1 Gy = 1 J/kg, and thus 1 rad = 0.01 Gy. Since the dose is proportional to the electron exposure, it is commonly used jargon to refer to the exposure as being the "dose." While this terminology is not strictly correct, the intended meaning becomes understandable in context.

Taking vitreous ice as an example, the LET for 300 keV electrons, divided by the mass density, is ~2.4 MeV cm^2/g. It follows that a dose of ~3.8×10^9 rad is deposited in a cryo-EM specimen as a result of an electron exposure of 10 e/$Å^2$. This estimate is too high, of course, because not all of the energy lost by incident electrons is actually deposited in a thin sample. Rather, as mentioned earlier, some of the energy escapes in the form of kinetic energy of secondary electrons. Depending upon the thickness of the specimen, the actual dose has been estimated to be reduced by half or more (Grubb, 1974). Unless one is concerned about making a very precise estimate of the radiation dose, however, it is not important to make a correction for this effect.

To further illustrate the linear relationship expressed in Eq. (1), the rad dose is plotted in Fig. 2 as a function of electron exposure. The specimen is again taken to be vitreous ice, and the energy of the electrons is assumed to be 300 keV. More is said below about each of the arrows shown in Fig. 2.

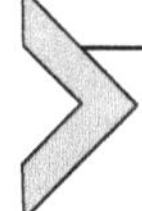

3. BIOLOGICAL MOLECULES BECOME STRUCTURALLY DAMAGED WHEN IRRADIATED

3.1 There Is a Large Literature of Radiation Chemistry and Radiation Biology

Extensive studies in radiation biology have described many effects caused by ionizing radiation. Much is known, for example, about how strand-breaks and mutations occur in DNA, even at relatively low doses (Hall & Giaccia, 2012). At ten thousand to a million fold higher doses, even enzymatic activity is destroyed (Kempner, 1993; Kempner & Schlegel, 1979). Ultimately, at doses rarely encountered outside the context of X-ray crystallography

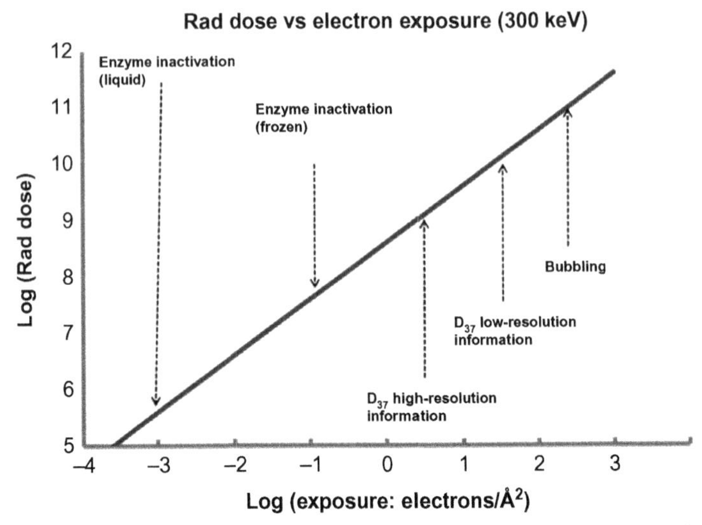

Fig. 2 Graph showing the linear relationship between electron exposure (300 keV electrons) and the rad dose deposited in a specimen. Five *annotated arrows* are included in the graph to indicate the general region of electron exposure at which various landmarks of radiation damage are incurred. The notation D_{37} indicates the dose/exposure at which the desired signal falls to 37% (e^{-1}) of its initial value.

and electron microscopy, radiation completely destroys protein structure (Howells et al., 2009).

The sensitivity of enzyme activity to radiation inactivation depends upon the size of the protein (Kempner & Schlegel, 1979), as well as upon its physical environment. As indicated in Fig. 2, for example, enzyme function may survive only up to doses $\sim 10^6$ rad when proteins are irradiated in liquid buffer, but it can survive doses as high as $\sim 10^8$ rads when frozen while being irradiated. The difference is understood to be due to the fact that many secondary chemical reactions occur in the liquid state, which cannot occur in the frozen state.

In the field of radiation chemistry, the radiolysis reaction

$$\text{Parent molecule} \xrightarrow{\text{Ionizing radiation}} \text{Products} \tag{2}$$

is often characterized in terms of the yield, or G-value for the reaction. The G-value is defined as the "number of molecules per 100 eV" that are either destroyed or produced. One therefore speaks of G-values for disappearance of parent molecules, and of various G-values for creation of any one of the product molecules. In general, multiple different reactions of the type indicated in Eq. (2) occur for any given parent molecule. These reactions often involve molecular fragmentation, although cross-linking and polymerization are also possible. G-values for some form of damage to the parent molecule are typically in the range of 2 or more for solid amino acids—see section 8.1 of Garrison (1968).

Biological molecules are also easily damaged by elastic events that literally knock an atom out of the molecule (Cosslett, 1970). While the large amount of momentum transfer that this requires is more favorable for a light atom than it is for a heavy atom, the Coulombic force is smaller when $Z=1$ than when $Z=6$. The net result is that knock-on events may be less likely for hydrogen atoms than they are for carbon, nitrogen, and oxygen atoms. In any case, the cross section for such "knock-on" events is very small compared to the cross section for inelastic scattering, and it is, instead, comparable to that for K-shell ionization mentioned in Section 2.2. As a result, the radiation damage caused by inelastic scattering, in particular that resulting from plasmon excitation and subsequent ionization, is far more important for biological materials than is knock-on damage. This remains the case even at very high electron energies (eg, 1–3 MeV), and even when account is taken of the displaced atom itself producing additional displaced atoms.

3.2 Fading of Diffraction Patterns Is a Convenient Indicator of Structural Damage

Electron diffraction patterns of hydrated protein crystals are easily observed in the electron microscope. These can go to quite high resolution if the crystals contain enough unit cells, and if the crystals are well ordered (Glaeser, Downing, DeRosier, Chiu, & Frank, 2007). As radiation damage begins to set in, however, the diffraction spots become progressively weaker, and the spots may also become broader. Fading of the (integrated) intensities of diffraction spots indicates that the contents of individual unit cells are becoming increasingly dissimilar, while broadening of the spots indicates that the crystal lattice is becoming less and less perfect. As is indicated in Fig. 2, high-resolution diffraction spots fade at a much faster rate than do the low-resolution spots (Baker, Smith, Bueler, & Rubinstein, 2010; Bammes, Jakana, Schmid, & Chiu, 2010; Taylor & Glaeser, 1976).

All of these phenomena are consistent with the model that radiation damage causes random structural changes in the contents of the unit cell, and the resulting structural differences first become noticeable at high resolution. An alternative interpretation would be that the crystal packing is sensitive to any change in molecular structure, even though the amounts of structural change within the molecules themselves may be quite minor. If, for example, molecules rotate and repack within a crystal, but change little—if at all—in the process, that too might explain the fading of diffraction intensities and the broadening of diffraction spots. To the extent that this second interpretation is correct, fading of the intensities of diffraction patterns would not be an accurate way to determine the structural sensitivity of biological molecules themselves. While one can, indeed, safely assume that repacking of molecular fragments, and even whole (undamaged) molecules does occur as the electron exposure progresses, prior studies in radiation chemistry make it unreasonable to suppose that the molecules themselves remain largely undamaged.

With the recent advent of superior electron-camera technology, Grant and Grigorieff were able to estimate the fading of signal from single-particle images of a large virus. They found that there was good agreement with earlier protein-crystal data as regards the radiation sensitivity of high-resolution features (Grant & Grigorieff, 2015). As Grant and Grigorieff pointed out, however, the structure of a virus—not unlike that of a protein-crystal—is based on the packing of identical subunits. As a result, this agreement does not completely rule out the possibility that changes in subunit packing, rather than changes in subunit structure alone, contribute to fading of the high-resolution signal. In addition, Grant and Grigorieff report that the lower-resolution signal in images of this virus particle is less sensitive than it was reported to be for images of protein crystals by Baker et al. (2010). They discuss a number of alternative hypotheses that might account for the results obtained at lower resolution, but whether there is a fundamental difference in the radiation sensitivity of single particles and of protein crystals is not yet clear.

3.3 Some Residues in Proteins Are Especially Sensitive to Radiation Damage

It is well known that a few, specific amino acid residues within a protein are likely to be damaged very early, well before the diffraction intensities of the crystal have changed to a significant degree. In this regard, fading of diffraction intensities is not a sufficiently sensitive way, rather than being a too sensitive way, to determine how much electron exposure can be safely used. The residues that are likely to be damaged first include those at the active site of an enzyme, solvent-exposed disulfide bonds, and side-chain carboxyl groups.

Much of the information on this point has come from the field of protein crystallography (Weik et al., 2000). It is reasonable to ask why one should expect the effects of X-ray radiation and of electron radiation to be the same. The explanation (Henderson, 1990) is that the primary event in inelastic scattering of X-rays is the production of an electron with almost the same energy as that of the X-ray photon—either through the photoelectric effect or through Compton scattering, depending upon the energy of the photon. After that, the (moderately) high-energy electron deposits energy in the sample by exactly the same inelastic-scattering processes as occur for incident electrons in the electron microscope.

3.4 Caging of Fragments and "Trapping" of Radicals Results in Cryo-Protection: This Helps Only to a Limited Extent

When proteins are irradiated at liquid nitrogen temperature, chemical attack by reactive species such as hydroxyl radicals, which are generated by radiolysis of the surrounding buffer, is greatly reduced. As a result, only the direct effects of radiation on the macromolecule itself remain a significant cause of radiation damage (Kempner, 1993). In addition, molecular fragments produced at liquid nitrogen temperature are themselves effectively caged by their surroundings, whether that consists of adjacent solvent molecules or of adjacent parts of the macromolecule. Indeed, even free radicals produced within irradiated organic crystals do not initially participate in secondary reactions. Instead they are trapped so well by their surroundings that each time a given radical is formed, it adopts the same orientation relative to its surroundings. We thus may conclude that, even when chemical bonds are broken, the two, previously bonded atoms may, initially, still remain in van der Waals contact. It is thought that both of these effects are responsible for the significant protection effect (Glaeser & Taylor, 1978; Hayward & Glaeser, 1979) observed for proteins in cryo-EM specimens.

The desirable effects of "cryo-protection" do not continue forever as the dose continues, however. As more and more radicals form and accumulate, it becomes more and more likely that a newly trapped radical comes in van der Waals contact with a preexisting one. When that happens, the two will react spontaneously, ie, there is no activation barrier to overcome for reactions between free radicals. The resulting heat released by this favorable reaction may be sufficient to activate local rotations and displacements, thus facilitating further reactions and consequent structural disorder. This series of events can occur even when the specimen is at helium temperature. It thus is understandable that helium temperature does not provide a major cryo-protection effect, as judged by the decay of high-resolution electron diffraction intensities, beyond that produced at nitrogen temperature (Chiu et al., 1986). A similar result has been reported for X-ray diffraction intensities of protein crystals irradiated at various temperatures, where it was found that the lifetime of high-resolution reflections did not change by more than 20% at temperatures below 100 K (see fig. 1A in Meents, Gutmann, Wagner, & Schulze-Briese, 2010). In this same study, however, specific radiation damage did show significantly greater cryo-protection, for example, that at solvent-exposed disulfide bonds (Meents et al., 2010).

3.5 Radiation Sensitivity of Enzyme Activity: Implications for Dynamic Studies in Liquid Samples

As indicated in Section 3.1, enzymes are inactivated by radiation doses of 10^6 rads or less in the liquid state. As seen in Fig. 2, this dose corresponds to an electron exposure of $\sim 2 \times 10^{-3}$ e/Å^2 or less. If two images are recorded at or above this exposure, and if the structure is observed to change between the first and the second such exposure, one can be sure that the changes are not due to some enzymatic function. Protein particles as small as ~5 nm, on the other hand, are calculated to first become detectable at an electron exposure of $\sim 2 \times 10^{-1}$ e/Å^2 under the most favorable circumstances, including the use of full phase contrast (Glaeser & Hall, 2011). There thus seems to be a large gap between what physics would allow for imaging biological specimens in the liquid state, using an environmental cell at room temperature and pressure (Ross, 2015), and what some would like to achieve.

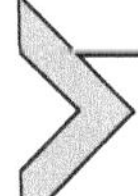

4. VITREOUS ICE ALSO BECOMES STRUCTURALLY DAMAGED BY IONIZING RADIATION

4.1 Water Molecules Are Easily Damaged by Ionizing Radiation

Radiolytic production of hydroxyl radicals, hydrogen peroxide, and other reactive species in liquid water plays a major role in radiation biology. As a result, there is an extensive literature on the radiolysis of water and aqueous solutions (Allen, 1961; Le Caër, 2011). Fig. 3 shows a summary of the main reactions that are currently believed to occur within the first picosecond after a water molecule has been ionized (Le Caër, 2011). The yield of the final, reactive intermediates shown in Fig. 3 is actually quite high, and it might be said that water is at least as radiation-sensitive as are proteins.

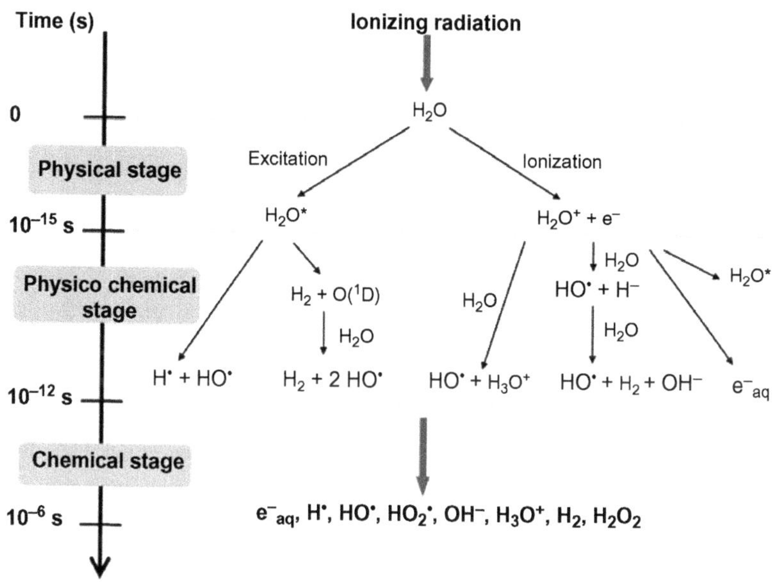

Fig. 3 Schematic progression of temporal events in the radiolysis of pure water. Products of the initial, "physical" stage of excitation and ionization evolve, within the first picosecond, to form multiple reactive intermediates such as hydrogen radicals, hydroxyl radicals, and hydrated electrons. Within a microsecond, hydrogen gas and hydrogen peroxide are among the products, as well as a number of the previously formed, reactive species. *Reproduced with permission from Le Caër, S. (2011). Water radiolysis: influence of oxide surfaces on H2 production under ionizing radiation.* Water, *3(1), 235.*

In pure water, many of the chemical intermediates shown in Fig. 3 have no alternative but to react with each other, ultimately returning nearly everything back to molecules of water. It is even believed that hydrogen radicals and hydroxyl radicals can convert hydrogen gas and hydrogen peroxide back to water (Le Caër, 2011). This very likely is why pure ice may appear to be unchanged when irradiated in the electron microscope.

4.2 Weak Thon Rings at High Resolution Show That Vitreous Ice Is Very Sensitive to Radiation Damage

It has long been thought to be strange that images of amorphous ice—in contrast to those for graphitic, amorphous carbon—show only very weak Thon rings, if any. It now seems that two factors are involved.

At low to medium resolution, the structure factor for vitreous ice should be very weak to begin with, as it is for liquid water. The reason for this is that density fluctuations for liquid water depend only on the temperature and the isothermal compressibility of the liquid—see eq. (1) in Clark, Hura, Teixeira, Soper, and Head-Gordon (2010). Amorphous carbon, on the other hand, is thought to have a somewhat granular, domain-like structure. In addition, the mass density of (predominantly) sp^2-bonded amorphous carbon films is two or three times greater than that of hydrogen-bonded amorphous ice. These differences all lead to much stronger structure factors, and thus stronger Thon rings, at low spatial frequencies for carbon compared to ice.

At high resolution, however, the diffraction pattern of amorphous ice, like that of liquid water, displays a strong feature known as the "water ring." One thus expects that Thon rings from ice should be especially strong at a resolution of 3–4 Å, but that is not the case.

Based on recent work by McMullan, Vinothkumar, and Henderson (2015), it now seems that radiation damage causes continual reorganization of the structural features responsible for the "water ring." As is illustrated in Fig. 4, proof that this is the case came from comparing the strength of Thon rings observed in the incoherent sum of power spectra (for successive frames

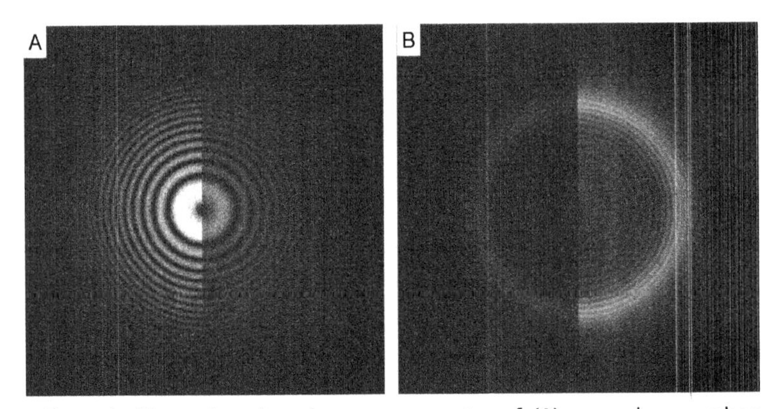

Fig. 4 Example illustrating that the power spectra of (A) amorphous carbon and (B) amorphous ice are dramatically different. In both cases images were obtained as dose-fractionated movies, using far greater electron exposures than could be tolerated by biological specimens, in order to improve the statistical definition of the power spectra. Each panel is, furthermore, split into two half-plane images in which the power spectrum of the coherent sum of frames is shown on the left half, and the "incoherent sum" of power spectra of individual frames is shown on the right half. *This figure was kindly prepared by Dr. Greg McMullan, using the same data published in McMullan, G., Vinothkumar, K. R., & Henderson, R. (2015). Thon rings from amorphous ice and implications of beam-induced Brownian motion in single particle electron cryo-microscopy.* Ultramicroscopy, 158, *26–32. doi: 10.1010/j.ultramic.2015.05.017. Republication of these data is with permission.*

of a dose-fractionated movie) to those observed in the power spectrum of the (aligned) sum of frames. If the specimen structure does not change from one frame to the next, one expects the incoherent sum of power spectra to have a lower signal-to-noise ratio (ie, the visibility of Thon rings should be less) and that was shown to be the case for a thin carbon film (see Fig. 4A). The opposite effect was seen for a thin vitreous-ice specimen; however (see Fig. 4B), leading to the conclusion that the high-resolution structure in one frame is not the same as (is not "fully coherent with") that in successive frames.

The resulting picture is that new, but equivalent structural features form in vitreous ice as the previous ones are lost due to radiation damage. In this picture, individual oxygen atoms (ie, water molecules) move significantly on the size scale of the structural organization responsible for producing the "water ring." While diffraction intensities (or power spectra) are insensitive to this reorganization, high-resolution image features are smeared out. In effect, there is a large Debye–Waller "temperature" factor (B factor) in the images, but not in the diffraction intensities.

McMullan et al. then went on to estimate that the root mean square displacement of water molecules may be ~5 Å after an exposure of 25 e/Å^2. These random, radiation-induced motions of water molecules must also jostle proteins, causing them to diffuse randomly within the vitreous ice, just as thermal motion of water molecules does in the liquid state. McMullan et al. conclude, however, that radiation-induced diffusion is unlikely to be a factor limiting the resolution attainable by single-particle cryo-EM except above ~2 Å, and then only for very small macromolecules.

4.3 Electron-Stimulated Desorption Progressively Thins Ice Specimens

It is commonly known that ice becomes progressively more transparent to the electron beam—ie, thinner, the longer that the same area is irradiated. This is actually a useful effect, as it provides a convenient way to see, when viewing the grid at low magnification (eg, in "Search" mode), where the electron beam was positioned on the specimen at the time when images were recorded at high magnification. If desired, even a complete hole can be "burned" through the ice, thereby providing a local area to accurately measure the incident electron intensity. This, in turn, makes it possible to estimate the ice thickness by making a quantitative measurement of the percent transmission of electrons in an area of interest (Agard et al., 2014).

Continuous removal of water molecules from the surface is an example of the well-known "electron-stimulated desorption" phenomenon (Ramsier & Yates, 1991). Radiation-induced thinning of ice at different temperatures was studied in detail by Heide (1984), using 100 keV electrons. He found that the rate of desorption increased steeply above 100 K, but at lower temperature, even down to 10 K, the rate remained constant. Heide reported that the rate of mass loss below 90 K corresponds to ~1 monolayer of water being removed from the surface for every 25 electrons/Å^2 of exposure. Heide also reported that the removal rate for microcrystalline cubic ice was higher than that for amorphous ice.

The amount of thinning expected for exposure to 300 keV electrons should be about half that for exposure to 100 keV electrons, ie, it should scale as the inelastic-scattering cross section. McMullan et al., on the other hand, quote a much greater rate, namely, 100 Å of ice for every 170 electrons/Å^2, attributing this estimate to Wright, Iancu, Tivol, and Jensen (2006). This discrepancy may be something that needs to be investigated further, but for most purposes, it should not affect the conduct of cryo-EM data collection.

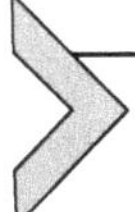

5. BUBBLING OF HYDRATED BIOLOGICAL SPECIMENS BECOMES APPARENT AT HIGH ELECTRON EXPOSURE

When hydrated biological materials are irradiated for a longer time than usual, microscopic bubbles begin to appear at random. The first noticeable bubbles appear after the accumulated exposure (for 300 keV electrons) is approximately 150 e/Å^2. At this high exposure, high-resolution features would long since be destroyed, of course, but the macromolecular particles might still be visible. At an earlier point, nascent bubbles or "nuclei" must have been present, but these are presumably too small to be seen in noisy images. The size of bubbles continues to grow with further exposure, soon leaving no trace of the biological macromolecule, but at some point the bubble growth finally stops. An example of the bubbling effect is shown in Fig. 5.

This bubbling effect occurs mainly for biological (and other organic) materials that are hydrated, and even for frozen solutions of small organic molecules such as glycerol. Bubbling is not observed in pure water at liquid nitrogen temperature, whether crystalline or amorphous. With only some

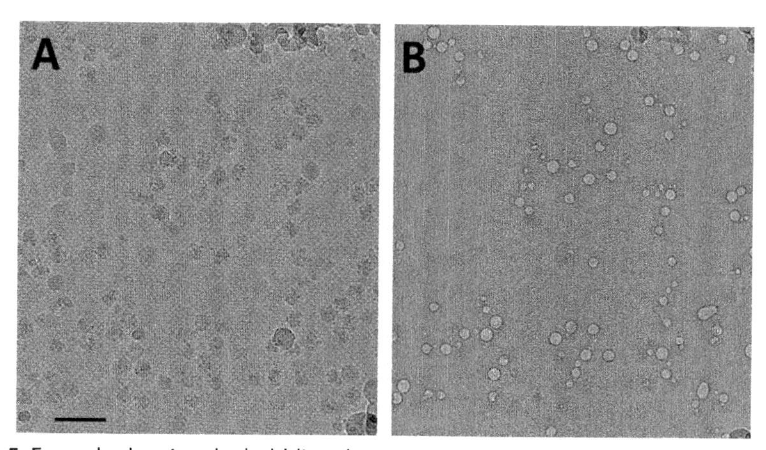

Fig. 5 Example showing the bubbling that occurs in a hydrated biological specimen at very high electron exposure. The specimen in this case consisted of biotinylated 70S ribosomes bound to a streptavidin monolayer crystal, which served as the support film spanning over the open holes of a holey carbon film. The bubbles have a variety of diameters, the largest of which may provide an estimate of the ice thickness. (A) High-resolution image recorded with an exposure of approximately 25 electrons/Å^2. (B) Image recorded after irradiating the sample with a high enough exposure that the growth of bubbles had ceased. Note that much of the globular surface contamination had sublimed at this point, as well. There is a clear correlation between the location of bubbles in (B), and the location of ribosome particles in (A), but not every ribosome particle nucleates a bubble. This image, taken from previously unpublished work using a sample provided by Dr. Arto Pulk, was prepared by Dr. Bong-Gyoon Han.

exceptions, bubbling is also not observed in dry biological materials. The exceptions include very thick organic specimens (eg, plastic sections ~0.5-μm thick), or relatively thick films of carbohydrate materials.

5.1 Bubbles Consist of Molecular Hydrogen

Leapman and Sun used electron energy-loss spectroscopy to identify H_2 gas as the radiolysis product that accumulates in these microscopic bubbles (Leapman & Sun, 1995). More recently, Meents et al. (2010) have also found that H_2 gas accumulates in protein crystals, and in a range of other test samples, when exposed to intense X-ray beams. In this case the gas forms macroscopically visible bubbles when the samples are warmed. Leapman and Sun also went on to estimate that the gas pressure can reach values as high as 1000 atm. At some point the growing bubble touches the surface, and the hydrogen gas suddenly escapes, leaving behind an empty hole in the vitreous ice.

It is not known whether the hydrogen gas is produced within protein, within water, or as a reaction product between the two. Since molecular hydrogen is a radiolysis product both for adjacent $-(CH_2)-$ groups and for pure water, both ought to contribute. What seems more certain is that biological materials appear to serve as nucleation sites for the formation of nascent bubbles, wherever the H_2 gas is first produced.

It is also worth noting that a greater electron exposure may be required in order to eventually observe bubbling at lower dose rates (Brilot et al., 2012; Chen et al., 2008), and the specimen may not show bubbling at all if it is thin enough. In order to account for these effects, it is hypothesized that slower production of H_2 gas, and thinner samples, both increase the chance that hydrogen can escape across the ice surface, and thus not contribute to the formation of bubbles.

5.2 Bubbling Can Be Used to Evaluate the Specimen Thickness

According to the model that bubbles grow until they finally touch the ice-vacuum surface, the largest bubbles would be ones that nucleate near the center of the sample. If this is the case, then the ice thickness can be estimated to be similar to the diameter of the largest bubbles. As an example, the largest bubbles in Fig. 4 are approximately 250 Å in diameter, which is a reasonable value for the thickness of ice-embedded ribosome particles.

5.3 Bubbling Can Be Used to Distinguish Regions with Different Chemical Composition (Bubblegrams)

It has been observed that bubbles form preferentially, and at unusually low electron doses, within the "inner body" protein components of bacterial phage particles (Cheng, Wu, Watts, & Steven, 2014; Wu, Thomas, Cheng, Black, & Steven, 2012). Images that show preferential formation of bubbles, called "bubblegrams," proved to be useful because this effect revealed the location of the inner body, which previously could not be distinguished from the surrounding DNA. In the case of the T7 phage particle, a secondary site of bubbling was also observed, for which bubbles seemed to appear later in the exposure series (Cheng et al., 2014). These studies have also provided information in support of the idea that allowing time for molecular hydrogen to escape the site where it was produced would delay the onset of bubble formation.

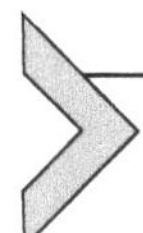

6. CRYO-SPECIMENS EXHIBIT COLLECTIVE BEAM-INDUCED MOVEMENT WHEN IRRADIATED

6.1 Radiation-Sensitive Specimens Show Beam-Induced Motion at Quite Low Electron Exposures

It has long been known that adjacent points within local areas of radiation-sensitive specimens move similarly (collectively) when samples are irradiated in the electron microscope. Perhaps the clearest example of collective motion is shown in the behavior of the bend contours exhibited by thin, organic crystals. These bend contours sweep across the face of such crystals well before the electron diffraction pattern itself has faded away. An example documenting such motion is included in a review of radiation damage and electron microscopy of organic polymers by Grubb (1974).

Movement of bend contours reflects a local change in the tilt angle for a given area of a crystalline specimen, such that different areas satisfy one or another Bragg-diffraction condition at different points in time. A dynamic change in the tilt angle does not necessarily mean, however, that there are associated translational shifts in the projected locations of the same points in the crystal.

It has also been suspected that the specimen height of irradiated areas may change relative to the plane of the EM grid. It is well known that the resolution in the image is degraded when a radiation-sensitive specimen is tilted, and this loss of resolution is always in a direction perpendicular to the tilt axis. Furthermore, the resolution is increasingly degraded, the higher the tilt angle. While these facts are easily explained by imagining that a thin crystal bows and flexes as it is irradiated, one cannot rule out the alternative hypothesis that electrostatic charging of the specimen causes the image to be deflected in a direction perpendicular to the tilt axis.

Another observation, which initially suggested the possibility of there being beam-induced specimen movement, was that the high-resolution image contrast is much less than it is expected to be (Henderson & Glaeser, 1985). Crystalline, radiation-sensitive specimens were used in these experiments in order to compare the strength of diffraction spots in the computed Fourier transforms of images to their corresponding strengths in electron diffraction patterns. Once again, the alternative possibility that charging might cause deflection of the image could not be completely ruled out. However, vermiculite, which is expected to become charged as easily as biological specimens, but which is far less prone to radiation damage, showed considerably higher image contrast, when adjusted for the values of its electron diffraction intensities.

In a recent follow-up to the earlier work, Glaeser, McMullan, Faruqi, and Henderson (2011) showed that images of paraffin could be greatly improved by using much thicker (carbon) support films. The explanation proposed was that the thicker carbon films were much stiffer, and thus they did not deform as easily in response to radiation damage.

The idea to illuminate only a small area of the specimen at a time, referred to as "spot-scan imaging," was introduced as a way to possibly reduce movement of the specimen (Bullough & Henderson, 1987; Downing & Glaeser, 1986). This idea proved to be quite successful, as long as specimens are prepared on continuous carbon films. When specimens are prepared as thin, self-supported films over open holes in the carbon films, however, charging effects severely limit the image quality when the size of the electron beam is smaller than the hole (Brink, Sherman, Berriman, & Chiu, 1998).

As is discussed in Chapter "Direct Electron Detectors" by Henderson and in Chapter "Processing of Cryo-EM Movie Data" by Rubinstein, the development of direct-detection cameras finally made it possible to unambiguously confirm the long-suspected phenomenon of beam-induced movement. More importantly, it also became possible to actually reduce the consequences of this movement by recording images as a series of dose-fractionated "movie frames."

6.2 Thin Cryo-Specimens Undergo Drum-Head-Like Flexing and Doming When Irradiated

One result that influenced thinking greatly, before the advent of direct-detection cameras, was the observation that particles could change their orientation by as much as 2 degree, while low-dose images were being recorded (Henderson et al., 2011). These changes in particle orientation occur as a collective behavior extending over local regions of the specimen, rather than being independent rotations of individual particles. The conclusion was that irradiation causes the thin-film specimen to tilt and bend relative to the plane of the EM grid.

Shortly thereafter, these studies led to recording images as a series of "movie frames" (Brilot et al., 2012). The results that were obtained added the crucial information that the particle rotations are accompanied by translations, both of which vary in magnitude and direction from one part of the field of view to another. In addition, both types of motion occur progressively, as the exposure continues, with the largest movements happening at the beginning of the exposure. As Brilot et al. then showed, alignment of the images in these frames is effective in compensating for the translational motion that occurs from frame to frame, resulting in better recovery of high-resolution detail.

In at least some cases, the observed particle rotations and translations suggested that the entire irradiated area of the specimen was moving like a drum-head, in effect forming a thin dome (Brilot et al., 2012). Exactly that same type of beam-induced distortion of the specimen had been demonstrated for paraffin crystals by Downing (1988). In the latter case, metal shadowing of the specimen was used to show that in-plane expansion must have occurred in the previously irradiated areas, which resulted in the formation of bulges, or "domes."

6.3 Images Can Be Corrupted Significantly by There Being Changes in *Z*-Height

As noted by Brilot et al. (2012), formation of domes can change the *Z*-height at the center of the field of view by as much as ~20 nm. In addition, we now know—see Chapter "Specimen Preparation for High-Resolution Cryo-EM" by Passmore—that thin-film specimens are susceptible to additional bending and warping, resulting in even greater changes in the *Z*-height within the field of view (Russo & Passmore, 2014). This additional movement, mainly perpendicular to the plane of the grid, happens if precautions are not taken to avoid crinkling of the support film as the sample is cooled, a phenomenon described by Booy and Pawley (1993). When changes in *Z*-height are comparable to the depth of focus for a given resolution, they will produce the same blurring of the image as that produced by idly changing the objective-lens focus, by the same amount, during an exposure. In this context, it is worthwhile to mention that the depth of focus is only 40 nm at a resolution of 4 Å, and just 10 nm at 2 Å resolution.

When the specimen is tilted, changes in *Z*-height (relative to the plane of the grid) also produce a significant component of image motion that is perpendicular to the tilt axis, as was mentioned in Section 6.1. Even for a tilt angle as small as 5 degree, which is difficult to avoid, a 200-Å change of in *Z*-height results in a component of motion perpendicular to the tilt axis that is more than 17 Å.

6.4 The Pattern of Beam-Induced Movement Can Be Quite Unpredictable

As was mentioned in Section 6.2, particle motions are locally correlated, but such motions may differ considerably from one area to another within the same image. In some cases the differences are symmetric enough to be explained by formation of a dome. In other cases, however, the variation from one area to another can be less simple to describe, almost as when there is turbulent flow. The latter type of locally correlated motion had been demonstrated previously by classifying the Fourier transforms of single particles, a technique introduced by Sander, Golas, and Stark (2003). These classes differed by the amount of motion, even its direction, and by the resolution to which Thon rings extended. When the locations of all particles within a given class was displayed, however, it turned out that members of different classes were grouped together in different areas of a micrograph (Glaeser & Hall, 2011).

Chapter "Processing of Cryo-EM Movie Data" by Rubinstein describes how movies can be processed in a way that accounts for such local correlations in particle motion. While compensating for beam-induced movement in this way is very effective, eliminating such movement to the greatest extent possible remains a top priority. Very significant progress, described in Chapter "Specimen Preparation for High-Resolution Cryo-EM" by Passmore, has been made in developing EM grids that exhibit much less beam-induced movement to begin with, and such grids thus leave less to be compensated by computational methods.

7. MORE THAN ONE MECHANISM MAY CONTRIBUTE TO BEAM-INDUCED MOTION

It is important to remember that the beam-induced particle motions described in Section 6 are actually motions that we observe to occur in the image. Such image motions might reflect mechanical motions that occur in the specimen, of course, but that is not the only possibility. It had long been believed, in fact, that the observed image motion was caused by charging of the specimen, which in turn caused deflection of the electrons as they formed the image. It seems very likely, in fact, that both mechanical motion and image-deflection do happen. Nevertheless, it is still not known how significant the contribution of charging actually is, when care is taken to avoid it as best as one can.

7.1 Cryo-EM Specimens, as Made, Are Expected to Be Under Considerable Stress

It seems certain that cryo-EM specimens, as prepared, are under considerable stress, just as is the case for any glass that has not been subjected to annealing. From a fundamental point of view, we know that the vitreous state is not one that is at thermodynamic equilibrium. Rather, we have to think of the water molecules as being "frustrated" in their current situation, prevented from getting to a lower-energy state by there being too little thermal energy to get out of the local minimum in which they are trapped.

The frustration referred to above occurs as the water solidifies during rapid cooling. The density of so-called low-density amorphous ice is about 0.94 g/cm^3, which is less than that of super-cooled water (~0.97 g/cm^3) at the temperature of vitrification. It thus is easy to imagine that the expansion that occurs as the solvent solidifies, while everything else is contracting, might result in stress within the vitrified sample.

After solidification occurs, further cooling is expected to generate additional mechanical stress in the sample. This is because the coefficient of thermal expansion for vitreous ice is much greater than that of the EM grid and its support film. As a result, a mismatch in the amount of contraction of these different materials must initially produce unbalanced stress, which, in turn, will deform the thin-foil sample until a static equilibrium is reached. There is also bound to be a mismatch in the thermal contraction behavior of biological samples and the surrounding, vitreous ice. This too might contribute to there being stress in the specimen at the time that it is first prepared.

7.2 Irradiation Can Relieve Mechanical Stress

As is described in Section 2.2, inelastic-scattering events deposit energy in packets that, for the most part, contain between 10 and 100 eV each. This amount of energy is enough to overcome activation barriers and allow local relaxation of stress in the close neighborhood to where the scattering event occurred. When the state of frustration is locally relieved, however, the overall balance of force (stress) changes, and the specimen as a whole can be expected to move by an amount sufficient to restore mechanical equilibrium. This much is only one part of the story, however, since the local input of large packets of energy can also drive a system out of equilibrium, as well as provide energy to overcome the activation barriers that initially serve to frustrate the system from reaching equilibrium.

7.3 Irradiation Can Generate (New) Mechanical Stress

There are multiple ways in which irradiation of cryo-EM specimens can generate stress, which in turn will cause movement (strain) in the specimen. As was pointed out by McBride, Segmuller, Hollingsworth, Mills, and Weber (1986), for example, when a chemical bond is broken, the two atoms involved move apart from a distance of ~1.5 Å to about 3.5 Å, ie, from a covalent-bond distance to van der Waals contact. More generally, the daughter-molecule products of radiolysis no longer fit into the cavity occupied by the parent molecule. In fact, McBride found that radiolysis of only about 5% of a sample causes the pressure to increase to about half the value that would convert graphite into diamond. Long before that point is reached, however, our thin, foil-like samples will easily bend and buckle in order to prevent the accumulating stress from becoming that large.

As was described in Section 5.1, one of the products of radiolysis in cryo-EM samples is hydrogen gas. Once a bubble of H_2 gas has nucleated, and other molecules of H_2 continue to be produced by further irradiation, the bubbles grow in size. While bubble growth causes obvious specimen motion by displacing adjacent material, visible bubbles only appear at much higher electron exposures than what are used to collect high-resolution data. What is not so obvious is the extent to which nucleation itself already contributes to specimen motion.

7.4 Which Comes First, Relaxation or Creation of Stress?

Beam-induced specimen motion is often observed to occur in two phases, an initial, rapid "burst" phase followed by a slower phase that continues almost indefinitely (see Chapter "Processing of Cryo-EM Movie Data" by Rubinstein and Chapter "Specimen Preparation for High-Resolution Cryo-EM" by Passmore for more about this). Little is known about why the rate of beam-induced movement changes in this way, but two hypotheses seem reasonable. One possibility is that the burst phase reflects the relaxation of stress that had been created during vitrification and subsequent (ie, further) cooling, and the second, slower phase reflects the continuous generation of stress due to radiation damage. A second possibility is that generation of damage-related stress dominates throughout, and any initial relaxation of stress has little to do with beam-induced movement. In this model, it might be hypothesized that beam-induced movement would still be greatest at the beginning, because that is when the specimen experiences the most damage. Later, after almost everything has been damaged, there would be little change in the mechanical stress within the sample.

8. IRRADIATION CAN PRODUCE ELECTROSTATIC CHARGING OF THE SPECIMEN

Irradiation of any specimen by high-energy electrons produces secondary electrons that escape from the surface. If the specimen is an insulator, as is the case for cyro-EM specimens, it will, as a result, become positively charged. Charging of specimens can, in principle, cause beam-induced movement to occur in two different ways.

One possible mechanism is that electrostatic forces might cause mechanical movement of the specimen. A spot of uncompensated charge on the specimen would be attracted to the nearest conducting surface, the objective aperture or the lens pole piece, for example. Then, since the specimen is a flexible foil, this attractive force would cause bending and warping to occur, just as was described in Section 6. The other possible mechanism is that charging might cause deflection of the image, rather than mechanical motion of the specimen. The two mechanisms are not mutually exclusive, of course, so both may happen simultaneously. It thus is appropriate to describe, a bit further, what is known about the effects of specimen charging, and how to minimize these effects.

8.1 A Buildup of Net Charge on the Specimen Can Be Easy to Detect

Very obvious electron-optical effects occur when an electrically insulating material is put into the electron beam. Examples of such specimens include thin plastic support films and uncoated plastic sections of tissue. Even uncoated biological specimens that are self-supported over holes in a carbon film show strong charging effects when the beam is confined to within the area of the hole.

The effect that is easiest to observe with such specimens is that it can become nearly impossible to focus the unscattered beam in the electron diffraction pattern (Brink, Sherman, et al., 1998; Curtis & Ferrier, 1969). Further, even when the unscattered beam is focused to the best extent possible, moving the specimen by a small amount again causes major distortion of the beam. Looking at the (focused) diffraction pattern is thus a good way to tell whether charging deflects the incident electron beam. In effect, specimen charging can produce an unwanted, additional lens, whose optical quality might be quite nonideal!

8.2 Evidence of Net-Charge Buildup Can Be Reduced in Several Ways

When specimens are prepared on a continuous carbon film, or when they are coated afterward with evaporated carbon, there is no longer any difficulty to focus the diffraction pattern (Brink, Gross, Tittmann, Sherman, & Chiu, 1998), and there is no obvious change in the focused, unscattered beam as the specimen is moved about. It thus appears that a compensating amount of negative charge is induced in the electrically conductive carbon, matching the positive charge left behind by the escape of secondary electrons (Glaeser & Downing, 2004). The result is to generate an electrostatic-dipole structure rather than an uncompensated, Coulombic charge on the specimen.

It has long been known, as well, that use of an objective aperture reduces specimen charging substantially. The explanation for this effect is that scattered electrons produce low-energy secondary electrons when they hit the aperture. These low-energy electrons are attracted to, and neutralize the positive charge on the surface of the specimen.

Another, commonly used way to minimize charging of uncoated specimens is to include part of the surrounding carbon film within the field that is illuminated by the electron beam. There is compelling evidence that low-energy secondary electrons are once again produced, this time by the surrounding carbon film, and these neutralize the positive charge on the surface of the specimen (Berriman & Rosenthal, 2012). There is also the possibility that the phenomenon of radiation-induced conductivity plays a role in preventing significant specimen charging (Curtis & Ferrier, 1969; Downing, McCartney, & Glaeser, 2004), provided that that the beam illuminates a continuous path between the area of interest and an adjacent part of the carbon support film.

8.3 Other Forms of Specimen Charging Are More Subtle to Detect

There are two additional effects, attributable to specimen charging, which become observable only at low magnification, and then only in highly defocused images. Since it has not been established whether either form contributes to degradation of high-resolution images, it is not known whether these phenomena are things that we need to be concerned about. Still, any form of charging is unwelcome, and thus—for completeness—brief descriptions are included here.

The first effect was initially reported by Dove (1964), who noticed that very low-magnification images of thin carbon films exhibit granular features that fluctuate rapidly in time. Dove coined the phrase "bee swarm" to convey the visual impression that is observed. This effect was then pursued by others including Curtis and Ferrier (1969), who attempted to give a quantitative theoretical explanation of the effect.

The second type of effect was initially described by John Berriman, who noticed that the electron beam can leave an erasable mark on thin carbon films. In the original observations, the irradiated area appeared darker than the surround when the low-magnification image was under-focused, and brighter than the surround when the image was over-focused. This contrast reversal is consistent with the projected Coulomb potential of the irradiated area being more positive than that of the surround. A hallmark of this behavior, sometimes called the "Berriman effect," is that the mark disappears again when an adjacent area of the specimen is irradiated (Downing et al., 2004). This behavior rules out the possibility that the beam-induced mark was due to the well-known buildup of hydrocarbon contamination that occurs when specimens are exposed to intense, focused electron beams at room temperature. A second hallmark of the Berriman effect is the fact that the amount of contrast in the marked location soon saturates. This behavior again differs from what happens in the case of beam-induced contamination, the thickness of which continues to increase the longer the specimen is irradiated.

Further work has found that the radiation-induced mark, while saturable (and thus not due to contamination) is not always erasable (Downing et al., 2004). Operationally, the charging pattern created by irradiating the sample cannot be reversed either by low-energy secondary electrons or by radiation-induced conductivity of adjacent areas.

To complicate things even further, the "Berriman mark" can show the opposite focus-dependent contrast, consistent with the projected Coulomb potential of the irradiated area being less positive (more negative) than that of the surround. The same thing is seen when evaporated carbon films are used to make a "Volta phase plate" for the electron microscope. In this last case it is suggested that the change in projected Coulomb potential reflects a change in the dipole moment per unit area in the region exposed to the intense electron beam, compared to the surround (Danev, Buijsse, Khoshouei, Plitzko, & Baumeister, 2014).

One other point that seems worth mentioning here is that very thin cryo-specimens often show a conventional, dark-contrast "Berriman effect" mark after some tens or hundreds of seconds of exposure, under the same conditions otherwise used for low-dose data collection. This mark on the specimen can be seen in the low-magnification "Search" mode that is used when identifying areas for further data collection. More commonly than not, this mark is seen as a dark ring at the perimeter of the irradiated area (Brink, Sherman, et al., 1998), rather than a uniform, dark disk, as is the case for the original Berriman effect. Thicker cryo-specimens, on the other hand, do not show such a mark, but instead show the progressive thinning of the irradiated area that is described in Section 4.3. One speculation is that, in regions of thicker ice, positive charge might be more completely carried off by electron-stimulated desorption of ionized water molecules, thereby avoiding the buildup of any significant, positive surface charge.

9. SUMMARY AND FUTURE DIRECTIONS

Vitreous ice and biological macromolecules both sustain extensive amounts of structural damage when irradiated in the electron microscope. It thus has been standard practice to limit the exposure to a "safe" value, ie, not much more than what destroys features of interest.

The ability to record images as a series of movie frames, has established, more clearly than ever before, that significant beam-induced motion occurs long before high-resolution features are fully destroyed. A large component of this beam-induced motion occurs in a direction perpendicular to the plane of the EM grid. A much smaller component is in the plane of the specimen, and—unless compensated by aligning successive movie frames, this smaller component is still large enough to severely limit the image resolution.

Our current ideas of what causes beam-induced movement include the relief of preexisting stress in the specimen as well as progressive generation of (new) stress as the exposure continues. Although specimen charging produces well-known corruption of images, at least the most severe of these effects can be avoided by a number of recommended practices. The extent to which residual specimen charging nevertheless contributes to beam-induced motion, either by generating electrostatic forces or by electron-optical deflection of the image, remains yet to be resolved.

As is described in other chapters of this volume, current research has made significant progress to reduce the amount of beam-induced motion to begin with, and to more effectively recover the signal that would otherwise be lost due to that motion. Nevertheless, mitigation of beam-induced motion remains at the frontier of cryo-EM methodology, and further improvements are bound to be welcome. The ability to prevent local changes in specimen tilt has not yet been addressed, for example. Beam-induced motion may thus continue to be an important issue, especially as the resolution of cryo-EM structures continues to press to ever better values.

ACKNOWLEDGMENTS

It is a pleasure to thank numerous colleagues, especially Kenneth H. Downing, Bong-Gyoon Han, and Richard Henderson, with whom many of the concepts in this chapter, as well as several research papers cited here, have been discussed over many years. Furthermore, I especially want to thank Dr. Greg McMullan for preparing Fig. 4, and Dr. Bong-Gyoon Han for preparing Fig. 5.

REFERENCES

Agard, D., Cheng, Y. F., Glaeser, R. M., & Subramaniam, S. (2014). Single-particle cryo-electron microscopy (cryo-EM): Progress, challenges, and perspectives for further improvement. In P. W. Hawkes (Ed.), *Advances in imaging and electron physics: Vol. 185* (pp. 113–137).

Allen, A. O. (1961). *The radiation chemistry of water and aqueous solutions*. Princeton, NJ: Van Nostrand.

Aronova, M. A., & Leapman, R. D. (2012). Development of electron energy loss spectroscopy in the biological sciences. *MRS Bulletin/Materials Research Society*, *37*(1), 53–62. http://dx.doi.org/10.1557/mrs.2011.329.

Baker, L. A., Smith, E. A., Bueler, S. A., & Rubinstein, J. L. (2010). The resolution dependence of optimal exposures in liquid nitrogen temperature electron cryomicroscopy of catalase crystals. *Journal of Structural Biology*, *169*(3), 431–437. http://dx.doi.org/10.1016/j.jsb.2009.11.014.

Bammes, B. E., Jakana, J., Schmid, M. F., & Chiu, W. (2010). Radiation damage effects at four specimen temperatures from 4 to 100 K. *Journal of Structural Biology*, *169*(3), 331–341. http://dx.doi.org/10.1016/j.jsb.2009.11.001.

Berger, M. J., & Selzer, S. M. (1964). Tables of energy-losses and ranges of electrons and positrons. *Studies in penetration of charged particles in matter* (pp. 205–268). Washington, DC: National Academy of Sciences—National Research Council.

Berriman, J. A., & Rosenthal, P. B. (2012). Paraxial charge compensator for electron cryomicroscopy. *Ultramicroscopy*, *116*, 106–114. http://dx.doi.org/10.1016/j.ultramic.2012.03.006.

Booy, F. P., & Pawley, J. B. (1993). Cryo-crinkling—What happens to carbon-films on copper grids at low-temperature. *Ultramicroscopy*, *48*(3), 273–280.

Brilot, A. F., Chen, J. Z., Cheng, A. C., Pan, J. H., Harrison, S. C., Potter, C. S., ... Grigoriefff, N. (2012). Beam-induced motion of vitrified specimen on holey carbon film. *Journal of Structural Biology*, *177*(3), 630–637. http://dx.doi.org/10.1016/j.jsb.2012.02.003.

Brink, J., Gross, H., Tittmann, P., Sherman, M. B., & Chiu, W. (1998). Reduction of charging in protein electron cryomicroscopy. *Journal of Microscopy*, *191*, 67–73.

Brink, J., Sherman, M. B., Berriman, J., & Chiu, W. (1998). Evaluation of charging on macromolecules in electron cryomicroscopy. *Ultramicroscopy*, *72*(1–2), 41–52.

Bullough, P., & Henderson, R. (1987). Use of spot-scan procedure for recording low-dose micrographs of beam-sensitive specimens. *Ultramicroscopy*, *21*(3), 223–229.

Chattarji, D. (1976). *The theory of auger transitions*. London/New York: Academic Press.

Chen, J. Z., Sachse, C., Xu, C., Mielke, T., Spahn, C. M. T., & Grigorieff, N. (2008). A dose-rate effect in single-particle electron microscopy. *Journal of Structural Biology*, *161*(1), 92–100. http://dx.doi.org/10.1016/j.jsb.2007.09.017.

Cheng, N. Q., Wu, W. M., Watts, N. R., & Steven, A. C. (2014). Exploiting radiation damage to map proteins in nucleoprotein complexes: The internal structure of bacteriophage T7. *Journal of Structural Biology*, *185*(3), 250–256. http://dx.doi.org/10.1016/j.jsb.2013.12.004.

Chiu, W., Downing, K. H., Dubochet, J., Glaeser, R. M., Heide, H. G., Knapek, E., … Zemlin, F. (1986). Cryoprotection in electron microscopy. *Journal of Microscopy, 141*(3), 385–391. http://dx.doi.org/10.1111/j.1365-2818.1986.tb02731.x.

Clark, G. N. I., Hura, G. L., Teixeira, J., Soper, A. K., & Head-Gordon, T. (2010). Small-angle scattering and the structure of ambient liquid water. *Proceedings of the National Academy of Sciences of the United States of America, 107*(32), 14003–14007. http://dx.doi.org/10.1073/pnas.1006599107.

Cosslett, V. E. (1970). Beam and specimen: Radiation damage and image resolution. *Berichte der Bunsengesellschaft für Physikalische Chemie, 74*(11), 1171–1175. http://dx.doi.org/10.1002/bbpc.19700741115.

Curtis, G. H., & Ferrier, R. P. (1969). The electric charging of electron-microscopic specimens. *Journal of Physics D: Applied Physics, 2*(7), 1035–1040. http://dx.doi.org/10.1088/0022-3727/2/7/312.

Danev, R., Buijsse, B., Khoshouei, M., Plitzko, J. M., & Baumeister, W. (2014). Volta potential phase plate for in-focus phase contrast transmission electron microscopy. *Proceedings of the National Academy of Sciences of the United States of America, 111*(44), 15635–15640.

Dove, D. B. (1964). Image contrasts in thin carbon films observed by shadow electron microscopy. *Journal of Applied Physics, 35*(5), 1652–1653. http://dx.doi.org/10.1063/1.1713709.

Downing, K. H. (1988). Observations of restricted beam-induced specimen motion with small-spot illumination. *Ultramicroscopy, 24*(4), 387–398.

Downing, K. H., & Glaeser, R. M. (1986). Improvement in high-resolution image quality of radiation-sensitive specimens achieved with reduced spot size of the electron-beam. *Ultramicroscopy, 20*(3), 269–278.

Downing, K. H., McCartney, M. R., & Glaeser, R. M. (2004). Experimental characterization and mitigation of specimen charging on thin films with one conducting layer. *Microscopy and Microanalysis, 10*(6), 783–789.

Garrison, W. M. (1968). Radiation chemistry of organo-nitrogen compounds. In M. Ebert & A. Howard (Eds.), *Current topics in radiation research*: *Vol. 4*. Amsterdam: North-Holand Publishing Company.

Glaeser, R. M., & Downing, K. H. (2004). Specimen charging on thin films with one conducting layer: Discussion of physical principles. *Microscopy and Microanalysis, 10*(6), 790–796.

Glaeser, R. M., Downing, K., DeRosier, D., Chiu, W., & Frank, J. (2007). *Electron crystallography of biological macromolecules*. Oxford, UK: Oxford University Press.

Glaeser, R. M., & Hall, R. J. (2011). Reaching the information limit in cryo-EM of biological macromolecules: Experimental aspects. *Biophysical Journal, 100*(10), 2331–2337. http://dx.doi.org/10.1016/j.bpj.2011.04.018.

Glaeser, R. M., McMullan, G., Faruqi, A. R., & Henderson, R. (2011). Images of paraffin monolayer crystals with perfect contrast: Minimization of beam-induced specimen motion. *Ultramicroscopy, 111*(2), 90–100. http://dx.doi.org/10.1016/j.ultramic.2010.10.010.

Glaeser, R. M., & Taylor, K. A. (1978). Radiation-damage relative to transmission electron-microscopy of biological specimens at low-temperature—Review. *Journal of Microscopy, 112*, 127–138.

Grant, T., & Grigorieff, N. (2015). Measuring the optimal exposure for single particle cryo-EM using a 2.6 Å reconstruction of rotavirus VP6. *Elife, 4*, e06980. http://dx.doi.org/10.7554/eLife.06980.

Grimm, R., Typke, D., Barmann, M., & Baumeister, W. (1996). Determination of the inelastic mean free path in ice by examination of tilted vesicles and automated most probable loss imaging. *Ultramicroscopy, 63*(3–4), 169–179.

Grubb, D. T. (1974). Radiation damage and electron microscopy of organic polymers. *Journal of Materials Science*, *9*(10), 1715–1736. http://dx.doi.org/10.1007/bf00540772.
Hall, E. J., & Giaccia, A. J. (2012). *Radiobiology for the radiologist*. Philadelphia: Wolters Kluwer Health/Lippincott Williams & Wilkins.
Hayward, S. B., & Glaeser, R. M. (1979). Radiation-damage of purple membrane at low-temperature. *Ultramicroscopy*, *4*(2), 201–210.
Heide, H. G. (1984). Observations on ice layers. *Ultramicroscopy*, *14*(3), 271–278. http://dx.doi.org/10.1016/0304-3991(84)90095-0.
Henderson, R. (1990). Cryoprotection of protein crystals against radiation-damage in electron and X-ray-diffraction. *Proceedings of the Royal Society of London. Series B, Biological Sciences*, *241*(1300), 6–8.
Henderson, R., Chen, S. X., Chen, J. Z., Grigorieff, N., Passmore, L. A., Ciccarelli, L., … Rosenthal, P. B. (2011). Tilt-pair analysis of images from a range of different specimens in single-particle electron cryomicroscopy. *Journal of Molecular Biology*, *413*(5), 1028–1046. http://dx.doi.org/10.1016/j.jmb.2011.09.008.
Henderson, R., & Glaeser, R. M. (1985). Quantitative analysis of image contrast in electron micrographs of beam-sensitive crystals. *Ultramicroscopy*, *16*(2), 139–150.
Howells, M. R., Beetz, T., Chapman, H. N., Cui, C., Holton, J. M., Jacobsen, C. J., … Starodub, D. (2009). An assessment of the resolution limitation due to radiation-damage in X-ray diffraction microscopy. *Journal of Electron Spectroscopy and Related Phenomena*, *170*(1–3), 4–12. http://dx.doi.org/10.1016/j.elspec.2008.10.008.
Kempner, E. S. (1993). Damage to proteins due to the direct action of ionizing radiation. *Quarterly Reviews of Biophysics*, *26*(1), 27–48.
Kempner, E. S., & Schlegel, W. (1979). Size determination of enzymes by radiation inactivation. *Analytical Biochemistry*, *92*(1), 2–10. http://dx.doi.org/10.1016/0003-2697(79)90617-1.
Le Caër, S. (2011). Water radiolysis: Influence of oxide surfaces on H2 production under ionizing radiation. *Water*, *3*(1), 235.
Leapman, R. D., & Sun, S. Q. (1995). Cryoelectron energy-loss spectroscopy—Observations on vitrified hydrated specimens and radiation-damage. *Ultramicroscopy*, *59*(1–4), 71–79.
McBride, J. M., Segmuller, B. E., Hollingsworth, M. D., Mills, D. E., & Weber, B. A. (1986). Mechanical stress and reactivity in organic solids. *Science*, *234*(4778), 830–835.
McMullan, G., Vinothkumar, K. R., & Henderson, R. (2015). Thon rings from amorphous ice and implications of beam-induced Brownian motion in single particle electron cryo-microscopy. *Ultramicroscopy*, *158*, 26–32. http://dx.doi.org/10.1010/j.ultramic.2015.05.017.
Meents, A., Gutmann, S., Wagner, A., & Schulze-Briese, C. (2010). Origin and temperature dependence of radiation damage in biological samples at cryogenic temperatures. *Proceedings of the National Academy of Sciences of the United States of America*, *107*(3), 1094–1099. http://dx.doi.org/10.1073/pnas.0905481107.
Ramsier, R. D., & Yates, J. T. (1991). Electron-stimulated desorption: Principles and applications. *Surface Science Reports*, *12*(6–8), 243–378.
Reimer, L., & Kohl, H. (2008). *Transmission electron microscopy physics of image formation*. New York: Springer-Verlag. http://site.ebrary.com/id/10274982.
Ross, F. M. (2015). Opportunities and challenges in liquid cell electron microscopy. *Science*, *350*(6267), 1490. http://dx.doi.org/10.1126/science.aaa9886.
Russo, C. J., & Passmore, L. A. (2014). Ultrastable gold substrates for electron cryomicroscopy. *Science*, *346*(6215), 1377–1380. http://dx.doi.org/10.1126/science.1259530.

Sander, B., Golas, M. M., & Stark, H. (2003). Automatic CTF correction for single particles based upon multivariate statistical analysis of individual power spectra. *Journal of Structural Biology, 142*(3), 392–401.

Taylor, K. A., & Glaeser, R. M. (1976). Electron microscopy of frozen hydrated biological specimens. *Journal of Ultrastructure Research, 55*(3), 448–456.

Weik, M., Ravelli, R. B. G., Kryger, G., McSweeney, S., Raves, M. L., Harel, M., … Sussman, J. L. (2000). Specific chemical and structural damage to proteins produced by synchrotron radiation. *Proceedings of the National Academy of Sciences of the United States of America, 97*(2), 623–628. http://dx.doi.org/10.1073/pnas.97.2.623.

Wright, E. R., Iancu, C. V., Tivol, W. F., & Jensen, G. J. (2006). Observations on the behavior of vitreous ice at ~82 and ~12K. *Journal of Structural Biology, 153*(3), 241–252. http://dx.doi.org/10.1016/j.jsb.2005.12.003.

Wu, W. M., Thomas, J. A., Cheng, N. Q., Black, L. W., & Steven, A. C. (2012). Bubblegrams reveal the inner body of bacteriophage phi KZ. *Science, 335*(6065), 182. http://dx.doi.org/10.1126/science.1214120.

CHAPTER THREE

Specimen Preparation for High-Resolution Cryo-EM

L.A. Passmore[1], C.J. Russo[1]
MRC Laboratory of Molecular Biology, Cambridge, United Kingdom
[1]Corresponding authors: e-mail address: passmore@mrc-lmb.cam.ac.uk; crusso@mrc-lmb.cam.ac.uk

Contents

Abstract

Imaging a material with electrons at near-atomic resolution requires a thin specimen that is stable in the vacuum of the transmission electron microscope. For biological samples, this comprises a thin layer of frozen aqueous solution containing the biomolecular complex of interest. The process of preparing a high-quality specimen is often the limiting step in the determination of structures by single-particle electron cryomicroscopy (cryo-EM). Here, we describe a systematic approach for going from a purified biomolecular complex in aqueous solution to high-resolution electron micrographs that are suitable for 3D structure determination. This includes a series of protocols for the preparation of vitrified specimens on various supports, including all-gold and graphene. We also describe techniques for troubleshooting when a preparation fails to yield suitable specimens, and common mistakes to avoid during each part of the process. Finally, we include recommendations for obtaining the highest quality micrographs from prepared specimens with current microscope, detector, and support technology.

Methods in Enzymology, Volume 579
ISSN 0076-6879
http://dx.doi.org/10.1016/bs.mie.2016.04.011

1. INTRODUCTION

In the past 3 years, spectacular progress has been made in the ability to determine structures by cryo-EM. But the method as a whole is still in its adolescence when compared with established methods like X-ray crystallography. Recent advances, including direct electron detectors with increased quantum efficiency (see chapter "Direct Electron Detectors" by McMullan et al.), easier-to-use microscopes with automated alignment and data collection (see chapter "Strategies for Automated CryoEM Data Collection Using Direct Detectors" by Cheng et al.), improved software for the classification and reconstruction of density maps (see chapters "Processing of Structurally Heterogeneous Cryo-EM Data in RELION" by Scheres, "Single Particle Refinement and Variability Analysis in EMAN2.1" by Ludtke, and "FREALIGN: An Exploratory Tool for Single-Particle Cryo-EM" by Grigorieff), and more stable and reproducible specimen support technology (Russo & Passmore, 2016a), have coalesced to bring the method out of its infancy. Still, for many projects, the major limiting factor for structure determination is specimen preparation.

The origin of this limitation is twofold in nature:

(1) During the creation of a thin layer of water for vitrification and imaging, specimens are exposed to surfaces and conditions which are very different from the inside of a test tube or cell. The effects of these on the molecules and complexes are not known a priori, and can be difficult to remedy if destructive to the specimen.

(2) Specimen preparation for cryo-EM is a delicate process that still requires skilled handling and careful technique through a number of detailed preparation steps. This often confounds novice and experienced microscopists alike by making it difficult to distinguish problems with the specimen from problems in technique and methods.

Together, these obstacles can slow the progress of a cryo-EM project and prevent structure determination, irrespective of other advances in cryo-EM technology. While an improved understanding of protein–surface interactions as well as new approaches and technology will be necessary to fully remove these obstacles to structure determination, there are a number of techniques currently available to improve specimen preparation in practice. Our intent here is to provide a systematic approach to the problem of specimen preparation and help novice microscopists to avoid many common pitfalls. We also provide protocols for various parts of the process and make recommendations for microscope and data collection settings to make this as efficient as possible. We do not provide the theoretical and historical background to cryo-EM, as there are several other recent reviews that address this (Cheng, 2015; Henderson, 2015; Nogales & Scheres, 2015; Vinothkumar & Henderson, 2016). The field of cryo-EM is rapidly evolving so methods will continue to advance as the technology improves. Still, the procedures provided here represent a snapshot of the current state-of-the-art, and the systematic approach to specimen preparation delineated below is likely to remain important even in the context of future disruptive advances in the field.

2. A SYSTEMATIC APPROACH TO SPECIMEN PREPARATION

Cryo-EM specimen preparation, as we consider it here, can be defined in one sentence: It is the process of taking an aqueous sample of a biological material (usually a purified protein complex), applying it to a support structure (grid), reducing its dimension to a layer that is as thin as possible (~100–800 Å depending on the size of the biological molecule), and then freezing this layer fast enough to prevent the water from crystallizing. This process is essentially the same now as when the method was first developed by Dubochet and colleagues in the 1980s (Adrian, Dubochet, Lepault, & McDowall, 1984). Still, as Dubochet recognized even then (Dubochet et al., 1988; Dubochet, Adrian, Lepault, & McDowall, 1985), many aspects of this process are problematic because we have a poor understanding, and poor control of the microscopic surfaces, materials, and dynamic changes that the purified complexes encounter during their journey from the test tube to the thin layer of vitreous ice. In practice, this means that what we see on the grid in the cryomicroscope often bears little resemblance to what we know was in the test tube.

To address this problem in the most efficient way possible, with respect to both the time of the scientist and the use of expensive resources like electron microscopes, each step of the sample preparation process should be assessed. This affords one a much better chance of determining where problems occur and hopefully, of systematically reaching the goal of an excellent specimen that yields a high-resolution structure with a minimum of data processing and effort. The process is outlined in Fig. 1 and described in detail below, followed by procedures to perform most of the steps listed.

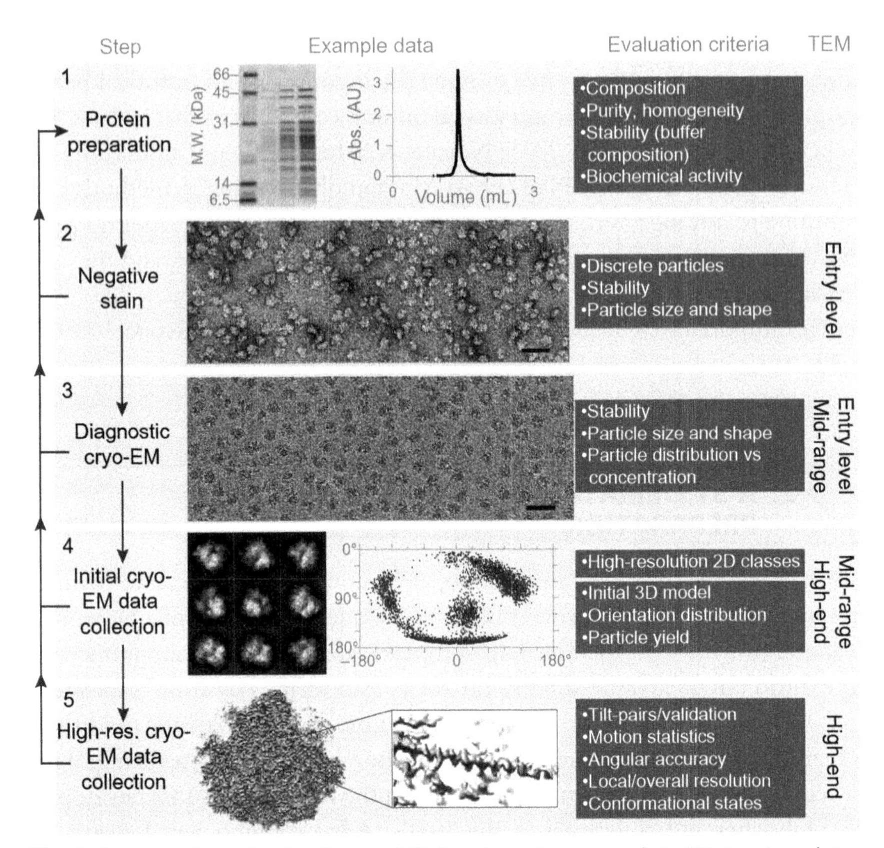

Fig. 1 *Structure determination by cryo-EM.* A systematic approach to 3D structure determination is shown. In the left column, the major steps are listed. Each step should be performed successively and only after one has been completed successfully should the scientist move onto the next step. In the second column, example data are shown for ribosomes (details in text). Scale bars on the micrographs are 500 Å. Each step should be evaluated with the criteria listed in the third column, returning to earlier steps for troubleshooting. The final column lists the class of electron microscope to be used, as defined in Table 1.

Table 1 General Classes of Transmission Electron Cryomicroscopes

Microscope Class	Typical Examples	~ Marginal Cost/Day (in 2016 £, Including Detectors)
Entry level	FEI T12/Spirit, JEOL 1400	250
Mid-range	FEI F20/Talos, JEOL 2100F	600
Upper-mid-range	FEI F30/Polara, JEOL 3200FS	1000
High-end	FEI Titan Krios	3000

2.1 Protein Preparation

The first step in a systematic approach is to evaluate protein composition and homogeneity thoroughly using biochemical methods (Fig. 1). Contaminating proteins or degradation products may interfere with complex stability and subsequent computational analysis of the particle images, wasting resources on the more time-consuming and expensive cryo-EM data collection and image processing steps. Specimen homogeneity can be evaluated using SDS-polyacrylamide gel electrophoresis (PAGE) and proteins can be identified using mass spectrometry. A Coomassie blue stained gel of purified ribosomes is shown in Fig. 1. It is important to ensure that only a single species is present and this can be judged using various techniques including native-PAGE, size exclusion chromatography (example chromatogram shown in Fig. 1), dynamic light scattering (DLS), and size exclusion chromatography coupled to multiangle light scattering (SEC-MALS). Using these methods to monitor subunit association, sample stability can then be optimized by changing buffer conditions (eg, salt, pH, and detergent). In addition, functional or biochemical assays should be used to test the activity of the protein in a given buffer condition. All of these steps should be done *prior* to making the first EM specimen, since no EM specimen will be better than the preparation from which it was made.

EM specimens are typically prepared using 3 μl protein solution at a concentration of 0.05–5 μM. Thus, it is essential for the protein complex to remain intact at these concentrations. If the dissociation constant (K_d) for the subunits is known, one can calculate whether it is expected to remain intact. Experimentally, one can run the protein complex on a size exclusion column repeatedly, at decreasing concentrations, to ensure it will not dissociate at the concentration required for cryo-EM. If the complex does dissociate, a few options are available. One can work at concentrations above the K_d and adjust the subsequent plasma and blotting conditions to achieve a very thin layer that only allows a monolayer of complex. Alternatively, a chemical cross-linking agent can be used to covalently link the subunits together, possibly preventing their dissociation after dilution (Plaschka et al., 2015; Weis et al., 2015). Cross-linking can be performed in solution or within a glycerol or sucrose gradient (GraFix) (Kastner et al., 2008; Stark & Grant, 2010). It is important to minimize (eg, using optimal protein and cross-linker concentrations) or remove (eg, using gradients or size exclusion chromatography) aggregates which can occur when multiple complexes are cross-linked to each other.

2.2 Negative Stain

After suitable protein is purified, the next step is to evaluate it using negative stain electron microscopy (Fig. 1). In this method, the protein is embedded in a layer of heavy metal salts (eg, uranium, molybdenum, or tungsten) which surrounds the protein like a shell (Brenner & Horne, 1959; Unwin, 1974). Since the density of the stain is about 3× higher than the protein, there is excellent contrast even with modest dose on an inexpensive microscope. A negative stain micrograph is shown in Fig. 1.

Negative staining is fast—making and imaging a specimen takes around the same time as running a gel—and is best used to rapidly assess various parameters in a preparation like pH, salt and buffer conditions, or the different fractions from a chromatography run. It reveals the solvent-excluded surface and shape of molecules and can therefore be used to evaluate homogeneity and size of proteins as well as the presence of binding partners. A homogenous sample will have discrete particles of uniform size. Averaging of a translationally (but not rotationally) aligned set of negatively-stained particles will provide an estimate of the diameter of the protein. This should be consistent with the size determined by solution methods like DLS or SEC-MALS. If the size and shape of the particles are not consistent, this is a clear sign that something is wrong and further troubleshooting should be undertaken before starting cryo-EM preparations. 3D structures can also be determined from negative stain and these can be used as a starting point for structure determination by cryo-EM, although in practice, this is often not required with newer algorithms for generating initial maps for 3D reconstruction (Elmlund, Elmlund, & Bengio, 2013; Tang et al., 2007). Methods for preparation of negatively stained specimens have been described previously and we refer the reader to these for the details (Ohi, Li, Cheng, & Walz, 2004). Even in the era of high-resolution cryo-EM, negative staining remains a valuable method for evaluating protein preparations quickly and cheaply.

2.3 Diagnostic Cryo-EM

After one is confident that a given protein preparation is monodisperse, stable in well-defined solution conditions, and yields a uniform distribution of discrete particles in negative stain, the specimen can be evaluated by cryo-EM (Fig. 1). Initially, cryo-EM should be performed at a diagnostic level. Three properties of the specimen should be evaluated at this point: (1) protein concentration and stability, (2) ice thickness and uniformity across the grid, and (3) phase of the ice (should be uniformly amorphous, not crystalline). All three should be consistent enough across the grid to yield several squares appropriate for data collection. An example of a micrograph of a ribosome specimen is shown in Fig. 1.

Standard substrates for cryo-EM consist of a 3-mm metal grid which supports a perforated foil (Fig. 2) (Russo & Passmore, 2016a). Foils are patterned with a regular array of holes (Ermantraut, Wohlfart, & Tichelaar, 1998; Quispe et al., 2007) and can be made from various materials, the most common of which is amorphous carbon (Table 2). We recently showed that

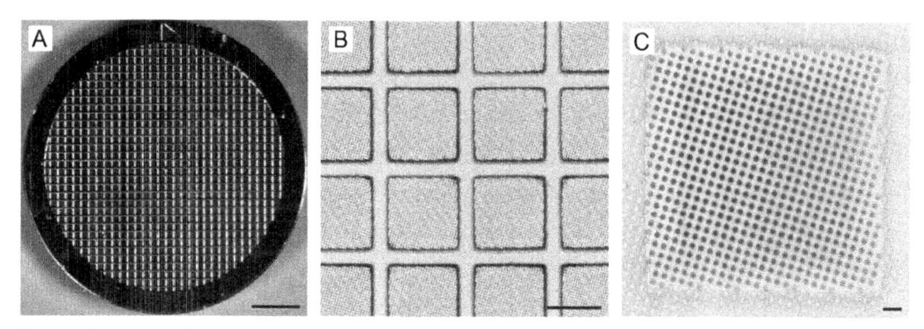

Fig. 2 *Supports for cryo-EM.* Here, we show a support comprising a 3-mm metal mesh grid (A) with a perforated gold foil covering the surface of the mesh (B and C). Thin films (~2–30 Å thick) can be added on top of the perforated foil. Scale bars are (A) 0.5 mm, (B) 50 μm, and (C) 5 μm.

Table 2 Specimen Support Geometries for Particular Applications

Mode	Grid	Foil	Film
Negative stain	400 mesh	None	50–100 Å am-C
Diagnostic cryo	300 mesh	1.2/1.3 μm	None/20 Å am-C
Medium-resolution cryo (≥3.5 Å)	300 mesh	1.2/1.3 μm	None/20 Å am-C
High-resolution cryo (<3.5 Å) >400 kDa	300 mesh	1.2/1.3 μm	None/20 Å am-C
High-resolution cryo (<3.5 Å) <400 kDa	300 mesh	1.2/1.3 μm	None/graphene
Very high-resolution cryo (<2.8 Å)	300 mesh	0.6/1.0 μm	None/graphene
Cryo-tomography: cellular (>30 Å)	200 mesh	2.0/2.0 μm[a]	None
Cryo-tomography: high-resolution subtomogram avg. (<15 Å)	300 mesh	1.2/1.3 μm	None/20 Å am-C

[a]Larger holes may be used for particularly large specimens.

making both the grid and the foil from the same material, gold, improves the stability of the support and reduces the movement of the specimen during imaging by more than an order of magnitude, thus improving image quality (Russo & Passmore, 2014c).

During vitrification, proteins are preserved in a hydrated, native state in the holes of the foil. The contrast of protein in vitreous ice is low so it is important to ensure that the ice is only slightly thicker than the particle diameter to maximize contrast. A water layer that is too thin will exclude proteins and this is usually apparent from the presence of a circular region of ice near the center of a hole where no particles are present. Given an appropriate concentration of stable complex in the droplet, the particle distribution and ice thickness should be optimized by changing one of a few parameters: (1) the blotting and plasma conditions, (2) adding a small amount of detergents, and (3) adding a thin film of amorphous carbon or graphene to the surface of the foil to induce adsorption of particles across the holes. Currently, this remains a trial and error process and so must be done in a way that minimizes uncontrolled changes.

Proteins often denature when they come into contact with surfaces (Seigel et al., 1997), and several are present during the vitrification process. These include the hydrophobic air–water interface, amorphous carbon support layers, copper, gold, and other metals used to create grids, and even the cellulose paper used to blot away the excess liquid. Of these, the air–water interface is likely the most problematic as the requirement for a thin layer of water ice means the particles must be brought in close proximity to one just before freezing. Its hydrophobic nature means it is likely to induce the adsorption, and possible denaturation, of many proteins and complexes (Seigel et al., 1997; Vogler, 2012).

For a given concentration of protein, one can calculate the number of particles that should be found in the thin layer of vitreous ice. For example, for a 1-MDa protein in 800 Å thick vitreous ice, a solution of 2 mg/ml should give approximately 100 particles/μm^2. A 250-kDa protein at the same concentration should give approximately 400 particles/μm^2. Often, the protein concentration in solution does not match the protein concentration on the support: protein density can be higher than expected if the particles tend to adsorb to the surface, or lower if they tend to repel the surface (or are attracted to the support or other surfaces like the blotting paper instead). Changing the buffer conditions (eg, changing pH or salt concentration, adding detergents, lipids, or other chemical modifying agents like cross-linkers) may change the interaction of proteins with surfaces (Cheung et al., 2013; Dubochet et al., 1985; Glaeser et al., 2016). A support that is too hydrophobic, for example because of insufficient plasma treatment or contamination with hydrocarbons, can strongly adsorb proteins to itself, making the concentration of particles in the holes much less than was in the applied droplet. This is a clear indication that there is a problem with the specimen. The opposite situation can also occur. That is, the particles can adhere to the air–water interface, artificially concentrating them prior to the removal of most of the water and the creation of the thin layer. This leads to a particle concentration in the holes that is much higher than it was in the droplet; also a clear indication that there is a problem with the specimen. Often, the particles adhered to the interface will have low contrast and will be partially or fully denatured. In addition, the particles will appear larger in size than expected as they unfold to varying degrees on the surface (Seigel et al., 1997). This should be resolved *before* moving on to collect large datasets in cryo.

2.4 Initial Cryo-EM Data Collection

Once protein density and ice thickness/quality have been optimized, an initial data collection should be performed on a mid-range electron microscope (Table 1). The goal is to obtain 2D reference-free class averages which have high-resolution features at the level of secondary structure (alpha-helices or better). Particles should be hand-picked, at least at first, to avoid the problems of template matching and "Einstein from noise" (Henderson, 2013). With modern field emission microscopes and direct detectors, only a few hundred particle images are needed to generate a 2D class of sufficient resolution to see alpha helices. Thus a dataset of a few thousand particles should be sufficient to yield a few high-resolution classes. High-resolution 2D classes are shown for ribosomes in Fig. 1. If high-resolution 2D classes are not obtained, the specimen should be reevaluated and optimized, including modification of protein preparations, buffers, detergents, and surface modifications, and the addition of support films like amorphous carbon or graphene. Particles that don't align well may be partially or fully denatured and could lead to incorrect structures if the subsequent computation and validation is not done with great care.

Many high-resolution structures to date have relied on discarding a large portion of the particles from the initial data collected. The proportion of discarded "junk" particles can vary by orders of magnitude from one specimen to the next. We suggest that many, if not most of these particles are damaged during the sample preparation process. By evaluating specimen preparation using the above criteria, one can optimize the specimen and improve particle yield. This is generally preferable to collecting large datasets and discarding most particles to reach a desired resolution. Improving particle yield will make the entire process more efficient and more likely to succeed.

Once a preliminary dataset is collected that generates suitable 2D classes, a larger dataset with a few hundred micrographs and several tens of thousands of particles is collected on a mid-range or high-end microscope, from which an initial 3D map can be calculated. Several important factors can then be evaluated. First, do the 2D classes show several distinct views of the particle, with self-consistent dimensions, each with secondary structural features? Second, is the orientation distribution sufficient to allow calculation of a 3D structure with isotropic resolution? Fig. 1 shows an equal area projection map of the orientation angles of ribosomes relative to an amorphous carbon substrate. Although these ribosomes exhibit preferred orientations, they cover Fourier space sufficiently for a high-resolution structure. If the orientation distribution is not suitable, one can alter buffers, add detergents, change plasma conditions, or use an alternative surface, eg, graphene or amorphous carbon to improve the distribution and promote additional views of the complex. Cross-linking can also alter particle orientations since modification of charged surface amino acids (often lysines) changes the surface properties of molecules and their interaction with the supporting surfaces and the air–water interface (Bernecky, Herzog, Baumeister, Plitzko, & Cramer, 2016). Collecting small datasets will be sufficient to determine if any changes in preparation conditions were effective at improving the orientation distribution.

2.5 High-Resolution Data Collection

If the resolution of the initial 3D reconstruction reaches a reasonable resolution for the number of particles and image acquisition settings, a larger data collection on a high-end electron microscope should be performed with the aim of obtaining a high-resolution structure (Fig. 1). What constitutes a "reasonable resolution" will be arbitrary and should ultimately depend on the resolution of the biological question and the state of the technology, but as a guideline, we suggest that currently, subnanometer resolution should be routinely possible from 10 to 50 thousand asymmetric particle images. With this in mind, a large dataset (500–1000 micrographs, ~24 h of microscope time) should be collected with the best available microscope: currently this is a 300 keV instrument with one of several commercially available direct electron detectors (Tables 1 and 3). The data collection strategy will depend on the size of the particle, the resolution desired, and the

Table 3 Current Electron Detectors for Cryomicroscopy

Detector Type	Current Examples	Pixel Pitch	Num. Pixels	Frame Rate (Hz)	Recommended Flux (Optimum [Range] e^-/px/s)
Phosphor–CCD	Gatan Orius 830	7.4	2048 × 2048	1	–
	Gatan US1000	14	2048 × 2048	1.5/15	–
Phosphor–CMOS	Tietz TemCam	15.6	4096 × 4096	1	–
	Gatan OneView	15	4096 × 4096	25	–
	FEI Ceta	14	4096 × 4096	32	–
Direct integrating	Direct El. DE16[a]	6.5	4096 × 4096	60	~ 240
	Direct El. DE20[a]	6.4	5120 × 3840	32	~ 100
	FEI Falcon 2[a]	14	4096 × 4096	18	50 [10–60] (300 keV) 40 [8–48] (200 keV)[b]
	FEI Falcon 3[a]	14	4096 × 4096	32	100 [20–120] (300 keV)
Direct counting	Gatan K2	5	3838 × 3710	400	5 [2–8] (300 keV)

[a]These detectors can also be operated in electron counting mode but current frame rates make this impractical for normal data collection.

[b]The flux, f_0 at one energy can be scaled with reasonable accuracy to other energies relevant to transmission electron microscopy using the equation $f_1 = f_0 \frac{\beta_1^2}{\beta_0^2}$, where β is the ratio of the electron velocity to the speed of light.

particular microscopes available to the microscopist. Our current recommendations for data collection on various specimen types are discussed in Section 8.

All biomolecular complexes exist in multiple conformational states, and for each to be resolved, even more high quality data are required. Many computational algorithms are available to help extract this information from a dataset and we refer the reader to chapters "Processing of Structurally Heterogeneous Cryo-EM Data in RELION" by Scheres, "FREALIGN: An Exploratory Tool for Single-Particle Cryo-EM" by Grigorieff, and "Single Particle Refinement and Variability Analysis in EMAN2.1" by Ludtke for more details. In some cases, ligands (eg, small molecules, binding partners, and nucleotides) can be added to stabilize particular conformations, or to interrogate a particular biological question related to how a ligand affects the conformational state of the complex. These usually require the collection of additional large datasets to achieve high-resolution after small datasets have been collected to confirm the change in conformation or presence of the ligand of interest.

Finally, but perhaps most importantly, all high-resolution datasets should be collected with validation in mind (see chapter "Testing the Validity of Single Particle Maps at Low and High Resolution" by Rosenthal). In particular, we recommend collecting a small set of tilt-pairs (~1–2% of the number of micrographs) with every dataset intended for determination of a previously unknown structure. They can be subsequently used to validate the reconstructed density map (Baker, Watt, Runswick, Walker, & Rubinstein, 2012; Henderson et al., 2011; Rosenthal & Henderson, 2003; Wasilewski & Rosenthal, 2014), measure the angular accuracy of the projection assignments (Russo & Passmore, 2014b), and determine the absolute hand of the structure (Rosenthal & Henderson, 2003).

By using the systematic approach described earlier, the microscopist will have the best chance of efficiently going from a protein in solution to a high-resolution structure. In the following section, we describe many of the practical methods and details of the processes that are important to successful specimen preparation. This includes a series of protocols that are intended to help guide the reader in the current state-of-the-art, and hone the techniques of new microscopists.

3. SUPPORT CHOICE, HANDLING, AND STORAGE

A fundamental but often overlooked factor in preparing high quality specimens for cryo-EM is correct handling and storage of the supports. They are delicate and can easily be damaged or contaminated, compromising later steps in the process and reducing reproducibility. Before starting, it is useful to check supports using an optical microscope and discard any with defects, large areas of broken squares or contamination. The time spent checking supports in a light microscope is trivial compared to subsequent time spent preparing and imaging the specimen. The supports should start and remain as flat as possible to prevent damage to the foil. Importantly, supports should always be handled with sharp tweezers and picked up by the rim only (Fig. 3). We like to use tweezers with a normally closed configuration (eg, Dumont type N5) as this applies a well-defined and reproducible force

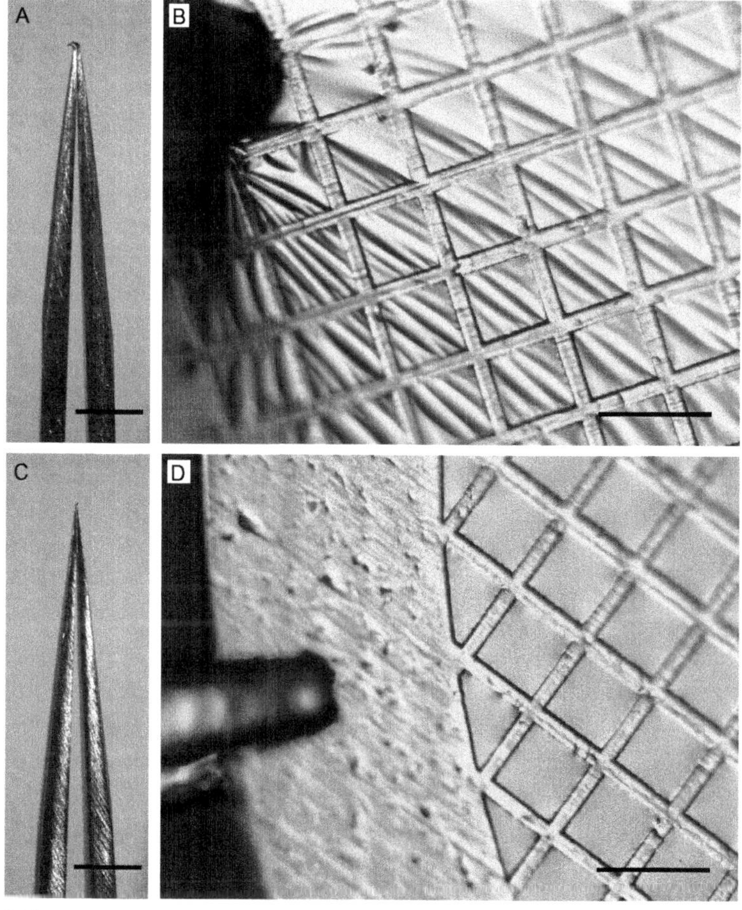

Fig. 3 *Tweezer damage to specimen supports.* Bent tweezers (A) or improper use (B) results in damage to the specimen support. For best results, sharp, straight tweezers (C) should be used and supports should be picked up by the rim only (D). For panels A and C, the scale bars are 1 mm. For panels B and D, the scale bars are 100 μm.

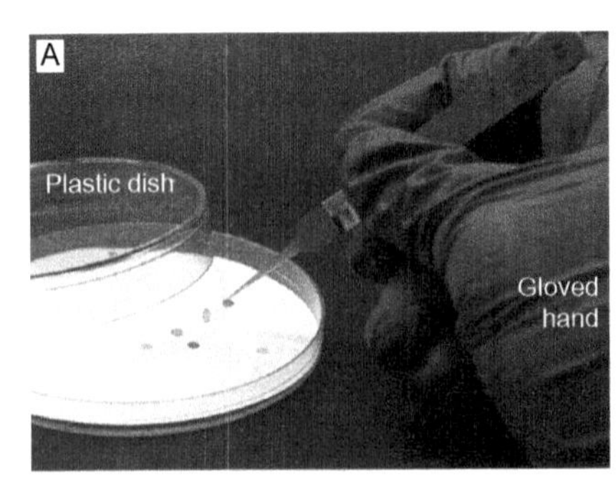

Fig. 4 *Storing grids to avoid contamination and static charge.* Panel A shows bad practice in grid handling: The use of gloves and plastic storage dishes results in the accumulation of static charge. In the image, a charged grid is standing on end. Panel B shows recommended handling procedures including glass containers, no glove on the hand holding the tweezers and a wrist grounding strap to prevent accumulation of charge.

to the grid during handling. Static discharge can also damage supports but this can be eliminated by wearing a wrist grounding strap whenever handling supports (Fig. 4). We do not recommend wearing a glove on the hand that holds the tweezers as this induces static buildup on the tweezer/support and reduces dexterity. After cleaning or modification, supports should be stored in glass, not plastic, petri dishes, and in clean, oil-free dry boxes to keep them clean and dust free.

To prevent carry-over of specimens and to remove contamination, grid handling implements, glass slides, and glass dishes should be thoroughly cleaned on a regular basis. Glass can be cleaned by sonicating in a high-purity detergent mixture (eg, 2% Micro-90) and rinsing with ultra-pure, 18 MΩ deionized water. Tweezers can be cleaned after routine use by sonication in alcohol (ethanol or isopropanol) or with more aggressive solvents like chloroform and acetone if they have been contaminated with plastic or oil residues. Note that tweezers used for grid plunging should be decontaminated with ethanol sonication after each distinct specimen, as cross-contamination from one grid to the next via the tweezers is common.

Supports with irregular arrangements (lacey or holey carbon) or regular arrays of holes are available but the latter, eg, Quantifoil, UltrAuFoil, C-flat, are more reproducible and simplify data collection methods. A detailed discussion of support types can be found in Russo and Passmore (2016a). Here, we provide recommendations for support geometries for various applications (Table 2).

As to the choice of material, we recommend amorphous carbon (am-C) on copper for negative stain, as they are commonly available and inexpensive, and all-gold supports for cryo-EM as they reduce the movement of specimens and improve image quality (Russo & Passmore, 2014c). All-gold supports can be made using the protocols described in Russo and Passmore (2016b) or are commercially available from Quantifoil (UltrAuFoil®).

4. CONTAMINATION AND CLEANING

Even under optimal storage conditions, supports may need to be cleaned prior to use. Contamination can arise from particulate matter including flakes of evaporated carbon, residual particles from manufacturing processes or accumulated dust from storage. If these are not removed, they will be resuspended upon application of the aqueous protein solution and may be visible in the vitreous ice, which can interfere with subsequent data processing and analysis. In addition, contamination can arise from organic residue deposited on the surface during handling or storage, and from residual photoresist or other plastics and solvents used during the manufacturing processes. Such surface residues can affect wetting and other material properties of the support and reduce reproducibility.

Here, we provide a method to clean supports by water and solvent washes prior to use (Protocol 1). Ultra-high purity (CMOS grade) solvents are required for cleaning without contamination. We do not recommend cleaning supports by heating as this can compromise the structural integrity of the support.

Warning: The protocols described here involve using high voltage, high temperatures, liquid nitrogen, hazardous and flammable chemicals/gases, and equipment under vacuum. They are for use by experienced scientists who know how to handle such equipment and have done all appropriate local safety training and risk assessments, etc. Wear safety glasses, appropriate protective equipment, and use at your own risk.

Protocol 1. *(Support cleaning)*

1. Fill a clean, glass crystallization dish (Pyrex, diameter 90 mm, height 50 mm) with deionized water (18 MΩ, filtered, UV treated). Overfill it to break the meniscus at the surface and remove any contaminating surface layers on the water (Fig. 5C).
2. Using small, clean glass test tubes, prepare three solvent washes (Fig. 5A): one with 1 ml chloroform (Sigma 650498), one with 1 ml acetone (Sigma 40289), and one with 1 ml isopropanol (Sigma 40301). Use glass pasteur pipettes as the solvents can dissolve and redeposit residue from most plastics.

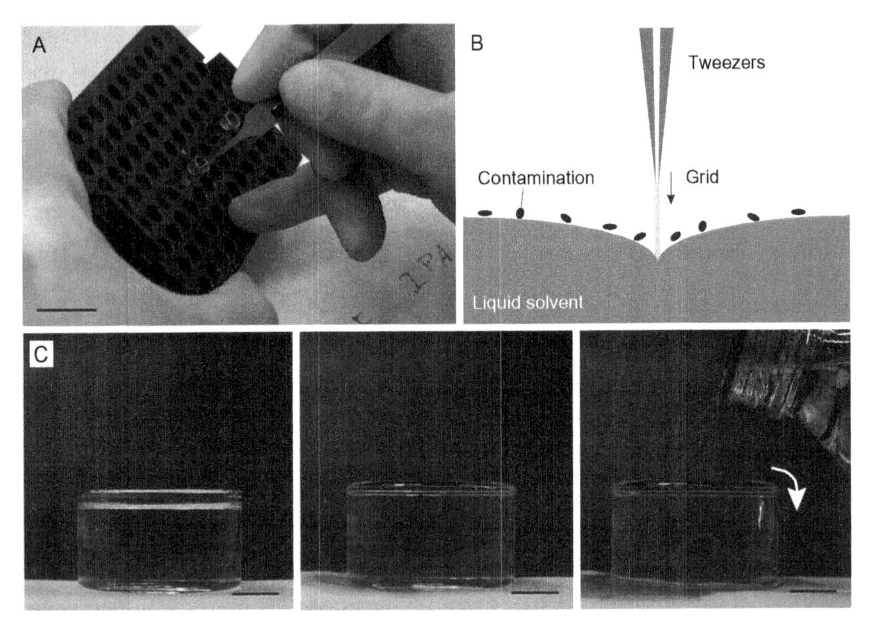

Fig. 5 *Removing surface contamination from specimen supports.* Supports can be washed sequentially (A) in chloroform, acetone, and isopropanol. Care must be taken to avoid deposition of contamination from the surface of water (or solvents) onto the support. A schematic is shown in panel B. Overfilling of containers can reduce surface contamination, shown in panel C. Scale bars are 20 mm.

3. Pick up a single support using clean tweezers (Dumont N5 or 5) and dunk it into the water at one side of the dish. As the grid enters the water, loose contamination can float off onto the surface of the water. Pulling the grid out in the same place will result in the contamination being redeposited on the grid surface. Move the grid through the water to remove it from the opposite side of the dish. (Ensure it is moved through the water with the thin edge leading to prevent damage to the perforated foil.) Touch edge of grid briefly to filter paper (Whatman No. 1) to remove excess water, but take care not to bend the grid. If not using anticapillary tweezers, blot between the tines.
4. Dunk grid sequentially in chloroform, acetone, and isopropanol (prepared above) for 10–20 s each. It is particularly important that the final rinse step is in the cleanest possible solvent as any residue from it will be deposited on the grid surface. If not using anticapillary tweezers, blot between the tines after each rinse to remove excess.
5. Touch edge of grid briefly to filter paper to remove excess isopropanol.

6. Place grid on filter paper (Whatman No. 1, 70 mm rounds) in clean glass petri dish (Schott, 70 mm), foil side up, to dry. Cover with glass lid to minimize dust accumulation.
7. Store in the covered glass petri dish after drying but use soon after cleaning.

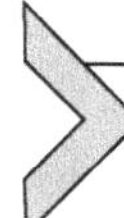

5. CONTINUOUS FILMS OF AMORPHOUS CARBON AND GRAPHENE

A continuous film of amorphous carbon or graphene provides an alternative surface for proteins to interact with. In some cases, this improves protein distribution or orientation within the vitreous ice. In addition, lower protein concentrations can sometimes be used because the particles adsorb to the surface before blotting away the excess liquid. Here, we provide a protocol for producing thin (20–50 Å) films of amorphous carbon for use as a surface for particle adsorption. The same protocol can be used for thicker films (50–100 Å) of carbon appropriate for transfer to bare grids (no perforated foil) for negative stain.

Protocol 2. *(Amorphous carbon deposition)*

This protocol is for depositing carbon onto a sheet of mica using an Edwards 306 Turbo coating system equipped with an Inficon crystal thickness monitor with water cooling, but can be adapted to other deposition systems.

1. Vent the chamber and remove the implosion guard and bell jar.
2. Put on clean gloves and handle everything inside the chamber with gloves.
3. Remove the carbon source apparatus, remove the shield and mount a new sharpened carbon rod (high purity graphite < 5 ppm impurity), under tension, in the chuck.
4. Use compressed nitrogen to blow out any bits of carbon or flakes as these can cause a short between the electrodes.
5. Replace the source in the chamber, with the rod positioned 125 mm from the stage. Finger tighten the nuts (no wrench).
6. Mica is a multilayered mineral crystal which is extremely flat (less than 1 nm RMS per mm^2) and easy to obtain in large sheets, and so is used as a template surface for carbon deposition. Cleave a piece of mica (eg, Agar G250-1) in half with a razor blade and position the two sheets, cleaved surface up, on the specimen stage directly beneath the source on top of a fresh piece of filter paper (Whatman no. 1). Mica should be cleaved immediately before coating as the freshly cleaved surface is

clean and hydrophilic, but it becomes contaminated (and thus hydrophobic) with time spent in air.

7. A clean penny or other small piece of metal can be used to hold the mica sheets and filter paper in place during evacuation.
8. Check the position of the shutter in the open and closed state: closed it should cover the solid angle from the source to the mica and the crystal thickness monitor. In the open state, the paths to both the mica and the crystal should be unobstructed and equal in length. Leave in the closed position.
9. Check the bell jar gasket and base plate for any dust, dirt, or flakes of carbon or metal. Clean with lint free paper and methanol if necessary.
10. Replace the bell jar and implosion guard.
11. Fill the liquid nitrogen cold trap.
12. Evacuate the chamber by pressing cycle on the panel.
13. When the gate valve opens, the pressure should drop rapidly into the 10^{-5} Torr range. Begin heating the source apparatus by turning on the low tension power supply (LT) and slowly increasing the current to about 1.0 Amp.
14. Wait 10 min and increase the current to 1.2 Amps.
15. Wait 5 min and increase to 1.4 Amps.
16. Leave at 1.4 amps for 20 min. Then shut off. Top up the liquid nitrogen.
17. Turn on the cooling water to the crystal thickness monitor and check the correct program for carbon is selected. Zero the thickness.
18. Vacuum should now drop to the 10^{-7} Torr range in about an hour.
19. Once vacuum is $< 5 \times 10^{-7}$ Torr, slowly ramp up the current again to 1.6–1.7 Amps over about 5 min, being careful not to let it spark. The pressure should not rise above $< 5 \times 10^{-6}$ Torr as the source is heated.
20. Once the carbon begins to deposit, the current will start to fluctuate. Open the shutter and deposit for 10–20 s, then close the shutter and turn the current back down slightly.
21. The crystal will initially go negative as it heats up but after closing the shutter, it will return to a positive value which is the accumulated thickness. This can be repeated until the desired thickness is reached.
22. When finished evaporating, close the shutter and turn off the power to the source.
23. Let the chamber cool for at least 30 min before venting.

24. Vent with dry nitrogen, remove your coated mica and store in a glass Petri dish.
25. Replace the bell jar and implosion guard, press cycle, wait for the vacuum to reach 200 mTorr and then press seal.

Notes: The crystal thickness monitor should be calibrated according to the manufacturer using an independent thickness measurement method. We have used atomic force microscopy in the past for this purpose (Russo & Passmore, 2014a). Carbon density can vary depending on the quality of your source material but should be approximately 2.2 g/cm^3. From start to finish, the process should take less than 4 h. Unless using a fully dry (no oil) pumping system, overnight pumps are not recommended as oil will backstream into the chamber when the liquid nitrogen trap is not cold and this will contaminate the chamber and mica with oil.

We previously provided a detailed protocol for transferring amorphous carbon onto supports (Russo & Passmore, 2016b) and we have had good results using this with Quantifoil and all-gold supports. This is reproduced here in Protocol 3.

Protocol 3. *(Amorphous carbon transfer onto supports)*

1. Wear wrist grounding strap to prevent static damage to supports during handling. Don't wear a glove in the hand that holds the tweezers, or ground the tweezers directly.
2. Inspect supports in dissecting microscope and discard any with defects. All grids should be flat, continuous, and without dust or lint.
3. If there is evidence of residual plastic from lithographic processing, they can be cleaned using Protocol 1.
4. Using gloves to prevent fingerprints and contamination, place filter paper (Whatman No. 1, diameter 70 mm) on a stainless steel mesh circle (diameter 65 mm, made with 0.7 mm wire with 3 mm mesh) inside a stainless steel ring (polished and beveled edges, 2 mm thick, height 10 mm, diameter 50 mm) in a glass crystallization dish (Pyrex, diameter 90 mm, height 50 mm) (Fig. 6) (Russo & Passmore, 2016b).
5. Fill dish with deionized water (18 MΩ, filtered, UV treated). Overfill it to break the meniscus at the surface and remove any contaminating surface layers on the water (Fig. 5C). Pour off excess water so the level is just below the rim of the glass dish.
6. Using tweezers (Dumont N5 or 5, cleaned by solvent rinse), carefully place grids, foil side up, onto the center of the filter paper. The grids should enter the water slowly and perpendicular to the surface.

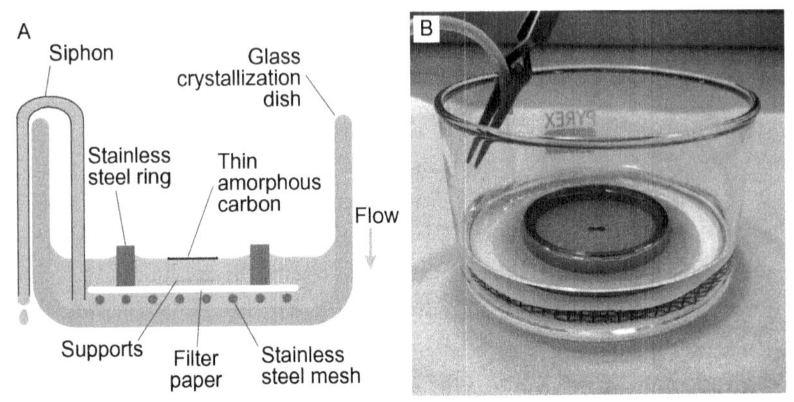

Fig. 6 *Apparatus for depositing thin films of amorphous carbon on supports.* Panel A shows a cross-sectional diagram of the float chamber. As the water level is lowered, the thin film of amorphous carbon is deposited onto the supports. The stainless steel ring makes a positive meniscus to help lower the carbon film down in the center of the ring. Panel B shows a photo of the apparatus in use. *This figure is reproduced from Russo, C. J., & Passmore, L. A. (2016b). Ultrastable gold substrates: Properties of a support for high-resolution electron cryomicroscopy of biological specimens.* Journal of Structural Biology, 193 *(1), 33-44. doi: 10.1016/j.jsb.2015.11.006. Drawings are available from the authors.*

7. Setup a siphon: attach tubing (length 0.6 m, outer diameter 3.2 mm, inner diameter 1.6 mm) to dish with normally closed forceps or a spring clip. Start flow of syphon with syringe and clamp off flow while floating the carbon (next step).
8. Float carbon off mica by slowly lowering a $\simeq 2.5 \times 2.5$ cm sheet of amorphous carbon coated mica (carbon-side up) into the water at a 20–30° angle. A light can be placed to shine off the surface of the water at a glancing angle, allowing you to see the carbon better.
9. Use the siphon to slowly lower the water level.
10. Monitor the position of the carbon with respect to the grids and ensure it stays centered by gently nudging with clean tweezers.
11. Once the carbon has been deposited on the grids and the water level is below the filter paper, carefully lift off the stainless steel ring. Remove the filter paper and mesh together. Place on a dry sheet of filter paper and cover with a clean glass petri dish or beaker, tilted slightly to leave room for evaporation. Allow several hours to dry.
12. When dry, store in a clean *glass* petri dish until ready to use.

Graphene is superior to amorphous carbon due to its defined structure and conductive properties (Geim, 2009; Pantelic, Meyer, Kaiser, Baumeister, & Plitzko, 2010; Pantelic et al., 2011; Russo & Passmore, 2014a). In addition, it is effectively invisible at the resolutions of interest for cryo-EM, whereas amorphous carbon adds additional background noise that is especially detrimental for smaller proteins. Here, we provide procedures for transferring graphene onto Quantifoil supports (Fig. 7, Protocol 4) (Regan et al., 2010; Russo & Golovchenko, 2012). Ultra-high purity (CMOS grade) solvents and acids are required for reliable transfers without contamination and a wrist grounding strap should be worn at all times to prevent static damage to supports during handling since graphene films are particularly sensitive to

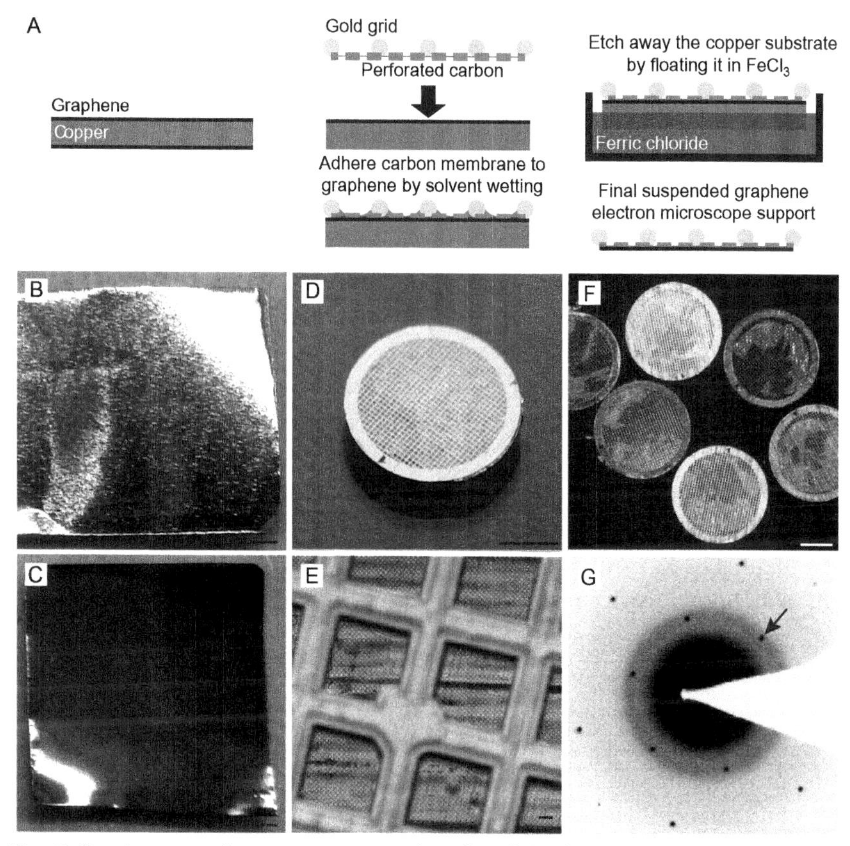

Fig. 7 *Graphene transfer onto supports with carbon foils.* The process is diagrammed in panel A. Panels B and C show copper heated to 150°C for 10 min in air, where B is fully covered in graphene so does not oxidize while C has no graphene and turns color due to oxidization, scale bars are 3 mm. This simple test is used to map the location of the graphene on the foil. Panels D and E show the grid–graphene–copper sandwich, scale bars are 1 mm and 10 μm, respectively. Panel F is the sandwich floating in the etchant, where the partially etched grains of copper are visible (scale 1 mm). Panel G is an electron diffraction pattern of suspended graphene with ice, where the arrow points to the 2.1 Å reflection from the graphene lattice. (See the color plate.)

static discharge. All glassware should be cleaned prior to use. A new transfer protocol for all-gold supports is currently under development, and we expect that graphene on all-gold supports will become commercially available.

Protocol 4. *(Graphene transfer onto supports)*

1. Inspect freshly cleaned supports (Quantifoil Au 300 1.2/1.3, see Protocol 1) in dissecting microscope and discard any with defects.
2. Graphene is grown on thin sheets of copper using chemical vapor deposition (CVD) and can be purchased from commercial vendors (Graphene Supermarket, Structure Probe Inc., etc.). Punch out 3.2 mm disks from copper/graphene source material where the copper has been checked for the presence of graphene (Fig. 7B, C). We use a custom-made mechanical punch but these are also available from Structure Probe, Inc.
3. Place a support, foil side down, on each disk.
4. Press together between two clean glass slides and then remove the top glass slide.
5. Using a clean pipet, add 7 μl of isopropanol (Sigma 40301) to the top of the grid–graphene–copper sandwich and let dry.
6. Inspect in high-resolution optical microscope to check for adherence of the carbon foil to the copper. The foil changes color when well adhered to the copper surface, (Fig. 7E) and only well-adhered regions will transfer successfully.
7. Layout and label seven borosilicate glass crystallization dishes (70 mm diameter) in fume hood.
8. Fill first crystallization dish 2/3 full with copper etchant containing buffered $FeCl_3$ (Sigma 667528).
9. Gently place the support disk sandwiches on the surface of $FeCl_3$ by sliding off glass slide. Cover with borosilicate glass Petri dish covers (80 mm) to keep out dust during the etch.
10. Etch for 20 min in $FeCl_3$ (for 25 μm thick copper).
11. Fill three dishes 2/3 with 20% HCl (Sigma 40233), 20% HCl and 2% HCl.
12. Transfer grids to each using a flamed platinum loop (homemade or EMS 70944). Leave floating for 10 min at each step.
13. Overfill three more dishes with cleanest possible water.
14. Transfer to each using platinum loop, and rinse for 3–5 min each.
15. Use loop to transfer and flip, graphene side up, onto filter paper (Whatman No. 1, 70 mm rounds). Let dry in clean borosilicate glass Petri dish (70 mm).

16. Dispose of acids in appropriate hazardous waste streams.
17. Store long term in glass in a dry environs free of any type of oils or hydrocarbons. Do not store in any type of plastic container.

6. SURFACE TREATMENTS

Support surfaces are often hydrophobic and this prevents efficient spreading of aqueous solutions. To control the hydrophilicity, treatment with low-energy plasmas is used (Fig. 8). Ions and radicals generated from a low-pressure gas interact with surfaces to remove residual organic contamination and react chemically with the surface to reduce their hydrophobicity. Although plasmas generated from residual air (glow discharging) can be used (Protocol 5), we recommend the use of plasmas with controlled composition as this is more reproducible (Protocol 8). Plasma treatment is typically performed with argon:oxygen mixtures but hydrogen can also be used (Protocol 7). In addition, other molecules (eg, amylamine) can be introduced to alter the surface and change the orientation distribution of particles adsorbed to the surface (Protocol 6) (da Fonseca & Morris, 2015; Miyazawa, Fujiyoshi, Stowell, & Unwin, 1999).

Protocol 5. *(Glow discharge treatment of supports)*

1. Inspect clean supports in dissecting microscope and discard any with defects. All supports should be flat, continuous, and without dust or lint.
2. Place supports in the chamber of a glow discharging apparatus (eg, Edwards S150B) on a clean glass slide, foil side up, in the center of the stage. Wear gloves when handling anything that will be placed in the chamber.

Ted Pella easyGlow (c. 2015)

Edwards S150B (c.1995)

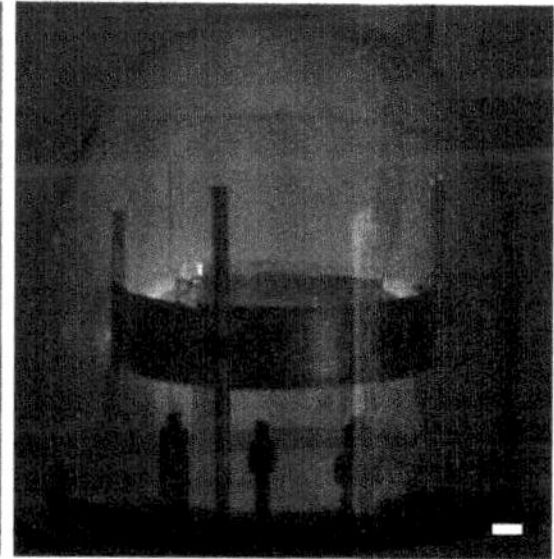
Edwards 12E6 (c.1962)

Fig. 8 *Plasmas generated by glow discharge.* Residual air plasma generation by three different instruments is shown. All are effective in increasing hydrophilicity of support surfaces but have varying degrees of reproducibility and can damage the supports. Note that the plasmas in glow discharge apparatuses are often nonuniform which can vary the exposure dose significantly, even in a single batch. Scale bars are 20 mm. (See the color plate.)

3. Pump the chamber to 200 mTorr.
4. Turn on the HT to 7 kV, current should be 28–30 mA.
5. Expose the supports for 30 s.
6. Turn off the HT and slowly vent the chamber.
7. Store the treated supports in a clean glass petri dish and use as soon as possible (within 1 h).

Protocol 6. *(Amylamine plasma treatment of supports)*

1. Inspect supports in dissecting microscope and discard any with defects. All supports should be flat, continuous, and without dust or lint.
2. Place supports in the chamber on a clean glass slide, foil side up, in the center of the electrode ring (Fig. 8).
3. A small glass vial containing 0.5 ml of amylamine is placed in the chamber at the edge of the stage.
4. Cover with the bell jar and pump the chamber to below 400 mTorr.
5. Turn on the HT and adjust the power to form a uniform plasma (~ 90 V, 1.5 A for our system).
6. Expose the supports for 30–60 s.
7. Turn off the HT and slowly vent the chamber. Discard the vial in an appropriate waste stream.
8. Store the treated supports in a clean glass petri dish and use within 1 h.

Protocol 7. *(Argon oxygen plasma treatment)*

1. Inspect supports in dissecting microscope and discard any with defects. All supports should be flat, continuous, and without dust or lint.
2. We use a Fischione 1070 plasma chamber with a custom suspension holder that holds up to 10 supports (Fig. 9). In this configuration, the grids are 15 ± 1 cm from the radio frequency coils. It is desirable to always position the grids in the same location relative to the plasma to improve the reproducibility of the exposures from batch to batch. Alternatively, the specimens can be placed directly in the chamber on a clean glass slide, foil side up, on the shelf in the chamber. Plasmas that are not well shielded or not sufficiently low in energy may damage gold supports by sputtering; this should be avoided. Load up to 20 supports into the plasma chamber. Everything that goes inside the vacuum chamber should be handled with gloves to prevent contamination and fingerprints.
3. Evacuate chamber to $\ll 10^{-4}$ Torr.
4. Admit high purity argon and oxygen (BOC 99.9999%) in a ratio of 9:1 to a pressure of 21 mTorr (31.0 SCCM gas flow).
5. Apply radio frequency plasma with 38 W of forward power and ≤ 2 W of reverse power (~70% setting on a Fischione 1070) for 10–60 s

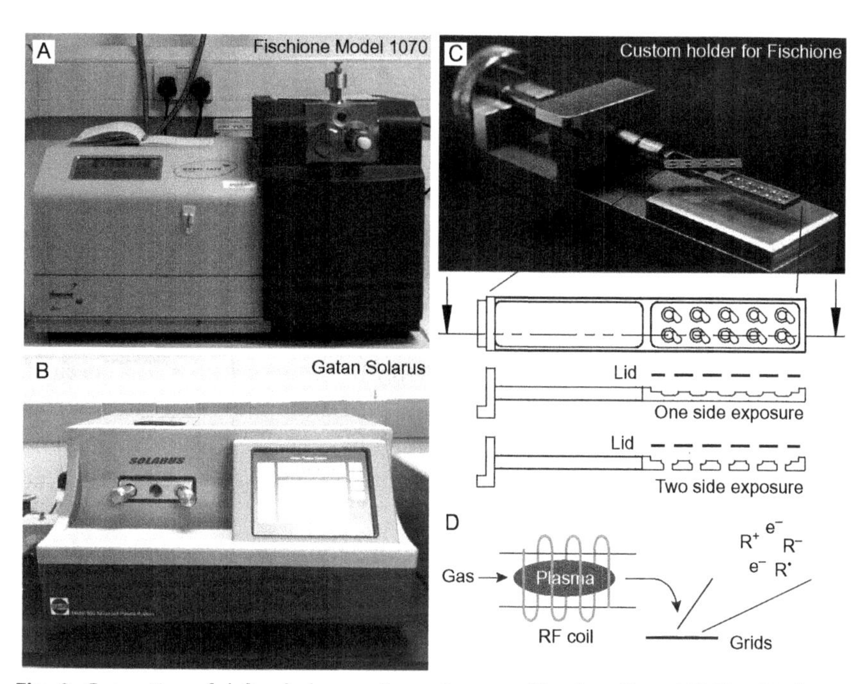

Fig. 9 *Generation of defined plasmas for surface modification.* (A and B) The Fischione Model 1070 and Gatan Solarus generate plasmas with defined compositions. (C) Photo (top) and diagram (bottom) of a custom specimen holder made at MRC LMB. Two different holder designs are shown—one is used for exposure of one side of a support and the other is used for exposure of both sides. The lid is used to prevent the supports from moving during the process. (D) Diagram of plasma generation.

(20–30 s works well for most supports). The time can be optimized for suitable hydrophilicity (spreading of aqueous solution and ice thickness). For amorphous carbon-containing supports, calibrate the carbon etch rate empirically by performing several plasma treatments of increasing dose and determining when a carbon layer of known thickness is fully etched.

6. For very thin films of continuous carbon, use a ratio of 19:1 argon: oxygen and 35 W forward power for 5–15 s.
7. Vent plasma chamber, remove supports and use within 1 h.

Using a low-energy hydrogen plasma, one can remove contamination and control protein adsorption onto graphene (Russo & Passmore, 2014a). Here, we provide the procedure for hydrogen plasma treatment using a Fischione 1070 with a Dominik Hunter model 20H-MD hydrogen generator, attached to one of the input ports of the plasma generator. Hydrogen plasma treatment is stable but should be performed immediately prior to use to minimize accumulation of contamination after modifying the surface.

Protocol 8. *(Hydrogen plasma treatment of graphene supports)*

1. Use only ultra-high purity hydrogen (>99.999% pure), all stainless steel tubing and fittings (no plastic), and ultra-high purity grade regulators.
2. The distance of the coils to the sample is important since it affects the energy of the hydrogen species. A plasma where atoms and ions strike the surface and recoil, delivering an energy greater than ~21 eV to the carbon atoms, might cause significant damage to the lattice rather than chemically modifying it. The energy of the plasma vs distance can be measured with a Langmuir probe to ensure that the energy is below the ~21 eV sputter threshold at the sample. Alternatively, one can test whether the energy is too high by imaging a graphene-covered support before and after increasing doses of hydrogen plasma to determine whether it remains intact.
3. Prior to hydrogen plasma treatment, the plasma chamber should be pretreated. Insert the holder and evacuate the plasma chamber to $\ll 10^{-4}$ Torr.
4. Burn the empty holder for 10 min in 100% pure hydrogen (70% power which is 35 watts forward, $\ll 2$ watts reverse power, 20 SCCM gas flow, pure hydrogen). Vent only just prior to loading.
5. Mount graphene supports in holder, load, and evacuate the plasma chamber to $\ll 10^{-4}$ Torr.
6. Treat with 5–40 s (typically 20 s) hydrogen plasma using the same settings.
7. Carefully remove the supports and use immediately.

7. VITRIFICATION

Vitrification was first developed by Jacques Dubochet and colleagues in the 1980s (Adrian et al., 1984; Dubochet et al., 1988). To successfully make vitreous ice, a thin layer of protein solution must be cooled quickly so that it remains in an amorphous, noncrystalline state. This is usually performed in an apparatus which plunges the specimen into a liquid cryogen (usually ethane or propane) such that it is cooled about 200 K in $< 10^{-4}$ s. For water to vitrify, the temperature has to drop faster than 10^5–10^6 K/s (Dubochet et al., 1988). Water is a poor thermal conductor so the sample must be less than 3 μm thick. Several designs for manual and semiautomated plungers are available for vitrification of biological specimens, including commercial models from FEI, Leica, EMS, and Gatan. These provide controlled temperature and humidity throughout the process, preventing evaporation, and making specimen preparation more reproducible. We note that after blotting, even a small amount of evaporation, equivalent to a few hundred monolayers of water, can concentrate the salt and change the pH of a suspended thin layer by a factor of two or more. For example, at 4°C and 90% relative humidity, the evaporation velocity is of order 100 Å/s so in the 2 s between the blot and the freeze, a 400 Å film can be concentrated by a factor of two, causing significant osmotic and conformational changes in the specimen. So we also recommend that the relative humidity surrounding the specimen support always be kept at 100% to prevent deleterious changes in the concentration of solutes just prior to freezing. This is easiest to achieve at 4°C, because much less water vapor per unit volume is required to bring the dewpoint to the air temperature, thus preventing evaporation. It is also important to work quickly once sample is applied to the support to minimize the interaction time of proteins with the often destructive, air–water interface.

A detailed discussion of vitrification procedures can be found in Dobro, Melanson, Jensen, and McDowall (2010) and we recommend consulting this review for further information. We include here the standard settings we use as a starting point and other recent work can be consulted for further advice (Thompson, Walker, Siebert, Muench, & Ranson, 2016), including more advanced techniques like time-resolved cryo-EM (Chen et al., 2015). The following procedure using a Vitrobot (FEI) works well for all-gold and standard Quantifoil supports:

Protocol 9. *(Standard vitrification procedure)*

1. Fill cryoplunger reservoir with fresh deionized water (18 MΩ) and equilibrate to 4°C and 100% relative humidity.
2. Ensure Vitrobot tweezers (FEI, Ted Pella 47000-500) are sharp by examining under an optical microscope. Check that they are not bent and symmetrically aligned in the following way: place on a flat surface and measure the tip to surface distance. Do this for all four sides; it should be the same (within 0.5 mm) for opposing sides.
3. Clean tweezers and blotting pads with ethanol; and clean the ethane cup, grid box holder, and foam dewar with a high purity detergent (2% Micro-90 in Milli-Q water); rinse with Milli-Q deionized water and dry completely.
4. Place new filter paper rounds (Whatman 595) on blotting pads.

5. After cryo-plunger is equilibrated (for at least 20 min to saturate the water in the filter paper) and just prior to specimen vitrification, cool down the plunging dewar with liquid nitrogen then fill the central cup with ethane (Dobro et al., 2010). *Warning* Liquid ethane can cause severe burns and blindness if splashed in the eye. *Always* wear safety glasses when handling.
6. Check that the temperature of the ethane is just above the melting point, 90 K.
7. Make supports hydrophilic by plasma treatment, as described earlier. We emphasize that only clean, flat, intact supports should be used (see earlier). Bent supports will be further damaged upon cryo-plunging (Fig. 10).
8. Mount a support in clean cryo-plunger tweezers. Use the foot pedal trigger on the Vitrobot to reduce the delay time between steps in the process.

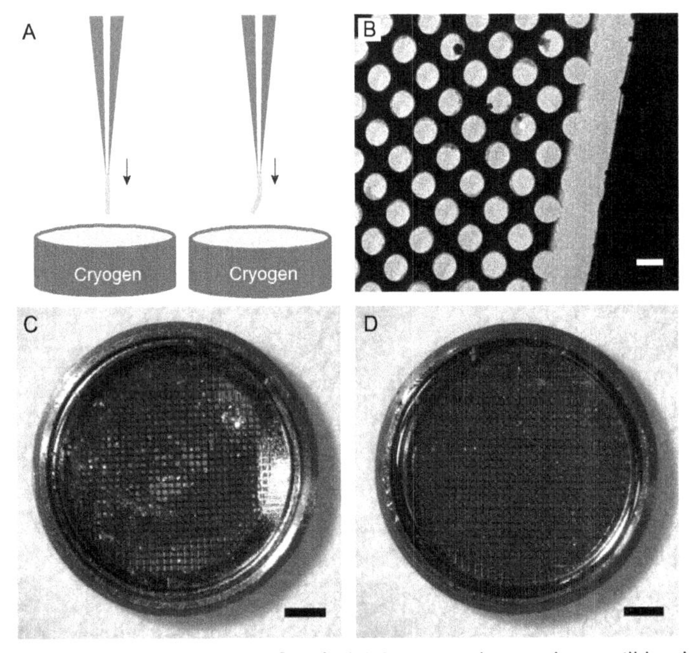

Fig. 10 *Vitrification and mounting of grids.* (A) Supports that are bent will be damaged upon cryo plunging, resulting in broken foils. An example of a broken gold foil is shown in panel B (scale bar 2 μm). Supports also need to be mounted correctly in microscope cartridges. Panel C shows a support that is incorrectly mounted, and so damaged, in a Krios cartridge. The support in panel D is correctly mounted. Scale bars in panels C and D are 500 μm.

9. Apply 3 μl protein solution (usually 10–5000 nM) to the foil side. *Make sure the pipet tip does not touch the support*—only touch the liquid droplet to the surface. Close access port.
10. Blot using the following settings:

Description	Manual Plunger	Vitrobot III	Vitrobot IV
Temperature	4°C	4°C	4°C
Relative humidity	100%	100%	100%
Wait time	0 s	0 s	0 s
Force setting	n/a	– 2	– 20
Blot time	2 + 1 s	4–6 s	2–4 s
Drain time	n/a	0 s	0 s

Settings here are for FEI Vitrobots but are easily adapted to other manual or semiautomated plunge instruments. We do not use a "wait time" or "drain time," in order to speed up the process and minimize contact time with the air–water interface. Note, changing the blot force setting to a more negative value will increase the force.

11. Plunge into liquid ethane.
12. Wait a minute for the tweezer to cool after the plunge into the ethane.
13. Carefully transfer the frozen specimen to a storage box via the cold vapor above the liquid nitrogen.
14. When releasing the grip of the tweezer, gently let the solidified ethane crack away from the support if present.
15. Store the supports in covered, labeled grid boxes in a liquid nitrogen storage dewar.

Take care not to bump the support into anything during transfers to storage boxes, mounting in stage cartridges, etc. Any support that is bumped or bent at any point should be discarded.

Generally, the only variables we change to optimize ice thickness are plasma exposure time and blot force. Blot time can also be varied but has a less reproducible effect on ice thickness, particularly for short blot times. If the support is not hydrophilic enough, one usually finds thicker ice in the center of the grid squares because the hydrophobic surfaces repel liquid, causing a drop of solution to accumulate in the center of the square.

8. DATA COLLECTION

As with most technological devices, there is a tradeoff between ultimate performance of the current microscopes and detectors and the amount of money and effort required to achieve this performance. But since the goal is ultimately to determine structures at resolutions which are sufficient to unambiguously answer biological questions, here we try to make reasonable compromises between the performance of the instrument and the amount and quality of data required to achieve a particular resolution that is appropriate for the stage of structure determination (Fig. 1). In general, we think it is better to spend more effort on preparing optimal specimens, as now, and even more so in the future, this is where the biggest differences in the resolution and interpretability of the resulting density maps are likely to come from.

Since biological imaging is ultimately limited by damage to the specimen (see chapter "Specimen Behavior in the Electron Beam" by Glaeser), currently the most important single factor governing the quality of the micrographs is the efficiency of the electron detector (see chapter "Direct Electron Detectors" by McMullan et al.). Using the current understanding of detective quantum efficiency (DQE) and published measurements of DQE for the various commercially available detectors (Chiu et al., 2015; Kuijper et al., 2015; McMullan, Faruqi, Clare, & Henderson, 2014), we have tabulated the parameters and recommended flux of current detectors in Table 3. These, in conjunction with the illumination and specimen movement considerations discussed later, form the basis for our recommended imaging conditions in Table 4. These can be considered a starting point for users to make trade-offs in the amount of data collected vs resolution or other factors like microscope time, and they will certainly change as the technology continues to develop in the coming years.

Here, we provide a standard protocol for data collection at high-resolution on a Krios with a Falcon 2 detector. The protocol is easily adapted to other specimen and microscope configurations using the data collection settings in Tables 3 and 4.

Protocol 10. *(High-resolution data collection using a Krios/Falcon 2 and all-gold supports)*

1. Load specimens in cartridges and mount in autoloader, including a calibration grid (PtIr, graphitized carbon, or similar). Discard any that are bent or broken during handling.
2. Load a specimen in the column and use low mag to check grid is intact and contains several squares with an appropriate thickness of ice.

Table 4 Currently Recommended Data Collection Settings at the MRC LMB

Mode	Source	Microscope type	Energy (keV)	Detector	Pixel size (Å/px)	Flux ($e^-/Å^2/s$)	Exp. Time (s)	Fluence ($e^-/Å^2$)
Negative stain	W-thermal	Entry level	80–120 [a]	CCD	3.3	9	2	18
Diagnostic cryo	W-thermal	Entry level	80–120 [a]	CCD	3.3	9	2	18
Diagnostic cryo	FEG	Mid-range	200	Falcon 2	2.1	10	2	20
Medium-resolution cryo (≥3.5 Å)	FEG	Mid-range/ high-end	300	Falcon 2	1.7	17	3	51
High-resolution (≤3.5 Å) ≥400 kDa	FEG	High-end	300	Falcon 2	1.3	28	2	56
High-resolution (≤3.5 Å) <400 kDa	FEG	High-end	300 (±5 eV)	K2	1.8	1.5	40	60
Very high-resolution (<2.8 Å)	FEG	High-end	300 (±5 eV)	K2	0.90	6.2	10	62
Cryo-tomography cellular (>30 Å)	FEG	High-end	300 (±5 eV)	K2	3.5 [b]	0.65	1.25 per tilt angle	100
Cryo-tomography high-resolution subtomogram avg. (<15 Å)	FEG	High-end	300 (±5 eV)	K2	2.2 [b]	1.5	1 per tilt angle	60

[a] DQE is maximum for a phosphor coupled CCD at approximately 80 keV but the effects of specimen charging and mean free path are lower at 120 keV.
[b] Pixel size is limited by flux instead of spatial resolution.

Ice thickness is best judged at low magnification by increasing the defocus to hundreds of micrometers to improve the contrast.

3. Once a grid is selected for imaging, load the calibration grid and perform standard microscope alignments in low dose exposure mode. On a well-aligned and stable Krios these include gun alignment, condenser and objective aperture alignments, beam tilt (coma free alignment), camera gain correction, condenser, and objective stigmation. Setup the exposure mode using the parameters in Table 4.
4. Reload the selected specimen and locate all the squares for data collection using low magnification and high defocus. Using automated software to create a montage of images at low magnification (atlas) can be useful for this (see chapter "Strategies for Automated CryoEM Data Collection Using Direct Detectors" by Cheng et al.).
5. All squares for data collection should be free from cracks, crystalline ice and contamination.
6. Collect a couple of test micrographs to confirm ice phase, thickness, and particle distribution.
7. Set focus in exposure mode using beam tilt or by looking at the fringes / bright spots at the edge of the hole (Russo & Passmore, 2016b).
8. Check the illumination geometry of the beam in exposure mode: it should be round, centered on the imaging axis, and centered on the hole while encompassing an annulus of the support around the hole (Fig. 11).
9. Begin collecting data, either manually or using automated data collection software (see chapter "Strategies for Automated CryoEM Data Collection Using Direct Detectors" by Cheng et al.).[1]
10. Check the focus and illumination every 20–30 holes. If mounted correctly, focus should vary by <1–2 μm across an 80 μm square.
11. Collect a few (1–2% of the total micrographs) tilt-pair micrographs for validation. Tilt angles and exposures of [0,15°] and [1,3 s], respectively, are reasonable for validation.

Every dataset collected using a direct electron detector should be checked to verify that the specimen preparation and illumination conditions are optimized to minimize particle motion. Currently, the simplest way to do this

[1] Notes on automated data collection on all-gold supports: For hole recognition using FEI's EPU program (hole selection), use a slightly smaller template hole then the actual hole size and then readjust to the true size before selecting the illumination geometry (template definition). This helps with hole locating because the gold foils have more contrast than the standard carbon foils. The "drift" step is not required on all-gold supports since they do not move during irradiation. Focusing is only required every 20–40 μm.

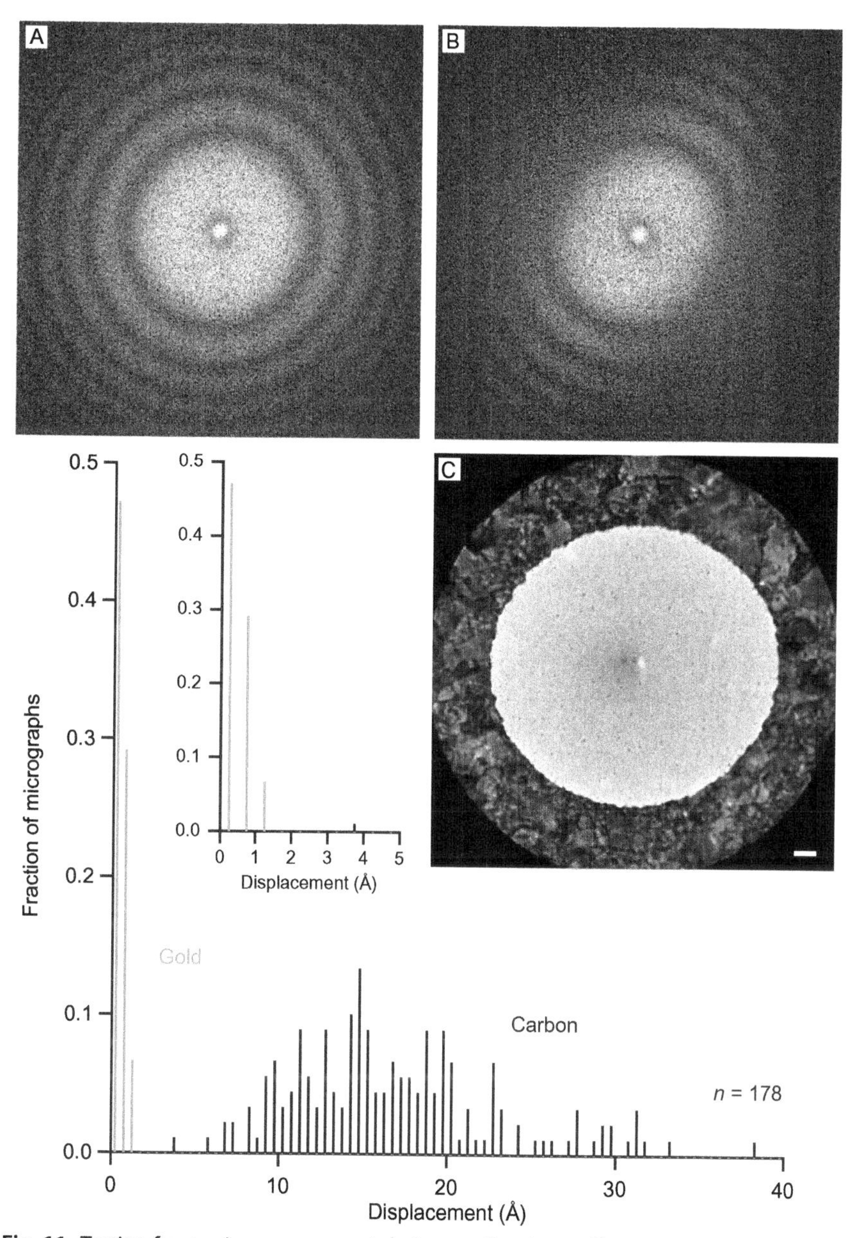

Fig. 11 *Testing for specimen movement during or after data collection.* Large amounts of specimen movement or stage drift can be detected during data acquisition using real-time fast Fourier transforms (FFTs) of the collected micrographs. FFTs of specimens are shown with no stage drift (A) and 10 Å/s temperature induced stage drift (B). Micrograph C shows the recommended, symmetric illumination of a frozen specimen suspended across a hole in an all-gold support foil. Histogram is the in-plane movement statistics for 1 s micrographs (16 $e^{-}/Å^{2}$) on all-gold supports vs amorphous carbon on gold (Quantifoil) under the same symmetric illumination conditions shown in C. Inset is enlargement of histogram near origin. (See the color plate.)

is using a per-micrograph motion correction program like motioncorr (Li et al., 2013; see chapter "Processing of Cryo-EM Movie Data" by Ripstein and Rubinstein). Each micrograph is collected as a movie where the total dose is subdivided into individual frames. The algorithm is used to determine the overall movement of the specimen in the micrograph with time. The program is fast, and therefore can be used to quickly check the overall movement of the specimen support in real time during imaging. Distributions of movement for micrographs in a 1 s exposure at 16 $e^{-}/Å^2$ (Titan Krios on a Falcon 2 direct electron detector with 1.7 Å pixels) are shown in Fig. 11. Both sets were collected with the recommended symmetric conditions. On all-gold supports, the movement is less than 1.5 Å, which is less than one pixel in the image. On standard Quantifoil supports under the same conditions, the per-micrograph movement is an order of magnitude larger. Thus, with ultra-stable supports and modern low-drift microscope stages, we use micrograph motion tracking algorithms to check for incorrect data collection settings, stage drift, or damage to the foil or grid since under normal conditions with ultra-stable supports, the overall movement of the specimen support should be essentially undetectable.

ACKNOWLEDGMENTS

The authors thank R. Henderson and G. McMullan for many helpful discussions, S. Scotcher for fabrication of custom instruments, the EM facility at MRC LMB (S. Chen, C. Saava), the Ramakrishan lab for the gift of ribosomes, S. Tan and C. Hill for a critical reading of the manuscript, and B. Carragher for the photograph of the Gatan Solarus. C.J.R. and L.A.P. are inventors on a patent filed by the Medical Research Council, UK related to this work, which is licensed to Quantifoil under the trade mark UltrAuFoil®. This work was supported by an Early Career Fellowship from the Leverhulme Trust (C.J.R.), the European Research Council under the European Union's Seventh Framework Programme (FP7/2007-2013)/ERC grant agreement no. 261151 (L.A.P.), and Medical Research Council grants U105192715 and U105184326.

REFERENCES

Adrian, M., Dubochet, J., Lepault, J., & McDowall, A. W. (1984). Cryo-electron microscopy of viruses. *Nature*, *308*(5954), 32–36.

Baker, L. A., Watt, I. N., Runswick, M. J., Walker, J. E., & Rubinstein, J. L. (2012). Arrangement of subunits in intact mammalian mitochondrial ATP synthase determined by cryo-EM. *Proceedings of the National Academy of Sciences*, *109*(29), 11675–11680.

Bernecky, C., Herzog, F., Baumeister, W., Plitzko, J. M., & Cramer, P. (2016). Structure of transcribing mammalian RNA polymerase II. *Nature*, *529*(7587), 551–554.

Brenner, S., & Horne, R. (1959). A negative staining method for high resolution electron microscopy of viruses. *Biochimica et Biophysica Acta*, *34*, 103–110.

Chen, B., Kaledhonkar, S., Sun, M., Shen, B., Lu, Z., Barnard, D., … Frank, J. (2015). Structural dynamics of ribosome subunit association studied by mixing-spraying time-resolved cryogenic electron microscopy. *Structure*, *23*(6), 1097–1105.

Cheng, Y. (2015). Single-particle cryo-EM at crystallographic resolution. *Cell*, *161*(3), 450–457.
Cheung, M., Kajimura, N., Makino, F., Ashihara, M., Miyata, T., Kato, T., … Blocker, A. J. (2013). A method to achieve homogeneous dispersion of large transmembrane complexes within the holes of carbon films for electron cryomicroscopy. *Journal of Structural Biology*, *182*(1), 51–56.
Chiu, P.-L., Li, X., Li, Z., Beckett, B., Brilot, A. F., Grigorieff, N., … Walz, T. (2015). Evaluation of super-resolution performance of the K2 electron-counting camera using 2D crystals of aquaporin-0. *Journal of Structural Biology*, *192*(2), 163–173.
da Fonseca, P. C., & Morris, E. P. (2015). Cryo-em reveals the conformation of a substrate analogue in the human 20s proteasome core. *Nature Communications*, *6*, 7573.
Dobro, M. J., Melanson, L. A., Jensen, G. J., & McDowall, A. W. (2010). Plunge freezing for electron cryomicroscopy. *Methods in Enzymology*, *481*, 63–82.
Dubochet, J., Adrian, M., Chang, J. J., Homo, J. C., Lepault, J., McDowall, A. W., & Schultz, P. (1988). Cryo-electron microscopy of vitrified specimens. *Quarterly Reviews of Biophysics*, *21*(2), 129–228.
Dubochet, J., Adrian, M., Lepault, J., & McDowall, A. (1985). Emerging techniques: Cryo-electron microscopy of vitrified biological specimens. *Trends in Biological Sciences*, *10*, 143–146.
Elmlund, H., Elmlund, D., & Bengio, S. (2013). PRIME: Probabilistic initial 3D model generation for single-particle cryo-electron microscopy. *Structure*, *21*(8), 1299–1306.
Ermantraut, E., Wohlfart, K., & Tichelaar, W. (1998). Perforated support foils with pre-defined hole size, shape and arrangement. *Ultramicroscopy*, *74*(1), 75–81.
Geim, A. K. (2009). Graphene: Status and prospects. *Science*, *324*(5934), 1530–1534.
Glaeser, R. M., Han, B.-G., Csencsits, R., Killilea, A., Pulk, A., & Cate, J. H. (2016). Factors that influence the formation and stability of thin, cryo-EM specimens. *Biophysical Journal*, *110*(4), 749–755.
Henderson, R. (2013). Avoiding the pitfalls of single particle cryo-electron microscopy: Einstein from noise. *Proceedings of the National Academy of Sciences*, *110*(45), 18037–18041.
Henderson, R. (2015). Overview and future of single particle electron cryomicroscopy. *Archives of Biochemistry and Biophysics*, *581*, 19–24.
Henderson, R., Chen, S., Chen, J. Z., Grigorieff, N., Passmore, L. A., Ciccarelli, L., … Rosenthal, P. B. (2011). Tilt-pair analysis of images from a range of different specimens in single-particle electron cryomicroscopy. *Journal of Molecular Biology*, *413*, 1028–1046.
Kastner, B., Fischer, N., Golas, M. M., Sander, B., Dube, P., Boehringer, D., et al. (2008). GraFix: Sample preparation for single-particle electron cryomicroscopy. *Nature Methods*, *5*(1), 53–55.
Kuijper, M., van Hoften, G., Janssen, B., Geurink, R., De Carlo, S., Vos, M., … Storms, M. (2015). FEIs direct electron detector developments: Embarking on a revolution in cryo-TEM. *Journal of Structural Biology*, *192*(2), 179–187.
Li, X., Mooney, P., Zheng, S., Booth, C. R., Braunfeld, M. B., Gubbens, S., … Cheng, Y. (2013). Electron counting and beam-induced motion correction enable near-atomic-resolution single-particle cryo-EM. *Nature Methods*, *10*(6), 584–590.
McMullan, G., Faruqi, A., Clare, D., & Henderson, R. (2014). Comparison of optimal performance at 300keV of three direct electron detectors for use in low dose electron microscopy. *Ultramicroscopy*, *147*, 156–163.
Miyazawa, A., Fujiyoshi, Y., Stowell, M., & Unwin, N. (1999). Nicotinic acetylcholine receptor at 4.6 Å resolution: Transverse tunnels in the channel wall. *Journal of Molecular Biology*, *288*(4), 765–786.
Nogales, E., & Scheres, S. H. (2015). Cryo-EM: A unique tool for the visualization of macromolecular complexity. *Molecular Cell*, *58*(4), 677–689.
Ohi, M., Li, Y., Cheng, Y., & Walz, T. (2004). Negative staining and image classification: Powerful tools in modern electron microscopy. *Biological Procedures Online*, *6*(1), 23–34.
Pantelic, R. S., Meyer, J. C., Kaiser, U., Baumeister, W., & Plitzko, J. M. (2010). Graphene oxide: A substrate for optimizing preparations of frozen-hydrated samples. *Journal of Structural Biology*, *170*(1), 152–156.

Pantelic, R. S., Suk, J. W., Magnuson, C. W., Meyer, J. C., Wachsmuth, P., Kaiser, U., ... Stahlberg, H. (2011). Graphene: Substrate preparation and introduction. *Journal of Structural Biology, 174*(1), 234–238.

Plaschka, C., Lariviere, L., Wenzeck, L., Seizl, M., Hemann, M., Tegunov, D., et al. (2015). Architecture of the RNA polymerase II-Mediator core initiation complex. *Nature, 518*(7539), 376–380.

Quispe, J., Damiano, J., Mick, S. E., Nackashi, D. P., Fellmann, D., Ajero, T. G., ... Potter, C. S. (2007). An improved holey carbon film for cryo-electron microscopy. *Microscopy and Microanalysis, 13*(05), 365–371.

Regan, W., Alem, N., Alemán, B., Geng, B., Girit, Ç., Maserati, L., ... Zettl, A. (2010). A direct transfer of layer-area graphene. *Applied Physics Letters, 96*, 113102–113104.

Rosenthal, P. B., & Henderson, R. (2003). Optimal determination of particle orientation, absolute hand, and contrast loss in single-particle electron cryomicroscopy. *Journal of Molecular Biology, 333*(4), 721–745.

Russo, C. J., & Golovchenko, J. A. (2012). Atom-by-atom nucleation and growth of graphene nanopores. *Proceedings of the National Academy of Science, 109*(16), 5953–5957.

Russo, C. J., & Passmore, L. A. (2014a). Controlling protein adsorption on graphene for cryo-EM using low-energy hydrogen plasmas. *Nature Methods, 11*(6), 649–652.

Russo, C. J., & Passmore, L. A. (2014b). Robust evaluation of 3D electron cryomicroscopy data using tilt-pairs. *Journal of Structural Biology, 187*(2), 112–118.

Russo, C. J., & Passmore, L. A. (2014c). Ultrastable gold substrates for electron cryomicroscopy. *Science, 346*(6215), 1377–1380.

Russo, C. J., & Passmore, L. A. (2016a). Progress towards an optimal specimen support for electron cryomicroscopy. *Current opinion in structural biology, 37*, 81–89. http://dx.doi.org/10.1016/j.sbi.2015.12.007.

Russo, C. J., & Passmore, L. A. (2016b). Ultrastable gold substrates: Properties of a support for high-resolution electron cryomicroscopy of biological specimens. *Journal of Structural Biology, 193*(1), 33–44. http://dx.doi.org/10.1016/j.jsb.2015.11.006.

Seigel, R. R., Harder, P., Dahint, R., Grunze, M., Josse, F., Mrksich, M., & Whitesides, G. M. (1997). On-line detection of nonspecific protein adsorption at artificial surfaces. *Analytical Chemistry, 69*(16), 3321–3328.

Stark, H., & Grant, J. (2010). GraFix: Stabilization of fragile macromolecular complexes for single particle cryo-EM. *Methods in Enzymology, 481*, 109.

Tang, G., Peng, L., Baldwin, P. R., Mann, D. S., Jiang, W., Rees, I., & Ludtke, S. J. (2007). EMAN2: An extensible image processing suite for electron microscopy. *Journal of Structural Biology, 157*(1), 38–46.

Thompson, R. F., Walker, M., Siebert, C. A., Muench, S. P., & Ranson, N. A. (2016). An introduction to sample preparation and imaging by cryo-electron microscopy for structural biology. *Methods, 100*, 3–15.

Unwin, P. (1974). Electron microscopy of the stacked disk aggregate of tobacco mosaic virus protein: II. The influence of electron irradiation on the stain distribution. *Journal of Molecular Biology, 87*(4), 657–670.

Vinothkumar, K. R., & Henderson, R. (2016). Single particle electron cryomicroscopy: Trends, issues and future perspective. *Quarterly Reviews of Biophysics*. in press.

Vogler, E. A. (2012). Protein adsorption in three dimensions. *Biomaterials, 33*(5), 1201–1237.

Wasilewski, S., & Rosenthal, P. B. (2014). Web server for tilt-pair validation of single particle maps from electron cryomicroscopy. *Journal of Structural Biology, 186*(1), 122–131.

Weis, F., Giudice, E., Churcher, M., Jin, L., Hilcenko, C., Wong, C. C., ... Warren, A. J. (2015). Mechanism of eIF6 release from the nascent 60S ribosomal subunit. *Nature Structural & Molecular Biology, 22*(11), 914–919.

CHAPTER FOUR

Strategies for Automated CryoEM Data Collection Using Direct Detectors

A. Cheng*[,1], Y.Z. Tan*[,†,1], V.P. Dandey*, C.S. Potter*[,†], B. Carragher*[,†,2]
*Simons Electron Microscopy Center, New York Structural Biology Center, The National Resource for Automated Molecular Microscopy, New York, NY, United States
†Columbia University, New York, NY, United States
[2]Corresponding author: e-mail address: bcarr@nysbc.org

Contents

Abstract

The new generation of direct electron detectors has been a major contributor to the recent resolution revolution in cryo-electron microscopy. Optimal use of these new cameras using automated data collection software is critical for high-throughput near-atomic resolution cryo-electron microscopy research. We present an overview of the practical aspects of automated data collection in the context of this new generation of direct detectors, highlighting the differences, challenges, and opportunities the new detectors provide compared to the previous generation of data acquisition media.

[1] These authors contributed equally to this work.

Methods in Enzymology, Volume 579
ISSN 0076-6879
http://dx.doi.org/10.1016/bs.mie.2016.04.008

1. INTRODUCTION

Image collection for cryo-electron microscopy (cryoEM) involves a series of tasks that require repetitively targeting areas of the grid at several different magnification scales to acquire a single final high magnification image. A cryoEM grid prepared using standard vitrification procedures is typically covered by a layer of sample suspended in vitreous ice of varying thickness and quality (Fig. 1A). The vitrified sample is usually suspended over a holey substrate and the goal is to acquire images, or a tilt series of images, of the sample suspended over these holes. If the sample is unevenly distributed across the hole, the area within the hole at which to acquire the target must also be identified. The grid is usually scanned at very low

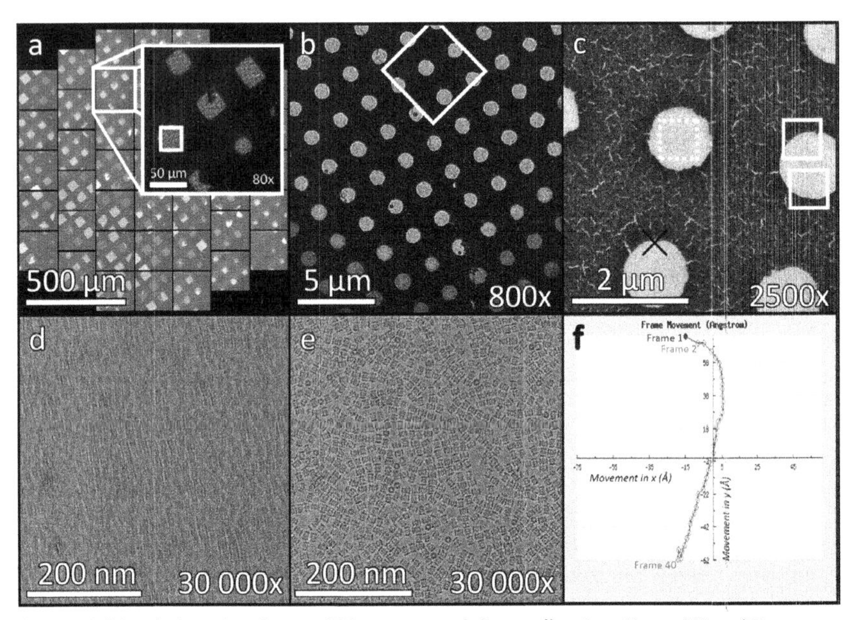

Fig. 1 Multiscale imaging in cryoEM automated data collection. From (A) to (D) are successively higher magnification images of the area of interest. The targets for imaging (*white squares*) in (C) are shifted to the center of the field of view using a combination of stage and image shifts (see text). Target for focus adjustment, drift monitoring, etc., is indicated by a *black cross* and this area is normally centered in the field of view using image shifts. (D) A micrograph obtained at the highest magnification (from target in (C) encompassed by *dotted white line*), before global frame alignment and (E) the same micrograph after frame alignment. The frame movement of the 40 frames acquired is plotted in (F). Images were acquired on a JEOL2100 equipped with a Gatan K2 direct detector camera operating in counting mode.

magnifications (Fig. 1A) to identify potentially suitable areas of vitreous ice of the right thickness. These areas are then scanned at a higher magnification to identify holes in the substrate (Fig. 1B and C). The holes are centered in the field of view and targets are selected for acquiring final high magnification images or tilt series, as well as for low dose focusing, astigmatism correction, and drift monitoring (Fig. 1D). This process is then repeated to acquire from 100s to several 1000s of images.

Automating these repetitive and exacting procedures has the benefit of freeing up the instrument operator and ensuring continuous operation of the instrument, and indeed automated data collection for cryoEM has a long history. Applications for acquiring tomographic tilt series were pioneered by the groups of Baumeister and Agard (Dierksen, Typke, Hegerl, Koster, & Baumeister, 1992; Koster, Chen, Sedat, & Agard, 1992) over 20 years ago. There are now many robust and established packages in routine use for automated tomographic data collection (including SerialEM (Mastronarde, 2005), UCSF Tomo (Zheng et al., 2007), TOM Toolbox (Nickell et al., 2005), Leginon (Suloway et al., 2005), EPU (FEI)), and few practitioners of this method would consider using manual methods for data collection. Automated single particle data collection applications (Potter et al., 1999) appeared a few years after the first tomography applications but were not generally and widely adopted until the advent of direct detectors (DDs) established that digital data collection was superior to acquiring images to film. Compared to films, which are typically scanned into images on the order of $\sim$6000 × 8000 pixels, previous generations of digital cameras were inferior in performance and had a very limited field of view ($\sim$10 times less area at an equivalent image pixel size). As a result, most users needing high-resolution data preferred to use film, but automated data collection (Suloway et al., 2005) using this medium was limited in impact—typically only a few hundred images per day could be acquired to film before the contamination of the vacuum brought a halt to data collection.

The performance of the new generation of DDs is now superior to film (see chapters "Processing of Cryo-EM Movie Data" by Ripstein and Rubinstein and "Direct Electron Detectors" by McMullan et al.) with the result that digital data collection has rapidly been established as the sole mean for acquiring high-resolution images. There are now almost a dozen different software packages available for automated acquisition of single particle data (for a recent review, see Tan, Cheng, Potter, & Carragher, 2016). Nevertheless, the field of view of these new digital cameras is still quite small, which means that a fairly large number of images must be acquired in order to accumulate enough particles to achieve a high-resolution reconstruction. For example, more than 1000 images were acquired to provide ~100,000 particles contributing to the 3.2 Å map of the *Plasmodium falciparum* 80S ribosome (Wong et al., 2014). Improvements in detectors and instrumentation (see chapter "Processing of Cryo-EM Movie Data" by Ripstein and Rubinstein) will likely reduce the number of particles required, but the opportunity to sort out highly heterogeneous datasets will drive the need for even higher numbers of images. While there is still some argument as to whether manual or automated data collection protocols are more efficient, there is no doubt that human beings are ill-suited to performing a repetitive and exacting task many 1000s of times without interruption. It has also now been established that automated methods are capable of acquiring data that results in maps of the highest resolution; as illustrated for example in the 2.8 Å map of the 20S proteasome (Campbell, Veesler, Cheng, Potter, & Carragher, 2015). It thus seems likely that automation of image acquisition will become firmly established for cryoEM. In this overview, we will discuss the practical aspects of automated data acquisition using the new generation of DDs and we will briefly describe the various strategies for data collection using a variety of microscopes and cameras.

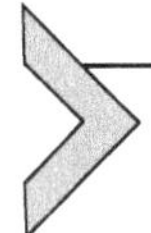

2. PRACTICAL CONSIDERATIONS IN INSTRUMENTATION CONFIGURATION

Before starting an automated data collection session it is important to understand how the DD camera is incorporated into the microscope in order to fully exploit the capabilities, as well as minimize the limitations, of these new cameras. The current generation of DD cameras can be separated into two categories: analog and electron counting. Analog DDs read out the amplified charge transferred directly and the variation of the charge arising from different numbers of electrons deposited produces the image. Electron counting DDs further process these signals under the condition of very low numbers of incident electrons and this allows the camera software to threshold the analog signal to a binary result; multiple frames of these binary images are summed together to produce the final image. Currently, only Gatan's K2 Summit camera is capable of recording in counting mode, but the other two DD manufacturers, FEI and Direct Electron, will soon offer their own counting cameras.

Near-atomic resolution cryoEM requires images with a pixel size in the range of 1 Å so that the physical Nyquist limit is in the range of 2 Å. When the camera is used for automated data collection, efficient searching and centering of final image targets are done at multiple scales of magnification. At each image magnification, the largest possible field of view is desired while still allowing for sufficient accuracy to bring the target to the center of the field of view at the next scale. Image shifts must be applied to keep the target centered across the magnification scales that are repeatedly accessed. To minimize image shift alignment hysteresis between the most frequently used magnifications, including the final exposure, it is desirable for these scales to be within the same objective and projection lens series. These three requirements set the upper and lower bound of the image pixel size at each multiscale imaging (MSI) level (Fig. 1), and the number of magnification levels required to view and target the entire specimen grid.

There are limited options to choose from in attempting to optimize image pixel sizes at all magnification scales of interest during automated data collection. For a specified camera and microscope, it may be possible to extend a lens series to enable the most frequently used magnifications to be used in the same series. If more than one camera is mounted on the microscope, the order of the camera stacking, which affects the post-magnification at the camera, needs to be considered. Before the advent of DDs, high-end TEM digital cameras typically had sensor pixel sizes in the range of 15–25 μm and an optimal set of image pixel sizes could be readily achieved. The DDs now available (see Tables 1 and 2) have sensor pixel sizes of 5 μm (K2), 14 μm (Falcon II), and 6.4 μm (DE20). The smaller sensor pixel sizes mean that the nominal magnification required for achieving an ideal image pixel size is now much lower than that typically used for film or CCDs and is frequently outside the optimal objective and projection lens series provided by TEM manufacturers. The need to use very low magnifications can also create issues like an obstructed field of view arising from optical components in the column, magnification

Table 1 DD Camera Specifications

Model	DE20	Falcon II	K2 Summit
Retrievable frame time unit (ms)	31	56	25
Physical pixel size (Å)	6.4 μm	14 μm	5 μm
Image dimension	3840 × 5120	4096 × 4096	3838 × 3710 (counting) 7676 × 7420 (super resolution)

Specifications for the three DD cameras that have been used at the New York Structural Biology Center, Simons Electron Microscopy Center.

Table 2 Examples of Parameters Used for High-Resolution CryoEM Structures on Various DD Cameras

Camera	DE20	K2 Summit	Falcon II
Sample	AAV-DJ with Arixtra	20S Proteasome	Yeast mitochondrial large ribosomal subunit
Symmetry	Icosahedral	D7	C1
Microscope	FEI Titan Krios	FEI Titan Krios	FEI Titan Krios
Magnification listed by microscope	29,000 ×	22,500 ×	104,478 ×
Pixel size	1.215 Å	0.66 Å (super resolution) 0.982 Å (processed)	1.34 Å
Total accumulated dose	66 $e^-/Å^2$	53 $e^-/Å^2$	25 $e^-/Å^2$
Number of frames	45	38	17
Exposure time per frame	31 ms	200 ms	1 s
Total exposure time	1.394 ms	7.6 s	17 s
Dose per frame	1.5 $e^-/Å^2$	1.39 $e^-/Å^2$	1.47 $e^-/Å^2$
Number of images acquired	1428	985; 196 used	1030
Total number of particles processed	120,166	87,066	135,949
Number of particles contributing to map	107,454	49,954	64,139
Resolution of map	2.8 Å	2.8 Å	3.2 Å
Reference	Spear et al. (2015)	Campbell et al. (2015)	Amunts et al. (2014)
EMPIAR ID	–	EMPIAR-10025	–

Raw data for the proteasome can be found in EMPIAR (Iudin, Korir, Salavert-Torres, Kleywegt, & Patwardhan, 2016).

anisotropy, and/or pin-hole distortion. It is therefore very desirable to place the camera with the smallest sensor pixel at the highest position possible; if this is not an option then a lens series extension of microscope magnifications may be required.

DD sensors are directly exposed to electrons and while they are radiation hardened, they are still more prone to radiation damage than scintillator-based digital cameras. This is especially a problem for analog DDs as a higher beam intensity is used to produce the desired signal in each movie frame. In order to minimize the exposure, a secondary scintillator-based CCD or CMOS camera is often installed along with the DD. Typically, the secondary camera is used during manual focusing and/or at very low magnification where the condenser lens strength limits the ability to expand the beam to reduce the electron dose rate on the detector. This camera is sometimes placed as a smaller sensor within the same camera body, which requires shifting the beam when acquiring images to the secondary camera. Alternatively, it may be installed as a retractable camera, requiring insertion every time it is used to acquire an image and then retraction to acquire to the DD camera. The switching of the cameras is handled by automated acquisition software so as to be transparent to the user but it is worth considering whether mechanical wear as a result of frequent DD insertion/retraction outweighs the benefit of using a secondary camera.

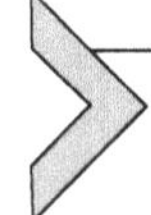

3. AUTOMATED DATA ACQUISITION USING DD CAMERAS

Most of the automated data acquisition applications have been updated (Tan et al., 2016) to accommodate the use of DDs; these include SerialEM (Mastronarde, 2005), Leginon (Suloway et al., 2005), UCSFImage4 (Li, Zheng, Agard, & Cheng, 2015), and EPU (FEI). Below we will focus on the use of Leginon to provide a specific example so that practical issues can be described in context. In general, the workflow for acquiring images on a DD closely follows that developed for the CCD (Suloway et al., 2005) and we will mostly focus on highlighting areas of importance or differences from that baseline in the description below.

3.1 Gain/Dark Reference Image Preparation

DDs are based on CMOS (complementary metal oxide semiconductor) technology, which means that the protocol for acquiring reference images is different from that used with a CCD. First, CMOS cameras are faster than CCDs but also have higher noise. Therefore, a higher accumulated exposure must be used on DDs during reference image preparation. For example, the recommended DE20 accumulated exposure dose for the gain reference is 1500–3000 electrons/pixel, while the Gatan software for K2 Summit accumulates 5000 electrons while preparing its gain reference. Second, while CCD dark images contain large bias and minimal exposure time-dependent dark current, CMOS sensors have negligible bias values but large dark currents. Dark current increases with the camera usage and thus dark images need to be acquired more frequently for DDs, ideally every day. Note that while the notion of a dark image is obvious for an analog DD, for counting cameras this dark image is the value associated with the threshold of the electron counting event, ie, the hardware dark reference image on the K2 Summit. Third, the charge response to electron dose rate is not as linear for a DD as a CCD. Thus, the dose rate used to obtain the gain reference should match that used to obtain subsequent images; for example, thick ice may offset the beam intensity measured by the detector, and a new gain reference may need to be acquired at a matching intensity.

3.2 Choosing Dose Rates

DD sensor saturation is determined within each frame, not the overall exposure dose of the movie. As a result, the dose rate should be adjusted to fit the individual requirements of the camera bearing in mind that the sensor can be permanently damaged by an intense beam. A DD camera is characterized by the length of time it takes for the rolling shutter to scan the sensor. This is called the base frame time to distinguish it from the movie frame time; movie frame time can be set to multiples of the base frame time in some DDs. Of the three DDs currently on the market, only the DE20 allows user control of the base frame time. The dynamic range for DDs is large, thus, analog DDs do not need to be used in a narrow dose rate range as long as the gain reference matches the value used for imaging. When using a counting DD, on the other hand, the effects of coincidence loss must be considered. For example, the K2 Summit requires 0.02 electrons/pixel per base frame to avoid 20% coincidence loss (Li et al., 2013); the exposure dose rate at the specimen is then dictated by the pixel size. With regards to automated data collection software, note that images acquired for purposes other than the final exposure do not need to take account of the optimal dose rate, except to avoid very high dose rates that may damage the camera sensor.

3.3 Choosing Total Exposure Time and Dose

The choice of the exposure time, previously carefully chosen in single-shot film or CCD exposures to give a total specimen dose typically in the range of 10–25 $e^-/Å^2$, is now only limited by the size of the movie file that can be accommodated by the available disk space and the power of the hardware for processing the movies. The higher dose frames are useful for enhancing contrast during particle picking steps and can always later be removed (Li et al., 2013) or downweighted (Grant & Grigorieff, 2015; Scheres, 2014) during map refinement steps.

3.4 Frame Saving and Drift Correction

All of the DD cameras support the ability to save individual frames acquired during the exposure. The principal benefit of collecting DD movies is the ability to align the frames to compensate for specimen drift and beam-induced movements (Brilot et al., 2012). Global, whole frame alignment can be performed using a variety of available packages, including, for example, those described in Grant and Grigorieff (2015) and Li et al. (2013) and software available from Direct Electron. It is also useful for users to be able to evaluate aligned movies during the data collection session; this is especially important for the long exposure times that are required for counting mode as the images are often so blurred before drift correction that it is not possible to evaluate the quality of the images or the particles (see Fig. 1D). Automated data collection systems can seamlessly integrate a global coarse movie correction algorithm into the data collection pipeline to provide the user with real-time feedback (Fig. 1E) and facilitate modification of the data collection strategy.

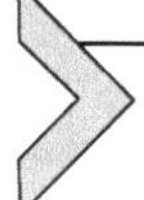

4. CHALLENGES ARISING FROM DDs AND HIGH-RESOLUTION BIOLOGICAL CRYOEM

The introduction of DDs has vastly improved the resolution achievable using cryoEM single particle analysis. They have also presented new challenges to efficient data collection. The three characteristics of the current generation of DDs that have the most important impact on automated data collection strategy are the smaller pixel sizes (in two of the three DDs available), the low dose rate requirement when using counting mode, and the need to save and display movies. These issues impose constraints on efficient automated data collection when seeking to acquire images of the highest possible resolution.

4.1 High-Resolution Data Collection Is Slower

The efficiency of automated data collection mostly depends on three factors: the speed of image acquisition, the time required for the specimen to settle after each movement of the stage, and the accuracy of targeting at multiple magnification scales. The higher the desired resolution, the longer the time required for each of these factors. The first factor is camera dependent; image acquisition using a DD with counting capability is slower than with analog models because of the extra processing required and the need to reduce coincidence loss (Li et al., 2013) by using a very low dose rate. The second factor depends on the method used to move the stage to the selected target and the mechanical design of the specimen holder. Specimens loaded using cartridges (eg, Polara, Krios, JEM3200) have better thermal isolation inside the microscope and are thus more stable than side-entry holders. The third factor is somewhat more complex. Most images are normally acquired from vitreous ice suspended over a hole in a supporting substrate. The sample is examined at low magnifications in order to select these holes and bring them to the center of the field of view (Fig. 1). As features are rarely perfectly centered when changing from one magnification to another, it is almost always necessary to move the area of interest to the center of the field of view when acquiring the final high magnification images. This movement can be done very rapidly and accurately using the image and beam shift coils or much more slowly using the specimen stage. The issue with using the first method is that the image phase shift imposed by beam-tilt coma increases as the cube of the resolution (Glaeser, Typke, Tiemeijer, Pulokas, & Cheng, 2011). A beam tilt of 0.167 mrad induced by 1 μm beam-image shift on a Titan Krios can result in a 45 degree phase change at 2.5 Å resolution (Fig. 2). As a result, when a specimen has the possibility of producing a 3D reconstruction at sub 3 Å resolution, centering the target using stage movement may be preferable to using the image shift coils. When using the stage for repositioning, targeting accuracy is improved by reducing the speed of the stage, incorporating backlash correction, and iteratively repeating the process until the feature of interest is centered to the required accuracy; all three actions add to the time required for image collection. Leginon takes ~15–30 s when using the stage for centering in order to achieve ~0.1 μm targeting accuracy, whereas using multiple image shifts to target can be achieved in less than a second. In practice even if the user prefers to choose speed over high-resolution data, some stage shifts will typically be used in order to achieve large movements required to bring holes far from the center of the image (Fig. 1C) to the center and then image shifts can be used to do the fine targeting within the hole.

The accuracy of the targeting is also dependent on the image shift hysteresis. While looping through target selection, target centering, and final imaging, the lens settings are cycled between low magnification

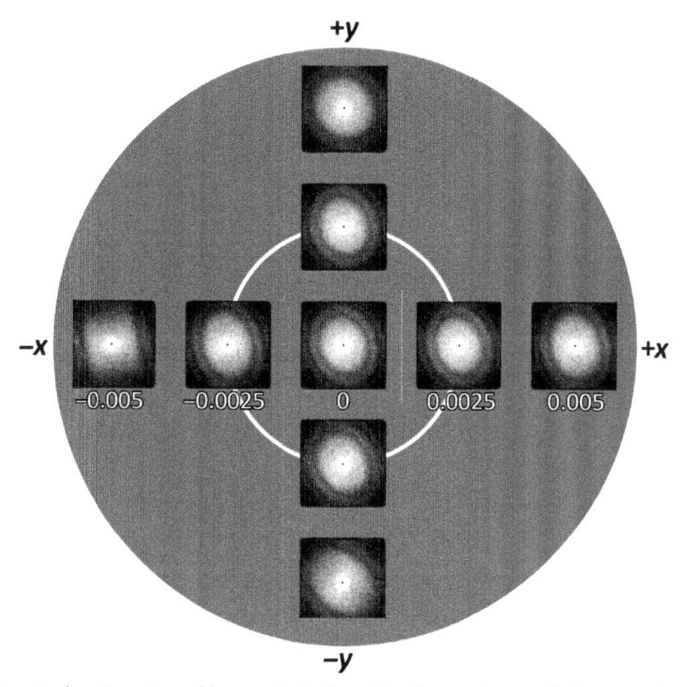

Fig. 2 Practical estimation of beam tilt induced by beam-image shift. A Zemlin tableau (Zemlin, Weiss, Schiske, Kunath, & Herrmann, 1978) is a series of digital diffractograms taken using a defined beam tilt in orthogonal directions. When arranged according to the amplitude and the direction of the tilts, a coma-free alignment results in a symmetrical pattern around the center. Estimation of beam tilt induced by beam-image shift can be obtained by increasing the shift in one direction until a symmetrical point is found. In this example, an 8 µm beam-image shift was applied and Leginon was used to acquire and display a two-layered tableau using a tilt amplitude of 0.0025 radians at each layer. The numbers displayed indicate the beam tilt in the *x* direction relative to a coma-free beam tilt when no beam-image shift is applied. The *red* (*light gray* in the print version) *line* indicates that along the *x*-axis symmetry is observed half way between 0 and +*x* 0.0025 radians. Therefore the coma-free beam tilt is displaced by 0.00125 radians from the beam-tilt value at the origin of the tableau. The tableau also shows that a smaller beam tilt needs to be applied in +*y* direction to achieve coma-free alignment.

(for targeting) and high magnification (for acquiring the final image). Normalization of the lenses is important to maintain relative stability of the image shift alignments between magnifications but is also time-consuming. Roughly 4 s are needed to normalize all lenses on the FEI Krios; an additional 4 s are spent on the JEOL JEM3200 to remove beam position hysteresis. If built-in normalization is absent or inadequate, cycling through all of the magnifications that are being used can achieve a similar effect, but it often takes longer as the cycling has to be done in a strict order through all magnifications every time a magnification change is required. In addition, the number of magnification steps that must be used to target from the grid overview atlas to the final high magnification depends on the targeting accuracy that can be achieved at each magnification setting. The limited DD sensor area may require adding in extra magnification steps to increase targeting accuracy and this again reduces the efficiency of data collection.

Efficiency of data collection is especially critical when using a DD in counting mode. In order to count single electron events, a dose rate is required that avoids loss due to coincidence. This translates to longer exposure times to accumulate a total electron dose on the specimen and there is also additional time required to process the counting images. These factors, in addition to the time required to save the frames to disk, could extend the idle time of the camera sensor up to 10 times that of the previous generation of CCD cameras. Time saving can be achieved by using the sensor idle time for moving to the next target, focusing, lens normalization, and waiting for stage drift to settle. This requires asynchronous functions in the software for automated data acquisition that was previously only relevant to cases where high-volume data collection on serial sections was acquired.

The longer exposure and collection times typically required for the movies also have an impact on providing feedback of the quality of the images to the user. Assessment of vitreous ice thickness and quality, as well as particle concentration, is essential to a decision of whether the images acquired will be useful. The long exposure times, of the counting cameras in particular, mean that the frame averaged images that are available immediately after image acquisition are frequently so blurred by specimen drift that these kinds of assessments are not possible (Fig. 1D). Thus, it is highly desirable that any automation software package have rapid on-the-fly frame alignment options and the ability to readily view the aligned images (Fig. 1E). Ideally, the frame alignment will keep pace with data collection and might require that the frame processing step takes advantage of parallelization.

4.2 Computing Requirements

The ability of DDs to collect in movie mode, combined with automated data collection, equates to an explosion of the amount of collected data that has to be stored. One frame stack from the K2 camera containing 50 frames is about 2 Gb in size—with automated data collection it is possible to collect thousands of these movies in a few days, resulting in an output of terabytes of data. The computing and networking architecture required to support state-of-art biological electron microscopy is thus an order of magnitude greater than when CCD or film cameras were the norm.

The first computational bottleneck is the communication of the camera with the storage cluster, which is relatively easily solved by the use of 10 Gb/s network connections. The second bottleneck stems from data storage during the entire project duration. After the raw frames have been globally aligned and summed into a single micrograph, this image is usually sufficient for downstream data processing until the final stages when per-particle motion correction is often used to improve the resolution. This final step requires reprocessing of the raw frames meaning that it is essential to have these raw frames available for the period of time taken to process the images. Thus, the raw frames may need to be stored for many months after the initial data collection. An alternative is to perform the per-particle alignment at the start of the processing and reconstruction process (Rubinstein & Brubaker, 2015). An additional factor is that new algorithms for particle and frame alignments are still being developed and that data may need to be reprocessed to see if the results can be improved.

Each data collection session potentially generates on the order of a few TB of raw frames, and different facilities have different strategies for addressing this storage issue. At facilities where no or minimal data processing capabilities are available to the user, the data will be immediately copied out for the user to bring back physically on hard disks at the end of data collection. In facilities that support local data processing, the raw frames are kept by the facility for several weeks to months, and the users have that period of time to backup their data to permanent storage before it is removed.

One way to reduce the amount of storage space required is compression. This is especially useful for movies collected in counting mode, as the pixels in these movie frames will be predominantly filled by zeros. Super resolution raw movies from the K2 Summit can be compressed losslessly with the bzip utility to 1/10 of their size in its 8-bit depth form. The analog DDs such as the DE and Falcon systems can also be compressed by two- to threefold. Also note that images prior to flat-field correction are stored as integers and compress better than the floating point images that result from flat-field and drift correction. Thus, uncorrected frames are the most efficient form for archiving movies.

The large movie files also mean that processing movies on-the-fly to provide user feedback is also a challenge. We believe this is a crucial aspect of automated data collection because it allows for identification and possible rectification of any problems with either the instrument or the specimen that arise during the process, especially since operator interaction is usually minimal during automated data collection. Parallelization is the main solution to dealing with on-the-fly tasks. For example, Appion (Lander et al., 2009) allows parallelization either by image or by separation of steps for gain/dark/defect correction and drift correction. With two GPU devices running in parallel, the throughput of image correction can match the speed of data collection when using the K2 Summit so that the user gets feedback about the quality of the images within a few minutes of their acquisition. Parallelization of the script for frame alignment provided by Direct Electron for the DE20 camera is achieved by aligning each movie in a separate CPU of a multi-CPU machine and this process is readily scalable. Particle picking and CTF estimation can also be done in parallel in Appion, usually running on individual images across multiple processors at the same time. Thus by the end of an automated data acquisition session using Leginon and Appion, the user should have globally aligned movies, an estimate of the CTF, an assessment of the quality of the image based on the CTF fitting (Sheth, Piotrowski, & Voss, 2015), an initial set of particles picked and an initial particle stack ready for reconstruction. Only a lack of computer power limits the speed with which the entire processing and reconstruction steps can be performed.

5. FUTURE OUTLOOK FOR DD INTEGRATION WITH AUTOMATED DATA COLLECTION

Development of DDs has until now largely been carried out with automated data collection as an afterthought—the goal had been to achieve atomic resolution. With that goal now reached by multiple groups over the past year, DD advances that aid in streamlining automated data collection will be valuable in making cryoEM accessible to a broader scientific community and in improving the overall efficiency of data acquisition. Commercial adoption of automated data collection is already well underway as both FEI and Gatan, Inc. are actively developing data packages to interface with their own DD cameras.

The influx of new projects into cryoEM, and the consequent pressure for time on very expensive instruments, means that increasing the throughput in data collection is increasingly important, and automation is likely to play a crucial part in this equation. Throughput can be increased in two ways: by increasing the data collected per unit time or by reducing the number of poor quality images collected. In the first instance, the new Direct Electron

DE64 cameras have a larger field of view (sensor size is 8k × 8k with 6.4 μm pixels) so that more particles are acquired in every image. Combining on-the-fly image alignment and CTF correction with automated algorithms may help improve throughput by identifying areas of the grid most likely to yield the highest quality data, or at least by avoiding areas of the grid that yield low quality images.

In the medium term, automated data collection will likely also become important for other nascent cryoEM techniques such as correlative light and electron microscopy followed by focused-ion-beam (FIB) milling and electron tomography (Villa, Schaffer, Plitzko, & Baumeister, 2013). We anticipate that the lessons learned from development of automated data collection with the DD will be used to improve the robustness and throughput of all forms of cryoEM for the general benefit of the biological community.

ACKNOWLEDGMENTS

We would like to acknowledge primary financial support from the National Institutes of Health (NIH); National Institute of General Medical Sciences (GM103310 to A.C., C.S.P., and B.C.); Simons Foundation (349247 to C.S.P. and B.C.); and Agency for Science, Technology and Research Singapore (to Y.Z.T.). We thank Melody Campbell and Scott Stagg for providing input for the chapter.

REFERENCES

Amunts, A., Brown, A., Bai, X. C., Llacer, J. L., Hussain, T., Emsley, P., … Ramakrishnan, V. (2014). Structure of the yeast mitochondrial large ribosomal subunit. *Science*, *343*(6178), 1485–1489. http://dx.doi.org/10.1126/science.1249410.

Brilot, A. F., Chen, J. Z., Cheng, A., Pan, J., Harrison, S. C., Potter, C. S., … Grigorieff, N. (2012). Beam-induced motion of vitrified specimen on holey carbon film. *Journal of Structural Biology*, *177*(3), 630–637. http://dx.doi.org/10.1016/j.jsb.2012.02.003.

Campbell, M. G., Veesler, D., Cheng, A., Potter, C. S., & Carragher, B. (2015). 2.8 Å resolution reconstruction of the Thermoplasma acidophilum 20S proteasome using cryo-electron microscopy. *Elife*, *4*, 1–10. http://dx.doi.org/10.7554/eLife.06380.

Dierksen, K., Typke, D., Hegerl, R., Koster, A. J., & Baumeister, W. (1992). Towards automatic electron tomography. *Ultramicroscopy*, *40*(1), 71–87. http://dx.doi.org/10.1016/0304-3991(92)90235-C.

Glaeser, R. M., Typke, D., Tiemeijer, P. C., Pulokas, J., & Cheng, A. (2011). Precise beam-tilt alignment and collimation are required to minimize the phase error associated with coma in high-resolution cryo-EM. *Journal of Structural Biology*, *174*(1), 1–10. http://dx.doi.org/10.1016/j.jsb.2010.12.005.

Grant, T., & Grigorieff, N. (2015). Measuring the optimal exposure for single particle cryo-EM using a 2.6 Å reconstruction of rotavirus VP6. *Elife*, *4*, e06980. http://dx.doi.org/10.7554/eLife.06980.

Iudin, A., Korir, P. K., Salavert-Torres, J., Kleywegt, G. J., & Patwardhan, A. (2016). EMPIAR: A public archive for raw electron microscopy image data. *Nature Methods*, *13*, 387–388. http://dx.doi.org/10.1038/nmeth.3806.

Koster, A. J., Chen, H., Sedat, J. W., & Agard, D. A. (1992). Automated microscopy for electron tomography. *Ultramicroscopy*, *46*(1–4), 207–227.

Lander, G. C., Stagg, S. M., Voss, N. R., Cheng, A., Fellmann, D., Pulokas, J., ... Carragher, B. (2009). Appion: An integrated, database-driven pipeline to facilitate EM image processing. *Journal of Structural Biology*, *166*(1), 95–102.

Li, X., Mooney, P., Zheng, S., Booth, C. R., Braunfeld, M. B., Gubbens, S., ... Cheng, Y. (2013). Electron counting and beam-induced motion correction enable near-atomic-resolution single-particle cryo-EM. *Nature Methods*, *10*(6), 584–590. http://dx.doi.org/10.1038/nmeth.2472.

Li, X., Zheng, S., Agard, D. A., & Cheng, Y. (2015). Asynchronous data acquisition and on-the-fly analysis of dose fractionated cryoEM images by UCSFImage. *Journal of Structural Biology*, *192*(2), 174–178. http://dx.doi.org/10.1016/j.jsb.2015.09.003.

Mastronarde, D. N. (2005). Automated electron microscope tomography using robust prediction of specimen movements. *Journal of Structural Biology*, *152*(1), 36–51. http://dx.doi.org/10.1016/j.jsb.2005.07.007.

Nickell, S., Forster, F., Linaroudis, A., Net, W. D., Beck, F., Hegerl, R., ... Plitzko, J. M. (2005). TOM software toolbox: Acquisition and analysis for electron tomography. *Journal of Structural Biology*, *149*(3), 227–234. http://dx.doi.org/10.1016/j.jsb.2004.10.006.

Potter, C. S., Chu, H., Frey, B., Green, C., Kisseberth, N., Madden, T. J., ... Carragher, B. (1999). Leginon: A system for fully automated acquisition of 1000 electron micrographs a day. *Ultramicroscopy*, 77(3–4), 153–161.

Rubinstein, J. L., & Brubaker, M. A. (2015). Alignment of cryo-EM movies of individual particles by optimization of image translations. *Journal of Structural Biology*, *192*(2), 188–195. http://dx.doi.org/10.1016/j.jsb.2015.08.007.

Scheres, S. H. (2014). Beam-induced motion correction for sub-megadalton cryo-EM particles. *Elife*, *3*, e03665. http://dx.doi.org/10.7554/eLife.03665.

Sheth, L. K., Piotrowski, A. L., & Voss, N. R. (2015). Visualization and quality assessment of the contrast transfer function estimation. *Journal of Structural Biology*, *192*(2), 222–234. http://dx.doi.org/10.1016/j.jsb.2015.06.012.

Spear, J. M., Noble, A. J., Xie, Q., Sousa, D. R., Chapman, M. S., & Stagg, S. M. (2015). The influence of frame alignment with dose compensation on the quality of single particle reconstructions. *Journal of Structural Biology*, *192*(2), 196–203. http://dx.doi.org/10.1016/j.jsb.2015.09.006.

Suloway, C., Pulokas, J., Fellmann, D., Cheng, A., Guerra, F., Quispe, J., ... Carragher, B. (2005). Automated molecular microscopy: The new Leginon system. *Journal of Structural Biology*, *151*(1), 41–60. http://dx.doi.org/10.1016/j.jsb.2005.03.010.

Tan, Y. Z., Cheng, A., Potter, C. S., & Carragher, B. (2016). Automated data collection in single particle electron microscopy. *Microscopy (Oxford, England)*, *65*(1), 43–56. http://dx.doi.org/10.1093/jmicro/dfv369.

Villa, E., Schaffer, M., Plitzko, J. M., & Baumeister, W. (2013). Opening windows into the cell: Focused-ion-beam milling for cryo-electron tomography. *Current Opinion in Structural Biology*, *23*(5), 771–777. http://dx.doi.org/10.1016/j.sbi.2013.08.006.

Wong, W., Bai, X. C., Brown, A., Fernandez, I. S., Hanssen, E., Condron, M., ... Scheres, S. H. (2014). Cryo-EM structure of the *Plasmodium falciparum* 80S ribosome bound to the anti-protozoan drug emetine. *Elife*, *3*, 1–20. http://dx.doi.org/10.7554/eLife.03080.

Zemlin, F., Weiss, K., Schiske, P., Kunath, W., & Herrmann, K. H. (1978). Coma-free alignment of high resolution electron microscopes with the aid of optical diffractograms. *Ultramicroscopy*, *3*, 49–60.

Zheng, S. Q., Keszthelyi, B., Branlund, E., Lyle, J. M., Braunfeld, M. B., Sedat, J. W., & Agard, D. A. (2007). UCSF tomography: An integrated software suite for real-time electron microscopic tomographic data collection, alignment, and reconstruction. *Journal of Structural Biology*, *157*(1), 138–147. http://dx.doi.org/10.1016/j.jsb.2006.06.005.

CHAPTER FIVE

Processing of Cryo-EM Movie Data

Z.A. Ripstein*,†, J.L. Rubinstein*,†,1
*Molecular Structure and Function Program, The Hospital for Sick Children, Toronto, ON, Canada
†University of Toronto, Toronto, ON, Canada
[1]Corresponding author: e-mail address: john.rubinstein@utoronto.ca

Contents

Abstract

Direct detector device (DDD) cameras dramatically enhance the capabilities of electron cryomicroscopy (cryo-EM) due to their improved detective quantum efficiency (DQE) relative to other detectors. DDDs use semiconductor technology that allows micrographs to be recorded as movies rather than integrated individual exposures. Movies from DDDs improve cryo-EM in another, more surprising, way. DDD movies revealed beam-induced specimen movement as a major source of image degradation and provide a way to partially correct the problem by aligning frames or regions of frames to account for this specimen movement. In this chapter, we use a self-consistent mathematical notation to explain, compare, and contrast several of the most popular existing algorithms for computationally correcting specimen movement in DDD movies. We conclude by discussing future developments in algorithms for processing DDD movies that would extend the capabilities of cryo-EM even further.

1. INTRODUCTION

The introduction of direct detector device (DDD) cameras for electron microscopes has led to significant recent advances in the field of single particle electron cryomicroscopy (cryo-EM) (Kühlbrandt, 2014;

Methods in Enzymology, Volume 579
ISSN 0076-6879
http://dx.doi.org/10.1016/bs.mie.2016.04.009

Smith & Rubinstein, 2014) and related fields such as electron cryo-tomography. The improved image signal-to-noise ratio (SNR) available from DDDs for ice-embedded protein particles has allowed for atomic models to be calculated directly from cryo-EM maps when protein complexes are sufficiently rigid (Li et al., 2013; Liao, Cao, Julius, & Cheng, 2013) and multiple conformations to be detected in dynamic protein complexes (Zhao, Benlekbir, & Rubinstein, 2015; Zhou et al., 2015). DDDs can improve image SNRs in four distinct ways. First, the inherent detective quantum efficiency (DQE) of the monolithic active pixel sensor is superior to that of photographic film or fiber-optic coupled charged coupled device (CCD) cameras (McMullan et al., 2009). Second, the high frame rate achievable due to the complimentary metal oxide semiconductor (CMOS) technology of the camera allows for individual electrons to be detected and counted in some camera designs, normalizing the contribution from each electron and further boosting the DQE of the camera (Li et al., 2013). Third, the high frame rate of the camera allows beam-induced movement and deformation of the specimen to be, at least partially, corrected by acquiring movies of specimens rather than integrated single exposures (Brilot et al., 2012; Campbell et al., 2012) (Fig. 1). Finally, the ability to collect movies rather than single integrated exposures allows high-resolution features in images to be recorded with low exposure and little radiation damage while low-resolution features, which are expected to be more radiation tolerant, may be imaged with higher total exposure (Baker & Rubinstein, 2010; Baker, Smith, Bueler, & Rubinstein, 2010).

At present, there are three commercial manufacturers of DDD cameras. Direct Electron LP produces the DE series of cameras with physical pixels that are between 6.0 × 6.0 μm and 6.5 × 6.5 μm and with frame rates

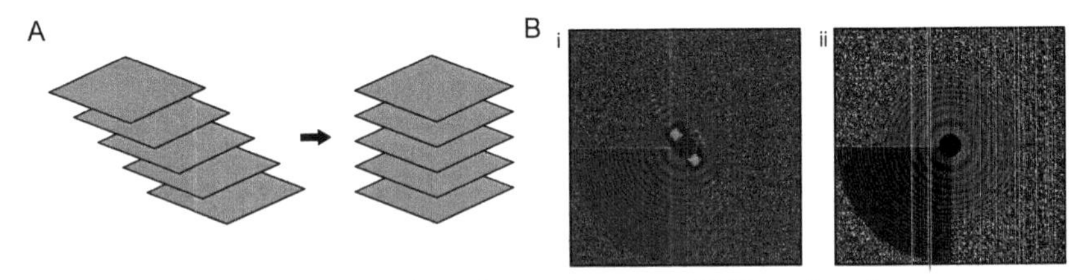

Fig. 1 (A) Cartoon depiction of whole-frame alignment. (B) Thon rings from an averaged DDD movie before (i) and after (ii) motion correction show that high-resolution information is restored by correcting for specimen movement. The bottom left corner of each power spectrum shows a CTF model fit to the Thon rings. CTF parameters were modeled and power spectra generated by CTFFIND4 (Rohou & Grigorieff, 2015).

between 30 and 60 frames per second (fps). FEI Company produces the Falcon cameras, which have 14.0 × 14.0 μm pixels and a thick sensitive layer on the detector, which lead to a high DQE, but currently provide a frame rate of only 18 fps. Electron counting with the current versions of these two cameras is not yet considered practical for single particle cryo-EM but the frame rate is sufficient to allow for correction of beam-induced movement. Gatan Inc. produces the K2 camera, which has 5.0 × 5.0 μm pixels and an internal frame rate of 400 fps, which allows for an approximate electron counting algorithm implemented on a dedicated field-programmable gate array (FPGA) computer. Movie frames that have been processed with the counting algorithm are added together and these sum images can be recorded at up to 40 fps. The different characteristics of the cameras suggest different optimal use strategies (McMullan, Faruqi, Clare, & Henderson, 2014). For integrating cameras such as the Direct Electron DE and FEI Falcon cameras, short movies with high frame rates are optimal. These short movies decrease the overall drift of the specimen holder during movie acquisition. For the Gatan K2 camera, the electron exposure rate must be limited in order to avoid coincidence loss during electron counting, where two or more electrons are counted as a single event due to their overlapping on the sensor during a single frame (Li et al., 2013; Li, Zheng, Egami, Agard, & Cheng, 2013). The slow exposure rate allows significant stage drift during the acquisition of the movie (Fig. 1). This review will describe several algorithms for whole-frame alignment. Cryo-EM of large virus particles with DDDs showed that, in addition to rigid body translation of entire movie frames, the specimen deforms during irradiation (Brilot et al., 2012). For large virus particles there is often sufficient signal in each movie frame to treat the frames of individual particles separately to account for this beam-induced displacement and rotation of particles during acquisition of the DDD movie. However, for smaller particles, algorithms are necessary to track particles during the movie using information available from considering the movie frames together. Therefore, in this review we also describe the current approaches for correcting this anisotropic movement of protein particles in images.

The objective of this chapter is to explain, compare, and contrast the current motion correction algorithms using self-consistent mathematical notation. At present, existing whole-frame and individual-particle motion correction approaches neglect any out-of-plane motion or rotation of the specimen during movie acquisition. Consequently, movement of the specimen or parts of the specimen between movie frames may be treated as a translation of each frame. Due to differences in the various algorithms described in this chapter, the best way to represent these translations mathematically varies. When referring to the translation of each frame relative to a single reference, for example the first frame or the middle frame of the movie, we write the translation as $-\vec{x}_t = (-\Delta x_t, -\Delta y_t)$ for frame t, where the opposite translation $\vec{x}_t$ is a vector that represents the best shifts, Δx_t and Δy_t, to apply to frame t to bring it into register with the frame that has been defined as stationary. Bringing these frames into register before averaging them optimizes the extraction of high-resolution information. Alternatively, for other algorithms, it is easier to describe the translations as $-\vec{s}_{mn} = (-\Delta x_{mn}, -\Delta y_{mn})$ for the movement from frame m to frame n. The aim of these algorithms is to identify the values of $\vec{s}_{mn}$ to apply to each frame to bring the frames into register with each other. The two different types of unknown are related by $\vec{s}_{mn} = \vec{x}_m - \vec{x}_n$.

Despite the relatively recent wide-spread adoption of DDD cameras in cryo-EM, a variety of different algorithms for aligning movie frames and frame regions have already been described. We begin this review by describing the algorithm in the program *Motioncorr* (Li et al., 2013), a whole-frame alignment approach. *Motioncorr* is intended to be used with the raw movie output of DDD cameras to begin the image analysis procedure. Next, we describe the algorithm used in *alignframes_lmbfgs* and *alignparts_lmbfgs* (Rubinstein & Brubaker, 2015), whole-frame and individual-particle motion correction programs, respectively, that are closely related to each other. *alignframes_lmbfgs* is intended for use with the raw movie output of a DDD camera to allow contrast transfer function (CTF) parameter determination and particle coordinate selection, two of the initial steps in single particle cryo-EM image analysis. *alignparts_lmbfgs* is then used to extract particles from the raw movie output of the camera while simultaneously correcting for whole-frame movement, individual particle movement, and accounting for radiation damage of the particles. Subsequently, we describe two programs intended for whole-frame alignment or alignment of large regions of frames: the Optical Flow method (Abrishami et al., 2015) and *Unblur* (Grant & Grigorieff, 2015), the latter also accounts for the radiation damage that occurs throughout the imaging process. Finally, we describe the particle polishing procedure in *Relion* (Scheres, 2014), an individual particle motion correction algorithm applied to prealigned movie frames after map refinement. We conclude by comparing the attributes of these different algorithms.

2. MOTIONCORR

In the process of implementing methods to use the K2 Summit electron counting DDD, Li and colleagues developed the first popular method for aligning whole movie frames (Li et al., 2013), which has become known as *Motioncorr*. For a movie that consists of T frames, the translations required to bring frames into register to extract high-resolution information are $\vec{s}_{mn}$ with $m \in \{1, T-1\}$ and $n = m + 1$. That is, one must know the translation from each frame to the next, $-\vec{s}_{mn}$, in order to apply their inverse, $\vec{s}_{mn}$. *Motioncorr*, which is currently the most widely used method in the field for whole-frame alignment, determines the translations by calculating the pair-wise cross correlation functions between all of the frames in a movie. To obtain these cross correlation functions, the Fourier transform of each frame is calculated, and in a pixel-wise fashion each Fourier component is multiplied by the complex conjugate of the equivalent Fourier component in the Fourier transform of another frame. The inverse Fourier transform of this product is calculated to give a cross correlation function for each unique pair of movie frames (Bracewell, 1965), yielding $T/2 \times (T-1)$ cross correlation functions for the T frames of the movie. To avoid the effects of fixed pattern noise, images are pretreated with a filter that has the form $exp\left(\frac{-B}{4d^2}\right)$, where B is a temperature factor and d is resolution. This filter down weights the high spatial frequencies that would cause fixed patterns in the image to dominate the cross correlation function. The maximum in each cross correlation function provides a vector $\vec{m}_{mn}$ that is a measured estimate of how that pair of frames should be aligned. In the absence of noise, the measured displacement vectors from cross correlation ($\vec{m}$ values) are equal to the desired translations ($\vec{s}$ values). For example, in the absence of noise $\vec{m}_{34}$ should equal $\vec{s}_{34}$ while $\vec{m}_{14}$ should equal $\vec{s}_{12} + \vec{s}_{23} + \vec{s}_{34}$. In the presence of noise, these relationships will not always hold true. The large number of $\vec{m}_{mn}$ vectors dramatically overdetermines the $T-1$ needed $\vec{s}_{mn}$ vectors that the method aims to discover. The measured $\vec{m}_{mn}$ values and desired $\vec{s}_{mn}$ values are therefore used to create a system of overdetermined linear equations as shown in Eq. (1), illustrated with an example movie that has four frames. Note that typically many more movie frames would be used, often between 7, the maximum available with a standard Falcon II detector from FEI, and 100. With 4 frames, 6 independent linear equations can be generated and recorded in the matrices shown:

$$\begin{bmatrix} 1 & 0 & 0 \\ 1 & 1 & 0 \\ 1 & 1 & 1 \\ 0 & 1 & 0 \\ 0 & 1 & 1 \\ 0 & 0 & 1 \end{bmatrix} \cdot \begin{bmatrix} \vec{s}_{12} \\ \vec{s}_{23} \\ \vec{s}_{34} \end{bmatrix} = \begin{bmatrix} \vec{m}_{12} \\ \vec{m}_{13} \\ \vec{m}_{14} \\ \vec{m}_{23} \\ \vec{m}_{24} \\ \vec{m}_{34} \end{bmatrix} \quad (1)$$

A least squares fit for the values of $\vec{s}_{mn}$ can be obtained from Eq. (1) using established methods from linear algebra (Li et al., 2013). Further improvement of the least squares solutions for the values of $\vec{s}_{mn}$ may be achieved using several approaches implemented in *Motioncorr*. First, cross correlation peaks close to the origin, where their precise location can be masked by a large peak at the origin due to residual fixed-pattern noise, can be avoided by ignoring equations from subsequent or nearly subsequent frames. Second, the maxima in cross correlation functions can be localized to subpixel accuracy using Fourier padding to interpolate the shape of the maximum from the cross correlation function. Third, equations from the matrix can be discarded when the residual of $\vec{m}_{mn} - \sum_{l=m}^{n-1} \vec{s}_{(l)(l+1)}$ exceeds a predefined amount. This least-squares whole-frame alignment method allowed high-resolution structures to be determined for important biological macromolecules (Cao, Liao, Cheng, & Julius, 2013; Li et al., 2013; Liao et al., 2013). Because of the large number of cross correlation functions that the method is often required to calculate, there is the potential for the method to be quite computationally expensive. Consequently, the original implementation of the program was on a GPU computer. It was pointed out by the authors that the method is not able to align image regions smaller than 2000 × 2000 pixels for movies acquired using typical conditions, suggesting that the method is not appropriate for aligning individual particle images in order to correct for deformation of the ice layer during imaging. While exploring different frame alignment methods and individual particle alignment methods, one of us (JLR) implemented the algorithm in standard modern Fortran for use with a desktop CPU computer, a program that we now make available online at https://sites.google.com/site/rubinsteingroup/direct-detector-align_lmbfgs. We used the linear algebra library LAPACK to find the least squares fit for values of $\vec{s}_{mn}$. We found the method to be impressively robust with the 30-frame movies we produced in the laboratory from a Gatan K2 camera. Because of how overdetermined the system of equations is with 30 frames, little gain was realized from subpixel maximum localization of the cross correlation function or removal of cross correlation functions above a minimum residual.

3. *alignframes_lmbfgs* and *alignparts_lmbfgs*

The programs *alignframes_lmbfgs* and *alignparts_lmbfgs* (Rubinstein & Brubaker, 2015) align whole movie frames and individual particles, respectively. In order to produce a robust and computational efficient method, Rubinstein and Brubaker pose the problem of finding $\vec{x}_t$ for the T frames as an optimization problem, either for whole movie frames or parts of frames corresponding to individual particle images. Optimization of functions, particularly where the partial derivatives of the objective function with respect to each of the parameters can be calculated analytically, is a well developed area in computer science. In this optimization approach, a single objective function that depends on all of the $\vec{x}_t$ values is minimized by adjusting the values of $\vec{x}_t$. Shifting of frames in Fourier space does not require interpolation and consequently the approach is performed with Fourier transforms of images rather than the images themselves. In Fourier space, image translation is equivalent to a phase change by ϕ_{jt}, and each Fourier component of the shifted frame is given by $F_{jt}(\cos\phi_{jt} + i\sin\phi_{jt})$ where F_{jt} is the jth Fourier component of the tth unshifted frame. Alternatively, this shifted Fourier component can be written $F_{jt}S_{jt}$ where $S_{jt} = (\cos\phi_{jt} + i\sin\phi_{jt})$. The amount of phase change is given by

$$\phi_{jt} = k_x(j) \cdot x_t \frac{2\pi}{N} + k_y(j) \cdot y_t \frac{2\pi}{N} \qquad (2)$$

where N is the extent in pixels in both the x and y direction of the $N \times N$ image, and $k_x(j)$ and $k_y(j)$ are the distance of the jth Fourier component from the origin in the k_x and k_y directions, respectively. The objective function used in the *alignframes_lmbfgs* and *alignparts_lmbfgs* programs is the negative of the sum of the unnormalized correlations of each shifted frame with the sum of the shifted frames:

$$O(\Theta) = -Re\sum_{t=1}^{T}\sum_{j=1}^{J}\left[F_{jt}^{*}S_{jt}^{*}\sum_{t'=1}^{T}F_{jt'}S_{jt'}\right] \qquad (3)$$

where O is a function of Θ, the translations of all frames, Re indicates the real part of the expression and * indicates the complex conjugate of a complex number. This objective function is useful because its partial derivatives with respect to x_t and y_t, the x- and y-components of the desired $\vec{x}_t$ vectors, can be calculated easily, giving

$$\frac{\partial O(\Theta)}{\partial x_a} = -Re\sum_{j=1}^{J}\frac{2\pi i k_x(j)}{N}\left[F_{ja}S_{ja}\sum_{t=1}^{T}F_{jt}^{*}S_{jt}^{*} - F_{ja}^{*}S_{ja}^{*}\sum_{t=1}^{T}F_{jt}S_{jt}\right] \quad (4)$$

and

$$\frac{\partial O(\Theta)}{\partial y_a} = -Re\sum_{j=1}^{J}\frac{2\pi i k_y(j)}{N}\left[F_{ja}S_{ja}\sum_{t=1}^{T}F_{jt}^{*}S_{jt}^{*} - F_{ja}^{*}S_{ja}^{*}\sum_{t=1}^{T}F_{jt}S_{jt}\right]. \quad (5)$$

Due to the availability of these derivatives, often called gradients in the field of optimization, the optimum in this objective function can be found in a computationally efficient manner with any of a variety of gradient-based optimization algorithms. The *alignframes_lmbfgs* and *alignparts_lmbfgs* programs make use of the limited memory Broyden–Fletcher–Goldfarb–Shanno (lm-bfgs) algorithm (Byrd, Lu, Nocedal, & Zhu, 1995) to optimize Eq. (3), giving the programs the "lmbfgs" part of their names.

The assumption that true frame or particle trajectories are unlikely to undergo sudden and dramatic changes in direction can be enforced by penalizing changes in $\partial x_t/\partial t$ and $\partial y_t/\partial t$. If $\partial x_t/\partial t$ and $\partial y_t/\partial t$ are constant ($\partial^2 x_t/\partial t^2$ and $\partial^2 y_t/\partial t^2$ are 0), the expected value for $(\vec{x}_t - \vec{x}_{t-1})$ is $(\vec{x}_{t-1} - \vec{x}_{t-2})$. Deviation from this expected linear trajectory can be penalized by an amount $\lambda\left(\left[\vec{x}_t - \vec{x}_{t-1}\right] - \left[\vec{x}_{t-1} - \vec{x}_{t-2}\right]\right)^2$. The overall penalty imposed on the objective function to encourage smoothness is then given by

$$P(\Theta) = \sum_{t=3}^{T}\lambda\left[(x_t - 2x_{t-1} + x_{t-2})^2 + (y_t - 2y_{t-1} + y_{t-2})^2\right] \quad (6)$$

where λ is a user selected weighting parameter. This penalty is known as second order smoothing because it penalizes finite difference approximations of the second derivatives of x_t and y_t with respect to t, $\partial^2 x_t/\partial t^2$ and $\partial^2 y_t/\partial t^2$. The penalty function described in Eq. (6) is added to the objective function in Eq. (3) to obtain the overall objective function that is minimized. The contribution to the penalty function in Eq. (6) from shifting of the ath frame when $a \in [3, T-2]$ is $\lambda[(\vec{x}_a - 2\vec{x}_{a-1} + \vec{x}_{a-2})^2 + (\vec{x}_{a+1} - 2\vec{x}_a + \vec{x}_{a-1})^2 + (\vec{x}_{a+2} - 2\vec{x}_{a+1} + \vec{x}_a)^2]$ and consequently the first derivative of Eq. (6) with respect to x_a is given by

$$\frac{\partial P(\Theta)}{\partial x_a} = \begin{cases} 2\lambda(x_a - 2x_{a+1} + x_{a+2}), & a=1, \\ 2\lambda(-2x_{a-1} + 5x_a - 4x_{a+1} + x_{a+2}), & a=2, \\ 2\lambda(x_{a-2} - 4x_{a-1} + 6x_a - 4x_{a+1} + x_{a+2}), & a \in [3, T-2], \\ 2\lambda(x_{a-2} - 4x_{a-1} + 5x_a - 2x_{a+1}), & a = T-1, \\ 2\lambda(x_{a-2} - 2x_{a-1} + x_a), & a = T. \end{cases} \quad (7)$$

And

$$\frac{\partial P(\Theta)}{\partial y_a} = \begin{cases} 2\lambda(y_a - 2y_{a+1} + y_{a+2}), & a=1, \\ 2\lambda(-2y_{a-1} + 5y_a - 4y_{a+1} + y_{a+2}), & a=2, \\ 2\lambda(y_{a-2} - 4y_{a-1} + 6y_a - 4y_{a+1} + y_{a+2}), & a \in [3, T-2], \\ 2\lambda(y_{a-2} - 4y_{a-1} + 5y_a - 2y_{a+1}), & a = T-1, \\ 2\lambda(y_{a-2} - 2y_{a-1} + y_a), & a = T. \end{cases} \quad (8)$$

The derivative of the smoothed objective function is therefore the sum of the values from Eqs. (4) and (7) for the derivative with respect to x_a, and the sum of the values from Eqs. (5) and (8) for the derivative with respect to y_a. Because of this second order smoothness constraint, *alignparts_lmbfgs* is best used to extract particle images from unaligned movies frames. Otherwise, changes in trajectory introduced into the frame-to-frame trajectories of individual particles by whole-frame alignment could be penalized by individual particle motion correction, turning a straight particle trajectory into one that has changes in direction or speed or otherwise altering the true trajectory.

For individual particle alignment, *alignparts_lmbfgs* makes use of the additional constraint that nearby particle images are unlikely to have significantly different trajectories, which is achieved by local averaging of trajectories. Once raw trajectories are determined, locally averaged trajectories are calculated according to

$$\vec{x}_{nt}' = \frac{\sum_{m=1}^{M} w_{nm} \vec{x}_{mt}}{\sum_{m=1}^{M} w_{nm}} \quad (9)$$

where $\vec{x}_{nt}'$ is the smoothed displacement vector for the nth particle in the tth frame and $\vec{x}_{mt}$ is the original displacement vector for the mth particle in the tth frame. The weight w_{mn} is given by

$$w_{mn} = exp\left(\frac{-d_{mn}^2}{2\sigma^2}\right) \quad (10)$$

where d_{mn} is the distance between the mth and n^{th} particles and σ is a user set parameter that determines the extent to which the smoothing is applied. This Gaussian weighting is equivalent to the local averaging used for fitting linear trajectories in *Relion* (Scheres, 2014). Because of the Gaussian form of Eq. (10), 95 % of the weight for a particle trajectory will come from the trajectories within 2σ pixels of that particle. With $\sigma=0$ there is no local averaging. As the value for σ increases the method forces all particle trajectories in an image to be the same, as they would be if only whole-frame alignment were performed. This method of individual particle alignment by gradient-based optimization with enforced smoothness and correlation in the trajectories of nearby particles allows for the movement of individual particles in movies to be determined quickly and accurately with minimal computational cost (Fig. 2A).

As proposed even before the widespread availability of DDDs (Baker et al., 2010), the *alignparts_lmbfgs* algorithm adjusts the contribution from each frame for the exposure and resolution dependent fading of information due to radiation damage. However, rather than the critical exposure information obtained from the study of 2D crystals (Baker et al., 2010), *alignparts_lmbfgs* uses the critical exposure curves determined from the study of icosahedral virus particles (Grant & Grigorieff, 2015).

4. UNBLUR

The program *Unblur* (Grant & Grigorieff, 2015) also uses a cross correlation approach for frame alignment. Similar to *alignframes_lmbfgs*, *Unblur* aligns noisy individual frames from a movie to the high SNR average of frames. Unlike *alignframes_lmbfgs a*, *Unblur* does not calculate a combined objective function or the gradients of the objective function. Instead, it iteratively aligns each individual frame to an average of frames that does not include the frame being aligned. Shifts for the individual frames are found by identifying the peak in a normalized cross correlation function, which for the ath frame is given by:

$$CC_a = \frac{Re\sum_{j=1}^{J}\left(F_{ja}^{*}\left(\sum_{t=1,\neq a}^{T}F_{jt}\right)\right)}{\sqrt{\sum_{j=1}^{J}\left(F_{ja}\right)^2\sum_{j=1}^{J}\left(\sum_{t=1,\neq a}^{T}F_{jt}\right)^2}}. \quad (11)$$

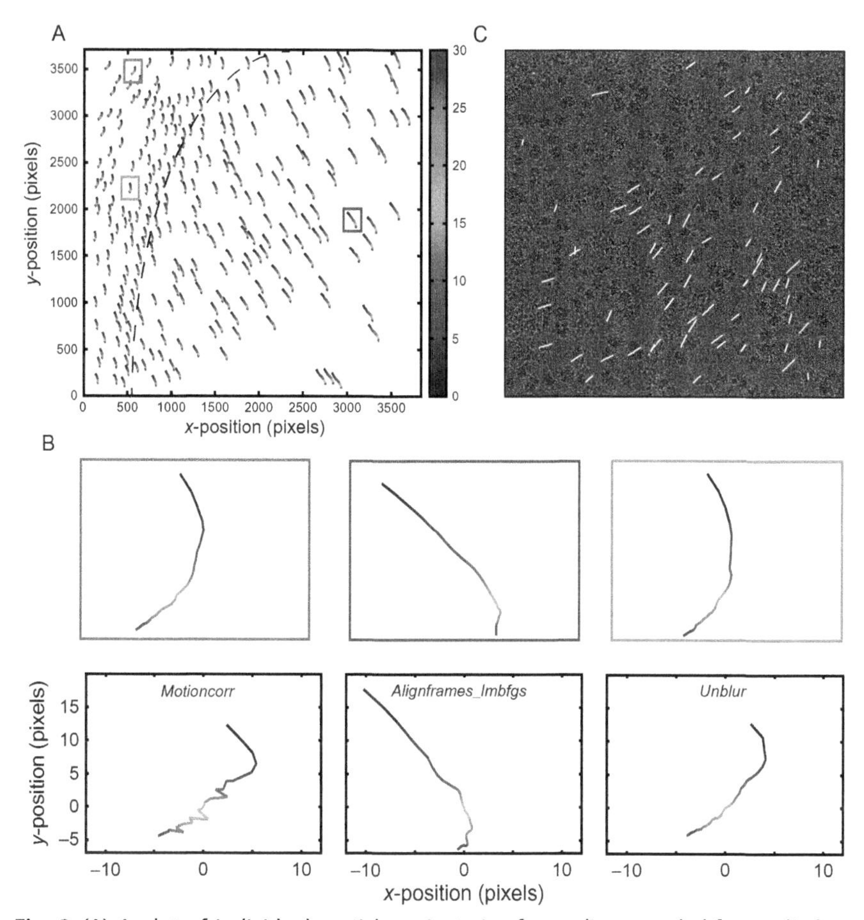

Fig. 2 (A) A plot of individual particle trajectories from *alignparts_lmbfgs* applied to unaligned movie frames. Each line in the plot indicates the trajectory of a single particle from frame 1 (*black*) to frame 30 (*blue*), exaggerated by a factor of 5. The broken line indicates the approximate edge of the carbon support film, with particles to the left of the line lying on the carbon support and particles to the right being in unsupported ice. (B) Inspection of individual particle trajectories from three regions of the micrograph shows smoothed particle trajectories that differ across the micrograph (three upper panels, *color* coded such that they correspond to the three small boxes in panel A). The whole-frame alignment programs *Motioncorr, alignframes_lmbfgs,* and *Unblur* all find somewhat different solutions to the alignment problem that produce similar power spectra from the average of aligned frames, each of which approximately matches the individual particle trajectories in different regions of the micrograph (lower panels). (C) Linear particle trajectories calculated using *Relion* particle polishing applied to whole-frame aligned movies. Particle trajectories start at the *green* dot and move to the *red* dot. Trajectories are from a different movie than used for parts A and B and are exaggerated by a factor of 50. *Source: Reproduced with permission from reference Scheres, S. H. (2014). Beam-induced motion correction for sub-megadalton cryo-EM particles.* Elife, 3, *e03665.* (See the color plate.)

The coordinates of the peak in this cross correlation function, $(\Delta x_a, \Delta y_a)$, are the components of the desired measured shift, $\vec{x}_a$, that would bring the frames into register with each other. Frames are assigned shifts sequentially and then the shifts are fit to either a spline or polynomial. This process of frame alignment by cross correlation, fitting of a smooth trajectory to the measured shifts, and application of smoothed shifts to each frame in order to calculate an improved average of frames is repeated until the estimates of shifts fall below a user-specified threshold, such as 0.5 pixels.

The fitting of the data to a spline or polynomial in *Unblur* smooths the trajectories, thereby removing sudden changes in trajectory that are due to noise in the images. Fitting of cubic splines is equivalent to minimizing the expression (Craven & Wahba, 1979):

$$\sum_{t=1}^{T}(\vec{x}_t - f(t))^2 + \lambda \int_{t=1}^{T} f''(t)dt \tag{12}$$

where the first term $\sum_{t=1}^{T}(\vec{x}_t - f(t))^2$ represents a least squares optimization of the fitted function $f(t)$ to the data $\vec{x}_t$. In this case $\vec{x}_t$ are the raw measured shifts, and $f(t)$ is the desired smoothed trajectory. In this model the measured values are assumed to come from the desired function plus noise, $\vec{x}_t = f(t) + \varepsilon_t$, where ε_t represents the noise. The second half of the equation accomplishes the same goal as the second order smoothing in *alignparts_lmbfgs* and *alignframes_lmbfgs*, which penalizes unlikely sudden changes in frame or particle trajectories, respectively. Here λ is a weighting factor for smoothing akin to that used in *alignparts_lmbfgs* and *alignframes_lmbfgs*. However, in *Unblur* the spline fitting algorithm implicitly calculates values for λ using a cross validation approach (Craven & Wahba, 1979), which takes into account the inherent noisiness of the signal. Finally, while averaging frames, *Unblur* performs exposure weighting to compensate for radiation damage (Baker et al., 2010) using values of critical exposures that were measured directly from single particle cryo-EM data, and a novel weighting scheme (Grant & Grigorieff, 2015).

5. OPTICAL FLOW

The Optical Flow approach (Abrishami et al., 2015) uses changes in pixel intensity to track the movement of frames relative to each other. Much like the least-squares approach implemented in the program *Motioncorr* (Li et al., 2013) and described in Section 3, the aim of the Optical Flow method is to find $\vec{s}_{mn}$ values, the frame-to-frame displacements in specimen that occur during movie acquisition. However, Optical Flow has been used to calculate $\vec{s}_{mn}$ values not only for whole frames but also for subregions of frames. Unlike the methods described above that have used cross correlation to measure the shift between frames, the Optical Flow implementation described by Abrishami and colleagues uses the difference in pixel intensities between subsequent frames to estimate translations. Specifically, the problem is phrased as a first order Taylor expansion of the change in pixel intensities I as a function of image shift. This formulation of the problem allows for computation of derivatives of the intensity of image pixels with respect to image translation:

$$I(x+\Delta x, y+\Delta y, t+\Delta t) \cong I(x,y,t) + \frac{\partial I}{\partial x}\Delta x + \frac{\partial I}{\partial y}\Delta y + \frac{\partial I}{\partial t}\Delta t \tag{13}$$

where $I(x,y,t)$ is the intensity of the pixel at coordinates (x,y) in the tth frame and Δx and Δy are the calculated shifts for this pixel. The assumption is made that the total intensity of frames does not change, such that $I(x + \Delta x, y + \Delta y, t + \Delta t) \cong I(x,y,t)$. Consequently, Eq. (13) becomes the set of linear equations:

$$\frac{\partial I}{\partial x}\Delta x + \frac{\partial I}{\partial y}\Delta y + \frac{\partial I}{\partial t}\Delta t = 0 \tag{14a}$$

$$\frac{\partial I}{\partial x}u + \frac{\partial I}{\partial y}v = -\frac{\partial I}{\partial t} \tag{14b}$$

where $u = \frac{\Delta x}{\Delta t}$ and $v = \frac{\Delta y}{\Delta t}$. In the case where two subsequent frames, m and n, are compared, $\Delta t = 1$ and the translation vector $\vec{s} = (\Delta x, \Delta y)$, which is equivalent to the $\vec{s}_{mn} = (x_{mn}, y_{mn})$ notation used elsewhere in this chapter. To calculate a translation vector for each pixel (x_0, y_0) one assumes the shifts to be locally correlated at that pixel, and Eq. (14b) is solved for a window of coordinates centered on the pixel. The problem of calculating a translation vector for each pixel in the window is similar to the problem solved by *Motioncorr*, as the large number of pixels in the window provides a large set of overdetermined linear equations of the form $A\vec{s}_{mn} = B$. This process is performed for each pixel (x,y) in the image, with the result being a vector field the size of the image, containing a vector $\vec{s}_{mn}$ for each pixel in the input image. For clarity we denote this optical flow shift measurement process here as $\circledast$ and the resulting vector field as OF_z.

Again, because of the noisy nature of the images, this alignment alone is insufficient to robustly calculate the shifts in a single step. Thus, as in the case of *Unblur*, the alignment is iterated to produce more reliable measured translation vectors. In the implementation of the Optical Flow approach, Abrishami et al. proposed iterative alignment in a pyramidal scheme of averaging frames. For example, with a 16 frame movie the alignment would proceed as follows:

1. Average all 16 unaligned frames: $I(x,y)=\sum_{t=1}^{16} I_t(x,y)$
2. Split the movie into 2^k groups of frames, where k is the current iteration, and average. That is: in the first iteration the frames are grouped and averaged as $I(x,y)=\sum_{t=1}^{8} I_t(x,y)$ and $I(x,y)=\sum_{t=9}^{16} I_t(x,y)$
3. Align these group averages with the global average by calculating optical flows, ie,

$$OF_{1-8}=\sum_{t=1}^{8} I_t \circledast \sum_{t=1}^{16} I_t \tag{15}$$

4. Use the measured OF to align each group of frames, then calculate the new aligned average of frames using the shifted frames
5. Repeat steps 2–4 until each group of frames has only 1 frame in it.

6. PARTICLE POLISHING IN *RELION*

"Particle polishing" is the individual particle motion correction procedure implemented in the program *Relion*. The name is derived from a catch phrase that became popular at the 2007 Gordon Research Conference on Three Dimensional Electron Microscopy. When Dr. Xuekui Yu of UCLA was asked what he meant when he said he used the "good" particles to calculate his 3.9 Å map of cytoplasmic polyhedrosis virus from a film dataset (Yu, Jin, & Zhou, 2008), he replied that he used particles that looked "shiny." After the identification of particle movement during imaging as a cause of information loss in single particle cryo-EM, particles that were not "shiny" could be made "shiny" by motion correction, which is analogous to "polishing" them. The method developed by Scheres (2014) differs from the other methods described in this review in that it aims to correct for particle movement using 2D projections of a 3D reference map to align individual particles in images, rather than performing this task using information from just the images. Consequently, unlike the methods described above, polishing is performed *after* a 3D map is calculated, rather than before, and is tightly integrated into the workflow of the *Relion* software package. During 3D map refinement, the Euler angles that relate the orientation of each particle image to the 3D map are estimated using particle images that are the averages of movie frames after whole-frame alignment. Once estimates of all of these orientations are obtained, particle polishing relaxes the assumption that the whole-frame alignment adequately describes the optimal alignment for individual particles in movies. Instead, it attempts to improve the translational alignment of particles in individual frames to projections of the 3D map. These translations are estimated by calculating cross correlation functions between map projections and particle images in movie frames to provide the appropriate values $-\vec{x}_t$. The low SNR in individual movie frames is accounted for in a few different ways. First, polishing is typically used with running averages of frames rather than individual frames, thus increasing the SNR over individual frames. In most cases, the trajectories for individual particle images estimated in this way are still noisy. Consequently, polishing makes the assumption that particle trajectories are linear and that particles move at a constant velocity. The initial estimate of each linear constant-velocity trajectory is obtained by performing regression of the estimates of $-\vec{x}_t$ to a straight line for the each particle. Finally, neighboring particles are assumed to have correlated trajectories. This correlation is enforced by replacing each particle trajectory with a weighted average of particle trajectories. The weighting function, $w_{p'}$, falls off as a Gaussian with the distance between particles, much as in *alignparts_lmbfgs*. This fit to a linear trajectory with a Gaussian weighting over neighboring particles is achieved by minimizing the function:

$$O(\Theta)_n = \sum_m^M w_m \sum_t^T \left(\vec{x}_{mt} - \left(\vec{\alpha}_n + \vec{\beta}_n t\right)\right)^2 \quad (16)$$

where $O(\Theta)_n$ is the overall objective function that describes the linear trajectory of particle n, M is the number of particles in the field of view, and $m \in \{1,M\}$ with $m \neq n$ denotes all neighboring particles. The Gaussian weights used for the neighboring particles, w_m, are set by having the user define a value for σ and are the same as the Gaussian weighting for local correlation in *alignparts_lmbfgs* shown in Eq. (10). The values of $\vec{x}_{mt}$ are the measured shifts for the neighboring particle m and α_n and β_n are the intercept and slope of the straight line trajectory for particle n, where $\vec{\alpha}_n = (x_0, y_0)$, and $\vec{\beta}_n = \left(\frac{dx}{dt}, \frac{dy}{dt}\right)$. It is important to note that in this version of particle polishing only x and y translations are considered, as opposed to earlier versions that accounted for particle rotation. In this way, each particle ends up with its own unique linear trajectory, with all of the trajectories tending to be the same when σ is large and $w_m \approx 1$. An example of these linear trajectories superimposed on a micrograph is shown in Fig. 2C.

In addition to particle motion correction, one of the key features of particle polishing is weighting each movie frame (or running averages of frames) in a frequency dependent manner to best reflect the information degradation caused by rapid movements and radiation damage. Rather than using estimates of information loss as a function of electron exposure (Baker et al., 2010), Polishing uses a data-driven approach with a filter estimated for each frame that has the form $exp\left(\frac{B_t}{4d^2} + C_t\right)$, where B_t is a relative temperature factor for frame t, d is the resolution, and C_t is the resolution independent component of the signal in the frame. Values of B_t and C_t are estimated for each frame by calculating 3D maps from individual frames. The advantage of this approach vs experimental measurement of radiation damage is that it is insensitive to inaccurate measurement of electron exposure in each frame and removes the assumption that all macromolecules suffer from radiation damage in the same way. A relative weighting for each frame is used according to:

$$\frac{W_t\left(\frac{1}{d}\right)}{W_\Sigma\left(\frac{1}{d}\right)} = \sqrt{\frac{FSC_t\left(\frac{1}{d}\right) - FSC_t\left(\frac{1}{d}\right)FSC_\Sigma\left(\frac{1}{d}\right)}{FSC_\Sigma\left(\frac{1}{d}\right) - FSC_t\left(\frac{1}{d}\right)FSC_\Sigma\left(\frac{1}{d}\right)}} \tag{17}$$

where $FSC_t\left(\frac{1}{d}\right)$ and $FSC_\Sigma\left(\frac{1}{d}\right)$ are the Fourier shell correlations between the two independently refined maps from halves of the individual frame datasets and the averaged frames dataset, respectively. A type of Guinier plot is produced by plotting $\frac{W_t\left(\frac{1}{d}\right)}{W_\Sigma\left(\frac{1}{d}\right)}$ against $\frac{1}{d^2}$, which reveals the decrease in signal of the individual frame maps relative to the map from the average of frames. From this relative Guinier plot, the slope B_t and intercept C_t are used to calculate relative B-factors for individual frames, and the following weighting scheme as a function of spatial frequency $\frac{1}{d}$ is applied:

$$w_t\left(\frac{1}{d}\right) = \frac{exp\left(\frac{B_t}{4d^2} + C_t\right)}{\sum_{t'} w_{t'}\left(\frac{1}{d}\right)}. \tag{18}$$

Consequently, if the signal from an individual frame map decreases faster than that of the map from the average of frames then the relative B-factor will be negative and that frame will be down weighted in a resolution dependent manner (note the $\frac{1}{d^2}$ term in Eq. (18)). In comparison, if the signal in an individual frame map is greater than the signal in the map from the average of frames then the relative B-factor will be positive and the frame will be up weighted relative to other frames. Note that in contrast to the form of most temperature factors, $exp\left(\frac{-B}{4d^2}\right)$, the relative B-factors used here have the opposite sign, and positive and negative relative B-factors up weight and down weight high spatial frequencies, respectively, while positive and negative B-factors in general do the opposite. This weighting scheme is normalized and consequently the sum of the weights for all spatial frequencies is unity, so that the weighted average has the same power as the unweighted average and the weighting does not sharpen or dampen the map (Fig. 3A). After these weights have been estimated for the frames, the interpolated linear trajectory shifts are applied to the frames, and the weighted average of frames is calculated. Note that the first few frames are often down weighted, which is most likely due to large movements of the sample upon initial irradiation that causes the loss of high-resolution information. After the first few frames the effect of radiation damage becomes dominant and the curve looks similar to the weighting scheme based on measurements of radiation damage used by *Unblur* and *alignparts_lmbfgs* (Fig. 3B). The utility of these weighting approaches and individual particle motion correction is shown in Fig. 3C and D, where the improvement in map quality is evident.

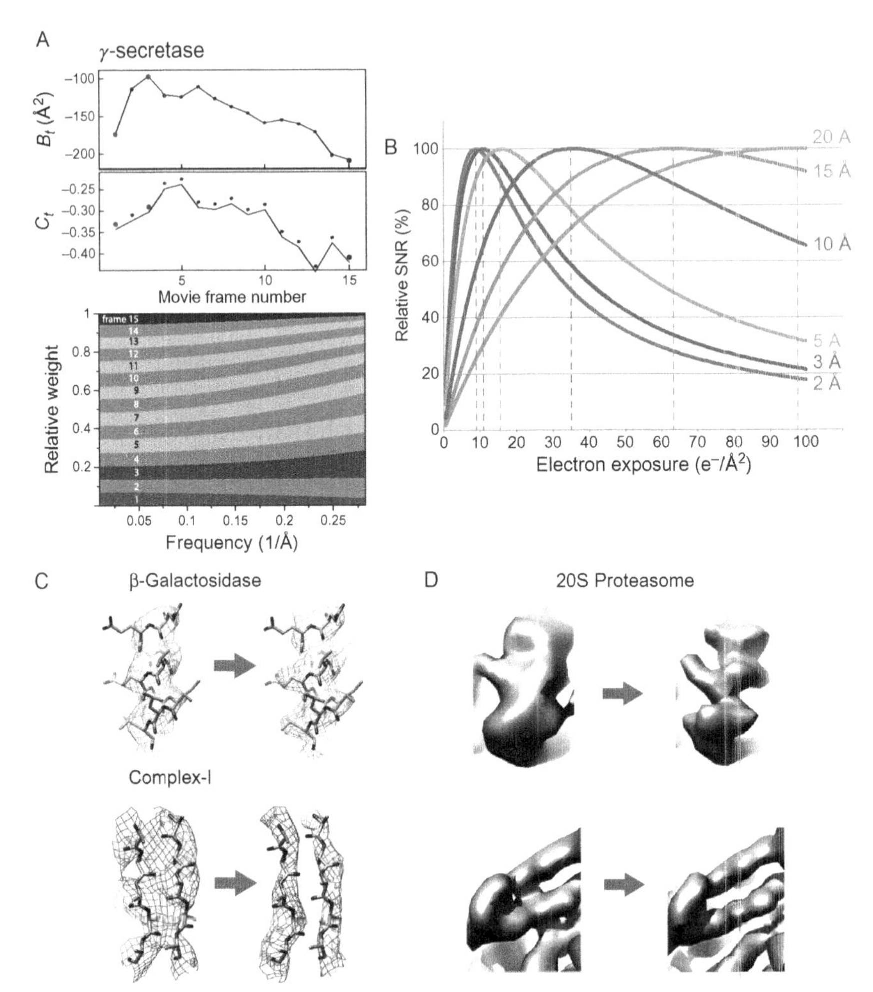

Fig. 3 (A) An example of measurements of B_t and C_t from *Relion* particle polishing. Lower panel shows the resulting weights of frames used. (B) Optimal exposure weighting curve used in *alignparts_lmbfgs* and *Unblur*. (C) Examples of map improvement upon motion correction and frame weighting with *Relion* particle polishing (Scheres, 2014). (D) Examples of map improvement upon motion correction and exposure weighting in *alignparts_lmbfgs*. *Source: (A)* and (C) *Reproduced with permission from Scheres, S. H. (2014). Beam-induced motion correction for sub-megadalton cryo-EM particles.* Elife, 3, *e03665. (B) Figure adapted from Baker, L. A., Smith, E. A., Bueler, S. A., & Rubinstein, J. L. (2010). The resolution dependence of optimal exposures in liquid nitrogen temperature electron cryomicroscopy of catalase crystals.* Journal of Structural Biology, 169 *(3), 431437. doi: 10.1016/j.jsb.2009.11.014 using the critical exposures measured by Grant and Grigorieff (2015) for virus particles at 300 kV.* (See the color plate.)

7. SUMMARY

Each of the different techniques described above has a number of advantages. *Motioncorr*, *alignframes_lmbfgs*, *Ublur*, and Optical Flow are all whole-frame alignment methods. *Motioncorr* benefits from highly overdetermined information from computing the pairwise cross correlation function between each pair of images in a movie. The disadvantage of the method is that noisy frames are compared to other noisy frames, which can make detection of signal difficult for low contrast images. The large number of calculations required makes CPU implementations of the least-squares algorithm somewhat slow, but this problem has been overcome by implementing the algorithm for GPU processing in *Motioncorr*. *alignframes_lmbfgs*, *Unblur*, and Optical Flow compare frames to the average of frames, the latter having a higher SNR than an individual frame and consequently these methods may perform somewhat better for low contrast images than *Motioncorr*. Alignment of a low SNR frame to an average of frames is expected to be more accurate than alignment of a low SNR frame to another low SNR frame, but the latter provides over-determination of parameters while the former does not. *Unblur* is arguably the simplest method of the three, repeatedly aligning frames to an average, which produces better alignments and a better average, while simultaneously forcing the trajectories of frames to be smooth. The Optical Flow method tackles this problem using a pyramidal scheme for aligning frames to the average of frames and the optical flow approach rather than a cross correlation approach. *alignframes_lmbfgs* makes use of the well-established field of derivative-based optimization, which allows it to quickly and robustly find the optimal alignment for all of the frames. *alignframes_lmbfgs* enforces a smooth trajectory for frames by including a penalty for nonsmooth trajectories in the objective function that is optimized. In comparing the output of these whole-frame alignment algorithms, it is evident that the calculated shifts measured by the different algorithms do not always agree for a given micrograph (Fig. 2B, lower panels). However, each approach produces a whole frame trajectory that is similar to a region of the individual particle trajectories (Fig. 2B, upper panel) showing that the algorithms pick up on different parts of the micrograph, thereby correcting some but not all of the particle motion. Consequently, it is clear that there is no correct answer for a whole frame trajectory when there is distortion of the specimen during imaging and the discrepancies between the different methods justifies the need for individual particle motion correction when attempting to obtain high-resolution structures from cryo-EM.

Particle polishing in *Relion* and *alignparts_lmbfgs* are the two methods that attempt to determine the trajectories for individual particles in the dataset. Their fundamental difference is that *alignparts_lmbfgs* translationally aligns the frames for particles only making reference to the average of frames for the particle (ie, at the beginning of the image analysis process) while polishing uses a map projection after estimating the pose or Euler angles for the particle (ie, at the end of the image analysis process). The map projection used by *Relion* will have a better SNR than the average of frames, which would provide polishing with an advantage over *alignparts_lmbfgs*. On the other hand, the average of frames used by *alignparts_lmbfgs* will always match exactly the appearance of the particle in the frames, while for polishing this agreement depends on accurate estimation of Euler angles, CTF parameters, particle conformation, and the absence of any contamination or signal from water in the raw image, which should provide an advantage to *alignparts_lmbfgs* over polishing. Both methods use a similar constraint that nearby particles must have similar trajectories. An additional difference between the methods is that polishing applies a linear and constant velocity trajectory for each particle while *alignparts_lmbfgs* allows variable direction and velocity, which if estimated accurately should be an advantage for *alignparts_lmbfgs*. However, polishing uses a powerful data-driven approach to determining the relative importance of each frame while *alignparts_lmbfgs* relies on physically measured fading of information for its weighting scheme. A final difference is in computation cost, with *alignparts_lmbfgs* being computationally inexpensive and polishing usually requiring significant computer resources.

8. FUTURE PROSPECTS

Since the advent of DDDs, the benefit of motion correction by alignment of movie frames has been apparent (Brilot et al., 2012). At present, the large transverse translations in the x–y plane are corrected by the movie processing algorithms described in this review, but challenges still remain. Chief among these would be the ability to correct for sample rotations and movement in the z-direction, which are known to occur during the complex flexing of samples in the electron beam. Additionally, one can conceive of different image acquisition schemes that enable motion correction during the initial period of rapid specimen movement, while maintaining the ability to align subsequent frames with less severe motion.

ACKNOWLEDGMENTS

We thank Sjors Scheres for providing the images used in parts of Figs. 2 and 3. We are grateful to Alexis Rohou, Niko Grigorieff, Timothy Grant, Xueming Li, Sjors Scheres, and Tony Crowther for a critical reading of this manuscript. Z.A.R. was supported by a scholarship from the Hospital for Sick Children and J.L.R. was supported by a Canada Research Chair. This work was supported by operating grant MOP 81294 from the Canadian Institutes of Health Research and the Natural Sciences and Engineering Research Council Discovery Grant 401724-12.

REFERENCES

Abrishami, V., Vargas, J., Li, X., Cheng, Y., Marabini, R., Sorzano, C. O. S., & Carazo, J. M. (2015). Alignment of direct detection device micrographs using a robust optical flow approach. *Journal of Structural Biology*, *189*(3), 163–176. http://dx.doi.org/10.1016/j.jsb.2015.02.001. http://www.sciencedirect.com/science/article/pii/S1047847715000313.

Baker, L. A., & Rubinstein, J. L. (2010). Chapter fifteen-radiation damage in electron cryomicroscopy. *Methods in Enzymology*, *481*, 371–388.

Baker, L. A., Smith, E. A., Bueler, S. A., & Rubinstein, J. L. (2010). The resolution dependence of optimal exposures in liquid nitrogen temperature electron cryomicroscopy of catalase crystals. *Journal of Structural Biology*, *169*(3), 431–437. http://dx.doi.org/10.1016/j.jsb.2009.11.014.

Bracewell, R. (1965). *The Fourier transform and its applications*. New York: McGraw-Hill.

Brilot, A. F., Chen, J. Z., Cheng, A., Pan, J., Harrison, S. C., Potter, C. S., … Grigorieff, N. (2012). Beam-induced motion of vitrified specimen on holey carbon film. *Journal of Structural Biology*, *177*(3), 630–637. http://dx.doi.org/10.1016/j.jsb.2012.02.003.

Byrd, R. H., Lu, P., Nocedal, J., & Zhu, C. (1995). A limited memory algorithm for bound constrained optimization. *SIAM Journal on Scientific Computing*, *16*(5), 1190–1208.

Campbell, M. G., Cheng, A., Brilot, A. F., Moeller, A., Lyumkis, D., Veesler, D., … Grigorieff, N. (2012). Movies of ice-embedded particles enhance resolution in electron cryo-microscopy. *Structure*, *20*(11), 1823–1828. http://dx.doi.org/10.1016/j.str.2012.08.026.

Cao, E., Liao, M., Cheng, Y., & Julius, D. (2013). TRPV1 structures in distinct conformations reveal activation mechanisms. *Nature*, *504*(7478), 113–118. http://dx.doi.org/10.1038/nature12823.

Craven, P., & Wahba, G. (1979). Smoothing noisy data with spline functions: Estimating the correct degree of smoothing by the method of generalized cross-validation. *Numerische Mathematik*, *31*, 377–403.

Grant, T., & Grigorieff, N. (2015). Measuring the optimal exposure for single particle cryo-EM using a 2.6 Åreconstruction of rotavirus VP6. *ELife*, *4*, 1–19. http://dx.doi.org/10.7554/eLife.06980. http://elifesciences.org/lookup/doi/10.7554/eLife.06980.

Kühlbrandt, W. (2014). The resolution revolution. *Science*, *343*(6178), 1443–1444.

Li, X., Mooney, P., Zheng, S., Booth, C. R., Braunfeld, M. B., Gubbens, S., … Cheng, Y. (2013). Electron counting and beam-induced motion correction enable near-atomic-resolution single-particle cryo-EM. *Nature Methods*, *10*(6), 584–590. http://dx.doi.org/10.1038/nmeth.2472.

Li, X., Zheng, S. Q., Egami, K., Agard, D. A., & Cheng, Y. (2013). Influence of electron dose rate on electron counting images recorded with the K2 camera. *Journal of Structural Biology*, *184*(2), 251–260.

Liao, M., Cao, E., Julius, D., & Cheng, Y. (2013). Structure of the TRPV1 ion channel determined by electron cryo-microscopy. *Nature*, *504*(7478), 107–112. http://dx.doi.org/10.1038/nature12822.

McMullan, G., Faruqi, A., Clare, D., & Henderson, R. (2014). Comparison of optimal performance at 300 keV of three direct electron detectors for use in low dose electron microscopy. *Ultramicroscopy*, *147*, 156–163.

McMullan, G., Faruqi, A., Henderson, R., Guerrini, N., Turchetta, R., Jacobs, A., & Van Hoften, G. (2009). Experimental observation of the improvement in MTF from backthinning a CMOS direct electron detector. *Ultramicroscopy*, *109*(9), 1144–1147.

Rohou, A., & Grigorieff, N. (2015). CTFFIND4: Fast and accurate defocus estimation from electron micrographs. *Journal of Structural Biology*, *192*(2), 216–221.

Rubinstein, J. L., & Brubaker, M. A. (2015). Alignment of cryo-EM movies of individual particles by optimization of image translations. *Journal of Structural Biology*, *192*(2), 188–195.

Scheres, S. H. (2014). Beam-induced motion correction for sub-megadalton cryo-EM particles. *Elife*, *3*, e03665.

Smith, M. T., & Rubinstein, J. L. (2014). Beyond blob-ology. *Science*, *345*(6197), 617–619.

Yu, X., Jin, L., & Zhou, Z. H. (2008). 3.88 å structure of cytoplasmic polyhedrosis virus by cryo-electron microscopy. *Nature*, *453*(7193), 415–419.

Zhao, J., Benlekbir, S., & Rubinstein, J. L. (2015). Electron cryomicroscopy observation of rotational states in a eukaryotic V-ATPase. *Nature*, *521*(7551), 241–245.

Zhou, A., Rohou, A., Schep, D. G., Bason, J. V., Montgomery, M. G., Walker, J. E., … Rubinstein, J. L. (2015). Structure and conformational states of the bovine mitochondrial ATP synthase by cryo-EM. *Elife*, *4*, e10180.

CHAPTER SIX

Processing of Structurally Heterogeneous Cryo-EM Data in RELION

S.H.W. Scheres[1]
MRC Laboratory of Molecular Biology, Francis Crick Avenue, Cambridge Biomedical Campus, Cambridge, United Kingdom
[1]Corresponding author: e-mail address: scheres@mrc-lmb.cam.ac.uk

Contents

Abstract

This chapter describes algorithmic advances in the RELION software, and how these are used in high-resolution cryo-electron microscopy (cryo-EM) structure determination. Since the presence of projections of different three-dimensional structures in the dataset probably represents the biggest challenge in cryo-EM data processing, special

Methods in Enzymology, Volume 579
ISSN 0076-6879
http://dx.doi.org/10.1016/bs.mie.2016.04.012

emphasis is placed on how to deal with structurally heterogeneous datasets. As such, this chapter aims to be of practical help to those who wish to use RELION in their cryo-EM structure determination efforts.

1. INTRODUCTION

Over the past two decades, statistical methods have become increasingly popular for the processing of cryo-electron microscopy (cryo-EM) images. In 2010, we wrote a review in this same series of *Methods in Enzymology* about a class of methods that are based on the optimization of a likelihood function (Sigworth, Doerschuk, Carazo, & Scheres, 2010). In that review, we explained how the averaging of two-dimensional (2D) projection images, or similarly their reconstruction into a three-dimensional (3D) map, can be considered as an incomplete data problem. The incompleteness of the data lies in the fact that the relative orientations of the 2D projection images are unknown, since the individual macromolecular complexes adopt random orientations on the experimental support. The way in which these unknown, or hidden, parameters are treated constitutes the main difference between the maximum-likelihood approach and alternative, least-squares methods that had dominated the field until then. In both refinement approaches one iteratively compares images from one or more 2D or 3D references in many different orientations with each experimental particle image, and references for the next iteration are calculated from averages over the experimental particles. In the least-squares approach one calculates a single, best orientation and class for each particle based on the squared difference between the reference and the particle, or the closely related cross-correlation coefficient. Each experimental particle then contributes only in its best orientation and class to the average that will form the reference for the next iteration. In the maximum-likelihood approach one employs a statistical noise model to calculate posterior probabilities for all possible orientations and classes, and each particle contributes to all references and in all orientations, but these are weighted according to the posterior probabilities. This treatment of the hidden parameters is called marginalization, and it results in a smearing out of each particle over multiple orientations and classes if the noise in the data is too strong to uniquely identify their correct assignment.

Because the statistical noise models are typically based on Gaussian distributions, the maximum-likelihood approach is closely related to the least-squares approach. In fact, in the absence of any noise in the data it becomes straightforward to identify the best orientation and class, and the two approaches become identical (Sigworth, 1998). However, because of the high noise levels in cryo-EM images, the two approaches typically behave rather differently on experimental data. These differences are most important during the initial iterations of the optimization, when low-resolution references contain too little information to unambiguously assign orientations and classes. Upon convergence, when high-resolution references provide much more information, the posterior probability distributions often approach delta functions, corroborating the selection of only the single most likely assignment for each particle.

By 2010, maximum-likelihood approaches had been used in a number of experimental studies, mostly for 2D and 3D classification tasks (Scheres, 2010). Since then, the use of likelihood-based approaches in cryo-EM has increased steeply. Existing implementations of 2D and 3D maximum-likelihood classification in the XMIPP package (Scheres, Nunez-Ramirez, Sorzano, Carazo, & Marabini, 2008) have remained in use, while Niko Grigorieff also implemented a new classification approach in FREALIGN that marginalizes over the class assignments but still treats the orientational assignments in a least-squares manner (Lyumkis, Brilot, Theobald, & Grigorieff, 2013). Probably the steepest increase in the use of likelihood-based methods has been due to the introduction of a new, empirical Bayesian approach (Scheres, 2012a) that was implemented in the RELION program (Scheres, 2012b). This approach differs from previously available likelihood optimization approaches in the introduction of a regularization term to the likelihood function. The resulting regularized likelihood optimization algorithm has proven useful for both high-resolution reconstruction as well as 2D or 3D classification in a wide range of experimental studies (Bai, McMullan, & Scheres, 2015).

This chapter reviews the algorithmic concepts that were implemented in RELION and that are new compared to the approaches existing in 2010. It also describes how this software may be used in high-resolution cryo-EM structure determination. One of the major challenges in many projects is the presence of multiple, different 3D structures in the data. Many macromolecules adopt a range of different conformations as an intrinsic part of their functioning, and many samples are not purified to absolute homogeneity. If left untreated, the presence of structural heterogeneity in the data will lead to blurred regions in the maps and possibly to incorrect interpretations. Because many different approaches to deal with structural heterogeneity exist, this probably represents the most challenging aspect of image processing in RELION. Therefore, this chapter is most detailed in its description of how to deal with structural heterogeneity. As such, it is also aimed as a useful resource for those who wish to use RELION in their cryo-EM structure determination efforts.

2. NEW ALGORITHMIC CONCEPTS

2.1 Regularization: The Empirical Bayesian Approach

The main algorithmic advance in RELION is the introduction of a regularization term to the likelihood target function. Previously available approaches optimized an unregularized likelihood function, which expresses the agreement between the experimental data and the reconstruction, while marginalizing over the hidden parameters. The problem with the unregularized target is that cryo-EM 3D reconstruction is severely ill-posed, ie, there are many possible noisy reconstructions that fit the data equally well. In order to define a unique solution one needs to introduce additional information, which is known as regularization. The key question then becomes, what information is available about the reconstruction in the absence of experimental data? A commonly employed approach to similar questions in machine learning is known as Tikhonov regularization, where one restrains the square of the Euclidean (or L2) norm of the solution (Tikhonov, 1943).

By describing both the signal and the noise components of the data using Gaussian distributions, Tikhonov regularization can also be understood from a Bayesian perspective. In this framework, the belief that the L2-norm of the reconstruction is limited is expressed as a prior on the solution. Optimization of the posterior distribution, which is proportional to the multiplication of the prior with the original likelihood function, is called maximum-a-posteriori (MAP) optimization and is mathematically equivalent to Tikhonov regularization. Whereas in standard Bayesian methods the prior is fixed before any data are observed, inside RELION parameters of the prior are estimated from the data themselves. This type of algorithm is referred to as an empirical Bayesian approach.

Both the likelihood and the prior are expressed in the Fourier domain, where all signal and noise components are assumed to be independent. In the likelihood function, the assumption of independent Gaussian noise in the Fourier domain is identical to a previously introduced, unregularized likelihood approach in the Fourier domain (Scheres, Nunez-Ramirez, et al., 2007).

Modeling the power of the noise as a function of spatial frequency in the Fourier domain allows for the description of nonwhite experimental noise and effectively results in a χ^2-weighting of the different spatial frequencies in alignment. In addition, the Fourier domain formulation permits a convenient description of the effects of microscope optics and defocusing (by the so-called contrast transfer function, or CTF). In the prior, the power of the Fourier components of the reconstruction is also restrained as a function of spatial frequency. Thereby, the prior basically acts as a Fourier filter that imposes smoothness on the reconstruction in real space.

In fact, optimization of the regularized likelihood target by the standard expectation-maximization algorithm (Dempster, Laird, & Rubin, 1977) results in an update formula for the reconstruction that shows strong similarities with previously introduced Wiener filters (Penczek, 2010a). Yet, while in 2010 it was still far from obvious how the Wiener filter for 3D reconstruction could be implemented to have its intended meaning (Penczek, 2010b), the empirical Bayesian perspective allowed an elegant and straightforward derivation of the correct implementation of the Wiener filter for 3D reconstruction. This filter depends on estimates for the power of both the signal and the noise as a function of spatial frequency. By constantly re-estimating these powers from the data themselves, RELION effectively calculates the best possible filter, in the sense that it yields the least noisy reconstruction, at every iteration of the optimization process.

Because optimal weights for the alignment and reconstruction are estimated without user intervention, the regularized likelihood optimization algorithm is intrinsically easy-to-use in the sense that is does not rely on an expert user to tune many critical parameters. This makes RELION suitable for automation and has likely contributed to its rapid uptake in the field.

2.2 Prevention of Overfitting: The Gold-Standard Approach to Refinement

Although the constantly updated Wiener filter explicitly dampens high-frequency components in the reconstruction, overfitting of the data is not entirely obviated. The reason for this lies in the empirical nature of the Bayesian approach. At every iteration, the widths of the Gaussians in the prior are estimated from the power of the reconstruction itself. Therefore, once one over-estimates the power of the true signal due to an inadvertent build-up of a small amount of noise in the reconstruction, one will allow even more noise in the next iteration. This can lead to overfitting, where noise in the model iteratively builds up.

By 2010, overfitting and over-estimation of resolution were still common problems in many cryo-EM structure determination projects, and this was recognized by a community-driven task-force for the validation of cryo-EM structures (Henderson et al., 2012). One of their recommendations was to prevent overfitting by a so-called gold-standard approach to refinement. In this approach one splits the data into two halves and one refines independent reconstructions against each half-set. The Fourier shell correlation (FSC) between the two independent reconstructions then yields a reliable resolution estimate, so that the iterative build-up of noise can be prevented. The original approach to estimate resolution from FSC curves was always intended to be used with two independently refined reconstructions (Harauz & van Heel, 1986). Nevertheless, common practices in the field had evolved toward the refinement of a single reconstruction, where the resulting angles would be used to make two (no longer independent) reconstructions from random half-sets at every iteration. The apparent rationale for this was that a reconstruction from all of the data would provide a better reference for alignment than the noisier reconstructions from only half of the data. However, evidence that this reasoning does not necessarily hold for cryo-EM refinements was obtained from tilt-pair experiments (Henderson et al., 2011). These experiments showed that alignments are dominated by the lower spatial frequencies, which are almost indistinguishable between reconstructions from all or half of the data. Further evidence came from a comparison of the two refinement approaches on simulated data and on experimental datasets for which a high-resolution crystal structure was available. Although the gold-standard approach gave nominally lower-resolution estimates, the resulting maps were actually better than when only a single reconstruction was refined (Scheres & Chen, 2012). Moreover, the previously introduced FSC = 0.143 threshold at which to interpret the resolution of the reconstruction (Rosenthal & Henderson, 2003) had been deemed as too optimistic when using the suboptimal refinement approach, and a more conservative FSC = 0.5 threshold had become the norm. The new experiments illustrated that the problem lies in the inflated FSC curves from the suboptimal refinement approach, and that the FSC = 0.143 threshold performs well when two independent reconstructions are used (Scheres & Chen, 2012).

These results prompted the implementation of a gold-standard refinement approach in RELION. In this implementation, the iterative build-up of noise is avoided by estimating the widths of the Gaussian priors from the FSC curve between two independently refined reconstructions.

Combined with a new approach to estimate the accuracy of alignment from the estimated signal-to-noise ratios in the data, this led to the implementation of "3D auto-refinement." In this approach, the estimated angular accuracies are used to automatically increase angular sampling rates during the refinement up to the point where the noise in the data prevents one from distinguishing smaller angular differences. This procedure allows the calculation of high-resolution reconstructions from structurally homogeneous datasets without intermediate user intervention (Scheres, 2012b).

2.3 Getting Clean, High-Resolution Maps: The Postprocessing Approach

In order to strictly prevent overfitting, the two independently refined reconstructions are not masked when calculating the FSC curves inside 3D auto-refinement. This leads to an underestimation of the true resolution of the reconstructed signal during refinement because the signal is restricted to a central region of the map and the surrounding solvent region merely contributes noise. Because orientational and class assignments are predominantly driven by the lower frequency content of the images, they are usually not noticeably affected by this underestimation of resolution. However, upon convergence the highest possible amount of information needs to be extracted from the reconstruction.

The noise in the solvent region may be removed by masking. Masking is a multiplication operation with a 3D map (the mask) that has values in the range of zero outside a region of interest to one inside the region. By masking out the solvent region from the two half-reconstructions, the noise gets reduced and the FSCs will increase. However, besides this beneficial effect of solvent removal, FSC values may also increase due to an undesirable convolution effect of masking. Multiplication with a mask in real-space corresponds to a convolution with the Fourier transform of the mask in the Fourier domain. A mixing of the stronger and better-correlating, low-frequency Fourier-components with the weaker and less-correlating, high-frequency components will cause an increase in FSCs that does not reflect the true signal-to-noise ratio in the maps. The more detailed features the mask has, the stronger this convolution effect is. Also this problem was recognized by the EM validation task-force (Henderson et al., 2012).

Adaptation of a method that was originally devised to estimate the amount of overfitting in refinement provided a solution to correct FSC curves for the convolution effects of masking (Chen et al., 2013). In this approach, the phases of Fourier components of the two half-reconstructions with spatial frequencies higher than a given cutoff are randomized. The cutoff for phase-randomization is chosen to be considerably lower than the resolution of the reconstructions without masking away the solvent. Then, a solvent mask is applied to the two phase-randomized half-reconstructions. In the absence of convolution effects, the resulting FSC curves should be zero beyond the phase-randomization cutoff. Therefore, any nonzero correlations are due to convolution effects and one can correct the FSC curve between the two masked half-reconstructions without phase randomization. The corrected FSC curve reflects the increase in resolution caused by removing the solvent noise, without being affected by mask convolution effects.

The phase-randomization approach to correct FSC curves was implemented in the postprocessing approach of RELION, where it was combined with a previously proposed method to sharpen the reconstruction (Rosenthal & Henderson, 2003). This is needed because high-frequency components in the reconstruction are dampened in the image formation process itself, in the detection process and in the image processing. This fall-off is often modeled by a Gaussian using a B-factor, in analogy to the temperature factor in X-ray crystallography. Application of a negative B-factor leads to sharpening of the map. The B-factor value may be estimated automatically for reconstructions that extend beyond 10 Å, or may be set by the user for lower-resolution reconstructions. Multiplication of the sharpened map with the corrected FSC curve typically leads to clean reconstructions with excellent high-resolution details.

2.4 Beam-Induced Motion Correction and Radiation-Damage Weighting: The Movie-Processing Approach

With the advent of direct-electron detectors in 2012/13, the possibility arose to collect movies during the exposure of the sample to the electron beam. This allowed for two improvements that were previously inaccessible: beam-induced motion correction and radiation-damage weighting. When the electrons hit the sample, inelastically scattered electrons deposit energy and chemical bonds in the sample are broken. The exact mechanisms of what happens are unknown, but one observation is that the sample starts to move upon exposure to the electron beam (see chapter "Specimen Behavior in the Electron Beam" by Glaeser). This beam-induced motion causes a blurring in the images that can be corrected by movie-processing, since each of the movie frames contains a sharper snapshot of the moving objects. Grigorieff and coworkers were the first to illustrate the potential of beam-induced motion correction for large rotavirus particles (Brilot et al., 2012; Campbell et al., 2012).

Correcting for beam-induced motions of smaller complexes is more difficult, since the movie frames contain only a fraction of the total electron dose each and are thus extremely noisy (see chapter "Processing of Cryo-EM Movie Data" by Rubinstein). Still, implementation of a movie-refinement approach, where one marginalizes over the orientations of running averages of multiple movie frames led to much improved maps. Using this approach, reconstruction of only thirty thousand ribosome particles led to a map in which side chain densities were clearly visible (Bai, Fernandez, McMullan, & Scheres, 2013). For even smaller particles, ie, with molecular weights less than 1 MDa, the running averages of multiple movie frames still contain too much noise to reliably follow beam-induced motions for individual particles. Therefore, the movie-processing approach was adapted based on the observation that neighboring particles often move in similar directions (despite the fact that particles further away on the same micrograph may move in very different directions). By fitting straight lines through the most likely translations from the original movie-processing approach, and by considering groups of neighboring particles in these fits, the high noise levels in the estimated movement tracks could be sufficiently reduced to allow beam-induced motion correction also for smaller complexes (Scheres, 2014).

The approach to simultaneously fit beam-induced motions for groups of neighboring particles was combined with a novel way of handling radiation-damage weighting. Radiation damage starts with the breakage of chemical bonds in the sample, which will rapidly destroy the high-resolution content of the images. Subsequently, secondary structure elements and protein domains are unfolded, and eventually the entire macromolecular complex will be destroyed. Consequently, low-resolution information in the images will persist longer than the high-resolution information. To model the dose- and resolution-dependent effects of radiation damage on the images, the calculation of a B-factor was proposed for each movie frame. In the resulting approach, the movie-frames of each particle are aligned according to the fitted motion tracks, and single-frame reconstructions are used to estimate a B-factor for each movie frame. Then, the aligned movie frames of each particle are averaged with frequency-dependent weights according to their relative B-factors. By weighting the different spatial frequencies in each movie frame differently, the useful information from each movie frame is retained. For example, whereas later movie frames may hardly contain any high-resolution information, they may still contribute constructively to the lower-resolution information content. Because the lower-resolution information may still be useful in particle alignment, this approach is more attractive than the alternative of throwing away later movie frames for high-resolution reconstruction. Because it results in particle images with improved signal-to-noise ratios this approach was called "particle polishing," a sideways reference to a quirk in the field whereby the crispest images are referred to as "shiny" (which was first used by Xuekui Yu at the 2007 Gordon research conference for 3D-EM).

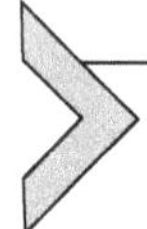

3. A TYPICAL HIGH-RESOLUTION STRUCTURE DETERMINATION PROCEDURE

Fig. 1 shows a flowchart of a typical high-resolution structure determination project from movies acquired on a direct-electron detector. Micrograph-wide beam-induced motions are first corrected by aligning the frames of the recorded micrograph movies, and an average micrograph is calculated for every movie. This step is often performed using the MOTIONCORR program (Li et al., 2013). Secondly, one uses another third-party program to estimate the parameters of the CTF for each average micrograph. Implementation of a wrapper in the RELION GUI makes it convenient to use CTFFIND (Mindell & Grigorieff, 2003; Rohou & Grigorieff, 2015) or Gctf (Zhang, 2016) for this task.

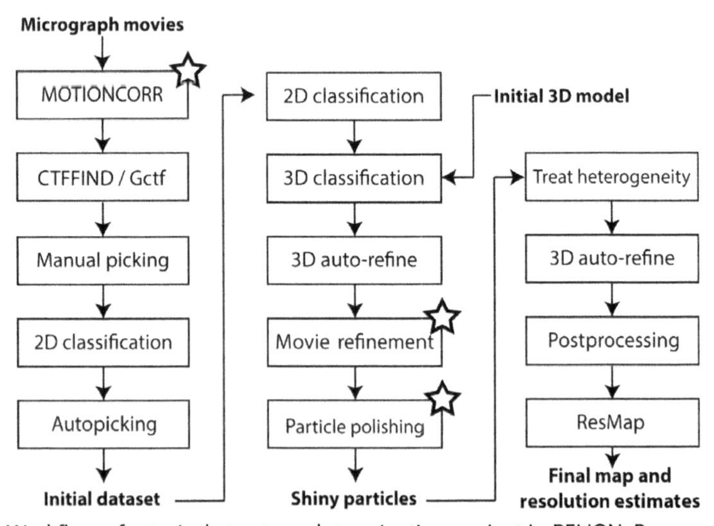

Fig. 1 Workflow of a typical structure determination project in RELION. Processes indicated with a star are skipped if movies are unavailable.

Next, one needs to identify the positions of individual particles in all micrographs. For this task the user first manually selects suitable particles from a few micrographs: typically around a thousand particles is enough. These particles are extracted into small square images, called boxes, where the size of the box is typically set to around two times the largest dimension of the complex of interest (see Fig. 2). Sometimes smaller boxes are used for highly elongated particles. Upon particle extraction, the individual particle images are normalized to have zero-mean and unity-variance in the background noise area, which is defined as the area outside a circular mask (see also Fig. 2). The diameter of this mask is set slightly larger than the largest dimension of the complex. A ramping background density may be modeled at this stage by fitting a 2D plane through the background pixels and subtracting this plane from the particle.

After extraction, the particles are subjected to a first round of reference-free 2D class averaging, which is called 2D classification in RELION. This task is performed using the same regularized likelihood optimization algorithm as described above, although in this case the references are 2D images and one marginalizes only over in-plane orientations. For cryo-EM data the number of classes for this run is typically set such that at least 50–100 experimental particles contribute on average to each class, so often one uses around 20 classes at this stage. From the resulting 2D class averages, one selects a few (usually not more than 5–10) representative views, which are used as templates for automated particle picking of all micrographs (Scheres, 2015). It is important to check that these templates are centered inside the particle box to prevent bias in the autopicked coordinates. In case the templates are not centered, they can be moved to their center-of-mass (assuming the signal is positive, ie, white) using the `relion_image_handler` program. Besides the templates, the autopicking algorithm has two additional parameters that are important: a threshold that expresses how restrictive the particle picking is (with higher threshold values being more restrictive, ie, picking fewer particles) and a minimum inter-particle distance. The minimum inter-particle distance is often set in the range of 50–70% of the longest dimension of the macromolecular complex of interest. Useful thresholds are often in the range of 0.1–0.5, where noisier micrographs generally require lower autopicking thresholds. For very noisy data, useful values as low as 0.05 have been observed, but in these cases one should be extra vigilant of template bias. Template bias leads to a dangerous situation where averages of experimental particle images reproduce the reference image used to pick the particles, even if the picked particles only

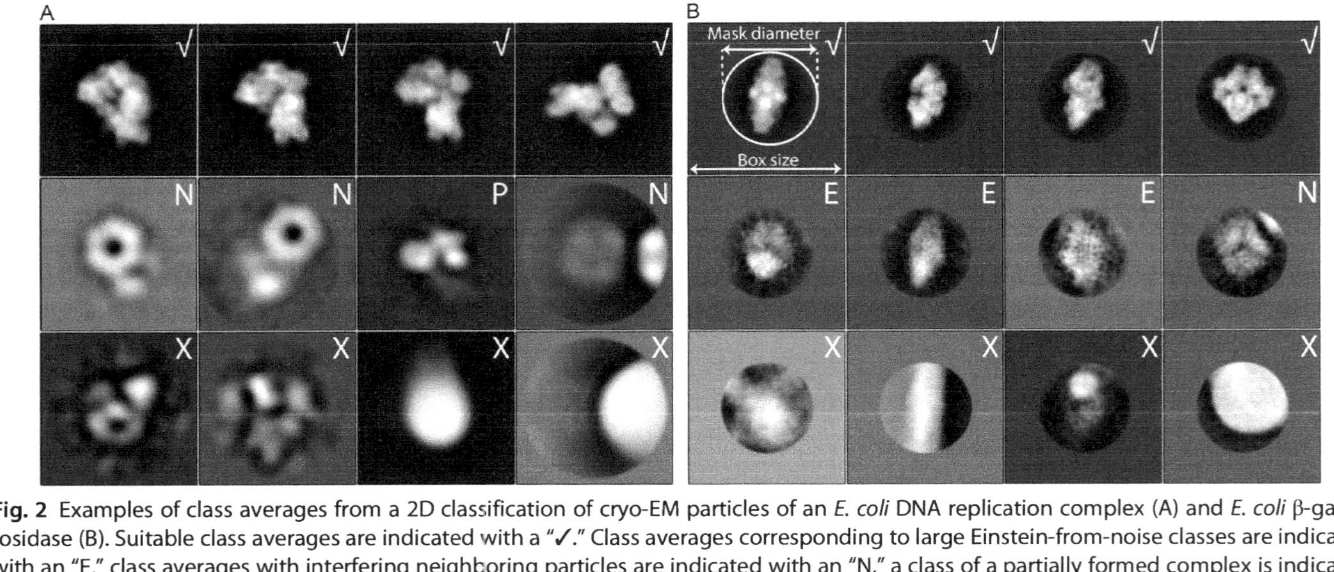

Fig. 2 Examples of class averages from a 2D classification of cryo-EM particles of an *E. coli* DNA replication complex (A) and *E. coli* β-galactosidase (B). Suitable class averages are indicated with a "✓." Class averages corresponding to large Einstein-from-noise classes are indicated with an "E," class averages with interfering neighboring particles are indicated with an "N," a class of a partially formed complex is indicated with a "P," and classes corresponding to otherwise unsuitable particles are indicated with an "X." The particle box size and the mask diameter used for particle extraction and normalization are indicated in the first class average in (B).

contain background noise. This is often referred to as "Einstein-from-noise," as a popular example to illustrate this effect uses a template picture from Einstein (Henderson, 2013; van Heel, 2013). To aid in the recognition of template bias and to prevent high-resolution bias, the template images are typically low-pass filtered to around 20–30 Å resolution. A more detailed description of how to recognize template bias is given below.

After extraction of the autopicked particles, one needs to identify those particles that are suitable for high-resolution structure determination. There are multiple reasons why particles may be unsuitable. The autopicking algorithm itself has been observed to be prone to selecting false positives in the form of high-contrast artifacts (Scheres, 2015), while noise-only images may result from low autopicking thresholds. In addition, particles may be partially unfolded during cryo-EM sample preparation, impurities in the sample may be mistaken for the particle of interest, or too closely neighboring particles may prevent their adequate alignment. A first, computationally cheap approach to identify false positives from the autopicking is called "particle sorting." This approach subtracts the corresponding templates from each experimental particle, and calculates a range of features on the difference images. Sorting all particles on the sums of their Z-scores over all features (Scheres, 2010) then may be more convenient to identify outliers in large datasets.

One of the most effective ways of selecting suitable particles is 2D classification. Typically, for large datasets one uses approximately 200 classes, while for smaller datasets, one restricts the number of classes such that at least 50–100 particles contribute to each class on average. Manual inspection of the resulting 2D class averages, often ordered on the number of particles contributing to each class, is then used to select suitable particles. Provided full CTF correction is performed, suitable particles will result in 2D class averages with strong, white density for the macromolecular complex and a black and featureless background. The macromolecular density should ideally not consist of white blobs for each protein domain, but contain clearly visible features of projected secondary structure elements throughout the complex. Overfitting of noise sometimes results in radially extending "hairy" artifacts in the solvent regions of the class averages. These undesired artifacts are indicative of bad alignments and typically only appear with very noisy data. Prevention of these artifacts may sometimes be achieved by limiting the resolution of the Fourier components that are used in the expectation step of the 2D classification. Often a resolution limit in the range of 15–20 Å is useful in initial runs, and this limit may be relaxed in later runs.

The selection of classes that are suitable for high-resolution reconstruction requires user supervision. High-contrast false positives from the autopicking and relatively small, low-resolution class averages are relatively easy to distinguish from the suitable classes. For the novice user, Einstein-from-noise classes are often harder to identify, yet their removal from the dataset is important. Sometimes, the class averages of Einstein-from-noise classes remain limited in resolution, and they can be readily identified by their blob-like aspect. In other cases, and especially for smaller macromolecular complexes, Einstein-from-noise classes may also give rise to relatively high-resolution averages. These class averages are often more grayish in their appearance than the suitable white-on-black classes, and high-resolution features often occur just as strongly in the surrounding background as in the apparent density of the complex. Fig. 2 shows examples of suitable and unsuitable 2D class averages of a 250 kDa DNA replication complex from *E. coli* (Fernandez-Leiro, Conrad, Scheres, & Lamers, 2015) and of the 450 kDa *E. coli* β-galactosidase complex (Scheres, 2015). Selection of the suitable classes may be performed conveniently through the GUI, and the selected particles from one 2D classification may be fed into an ensuing one. Sometimes, one also performs a subsequent 2D classification on particles that were selected from classes that did not look optimal in an attempt to rescue suitable particles that would otherwise be discarded. Most often, not more than three subsequent 2D classifications are performed. Strongly depending on the quality of the original micrographs and the threshold used in the autopicking, 20–80% of the autopicked particles may be discarded at this stage.

Next, one typically also uses an initial 3D multireference refinement run (called 3D classification) to continue the selection of suitable particles for high-resolution structure determination. The initial 3D reference that is required for 3D classification or 3D auto-refinement cannot be generated inside RELION itself. Popular programs to calculate such initial models from the particle images themselves are EMAN2 (Ludtke, Baldwin, & Chiu, 1999; Tang et al., 2007; see also chapter "Single Particle Refinement and Variability Analysis in EMAN2.1" by Ludtke) and SIMPLE (Elmlund, Davis, & Elmlund, 2010). Alternatively, one may use maps of similar complexes from the EMDB, maps generated from related PDB entries, or maps obtained by random conical tilt reconstruction (Radermacher, Wagenknecht, Verschoor, & Frank, 1987) or subtomogram averaging (Walz et al., 1997). In some cases, in particular when the complex of interest has a known point-group symmetry, refinements may even be started from a spherical blob. The number of classes to be used in the initial 3D classification run depends on the size of the dataset and the available computer power. For smaller datasets one uses on average at least 5000 particle images per class, while for larger datasets one rarely uses more than 8–10 classes to reduce computing costs. The initial 3D classification is run with exhaustive angular searches, and typically without imposing symmetry. As in the case of 2D classification, the resolution of the Fourier components that is taken into account in the expectation step can again be limited to prevent overfitting. More details about the 3D classification are given in the next section. Again, suitable classes will show strong protein-like features, for example, rod-like densities for α-helices, while unsuitable classes will be of lower resolution or may not resemble the complex of interest. At this stage, reconstructions that correspond to the complex of interest in different conformational states are usually still kept together for subsequent movie-refinement.

One problem that may arise at this point is that there are not enough different views for 3D reconstruction because the particles adopted a strongly preferred orientation on the experimental support. This problem, which is often already detectable from a shortage of different 2D class averages, may manifest itself in streaky reconstructions, where densities are smeared out in the direction of the predominant view. A related problem may be that different classes become streaky in different directions, which is an indication that the classification converged to separate different views rather than different structural states. Such classifications are typically not useful. As long as some of the minority views are available in the dataset, throwing away particles from the predominant view may help to balance the orientational distribution and thereby get better reconstructions and classifications. This was, for example, crucial in the structure determination of the dynactin complex (Urnavicius et al., 2015). In many cases it is probably more efficient to tweak the cryo-EM sample preparation procedure, for example, by adding small amounts of detergents or by changing the type of experimental support, in order to get more different views.

The selected particles after the initial 3D classification are then used for movie-processing. For this, one first runs a 3D auto-refinement of the selected particles. Some degree of structural heterogeneity in the data at this point is acceptable. The objective of this refinement is to determine the orientations of all particles with respect to a single consensus reference. If at this point the consensus map does not extend well beyond 10 Å resolution, then movie-processing will probably not be effective and could be skipped. Otherwise, the consensus refinement is continued for a single iteration, where running averages of the movie frames of each particle are aligned to reference projections of the consensus reconstruction. The user has to define the width of the running averages, which depends on the dose per movie frame and on the size of the complex of interest. For large complexes (eg, larger than 1 MDa) an accumulated dose of 3–6 electrons (e^-) per $Å^2$ in the running averages is usually sufficient. Smaller complexes may benefit from wider running averages, eg, with an accumulated dose in the range of 7–10 $e^-/Å^2$. In the subsequent particle polishing step, the translations from the movie refinement are used to fit straight tracks that describe the beam-induced motion for each particle. How many neighbors contribute to the fitting of each particle depends on a user-controlled standard deviation of a Gaussian weighting function. Larger standard deviations are needed for smaller particles, as their individual tracks will be noisier and more neighbors are needed to get reliable fits. Useful values for the standard deviation therefore also depend on how many particles there are on each micrograph. For particles with molecular weights less than 0.5 MDa useful values are often in the range of 500–1500 Å, while values below 500 Å are often used for larger particles. Reconstructions with individual movie frames are then used to determine the power of the signal at every spatial frequency for radiation-damage weighting. This often represents a critical step, and inspection of plots of both the B-factors and the linear scale factors is highly recommended. In the weighting process, the absolute values of the scale and B-factors of the movie frames are not important, only the differences between them matter. The B-factors are typically relatively high negative numbers in the first few frames (eg, with an accumulated dose of 1–3 $e^-/Å^2$) due to rapid initial beam-induced motions; they become smaller for intermediate frames (4–10); and then increase again for later frames due to radiation damage (eg, from an accumulated dose of 10–15 $e^-/Å^2$ onward). The linear scale factors typically decrease in a fairly linear manner throughout the movie, possibly with the exception of the first one or two frames, which may be somewhat lower. As a typical example the resulting plots and frequency-dependent weights for each movie frame are shown for a dataset on bovine complex-I in Fig. 3. If these plots do not look as expected, then the user may change the widths of the running averages, or the standard deviation of the Gaussian that expresses how many neighboring particles influence the fitted motion tracks. Also changing the resolution limits for the B-factor calculations or the tightness of the mask may help to get better plots. Visual inspection of the fitted beam-induced motion tracks and the Guinier plots that are used to estimate the B-factors may be

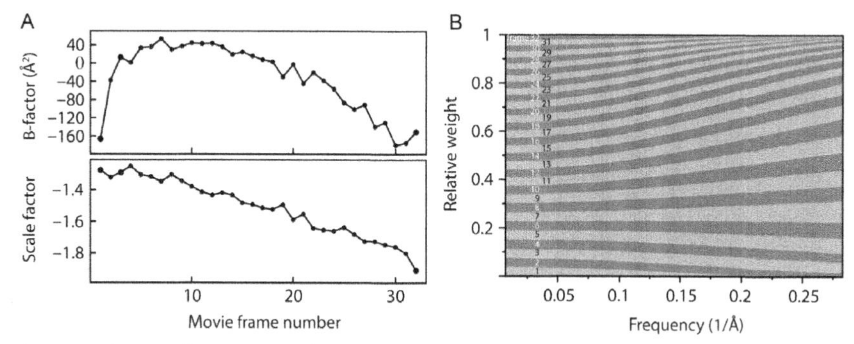

Fig. 3 Radiation-damage weighting. (A) B-factors and linear scale factors as estimated for all 32 movie frames of a cryo-EM dataset on bovine complex-I. (B) The resulting frequency-dependent weights for each movie frame. The width of each band indicates the relative weight at each frequency in each frame. At each frequency the sum over frames of the weights is unity. Note that the early and the later frames hardly contribute to the highest frequencies, whereas they do contribute to the lower frequencies. *This figure was adapted from Scheres, S. H. W. (2014). Beam-induced motion correction for sub-megadalton cryo-EM particles.* eLife 3, *e03665.*

helpful in identifying how these parameters may be changed. In addition, instead of performing reconstructions from individual movie frames to estimate the B-factors, one may perform reconstructions from running averages of multiple movie frames. The drawback of this approach is that large differences in the B-factors of sequential movie frames, as for example, observed in the first few frames in Fig. 3, cannot be modeled accurately.

After particle polishing, the resulting "shiny" particles may be used for further treatment of structural heterogeneity as outlined in more detail in the next section. Once a structurally homogeneous subset has been identified, a final 3D auto-refinement is performed, which is followed by postprocessing to sharpen the map and to estimate its final resolution. The solvent mask for the postprocessing procedure can be calculated automatically from the reconstructed density, but the user should check that the resulting mask does not cut off part of the complex and that the edges of the mask are sufficiently wide. Too sharp edges on the mask will lead to strong convolution effects on the FSC curves, and unreliable (typically too low) resolution estimates. Therefore, it is important that the user inspects the FSC curves from the phase-randomization procedure. In particular, FSC values between the masked phase-randomized maps should be close to zero at the estimated resolution of the map. RELION uses cosine-shaped edges on masks, and useful widths of the mask edges are often in the range of 3–10 pixels. Masks around specific parts of the complex may be used to estimate variations in local resolution. This may be useful to calculate better-resolved parts of the map to higher resolution, and more flexible parts to lower resolution. The overall reported resolution should however be calculated around the entire complex, and not merely reflect the resolution of the best ordered region. Finer variations in local resolution are often estimated using a wrapper from the RELION GUI to the RESMAP program (Kucukelbir, Sigworth, & Tagare, 2014). The same mask that encompasses the entire complex from the postprocessing approach should be provided to RESMAP in order to obtain reliable resolution estimates.

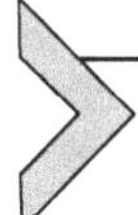

4. DEALING WITH STRUCTURAL HETEROGENEITY

4.1 3D Classification with Exhaustive Angular Searches

The main approach to distinguish projections from different 3D structures in RELION is 3D classification. Inside this approach, multiple reconstructions are refined simultaneously and one marginalizes over both the orientational and class assignments of the particle images. As demonstrated by the previously introduced unregularized likelihood approach to 3D classification in real-space (ML3D) (Scheres, Gao, et al., 2007), marginalization over both orientations and classes allows classification without the need for prior knowledge of the differences between the structures present in the data. This unsupervised classification is achieved by initializing multireference refinements from a single, low-resolution consensus model and assigning a random class to each particle in the first iteration. In this way, apart from the consensus model, the user only provides the number of desired classes. The actual number of different structures in the data is typically unknown, but in practice different runs with varying numbers of classes generate useful results. Because the user does not provide different references, against which to match the structural heterogeneity in the data, in machine-learning terminology this is called clustering instead of classification.

The simultaneous refinement of multiple references precludes the application of the gold-standard refinement approach, as the randomly seeded classification convergence may be very different for the two halves of the data. Therefore, inside the 3D (and also the 2D) classification approach in RELION, the resolution of each of the models is estimated directly from their power spectrum. This could in principle lead to overfitting, and the user has control over this by adjustment of the regularization parameter T. Lower values of T will prevent higher-resolution features to appear in the reconstructions and thus keep overfitting at bay. In practice, one does not change T very much. Most 2D classifications are performed with

$T=2$, while 3D classifications typically use values around $T=4$. One important exception to this is masked classification, which is described in more detail below.

If one starts without any knowledge of the relative orientations of the particles, then one needs to marginalize over all orientations in the likelihood function, or in other words one needs to perform exhaustive angular searches. To limit the associated computational costs, exhaustive angular searches in 3D classification are typically performed using a relatively coarse angular sampling. Because RELION uses the Healpix algorithm (Gorski et al., 2005) to sample the first two Euler angles on the sphere, only certain angular sampling rates are allowed. A sampling rate of 7.5 degrees is most often used for exhaustive searches. For highly symmetrical structures, eg, with icosahedral or octahedral symmetry, a 3.7 degree sampling may be better, although 3D classifications with exhaustive searches are typically performed without imposing symmetry, even if the complex of interest is known to have symmetry. In these asymmetric classifications one aims to remove junk or otherwise unsuitable particles from the dataset, and imposing symmetry on junk particles might make them look more like the structures of interest. Because each junk particle is probably different from all other junk particles, the concept of a "junk class" does not necessarily exist. Nevertheless, suitable particles tend to group together in one or more classes that yield reconstructions with better protein features than classes with junk particles. The latter tend to yield reconstructions with lower resolutions and without apparent symmetry.

If large conformational or compositional differences exist between the particles in the dataset, then 3D classification with exhaustive angular searches may also separate these out. Sometimes, this initial stage of separating out junk particles and very large conformational differences is repeated several times, where the selected subset from a previous run is used as input for the next.

4.2 Detection of Remaining Structural Heterogeneity

Once a subset of non-junk particles has been selected, all relative orientations of the selected particles may be determined with improved accuracy in a 3D auto-refinement run. If multiple different structures were encountered in the first 3D classification round, then separate 3D auto-refinements may be performed for each of the corresponding subsets. After 3D auto-refinement, postprocessing may then be used to sharpen the refined maps and to calculate their true resolution after solvent masking. Indications whether remaining structural heterogeneity is still present may then be obtained from visual inspection of the resulting reconstructions (both before and after postprocessing) for blurry parts of density. Often, looking at the maps in 2D slices, for example, in RELION itself, is highly complementary to looking at 3D rendered maps in a program like UCSF Chimera (Goddard, Huang, & Ferrin, 2007; Pettersen et al., 2004). In addition, the calculation of a map with local resolution variations by the wrapper to the ResMap program (Kucukelbir et al., 2014) is useful, since large variations in local resolution are a strong indicator of remaining structural heterogeneity.

4.3 3D Classification with Finer, Local Angular Searches

Small conformational or compositional differences may not be separated well using the relatively coarse angular samplings that are typically used in 3D classification with exhaustive angular searches. One option to separate structures with smaller differences is to use the refined orientations from a previous (consensus) 3D auto-refinement as input for a 3D classification with finer angular samplings. In order to reduce the computational costs of this classification, one performs only local searches of the angles around the angles from the consensus refinement. This approach implicitly sets the prior probabilities of angles that deviate much from the input ones to zero, ie, one assumes that the angles from the auto-refinement are close to the true angles. This assumption probably holds reasonably well when conformational changes are relatively small, and thereby this approach is highly complementary to 3D classification with exhaustive searches, which is better suited for large conformational differences.

An extreme, and computationally even cheaper, alternative to 3D classification with local angular searches is to keep the orientations from the consensus refinement completely fixed. In this type of 3D classification, one only marginalizes over the class assignment and assumes that the orientations from the consensus refinement are the correct ones. Because in this calculation one only needs to compare each particle image with a single projection for each of the references, these calculations can be performed very rapidly. Most often, these 3D classifications without angular searches are performed within masks, as explained below.

4.4 Masked 3D Auto-Refinement

In many cases, the structural heterogeneity of interest is of a continuous nature, for example, when a particular subunit rotates relative to the rest of the complex. In other cases, multiple subunits all move independently from each other. In such cases, masked 3D auto-refinement may be a suitable tool to deal with the multitude of different 3D structures in the data.

In this approach, one applies a 3D mask to the reference at every iteration. By masking out part of the complex, the experimental particle images are aligned only with respect to that part of the structure that lies within the mask. Thereby, the refinement becomes effectively insensitive to what happens in the regions outside the mask. Masks may be derived from fitted PDB models, or generated using the `relion_mask_create` program. Interactive editing of masks may be performed using the "Volume eraser" tool in UCSF Chimera (Pettersen et al., 2004). In order to prevent artifacts in the Fourier domain due to the convolution effects of masking, one again needs to use soft edges on the masks. Cosine-shaped soft edges may be added to masks using the `relion_mask_create` program. Often, useful widths of the soft edges are in the range of 3–10 pixels.

An insightful example of how masked 3D auto-refinement may be useful to deal with continuous motions was performed for the cytoplasmic ribosome of the *Plasmodium falciparum* parasite (Wong et al., 2014). As often observed for cryo-EM datasets of ribosomes, the ribosomes in this sample adopted multiple different ratcheted states, where the small subunit rotates relative to the large subunit. Because many intermediate rotated states exist, the structural heterogeneity can be interpreted as an almost continuous inter-subunit rotation. The resulting reconstruction from all particles showed relatively well-defined density for the large subunit (which dominates the alignment), while density for the small subunit was more blurred. Separation of this rotation in a discrete number of classes, like one would do in 3D classification, would be difficult. One would need to use a very large number of classes to describe the near-continuous motion, and one would end up with very few particles per class.

However, by applying a mask around the large subunit and aligning all particles to the mass within the mask, one can keep the entire dataset together. Conversely, one can do the same for the small subunit. For the *P. falciparum* dataset, this approach resulted in much improved reconstructions for both subunits. In addition, as one obtains two sets of angles for all particles, the differences between these sets carry information about the ratcheted state of each particle, although this information was not used in the *P. falciparum* study. One problem with these masked refinements is that one ends up with two separate (masked) reconstructions for the two subunits, in which information about the interface between the subunits may not be well defined. Therefore, apart from the two separately masked reconstructions, the map from the unmasked refinement of the entire ribosome remained useful to inspect the subunit interface. The reconstruction from the unmasked refinement was also used to report on a single resolution estimate.

4.5 Masked 3D Classification

Masked 3D classification is the multireference equivalent of masked 3D auto-refinement. By masking out a region of interest from all references at every iteration, the classification can be focused on a specific region of interest, while structural variability in other regions is ignored. This is useful for many types of structural heterogeneity. For example, it may be used to separate complexes from which parts are missing in the case of non-stoichiometric complex formation, or one may separate conformational differences in one part of the structure while ignoring differences in other parts. Masked 3D classification is also useful to describe continuous motions in parts of the structure by dividing the data into subsets. Whereas 3D auto-refinement could in principle describe continuous motions without subdivision into a discrete number of subsets, auto-refinement cannot be applied to cases where the flexible region is very small. In such cases, finding the orientations of the particles with respect to that small region will be difficult. This is because projections of the masked references will only contain a small amount of mass, and the corresponding signal in the particle images will be drowned in the experimental noise and the projected mass of the rest of the complex. As a rule of thumb, in order to be able to apply masked 3D auto-refinement, one needs to have at least as much molecular mass inside the mask as one would need for the structure of an isolated complex, say at least 150–200 kDa. Many macromolecular complexes have flexible parts that are much smaller than this, and for these cases masked 3D classification provides a useful alternative.

In the masked 3D classification approach one typically provides the angles as determined in a consensus refinement as input and one does not perform any orientational searches. Apart from strongly reduced computational costs, this approach has the added benefit that particles cannot rotate or translate their region of interest outside of the mask, which would require even more classes. The result of these masked 3D classifications without angular searches is that the potentially continuous motion of the region of interest is divided into a discrete number of subsets. Each of these subsets will naturally contain fewer particles than the original dataset, but the density of the subset reconstruction may be significantly improved, in particular when certain conformational states are more recurrent than others. If the size of the mask is a borderline case between masked auto-refinement or masked classification, then one may also employ local angular searches around the angles found in the consensus refinement. Because the 3D classification approach estimates the resolution from the power spectrum of each of the reconstructions, and because these reconstructions are now masked to select a potentially very small region, one typically uses higher values of the regularization parameter T in masked classifications. Whereas in unmasked 3D classifications T is typically kept around 4, values of 10–40 have been used in masked classification. Smaller regions require larger values of T. Interpretation of the maps by the user is required to find a useful value of T that balances the build-up of noise with the ability to visualize high-resolution features and separate small structural differences.

4.6 Masked 3D Refinement/Classification with Partial Signal Subtraction

The masked 3D auto-refinement and classification approaches suffer from a fundamental inconsistency. Inside these calculations, experimental projections of the entire complex are compared with projections of masked references, which contain only part of the complex. Therefore, the density in the experimental particle image that comes from the part of the complex that lies outside the mask will effectively act as noise in the comparisons. This noise will deteriorate the orientational and or class assignments. In the *P. falciparum* ribosome case described above, the part of the experimental particle images that corresponded to the large subunit caused larger inconsistencies in the masked refinement of the small subunit than vice versa. Consequently, the resolution in the masked refinement of the small subunit was considerably lower than for the large subunit.

The inconsistency in comparing experimental projections with projections from masked references can be reduced by subtraction of part of the signal from the experimental images (Fig. 4). In this approach, one first performs a consensus 3D auto-refinement of the entire dataset. Then, one designs two masks. The first mask encapsulates the region of interest on which one will perform a masked 3D classification (or a masked 3D auto-refinement). The second mask comprises the entire complex except for the region that lies inside the first mask. This second mask is applied to the reconstruction from the consensus refinement, and projections of this

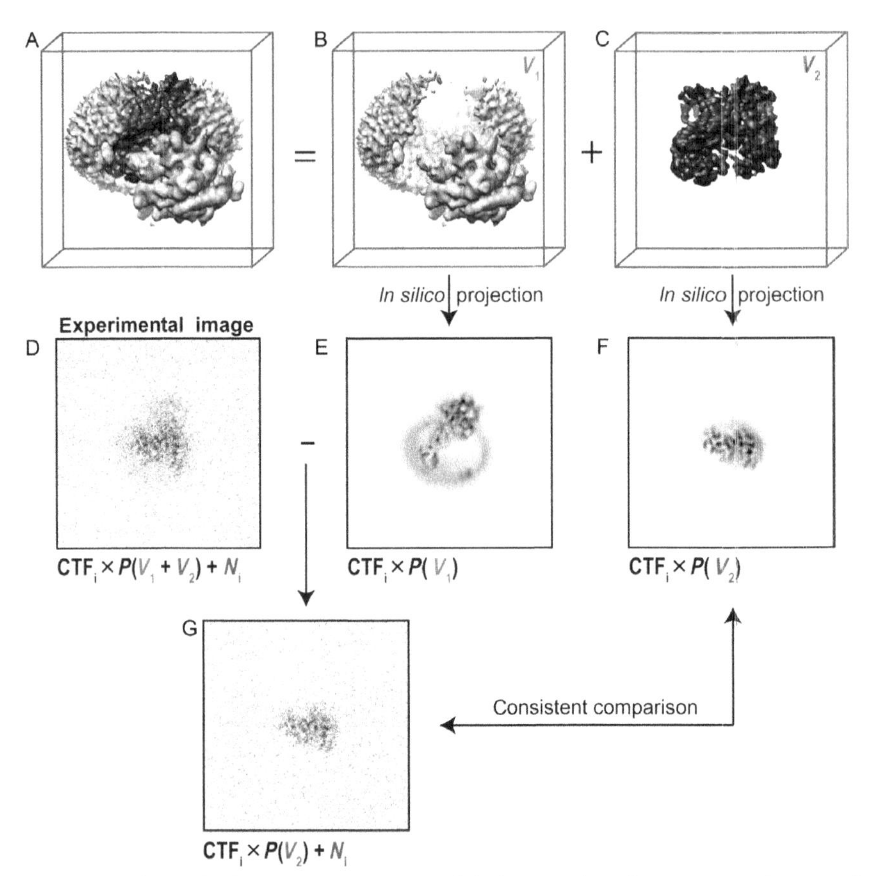

Fig. 4 Partial signal subtraction. (A) A 3D model of a complex of interest. (B) The part of the complex one would like to ignore (V_1). (C) The part of the complex one would like to focus classification on in a masked classification or 3D auto-refine (V_2). (D) An experimental particle image is assumed to be a noisy, CTF-affected 2D projection of the entire complex in (A). (E) A CTF-affected 2D projection of V_1. (F) A CTF-affected 2D projection of V_2. (G) Partial signal subtraction consists of subtracting the CTF-affected 2D projection of V_1 (E) from the experimental particle (D). This results in a modified experimental particle image (G) that only contains a noisy and CTF-affected projection of V_2. *This figure was adapted from Bai, X., Rajendra, E., Yang, G., Shi, Y., Scheres, S.H. (2015b). Sampling the conformational space of the catalytic subunit of human γ-secretase.* eLife 4. *doi:10.7554/eLife.11182.*

masked reconstruction are subtracted from all experimental particle images in the dataset. For this subtraction, one takes the effects of the CTF for each particle into account, and one uses the orientations as determined in the consensus refinement. This creates a new dataset of experimental images from which part of the signal was subtracted. The `relion_reconstruct` program may be used to calculate a 3D reconstruction of the subtracted particles in order to verify the subtraction. The subtracted dataset can then be used for masked 3D classification or auto-refinement. Again, multiple scenarios are possible. Orientational searches can be performed exhaustively, restricted to local searches, or kept fixed at the orientations from the consensus refinement. If the remaining part of the complex is small, masked 3D classifications may be used to divide the data into a discrete number of subsets, while continuous motions in larger parts of the complex may be described using masked 3D auto-refinements.

Early applications of the partial signal subtraction approach were used by Michael Rossmann and colleagues to study symmetry mismatches in bacteriophage ϕ29 (Morais et al., 2003) and flaviviruses (Zhang, Kostyuchenko, & Rossmann, 2007). Because their reconstruction algorithm did not maintain the correct absolute grayscale of the experimental projections, it was necessary to optimize a scale factor between the reconstruction and the experimental particles. This is not necessary inside RELION, where the original grayscale is maintained in the reconstruction. Moreover, whereas the early work on viruses was done with CTF phase-flipped experimental images, inside RELION both the phases and amplitudes of the CTF of each particle are taken into account in the subtraction process.

Implementation of the subtraction approach inside RELION (Bai, Rajendra, Yang, Shi, & Scheres, 2015) recently allowed separation of three different conformations of the human gamma-secretase complex. In this case, a masked 3D classification was performed on the catalytic subunit, while projections of the other three subunits were subtracted from the experimental particle images. Because the region with the mask of the catalytic subunit only comprised approximately 30 kDa, orientations were kept fixed at those obtained in a consensus refinement of the entire (170 kDa) complex. After identification of three distinct conformations of the catalytic subunit, unmasked 3D auto-refinement of the original experimental particle images led to three reconstructions of the entire complex to near-atomic resolution. The observation that these structures differed only in the orientation of a few alpha-helices illustrates the exciting potential of masked classification with partial signal subtraction in separating cryo-EM images which differ only at the secondary structure level. The same approach was also used by Ilca and colleagues to subtract viral capsid densities in order to visualize an RNA polymerase bound inside double-stranded RNA bacteriophage φ6 (Ilca et al., 2015), while Zhou and colleagues used their own modified version of RELION to subtract projections of NSF rings to analyze structural variability in the SNAP–SNARE complex (Zhou, Huang, et al., 2015).

4.7 Dealing with Pseudo-Symmetry

Many macromolecular complexes adopt some form of symmetry, but often this symmetry is only true for part of the complex. A generally applicable approach to deal with pseudo-symmetric complexes was recently used to solve the structure of a human apoptosome complex (Zhou, Li, et al., 2015). In this case, seven copies of the central protein form a ring-like hub with C7 symmetry. However, this symmetry is broken due to the dynamic nature of seven protruding spokes. Because each of the seven spokes seems to flex with respect to the central hub in an independent manner from the other six spokes, none of the complexes is truly symmetric. Consequently, a consensus refinement with imposed C7 symmetry led to a map with an overall resolution of 3.8 Å, where the central hub was well-resolved but the resolution for the spokes was considerably lower.

In order to deal with the pseudo-symmetry, the dataset was artificially expanded according to the pseudo-symmetric C7 point group. In this case, the dataset was enlarged sevenfold, by replicating each particle and adding $(n/7)*360°$ with $n=1,\ldots,7$ to the first Euler angle of every particle from the C7 consensus refinement. In this way, each of the seven spokes of every particle is oriented onto every position on the C7-symmetric ring. Masked classification within a single spoke and without angular searches then resulted in the identification of a subset of all spokes that were in a comparatively highly populated conformation relative to the symmetrical ring. Subsequent masked 3D auto-refinement in C1, where the mask included the entire central hub plus the single protruding spoke, yielded a reconstruction in which the spoke was determined to a resolution of 5 Å, which allowed docking of available crystal structures with confidence. Note that this C1 refinement was done with only local angular searches around the expanded set of orientations in order to prevent copies of the same particle from contributing to the reconstruction more than once in the same orientation. Because symmetry expansion is not as self-evident for noncyclic space groups, a stand-alone command line program was implemented to expand a dataset for any given symmetry point group. This program, which is called `relion_particle_symmetry_expand`, is already available upon request and will be part of RELION-2.0.

It should be noted that this approach shows similarities with an earlier approach to deal with pseudo-symmetric structures that was described by

Briggs et al. (2005). In their study of Kelp fly virus capsids, they identified two different types of fivefold vertices in the pseudo-icosahedral capsids. They used the orientations from an icosahedral consensus refinement to extract subimages of all vertices of each capsid, and used the known orientations from the icosahedral refinement to select similarly oriented vertices for multivariate statistical analysis and classification. Refinements of the identified classes were then performed exploiting the known orientations of the vertices with respect to the entire capsid.

Although not used for the apoptosome structure, the symmetry-expansion approach can also be combined with the partial signal subtraction approach. One could, for example, expand the set of orientations according to the broken point-group symmetry, and then subtract the consensus density for the symmetric part of the structure plus all-but-one of the asymmetric features from the experimental particle images. Then, one would perform a masked 3D classification or auto-refinement on the remaining asymmetrical feature, possibly followed by a masked 3D auto-refinement that includes both the symmetrical part of the complex and the single asymmetrical feature.

4.8 Multibody Refinement

At an early stage of the structure determination process of the spliceosomal U4/U6.U5 tri-snRNP particle from the Nagai lab (see also below) an automated approach to iterative masked refinements with partial signal subtraction from multiple regions was explored (Nguyen et al., 2015). The spliceosome has been an example of a notoriously difficult cryo-EM sample for many years due to its extremely dynamic nature (Lührmann & Stark, 2009). The 1.4 MDa Y-shaped tri-snRNP complex may be considered to consist of four more-or-less independently moving regions: a central "body"; a more flexible "foot" and "head"; and an extremely flexible "arm." In the automated approach, four regions were refined in parallel using masked 3D auto-refinement with partial signal subtraction. At every iteration, a mask around each of the four regions was applied in turn, and the signal of the other three regions was subtracted from every particle. Masked alignment of the four independent regions at every iteration led to four sets of optimal orientations for every particle, one orientation for each region. These optimal orientations were then used to perform a better signal subtraction in the next iteration. This still experimental implementation, which was called multibody refinement, holds the promise to automatically refine multiple independently moving regions, or bodies, almost without user-intervention. However, the current implementation still suffers from strong artifacts at the boundaries between the regions, and thus requires further development to make it generally applicable.

4.9 A More Elaborate Example

A recent near-atomic resolution structure of the spliceosomal U4/U6.U5 tri-snRNP complex from the Nagai lab (Nguyen et al., 2016) represents an insightful example of how several of the approaches described above may be combined to describe molecular motions in a highly dynamic complex. Despite the extensive amount of flexibility in this complex, maps in which the majority of the complex could be resolved to near-atomic resolution could be obtained using the procedures that are outlined in Fig. 5.

After following the standard procedures as outlined in Fig. 1, a refinement of 140 thousand shiny particles led to a consensus map with an overall resolution of 3.7 Å. However, due to apparent flexibility in the complex, the reconstructed density in large portions of the map was not of sufficient quality to allow building of an atomic model. Separate masked 3D auto-refinements on the body, the foot and the head, with subtraction of the signal corresponding to the other regions, resulted in improved reconstructions of these three regions to local resolutions of 3.6, 3.7, and 4.2 Å, respectively. The improvement in density was largest for the head region, presumably because of its very dynamic attachment to the body. Whereas large parts of the head domain were only resolved to 8–10 Å resolution in the consensus map, the 4.2 Å reconstruction from the masked refinement with partial signal subtraction allowed the building of an atomic model for most of the protein and RNA in this region.

Because the arm region is even more flexible than the head, and because of its relatively small size, masked 3D auto-refinement of this region did not work well. Instead, masked 3D classification with subtraction of the signal from the other three regions was performed on this region. In an initial classification, a mask around only the arm region was used to divide the data according to three different relative orientations of the arm with respect to the body. Subsequent subtraction of only the head and the foot region, combined with masked 3D auto-refinement of the body and the arm regions together yielded three overall resolutions of 4.5–4.6 Å. Local resolution estimates indicated that the arm regions in these maps were resolved to resolutions in the range of 6.2–7.5 Å.

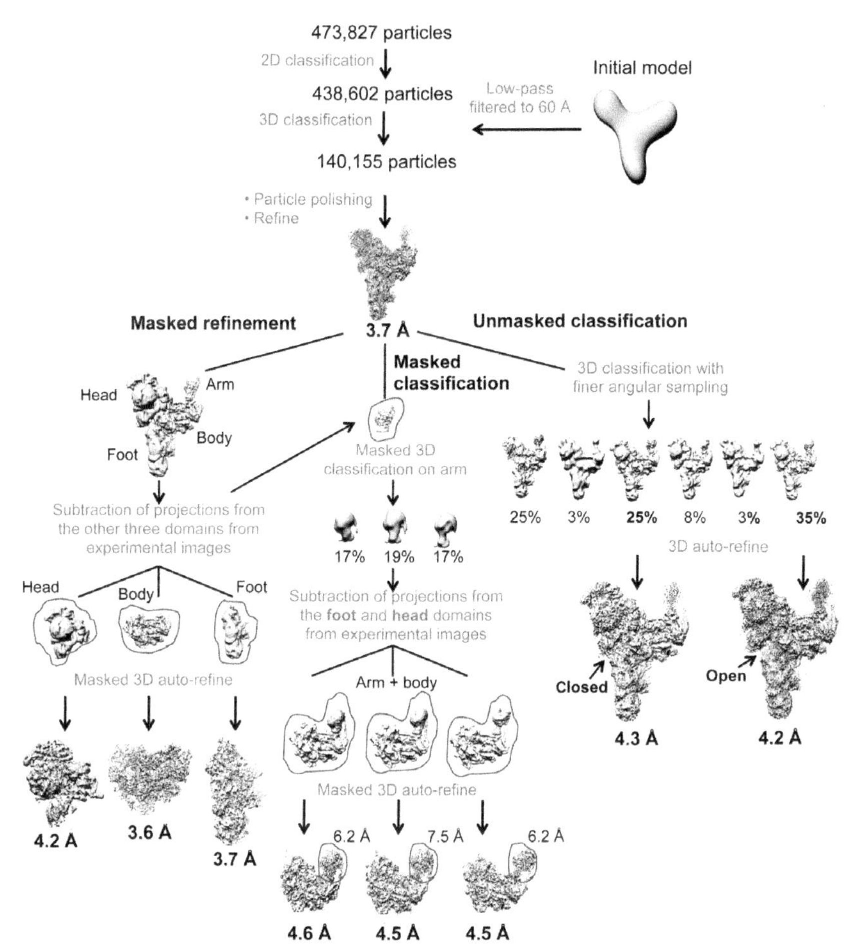

Fig. 5 An example of how to deal with structural heterogeneity is shown in the form of a flowchart for the processing of cryo-EM data on the spliceosomal U4/U6.U5 tri-snRNP complex. The various steps are described in more detail in the main text. *This figure was adapted from Nguyen, T.H.D., Galej, W.P., Bai, X.-C., Oubridge, C., Newman, A.J., Scheres, S.H.W., et al. (2016). Cryo-EM structure of the yeast U4/U6.U5 tri-snRNP at 3.7 Å resolution.* Nature, 530, *298–302. doi:10.1038/nature16940.*

In addition to the masked 3D classifications and auto-refinements, an unmasked 3D classification with 1.8 degree local angular searches centered around the orientations from the consensus refinement of the entire complex was also useful to identify an open and a closed conformation of the body and the head regions.

5. OUTLOOK

RELION is still under active development. Its high computational costs (typically ~100–200 thousand CPU hours to process a single dataset) are currently an important bottleneck, especially for those labs that do not have access to large computer clusters. Ongoing efforts in vectorization of the code and its implementation for graphical processing units (GPUs) will probably lead to significant reductions in these costs in the near future. This will allow on-the-fly processing of micrographs during data acquisition, and procedures to automatically execute a predefined workflow of individual tasks are currently being developed. In addition, new algorithmic developments to handle helical assemblies; new approaches for electron tomography (Bharat, Russo, Löwe, Passmore, & Scheres, 2015); and further developments of multibody refinement and other classification tools for flexible complexes are all active areas of research. Hopefully, these developments will continue to push the boundaries of cryo-EM structure determination.

ACKNOWLEDGMENTS

I thank Drs. Xiao-chen Bai, Anthony Fitzpatrick, and Titia Sixma for critical comments on the chapter. This work was funded by the UK Medical Research Council (MC_UP_A025_1013).

REFERENCES

Bai, X.-C., Fernandez, I. S., McMullan, G., & Scheres, S. H. (2013). Ribosome structures to near-atomic resolution from thirty thousand cryo-EM particles. *eLife*, *2*, e00461. http://dx.doi.org/10.7554/eLife.00461.

Bai, X., McMullan, G., & Scheres, S. H. W. (2015). How cryo-EM is revolutionizing structural biology. *Trends in Biochemical Sciences*, *40*, 49–57. http://dx.doi.org/10.1016/j.tibs.2014.10.005.

Bai, X., Rajendra, E., Yang, G., Shi, Y., & Scheres, S. H. (2015). Sampling the conformational space of the catalytic subunit of human γ-secretase. *eLife*. *4*. http://dx.doi.org/10.7554/eLife.11182.

Bharat, T. A. M., Russo, C. J., Löwe, J., Passmore, L. A., & Scheres, S. H. W. (2015). Advances in single-particle electron cryomicroscopy structure determination applied to sub-tomogram averaging. *Structure*, *23*, 1743–1753. http://dx.doi.org/10.1016/j.str.2015.06.026.

Briggs, J. A. G., Huiskonen, J. T., Fernando, K. V., Gilbert, R. J. C., Scotti, P., Butcher, S. J., et al. (2005). Classification and three-dimensional reconstruction of unevenly distributed or symmetry mismatched features of icosahedral particles. *Journal of Structural Biology*, *150*, 332–339. http://dx.doi.org/10.1016/j.jsb.2005.03.009.

Brilot, A. F., Chen, J. Z., Cheng, A., Pan, J., Harrison, S. C., Potter, C. S., et al. (2012). Beam-induced motion of vitrified specimen on holey carbon film. *Journal of Structural Biology*, *177*, 630–637. http://dx.doi.org/10.1016/j.jsb.2012.02.003.

Campbell, M. G., Cheng, A., Brilot, A. F., Moeller, A., Lyumkis, D., Veesler, D., et al. (2012). Movies of ice-embedded particles enhance resolution in electron cryo-microscopy. *Structure, 20,* 1823–1828. http://dx.doi.org/10.1016/j.str.2012.08.026.

Chen, S., McMullan, G., Faruqi, A. R., Murshudov, G. N., Short, J. M., Scheres, S. H. W., et al. (2013). High-resolution noise substitution to measure overfitting and validate resolution in 3D structure determination by single particle electron cryomicroscopy. *Ultramicroscopy, 135,* 24–35. http://dx.doi.org/10.1016/j.ultramic.2013.06.004.

Dempster, A. P., Laird, N. M., & Rubin, D. B. (1977). Maximum-likelihood from incomplete data via the EM algorithm. *Journal of the Royal Statistical Society, Series B, 39,* 1–38.

Elmlund, D., Davis, R., & Elmlund, H. (2010). Ab initio structure determination from electron microscopic images of single molecules coexisting in different functional states. *Structure, 18,* 777–786.

Fernandez-Leiro, R., Conrad, J., Scheres, S. H., & Lamers, M. H. (2015). cryo-EM structures of the E. coli replicative DNA polymerase reveal its dynamic interactions with the DNA sliding clamp, exonuclease and τ. *eLife. 4.* http://dx.doi.org/10.7554/eLife.11134.

Goddard, T. D., Huang, C. C., & Ferrin, T. E. (2007). Visualizing density maps with UCSF Chimera. *Journal of Structural Biology, 157,* 281–287. http://dx.doi.org/10.1016/j.jsb.2006.06.010.

Gorski, K. M., Hivon, E., Banday, A. J., Wandelt, B. D., Hansen, F. K., Reinecke, M., et al. (2005). HEALPix—A framework for high resolution discretization, and fast analysis of data distributed on the sphere. *The Astrophysical Journal, 622,* 759–771. http://dx.doi.org/10.1086/427976. ArXivastro-Ph0409513.

Harauz, G., & van Heel, M. (1986). Exact filters for general geometry three dimensional reconstruction. *Optik, 73,* 146–156.

Henderson, R. (2013). Avoiding the pitfalls of single particle cryo-electron microscopy: Einstein from noise. *Proceedings of the National Academy of Sciences of the United States of America, 110,* 18037–18041. http://dx.doi.org/10.1073/pnas.1314449110.

Henderson, R., Chen, S., Chen, J. Z., Grigorieff, N., Passmore, L. A., Ciccarelli, L., et al. (2011). Tilt-pair analysis of images from a range of different specimens in single-particle electron cryomicroscopy. *Journal of Molecular Biology, 413,* 1028–1046. http://dx.doi.org/10.1016/j.jmb.2011.09.008.

Henderson, R., Sali, A., Baker, M. L., Carragher, B., Devkota, B., Downing, K. H., et al. (2012). Outcome of the first electron microscopy validation task force meeting. *Structure, 20,* 205–214. http://dx.doi.org/10.1016/j.str.2011.12.014.

Ilca, S. L., Kotecha, A., Sun, X., Poranen, M. M., Stuart, D. I., & Huiskonen, J. T. (2015). Localized reconstruction of subunits from electron cryomicroscopy images of macromolecular complexes. *Nature Communications. 6.* http://dx.doi.org/10.1038/ncomms9843. article number 8843.

Kucukelbir, A., Sigworth, F. J., & Tagare, H. D. (2014). Quantifying the local resolution of cryo-EM density maps. *Nature Methods, 11,* 63–65. http://dx.doi.org/10.1038/nmeth.2727.

Li, X., Mooney, P., Zheng, S., Booth, C. R., Braunfeld, M. B., Gubbens, S., et al. (2013). Electron counting and beam-induced motion correction enable near-atomic-resolution single-particle cryo-EM. *Nature Methods, 10,* 584–590. http://dx.doi.org/10.1038/nmeth.2472.

Ludtke, S. J., Baldwin, P. R., & Chiu, W. (1999). EMAN: Semiautomated software for high-resolution single-particle reconstructions. *Journal of Structural Biology, 128,* 82–97. http://dx.doi.org/10.1006/jsbi.1999.4174.

Lührmann, R., & Stark, H. (2009). Structural mapping of spliceosomes by electron microscopy. *Current Opinion in Structural Biology, 19,* 96–102. http://dx.doi.org/10.1016/j.sbi.2009.01.001.

Lyumkis, D., Brilot, A. F., Theobald, D. L., & Grigorieff, N. (2013). Likelihood-based classification of cryo-EM images using FREALIGN. *Journal of Structural Biology, 183*, 377–388. http://dx.doi.org/10.1016/j.jsb.2013.07.005.

Mindell, J. A., & Grigorieff, N. (2003). Accurate determination of local defocus and specimen tilt in electron microscopy. *Journal of Structural Biology, 142*, 334–347. http://dx.doi.org/10.1016/S1047-8477(03)00069-8.

Morais, M. C., Kanamaru, S., Badasso, M. O., Koti, J. S., Owen, B. A. L., McMurray, C. T., et al. (2003). Bacteriophage φ29 scaffolding protein gp7 before and after prohead assembly. *Nature Structural & Molecular Biology, 10*, 572–576. http://dx.doi.org/10.1038/nsb939.

Nguyen, T. H. D., Galej, W. P., Bai, X.-C., Oubridge, C., Newman, A. J., Scheres, S. H. W., et al. (2016). Cryo-EM structure of the yeast U4/U6.U5 tri-snRNP at 3.7 Å resolution. *Nature, 530*, 298–302. http://dx.doi.org/10.1038/nature16940.

Nguyen, T. H. D., Galej, W. P., Bai, X., Savva, C. G., Newman, A. J., Scheres, S. H. W., et al. (2015). The architecture of the spliceosomal U4/U6.U5 tri-snRNP. *Nature, 523*, 47–52. http://dx.doi.org/10.1038/nature14548.

Penczek, P. A. (2010a). Image restoration in cryo-electron microscopy. In G. Jensen (Ed.), *Methods in enzymology. Cryo-EM, Part B: 3-D reconstruction* (pp. 35–72): Academic Press.

Penczek, P. A. (2010b). Fundamentals of three-dimensional reconstruction from projections. In G. Jensen (Ed.), *Methods in enzymology. Cryo-EM, Part B: 3-D reconstruction* (pp. 1–33): Academic Press.

Pettersen, E. F., Goddard, T. D., Huang, C. C., Couch, G. S., Greenblatt, D. M., Meng, E. C., et al. (2004). UCSF Chimera—A visualization system for exploratory research and analysis. *Journal of Computational Chemistry, 25*, 1605–1612. http://dx.doi.org/10.1002/jcc.20084.

Radermacher, M., Wagenknecht, T., Verschoor, A., & Frank, J. (1987). Three-dimensional reconstruction from a single-exposure, random conical tilt series applied to the 50S ribosomal subunit of Escherichia coli. *Journal of Microscopy, 146*, 113–136.

Rohou, A., & Grigorieff, N. (2015). CTFFIND4: Fast and accurate defocus estimation from electron micrographs. *Journal of Structural Biology, 192*, 216–221. http://dx.doi.org/10.1016/j.jsb.2015.08.008.

Rosenthal, P. B., & Henderson, R. (2003). Optimal determination of particle orientation, absolute hand, and contrast loss in single-particle electron cryomicroscopy. *Journal of Molecular Biology, 333*, 721–745.

Scheres, S. H. W. (2010). Classification of structural heterogeneity by maximum-likelihood methods. In G. Jensen (Ed.), *Methods in enzymology. Cryo-EM, Part B: 3-D reconstruction* (pp. 295–320): Academic Press.

Scheres, S. H. W. (2012a). A Bayesian view on cryo-EM structure determination. *Journal of Molecular Biology, 415*, 406–418. http://dx.doi.org/10.1016/j.jmb.2011.11.010.

Scheres, S. H. W. (2012b). RELION: Implementation of a Bayesian approach to cryo-EM structure determination. *Journal of Structural Biology, 180*, 519–530. http://dx.doi.org/10.1016/j.jsb.2012.09.006.

Scheres, S. H. W. (2014). Beam-induced motion correction for sub-megadalton cryo-EM particles. *eLife, 3*, e03665.

Scheres, S. H. W. (2015). Semi-automated selection of cryo-EM particles in RELION-1.3. *Journal of Structural Biology, 189*, 114–122. http://dx.doi.org/10.1016/j.jsb.2014.11.010.

Scheres, S. H. W., & Chen, S. (2012). Prevention of overfitting in cryo-EM structure determination. *Nature Methods, 9*, 853–854. http://dx.doi.org/10.1038/nmeth.2115.

Scheres, S. H. W., Gao, H., Valle, M., Herman, G. T., Eggermont, P. P. B., Frank, J., et al. (2007). Disentangling conformational states of macromolecules in 3D-EM through likelihood optimization. *Nature Methods, 4*, 27–29. http://dx.doi.org/10.1038/nmeth992.

Scheres, S. H. W., Nunez-Ramirez, R., Gomez-Llorente, Y., San Martin, C., Eggermont, P. P. B., & Carazo, J. M. (2007). Modeling experimental image formation for likelihood-based classification of electron microscopy data. *Structure*, *15*, 1167–1177.

Scheres, S. H. W., Nunez-Ramirez, R., Sorzano, C. O. S., Carazo, J. M., & Marabini, R. (2008). Image processing for electron microscopy single-particle analysis using XMIPP. *Nature Protocols*, *3*, 977–990.

Sigworth, F. J. (1998). A maximum-likelihood approach to single-particle image refinement. *Journal of Structural Biology*, *122*, 328–339.

Sigworth, F. J., Doerschuk, P. C., Carazo, J.-M., & Scheres, S. H. W. (2010). An introduction to maximum-likelihood methods in Cryo-EM. In G. Jensen (Ed.), *Methods in enzymology. Cryo-EM, Part B: 3-D reconstruction* (pp. 263–294): Academic Press.

Tang, G., Peng, L., Baldwin, P. R., Mann, D. S., Jiang, W., Rees, I., et al. (2007). EMAN2: An extensible image processing suite for electron microscopy. *Journal of Structural Biology*, *157*, 38–46. http://dx.doi.org/10.1016/j.jsb.2006.05.009.

Tikhonov, A. N. (1943). On the stability of inverse problems. *Doklady Akademii Nauk SSSR*, *39*, 195–198.

Urnavicius, L., Zhang, K., Diamant, A. G., Motz, C., Schlager, M. A., Yu, M., et al. (2015). The structure of the dynactin complex and its interaction with dynein. *Science*, *347*, 1441–1446. http://dx.doi.org/10.1126/science.aaa4080.

van Heel, M. (2013). Finding trimeric HIV-1 envelope glycoproteins in random noise. *Proceedings of the National Academy of Sciences of the United States of America*, *110*, E4175–E4177.

Walz, J., Typke, D., Nitsch, M., Koster, A. J., Hegerl, R., & Baumeister, W. (1997). Electron tomography of single ice-embedded macromolecules: Three-dimensional alignment and classification. *Journal of Structural Biology*, *120*, 387–395. http://dx.doi.org/10.1006/jsbi.1997.3934.

Wong, W., Bai, X.-C., Brown, A., Fernandez, I. S., Hanssen, E., Condron, M., et al. (2014). Cryo-EM structure of the Plasmodium falciparum 80S ribosome bound to the anti-protozoan drug emetine. *eLife*, *3*, e03080. http://dx.doi.org/10.7554/eLife.03080.

Zhang, K. (2016). Gctf: Real-time CTF determination and correction. *Journal of Structural Biology*, *193*, 1–12. http://dx.doi.org/10.1016/j.jsb.2015.11.003.

Zhang, Y., Kostyuchenko, V. A., & Rossmann, M. G. (2007). Structural analysis of viral nucleocapsids by subtraction of partial projections. *Journal of Structural Biology*, *157*, 356–364. http://dx.doi.org/10.1016/j.jsb.2006.09.002.

Zhou, Q., Huang, X., Sun, S., Li, X., Wang, H.-W., & Sui, S.-F. (2015). Cryo-EM structure of SNAP-SNARE assembly in 20S particle. *Cell Research*, *25*, 551–560. http://dx.doi.org/10.1038/cr.2015.47.

Zhou, M., Li, Y., Hu, Q., Bai, X.-C., Huang, W., Yan, C., et al. (2015). Atomic structure of the apoptosome: Mechanism of cytochrome c- and dATP-mediated activation of Apaf-1. *Genes & Development*, *29*, 2349–2361. http://dx.doi.org/10.1101/gad.272278.115.

CHAPTER SEVEN

Single-Particle Refinement and Variability Analysis in EMAN2.1

S.J. Ludtke[1]
National Center for Macromolecular Imaging, Baylor College of Medicine, Houston, TX, United States
[1]Corresponding author: e-mail address: sludtke@bcm.edu

Contents

Abstract

CryoEM single-particle reconstruction has been growing rapidly over the last 3 years largely due to the development of direct electron detectors, which have provided data with dramatic improvements in image quality. It is now possible in many cases to produce near-atomic resolution structures, and yet 2/3 of published structures remain at substantially lower resolutions. One important cause for this is compositional and conformational heterogeneity, which is both a resolution-limiting factor and presenting a unique opportunity to better relate structure to function. This manuscript discusses the

Methods in Enzymology, Volume 579
ISSN 0076-6879
http://dx.doi.org/10.1016/bs.mie.2016.05.001

canonical methods for high-resolution refinement in EMAN2.12, and then considers the wide range of available methods within this package for resolving structural variability, targeting both improved resolution and additional knowledge about particle dynamics.

1. INTRODUCTION

1.1 History

The first version of EMAN (Ludtke, Baldwin, & Chiu, 1999) was developed in 1998. At that time SPIDER (Frank et al., 1996; Shaikh et al., 2008) and IMAGIC (van Heel, Harauz, Orlova, Schmidt, & Schatz, 1996) were the primary nonicosahedral single-particle reconstruction packages in use, both of which had a steep learning curve, and required substantial manual effort. EMAN was the first of these packages to offer a substantial graphical interface (GUI) and gave the end user a straightforward path to follow toward 3-D reconstruction. It is important to note that at this time subnanometer resolution was just emerging, and near-atomic resolution, while theorized as possible (Henderson, 1995), was still in the distant future. EMAN's contributions at the time, in addition to its ease of use, was its integrated CTF correction strategy (Ludtke, Jakana, Song, Chuang, & Chiu, 2001), and a single-integrated refinement program which automated all of the steps in the reconstruction process. The combination of supervised classification and iterative class averaging to reduce model bias and speed convergence were also key features leading to its popularity.

The first version of EMAN2 (Tang et al., 2007) was developed in the mid-2000s and was a complete refactoring of EMAN1 into a hybrid C++/Python environment with a modern object-oriented design. One of the motivations in this development was a collaborative effort to make EMAN2 work with the PHENIX X-ray crystallography package, under an effort led by Dr. Robert Glaeser. This same effort brought Dr. Pawel Penczek into the collaborative effort, which allowed EMAN2 and SPARX (Hohn et al., 2007) to share a common C++ image processing core and Python library, which continues to this day. While the collaboration with PHENIX proved to be a bit premature at that time, since few CryoEM projects were achieving resolutions where X-ray tools were useful, it was an accurate predictor of the future, as PHENIX is now a frequently used tool for performing real-space refinement of CryoEM-derived models and other tasks. SPARX was then ceded to Penczek's group, which continues active development.

The development of EMAN2.1, first released in 2013, transitioned from the unpopular BDB database storage system used in EMAN2.0 to the interdisciplinary HDF5 image format. In addition, EMAN2.1 included automatic resolution estimation during each refinement iteration using the "gold-standard" FSC, automatic refinements using a system of heuristics, which could provide virtually all of the refinement parameters automatically, a new optimized orientation determination system an order of magnitude faster than the methods in EMAN1, as well as many other improvements.

The EMDatabank (Lawson et al., 2016) now identifies over 50 software packages which have some use in CryoEM reconstruction. The seven most-used of these packages, EMAN, RELION (Scheres, 2012), SPIDER (Shaikh et al., 2008), SPARX (Hohn et al., 2007), IMAGIC (van Heel et al., 1996), FREALIGN (Grigorieff, 2007), and XMIPP (de la Rosa-Trevín et al., 2013), account for over 97% of the published single-particle reconstructions in 2015.

1.2 The Scope of the Problem

Direct detectors and current generation microscopes are producing data of unprecedented quality, and indeed over 20% of the maps published in 2015 were at a resolution of 4 Å or better, with another 14% at 6 Å or better. Given the state of the art, one might ask why 2/3 of structures are still published when they are at comparatively low resolution. One theory for this phenomenon might be insufficient access to current generation hardware; however, while the statistics for improved microscopes and detectors are skewed slightly toward high resolution, only half of the structures published using a FEI Krios in 2016 surpassed 6 Å resolution. Similar statistics apply for use of Gatan K2 Summit detectors. So, even with current generation hardware it seems near-atomic resolution is not assured. While there may be specimen preparation issues, such as ice thickness or low contrast due to buffer constituents (detergent being a particular problem), it is hoped that these problems are largely being resolved on far less expensive equipment, similar to the use of tabletop crystallography machines to verify crystal quality prior to using time on a beamline. The most likely remaining resolution-limiting factor is conformational and compositional variability in the specimen.

An assumption fostered by the long success of crystallography over recent decades is the implicit assumption that structures are rigid, or at least fairly rigid, such that a single structure exists to determine. Even with the recognition of many examples where large-scale motion is an integral part of the function, it is still common to refer to *the* structure of a macromolecule, and deal with local motion in terms of B-factors describing Gaussian deviations from the true location of each atom, though use of the more extensive TLS analysis (Chaudhry, Horwich, Brunger, & Adams, 2004) is becoming more common. Since the vitreous state used in CryoEM mimics a native aqueous environment, and lacks the crystal packing forces in crystallography, it is not surprising that substantial flexibility is observed. Characterizing this motion is not just a mechanism for improving resolution, but can reveal fundamental insights into the structure–function relationship.

Even if a user's first experience with CryoEM turns out to be straightforward and rapidly achieves high resolution, it is important to remember that this is relatively uncommon, and it will frequently be necessary to invoke a variety of experimental methods and software tools. In the case of a specimen which is extremely rigid in solution, and data have been collected on high-end equipment, in our experience the various algorithms and software packages used in CryoEM tend to produce equivalent structures, in terms of both measured resolution and visible features. However, if the specimen has either compositional or conformational variability, then it is very common for various algorithms to explore the variability in different ways and produce visibly different structures. Understanding the nature of the variability through the differences among software packages, algorithms or even additional experiments are frequently required to obtain a complete and correct understanding of the system.

1.3 EMAN2.1 Philosophy

The overall philosophy of processing in EMAN2 would be best termed "guided flexibility." EMAN2 offers tools at every level ranging from GUIs to low-level image processing algorithms to encourage and support detailed exploration of experimental CryoEM and CryoET data. There is a simple to use workflow interface (*e2projectmanager*) to guide users through a range of different tasks from evaluation of micrograph quality through high-resolution 3-D reconstruction and heterogeneity analysis. Tools like *e2refine_easy* (Bell, Chen, Baldwin, & Ludtke, 2016) and *e2refine2d* provide quick and automatic processing of single-particle data (in 3-D and 2-D), similar in function to the refinement programs offered in packages focusing on a single algorithm, such as RELION (Scheres, 2012) (and chapter "Processing of Structurally Heterogeneous Cryo-EM Data in RELION" by Scheres) and FREALIGN (Grigorieff, 2007) (and chapter "Frealign: An Exploratory Tool for Single Particle Cryo-EM" by Grigorieff).

e2refine_easy has an extensive set of heuristics, such that for well-behaved projects, almost no options need be specified by the user. However, when projects prove to be more complicated than initially anticipated, it also offers a diverse set of options and alternative workflows, which can be used to gain more insight into the behavior of the specimen. Since there can be considerable value in comparative refinements, there is also an extensive set of interoperability tools making it straightforward to interoperate/interconvert among EMAN, RELION, FREALIGN, and CTFFIND3/4 (Mindell & Grigorieff, 2003; Rohou & Grigorieff, 2015). Since its inception, EMAN2 and SPARX have been jointly distributed, and while they have evolved different philosophies, different end-user programs, and different pipelines, they share a common core image processing system, Python programming environment, and image access. It is also possible to seamlessly exchange image data between EMAN2 and the popular NumPy/SciPy environments. Finally, EMAN2 has committed to supporting EMX (Marabini et al., 2016), the new electron microscopy exchange format for metadata interchange among CryoEM software packages.

At the highest level the graphical user interface *e2projectmanager* can guide the user through a canonical single-particle refinement starting with micrographs, particle locations, particles, or even particles with predetermined CTF information. This pipeline includes several 2-D analysis steps including identification of bad particles and de novo initial model generation. These pipelines are designed such that each step in the process should be straightforward, and in most cases alternative algorithms can be trivially substituted into the workflow. If a user believes CTFFIND3/4 will compute more accurate astigmatism values than the built-in particle-based approach, there is a tool for importing these values and automatically computing the additional particle-based values EMAN2 requires, while retaining defocus and astigmatism from the external software. If a user finds they prefer RELION's 2-D class-averages to EMAN2's, they can easily take that diversion. In short, EMAN2 not only offers its own complete pipeline but also endeavors to make it simple to compare tools among multiple packages.

When tasks are initiated in the graphical workflow interface, in reality the interface is simply building commands then executing them on the local machine. Any of these commands can, instead, be copied from the GUI for execution on a Linux cluster or other high-performance computer, as desired. The majority of the processing in EMAN2.1 can be completed on a mid-range or high-end workstation, with the exception of final high-resolution refinements for large projects. The current EMAN2.1 tutorial makes use of a subset of ~4200 beta-galactosidase particles drawn from the 2015 EMDatabank Map Challenge (http://challenges.emdatabank.org), which can be processed to ~4 Å resolution on a workstation in a few hours. Clearly there are larger projects which require clusters for high-resolution refinement, particularly in cases of variability, where very large data sets must be divided into several subpopulations. For such uses, the system supports both MPI and threaded parallelism. All executed EMAN2 commands are logged when executed, and processes like *e2refine_easy* further produce an output report with resolution plots as well as a detailed description of the methodologies employed during the refinement, and often suggestions for proceeding.

While there is a complex system of heuristics in place for many of the high-level tasks in the system, all of these can be overridden by the user, should the need arise. For example, a canonical *e2refine_easy* refinement run requires only: starting model, input particle stack, particle mass, target resolution, symmetry, number of iterations, "speed," and parallelism settings. All other settings can be determined automatically based on the properties of the particle. However if desired, this command offers ~40 additional documented options which can be specified should the user elect to do so. It must also be noted that the automatic system records all of the options selected by the heuristics, so one can easily see which options were selected automatically as a starting point for manual adjustments. For the majority of structures, this detailed option setting will not be useful, and should the user find it necessary to do so to achieve optimal results, they are encouraged to report what they changed and why, for use in improving future versions of the heuristics.

The system has a highly modular core, with methods for specific tasks organized by name and categorized by task (Tang et al., 2007). If someone were to design, for example, a new 3-D reconstruction algorithm, it would be added to the system as a new named option, and would immediately be usable in any program within the system, with no additional reprogramming required. The built-in *e2help* command provides documentation for each method of each type, as does the built-in help system in *e2projectmanager.*

2. SINGLE-PARTICLE RECONSTRUCTION

We begin with a brief summary of EMAN2.1's approach to a typical single-particle reconstruction, then move on to assessing the structure, and finally, what to do when structures prove to be less tractable than hoped.

Details on the internal workings of the processes invoked in this workflow can be found in other sources (Bell et al., 2016). Here we consider the broader perspective of the diverse set of situations which may be encountered in single-particle analysis. Normally the most convenient way to complete a standard refinement is to use the *e2projectmanager* interface (shown later in Fig. 4). This interface is useful for both beginners and experienced users, but exposes only the commonly used options in each program. Exactly the same process can be followed via a sequence of command-line programs. Detailed tutorials, both textual and video, are available online and are updated regularly (http://wiki.eman2.org).

The general sequence of steps in a simple single-particle refinement is:

Movie alignment (if applicable): If the data were collected on a direct detector in movie mode, whole-frame movie alignment is the first step. Tools for optimal alignment remain an active area of research in the community (see chapter "Processing of Cryo-EM Movie Data" by Rubinstein). EMAN2.1 has *e2ddd_movie* and *e2ddd_particles*, but any other alignment tool can also be used for this task.

Micrograph assessment: This includes determining defocus and (optionally) astigmatism as well as assessing the suitability of the image. EMAN2.1 has *e2evalimage* and *e2rawdata* for interactive or noninteractive processing. CTFFIND or EMX files can also be imported.

Particle picking: Many tools exist for this. EMAN2.1 includes *e2boxer* for both manual and semiautomatic picking. SPARX has a module for fully automatic picking. Any other tool can be used as well. EMAN1 .box files and EMX files can be imported, as can already extracted particle stacks. Note that it is absolutely critical for EMAN2.1 that box-size recommendations be followed. If not, results are guaranteed to be suboptimal (http://www.eman2.org/emanwiki/EMAN2/BoxSize).

Particle-based CTF and SSNR: If the defocus and astigmatism have not already been determined in a previous step, EMAN2.1 uses particle-based determination for this process. If existing values are provided, they may be used without modification, in which case this step determines the radial SSNR per-micrograph and an estimated radial structure factor. This step must be performed in EMAN2.1, using *e2ctf*, and is performed automatically by several of the particle import procedures.

Set building: In EMAN2.1, a set is a text file containing references to an arbitrary fraction of the particles the project. Conceptually this is somewhat similar to the STAR files used in some of the other CryoEM packages, but a major difference is that the .lst files can be used directly as if they were an image stacks by any EMAN2.1 program. This mechanism can be also used to coordinate multiple versions of particles, for example, one set of particles based on only the first few movie frames for reconstruction and another set using all frames (high dose) for orientation determination. Implemented in *e2buildsets*.

2-D unsupervised class-averaging: This process, implemented in *e2refine2d*, is used for two purposes. First, it can identify specific types of bad particles and mark them as such so they are excluded from future sets. Second, it can be used to assess the amount of compositional and conformational variability in the specimen. Since this process does not require the 2-D averages to be based on a consistent 3-D map, they are free to express the full range of variations present among the particles.

Initial model generation: The programs *e2initialmodel* and *e2initialmodel_hisym* normally serve this purpose, but a starting model can be drawn from any other source as well. See later for a discussion of model bias.

3-D refinement: As described earlier, *e2refine_easy* is the canonical program for this, assuming a highly homogeneous data set. This is also a possible branch point for running comparative refinements in other packages.

Evaluate 3-D reconstruction: Validation and considering possible compositional or conformational variability are critical. These issues are discussed in the remainder of this chapter.

It is important to note that EMAN2.1 uses a standard directory structure for all projects (http://www.eman2.org/emanwiki/EMAN2/DirectoryStructure), and the user has limited freedom to deviate from this during canonical workflows. This is not as onerous as it sounds. It is simply a standard naming convention for files and folders. Each macromolecular system should have its own project directory, and all *e2** programs should normally be run from this directory, not from any of the subdirectories. However, when using EMAN2.1 as a utility, rather than following a workflow, programs can be run from any location on the hard drive. Utilities such as *e2display*, *e2proc2d*, *e2proc3d*, *e2procpdb*, *e2proclst*, *e2procxml*, *e2filtertool*, and *e2iminfo* can be useful even if using another software package for 3-D refinement.

Internally, EMAN2.1 uses the HDF5 file format as a standard for all images and volumes. This format supports stacks of 1–3-D images with arbitrary tagged metadata in the header of each image. It is also compatible with UCSF Chimera, so no conversion is required before visualization. EMAN2.1 still supports all documented file formats in CryoEM, and any program can transparently read/write any format. However, specific programs often require specific metadata, such as CTF information, to be stored in the image header, so running *e2refine_easy* directly on an MRCS stack, for example, will result in disabling CTF correction.

Nonheader metadata is now stored in JSON format, which is a human-readable format popular for web programming. Note that some JSON files may contain binary image data encoded as text, which makes these files difficult to read with a text editor. The "Info" button in the EMAN2.1 file browser can be used in these cases to interactively browse JSON files.

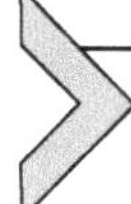

3. ASSESSING A REFINEMENT AND IDENTIFYING VARIABILITY

3.1 Resolution

In recent years, the "gold-standard" FSC (Henderson et al., 2012; Scheres, 2012) has become a fairly universal standard in the CryoEM field, though it is not without its detractors. The concept is quite simple, the particle data are split into two sets at the very beginning of processing, and different starting models are generated for each half of the data. Refinement then proceeds completely independently for the two halves, and only at the end of refinement is the two independent maps compared and averaged. In EMAN 2.1, the independent starting models are generated by randomizing phases beyond a resolution 1.5–2 × worse than the target resolution, insuring that any model bias will be completely different in the two independent maps. This methodology is an intrinsic part of *e2refine_easy* to the extent that it is no longer possible to run a refinement without following these practices.

The primary objections to the "gold standard," as far as we are aware, are that it does not prevent model bias (Stewart & Grigorieff, 2004) from occurring and is not immune to algorithmic bias. What the "gold standard" does do is to prevent resolution exaggeration due to model bias. In addition, since the FSC is also used to filter the map between iterations, it may help to retard the actual noise bias. *e2refine_easy* also has an independent strategy for minimizing bias through iterative 2-D class-averaging integrated into the 3-D refinement loop (Ludtke et al., 1999). The "gold standard" also does nothing to prevent resolution exaggeration due to mask correlation, algorithmic bias, or a range of other effects, which can cause resolution to be overestimated. Unfortunately, achieving near atomic resolution has done little to minimize problems with resolution exaggeration. While there are landmarks at specific resolutions to validate claimed results, such as helix separation at 7–9 Å resolution and strand separation at 4–5 Å resolution. It does nothing to validate that apparent side chains appearing at 3–5 Å resolution range are real. It is unreasonable to expect that people will not try to claim the best resolution they believe they can justify for their reconstructions, so it remains critical that the field develops strong validators and mandates their use to counter this tendency toward exaggeration. The "gold standard" represents one critical step in this direction, which is mandatory in EMAN2.1 refinements, but is clearly an incomplete solution.

3.2 Map Accuracy, Variability, and Symmetry

While a structure may be a homomultimer, it should be clear that no real molecule or assembly will have perfect geometric symmetry at infinitely high resolution. At some resolution the symmetry will be broken. As we approach true atomic resolution, this symmetry breaking will begin being observed even for the most rigid structures. In X-ray crystallography, the crystal packing forces act to restrain this variability somewhat, and what remains is typically expressed as B-factors. Since prior to vitrification, CryoEM molecules are in solution, typically at near-room temperature, much larger-scale variability should be expected. To assess the reliability of high-resolution maps, we must consider a variety of methods.

One powerful test, which is necessary, but not sufficient, to demonstrate a correct structure, is agreement among projections, projection-based class-averages, unsupervised class-averages, and the particles themselves (Murray et al., 2013). Clearly there are limits to comparisons with individual particles due to their low information content. However, both types of class-averages should agree with a corresponding projection to the limits of the variation between two angular steps. These comparisons can be made in EMAN2.1 using *e2classvsproj* and *e2ptclvsmap*. When any significant disagreement exists, even locally, this is an indication of imperfect agreement of the data and the 3-D map, which should be investigated further.

While largely an issue of low-, not high-, resolution reconstruction, if there is a question of the accuracy of the quaternary structure in a reconstruction, *e2RCTboxer* and *e2tiltvalidate* can be used together to validate the reliability of the map (Murray et al., 2013; Rosenthal & Henderson, 2003). As also discussed later, there is an extensive set of subtomogram averaging tools, which can also be applied for validation of quaternary structure (Galaz-Montoya, Flanagan, Schmid, & Ludtke, 2015; Shahmoradian et al., 2013).

Another potential difficulty is preferred orientation. Particularly in cases where a continuous carbon substrate is used, but even in some cases where particles are frozen over holes, macromolecules may exhibit preferred orientation on the grid. The extent to which this causes a problem for the reconstruction is highly dependent on both the shape and symmetry of the structure as well as the nature of the preferred orientation. Let us consider two examples. First, GroEL (Braig et al., 1994), a homo 14-mer, consisting of two back-to-back rings forming a cylindrical shape. GroEL on the grid, even in the absence of continuous carbon, is found predominantly in end-on views, with projection direction near parallel to the sevenfold symmetry axis, and in side views within a few degrees of the set of twofold symmetries relating the two rings (Ludtke et al., 2001). Very few particles are found in orientations between these extremes. While this might seem to be a very poor distribution, in fact, this represents precisely the minimum information required for a complete reconstruction in Fourier space. While the preferred orientation does cause some resolution anisotropy, this occurs in a pattern difficult to even track by eye, and generally speaking reconstructions are excellent to near-atomic resolution.

The second case we consider is the ribosome (either 70S or 80S). Ribosomes are almost always frozen on a continuous carbon substrate, and in this environment tend to sit with one of two specific points on the surface stuck to the carbon. While there is some angular variation near these specific orientations, and there is a fraction of particles present in truly random orientations, this preferred orientation leads to some severe artifacts visible in many published ribosome structures. Specifically, this causes a highly anisotropic resolution in the CryoEM reconstruction, forcing a choice. The entire map could be filtered to a resolution achievable isotropically. However, this is unsatisfying since clearly the isotropic resolution is substantially worse than the achievable resolution over much of the unit sphere. Failing to do this, however, can lead to a pattern of semilocalized "streaking" visible in the structure, which is clearly an artifact. For the ribosome, this problem is further complicated by the presence of multiple ratcheting states, which can lead to additional local streaking/blurring artifacts.

While anisotropic resolution poses the major problem caused by preferred orientation, there are also two other potential problems. First, if one particular orientation is massively overpopulated, for example, with 10–100 × more particles than typical orientations, then the very small fraction of particles which are inevitably assigned to the incorrect orientation may wind up contaminating the less populated orientations and cause significant structural distortions. The main solution to this problem is to eliminate some of the excess particles in the preferred orientation such that there are no more than $\sim 10\times$ as many particles in these orientations than in the typical orientation. Second, to achieve a complete (but not isotropic) reconstruction, it is necessary to have all orientations along a great circle around the unit sphere populated with number of particles. Otherwise the preferred orientation may lead to an artifact similar to the missing cone from random-conical tilt reconstructions. If information is completely absent in some directions in Fourier space, then the reconstruction algorithm may fill these missing areas with anything that is otherwise difficult to fit into the structure, such as contamination or particles in a different conformation, which again may lead to significant artifacts. There is no satisfactory computational solution to this problem, meaning alterations to surface chemistry should be considered to try and eliminate or reduce the preferred orientation experimentally. One common trick is to include a very tiny amount of a weak detergent to block the air–water interface and to exclude the continuous carbon substrate (Zhang et al., 2010), but many other such tricks also exist (Glaeser et al., 2016).

The problem of broken symmetry or pseudosymmetry can be more difficult to address quantitatively. The first question to address is whether a pseudosymmetry is broken in a consistent way, such as with the unique vertex present in many icosahedral viruses, or whether it is broken by independent motion among the subunits. If the former, then the correct solution is to refine the structure with no symmetry imposed at all, or potentially, with symmetry imposed only among subunits expected to be identical. The difficulty is that there must be sufficient information present in each individual particle to discriminate among the possible symmetry-related orientations unambiguously. In the case of a large portal on a virus particle, this is unlikely to be an impediment. However, in a case like CCT, a chaperonin with two copies of eight different proteins arranged in two back-to-back rings, the question can become intractable (Cong et al., 2010; Leitner et al., 2012). In CCT the eight subunits are so highly homologous, that differences among them only become significant at near-atomic resolution, where relatively little signal is present in the image data. In these cases, counting-mode direct detectors, which provide a significant contrast boost at high resolution, may be the difference between success and failure, as they increase the information available about each individual particle.

If the symmetry is broken due to independent Brownian-like motion among the subunits, then this problem is similar to the general issue of structures with conformational variability discussed in Section 4.5. The problem of self-consistent pseudosymmetric structures is discussed in Section 4.6.

3.3 Model-Based Refinement

As atomistic models based solely on CryoEM maps are now becoming more commonplace, it is natural for our field to consider adaptations of refinement methods long used in X-ray crystallography (see chapter "Refinement of Atomic Structures Against cryoEM Maps" by Murshudov). In this case, specifically, using models as a tool to improve the CryoEM maps themselves. The basic concept is to refine a structure to high resolution, build a model for that map, and then use that model to generate a starting map for another round of refinement.

Naturally there is a significant risk of producing a self-fulfilling prophecy via this approach. That is, side chains are modeled, then model bias amplifies the modeled side chains during refinement, whether they are accurate or not. To avoid this trap, we need a mechanism for validating any improved results. The simplest approach, which represents a trade-off between validation and potential structure improvement, is to strip the side chains from the model-based starting map. Then if side chains reappear, particularly if they agree among the even/odd halves of the data, clearly they must have resulted from the data itself, and not the side-chain-free model. However, this also means side chains are not present to help anchor the orientation of the structure. A slightly more aggressive method would be to, for example, remove all of the even numbered side chains from the initial map used for the even particles, and the odd numbered side chains from the initial map used for the odd particles. Then, once again, any correlations among the side chains between the two maps clearly could not have resulted from model bias. Alternatively, we can require two completely independent reconstructions with independent modeling, akin to a suggested method for model validation (DiMaio, Zhang, Chiu, & Baker, 2013), but this requires that only half of the data be available for the modeling/remodeling process, which may be a limiting factor in this method.

These tasks can easily be experimented with using the *e2pdb2mrc* program to convert models to maps. Further, we have developed a new version of *e2pathwalker*, which can generate acceptably accurate backbone traces of high-resolution CryoEM maps fully automatically. This eliminates the need for a human to be involved in the model-building process after each refinement cycle and paves the way for possible iterative model-dependent refinement in the future.

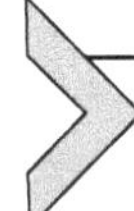

4. CONFORMATIONAL AND COMPOSITIONAL VARIABILITY

It should not be surprising that a large fraction of large macromolecules undergo some amount of motion in solution, or that even for ligands with measured high affinity, a significant amount of compositional variability is frequently observed. The underlying assumption in all single-model CryoEM refinements is that the particles being reconstructed are identical at some target resolution. When this is not the case, the impact on the structure may be modest or severe depending on quaternary structure and the nature of the variability. At present we do not believe there is any single strategy to unambiguously treat all of these situations. Indeed, resolving certain types of structural variability may be a mathematically ill-posed problem and require specialized data collection to resolve. EMAN2.12 offers several different strategies oriented at resolving these issues in different specific situations.

4.1 Variance Analysis

Before proceeding with the more detailed analyses below, it is useful to assess the variance of the map and locate any hotspots that indicate variability in the underlying structure. EMAN2.1 includes *e2refinevariance*, a slightly modified implementation of the bootstrapping method described in Penczek, Yang, Frank, and Spahn (2006). In the original published method the entire population of particles is resampled with replacement to build a set of N maps from which the variance could be computed. It was later determined that there were weaknesses in this approach for systems with nonuniform particle orientations and that this effect should be compensated for when resampling (Zhang, Kimmel, Spahn, & Penczek, 2008). Since *e2refine_easy* makes use of reference-based class-averages, these already roughly account for orientation anisotropy, so rather than resampling the particle population directly, *e2refinevariance* performs resampling with replacement on each class-average independently, then reconstructs the class-averages as usual (Chen, Luke, Zhang, Chiu, & Wittung-Stafshede, 2008).

This procedure works well, but one important caveat for all variance maps, regardless of source, is that a low-pass-filtered variance map is not equivalent to a variance map computed from a low-pass-filtered data set. That is, to observe low-resolution variances in the structure, it is necessary to filter the original independent reconstructions prior to computing the variance. It is not equivalent to low-pass filter the variance computed from the full resolution data. Most interesting variances are low-resolution features associated with large-scale motion or variability. For this reason, and for speed, it is typical to low-pass filter and downsample the class-averages (or alternatively the particles), an option provided directly within the program.

Once computed, variance maps are typically displayed as a color pattern overlaid on an isosurface. While *e2display*'s 3-D widget does have support for "color by map," UCSF Chimera (Goddard, Huang, & Ferrin, 2007) has a much more complete interface and is recommended for this purpose. It is important to note that due to differences in sampling, higher variance will always be observed on and near symmetry axes. While real, this is a mathematical effect and does not actually indicate a higher level of variability along the axis.

4.2 Multimodel Refinement

This general method, implemented as *e2refinemulti*, is widely used in multiple software packages and has been in common use for well over a decade (Brink et al., 2004; Chen, Song, Chuang, Chiu, & Ludtke, 2006). The concept is straightforward. Normal refinement uses projections of an initial model to determine individual particle orientations. In multimodel refinement, N sets of projections from N different initial models are used to determine which of the models each particle matches the best in addition to orientation. After reconstructing N new models, the process is iterated in the same way as normal single-model refinement. Procedurally, the only change is that multiple initial models must be provided. These N models can come from any source. They must not be perfectly identical, or the refinement will fail. However, perturbing the same starting model N times by adding a small amount of noise is a perfectly acceptable and unbiased strategy, though convergence could take as many as 15–20 iterations if the starting models are nearly identical. Alternatively, with some knowledge about the system, such as a ligand which may be present or absent, two starting models could be the single-model refined map and the same map with the ligand manually masked out. In this case, convergence will normally be achieved much more rapidly (3–5 iterations).

This algorithm is guaranteed to converge to produce N final maps, and it will tend to converge to maps as different as possible given the provided data. If the data are completely homogeneous apart from noise, N different maps will still emerge, and the observed differences will then be due to model bias. When structural variability is present in the data, providing that the variability is detectable in individual particles, this real variability will tend to dominate the classification process, rather than any sort of model bias. Nonetheless, it is important to follow this multimodel refinement with additional validations to insure the results are reliable.

The first recommended step is to perform a cross validation. The particles associated with each of the N models are automatically split into N new .lst files at the end of the *e2refinemulti* job. During the job, particles are free to migrate among the N maps during each iteration. The final particle separation is static and should be used as input for a new *e2refine_easy* refinement to achieve better convergence. As one validation, the starting model for each of these single-map refinements should be the final map from one of the other N-1 results, rather than the final result associated with the same particle set. If the refinement converges back to the *e2refinemulti* result associated with the particle set (away from the starting model), then it implies that this map is truly discriminated from the other N-1 maps by the particles. This, in itself does not prove that the results are meaningful, just that the particle sets are sufficiently well discriminated to overcome initial model bias.

The other obvious validation would be to rerun *e2refinemulti* using a different set of N starting models. If noise was used to discriminate among starting models in the first run, then this could be repeated with different noises. If deterministic starting models were used initially, then some strategy for perturbing them significantly could be applied, such as randomizing high-resolution phases using the filter.lowpass.randomphase processor.

It is critical to realize that in all least-squares classification methods, including methods such as maximum likelihood, while the N, 3-D maps produced by the method may be quite representative of the actual specimen variability, the accuracy of the classification of individual particles may be quite low. This is true for basically the same reason that in statistics the standard deviation of the mean is much smaller than the standard deviation of the distribution that formed it. Individual particles are noisy and difficult to classify, so, while we speak of the particles associated with a particular map, this subset is often highly contaminated with particles which should be in other subsets. Some of the methods described later strive to address this problem by stepping outside the least-squares framework, through use of masks, thresholds, exact subtraction, and other methods, to help improve the statistical accuracy of the classification of each individual particle.

4.3 Splitting 3-D Refinements Using 2-D Principal Component Analysis

Based on the above discussion, it would be desirable to have a more accurate mechanism for accurate classification of particles and/or for identifying important structural variabilities. EMAN2.1 includes *e2refine_split*, for unbiased binary separation of a single-model 3-D refinement. This method divides the data into two subsets and produces a new 3-D reconstruction for each state.

Consider the set of particles going into a single class-average. Principal component analysis (PCA) can be applied to the aligned particles to identify the most prominent differences, and split the particles into two sets, thus producing two class-averages from one. Applying this to an entire data set leads to two class-averages for each orientation, but does not identify whether each average should be associated with map A or map B.

The approach used by *e2refine_split* to perform this assignment (Fig. 1) is to begin with the class-average with the largest number of particles and arbitrarily assign average A to map X and B to map Y. Next, the next most populous class-average is compared to the 3-D Fourier volume after the first assignment. The two possibilities, A→X, B→Y or A→Y, B→X, are compared, the best match is selected and the class-averages inserted into the corresponding Fourier volumes. The process is then repeated for each

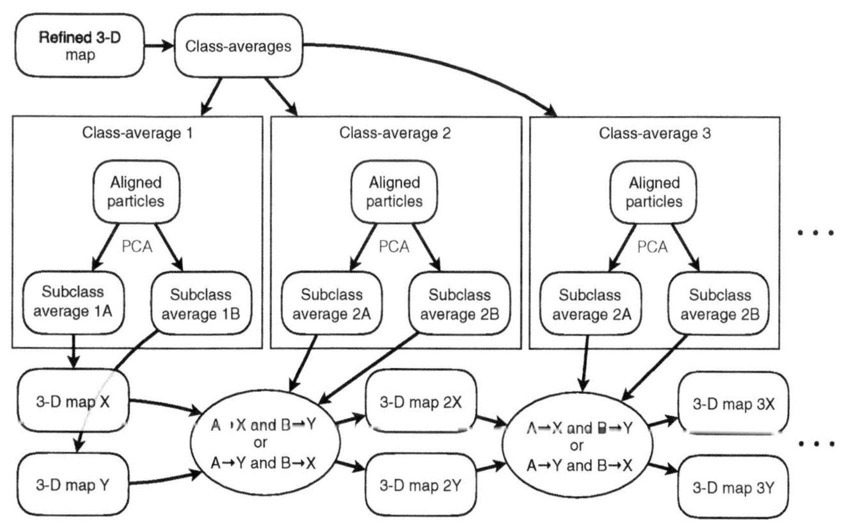

Fig. 1 A diagram showing how *e2refine_split* separates a single-3-D refinement into two distinct maps, by subclassifying the particles within each class-average.

pair of averages in particle count order, with Fourier space gradually becoming more and more complete, thus providing more information for each subsequent decision. That is, at the beginning of the process, there are few overlaps in Fourier space, but each inserted class-average has a comparatively large number of particles, and thus provides a high SSNR for A/B classification. Later in the process, where the tested class-averages are noisier due to low particle counts, Fourier space is more completely filled providing more existing information to make the decision. At the end of the process, two 3-D maps and two sets of particles are produced where only one existed before.

There are several advantages of this algorithm over *e2refinemulti*. First, it is unbiased, requiring no starting models. Second, it is very fast, since no additional 3-D orientation determination is required, and class-averaging is a fast process. Third, the effective noise reduction provided by PCA leads to more accurate classification.

There are also some weaknesses of the method: First, if the data include a wide range of defocus values and the structural variability is modest, then the first PCA component may represent defocus variation rather than a structural difference. Second, if there are multiple independent motions present in a 3-D map, this method may identify a different variation as being the most prominent in one orientation than the variation identified for another orientation. In this situation, the algorithm will simply do the best it can, and it is hoped that a second round of binary splitting, applied to each of the maps in the first round, will identify the next strongest remaining inhomogeneity among the data, and thus effectively separate the two motions. The user also has the option to select which PCA vector is used in the classification, though, so it may be possible to combat problems such as defocus dominating the result.

While in principle it would be possible to develop a method which splits each class-average into N groups rather than only two groups, this would increase the number of permutations to consider exponentially as well as increase sensitivity to misassignment of class-averages, so at present we limit the method to two states at a time. A related technique (Zhang, Minary, & Levitt, 2012) proposes to use individual class-averages in conjunction with simulation methods to derive more continuous dynamics information. It is possible that a future algorithm could hybridize these approaches and yield a data-based solution for rigorous separation of states.

4.4 Model-Based Binary Separation in 3-D

In the approach above, classification is performed in 2-D and is bias free. However, if more information about the specific variation is available, this

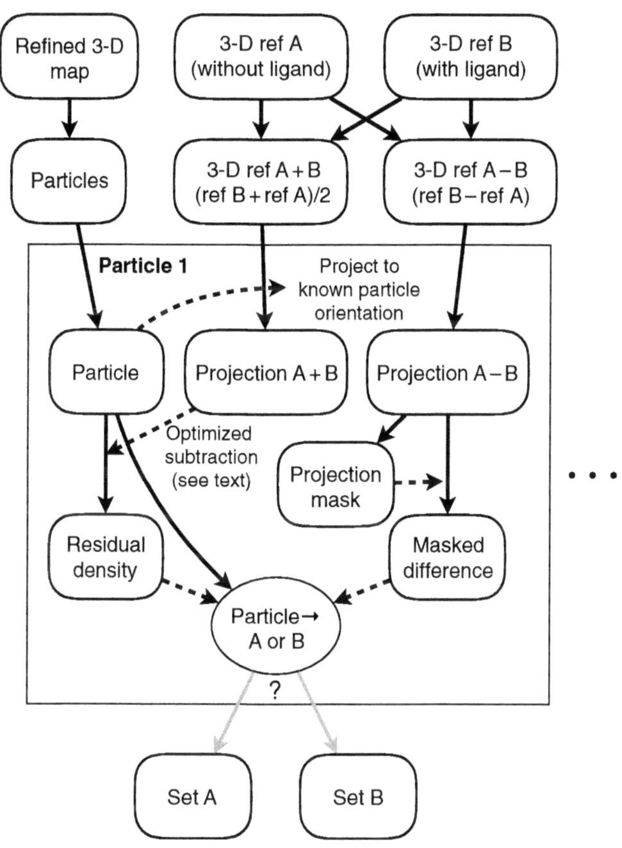

Fig. 2 A diagram showing how particles can be accurately classified between two 3-D reference maps. The classification performed using this process provides demonstrably better statistical separation than standard multimodel refinement (Section 4.2), provided the major differences between the two references are fairly localized.

can be used to improve per-particle classification accuracy (Fig. 2). For example, with ligand binding, or with a localized discrete motion, the voxels which vary the most in 3-D between the two states can be identified, then the voxels which do not profitably contribute to classification can be excluded or aggressively downweighted. This method is implemented in *e2classifyligand* (Park et al., 2014).

This algorithm requires two starting maps representative of the two states to be separated. The average of the two maps is computed to serve as a neutral standard, and the difference of the two maps is computed to construct a mask and a classification vector. For each particle a projection of the sum and difference maps is generated in the already known orientation. A mask, identifying the most important pixels for classification, is constructed and applied to the difference projection. The neutral standard projection is used to eliminate as much irrelevant information from the particle as possible. This subtraction includes appropriate CTF imposition, filtration, and density matching, similar to the method described in Section 4.5. Ideally, this remaining density would include only noise and the information relevant for classification. This residual density is then used in conjunction with the masked classifier to separate the particles. In addition to A↔B separation, ambiguous particles can also optionally be split into a separate class, retaining only the particles most clearly representative of each state. In addition, the per-particle classification values are stored and can be used to assess the statistics of the separation.

4.5 Robust Extraction of Particle Fragments

In some instances one particular domain of a particle may undergo some motion independent from any other structural variability in the particle. In such situations, it is desirable to focus classification and in some cases, alignment, on only this region of interest (ROI). However, to isolate only one portion of the particle from a projection image, the remaining "constant" portion of the particle must be very accurately subtracted from the image (Fig. 3). This process is very similar to the process used in Section 4.4 for 2-D classification. After completion of a standard 3-D refinement of the data, the final 3-D reconstruction is masked, such that the region being targeted for variability analysis is excluded from the volume, creating an exclusion map, which contains only the density we are *not* interested in. For each particle a projection of this masked map is generated in the same orientation as the particle, modified for CTF, filtered to match, and with densities rescaled to best match the projection for optimal subtraction. After subtracting the masked projection, what is left behind is ostensibly only the ROI, still in the same orientation with respect to the refined map. While there may also be some structural variability present outside the ROI making the subtraction imperfect, this operation still isolates the ROI far better than would, say, a simple mask.

If the particle possesses symmetry or pseudosymmetry, it may be possible to impose this symmetry during the masking process to extract individual monomers as independent particles, each in a different orientation. For example, if a homotetramer is believed to undergo independent motion in each subunit within a specific domain, that domain could be extracted from each of the four subunits, providing four times as many extracted particle fragments.

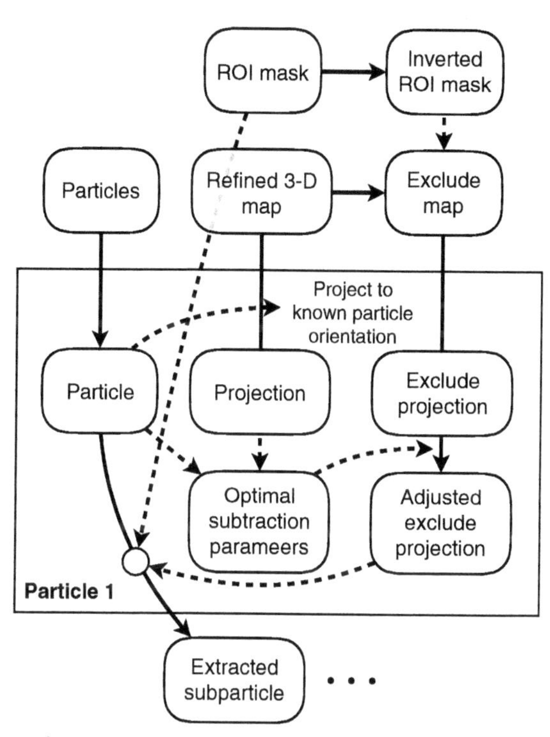

Fig. 3 A diagram showing how local fragments can be accurately isolated from 2-D images of individual particles based on a 3-D mask and a refinement of the complete data set.

If the extracted particle fragments are very small, it may not be feasible to perform any further alignment on them, and simply classifying into subsets based on their already known orientations with respect to the rest of the structure may be appropriate. If the extracted fragments are sufficiently large, they may be treated as single particles in their own right and be refined using a process like *e2refine_easy*.

This approach was actually the precursor to the classification methodology described in Section 4.4 and has been available in EMAN1 (as *extractmonomers*) and EMAN2 (as *e2extractsubparticles*) for over a decade. While it has been used for preliminary analysis of data frequently over that time, until the development of direct detectors, the lower size limit for particles (or fragments) in CryoEM was generally regarded to be ~200 kDa. This limit put practical constraints on the use of this methodology, since the extracted fragments needed to be large enough to be processed somewhat independently as particles. With the dramatic improvement in DQE extending to high-resolution provided by direct electron detectors, this and similar (Nguyen et al., 2015) (and see chapter "Processing of Structurally Heterogeneous Cryo-EM Data in RELION" by Scheres) methods have become practical for a wide range of purposes.

4.6 Breaking Symmetry

The previously described approach is appropriate for symmetric particles in which each symmetry-related copy is undergoing some form of variability independently. However, if this is not the case, and the particle possesses some form of self-consistent symmetry breaking, then alternative methods are required. The standard example is an icosahedral virus particle with a portal at one or more vertices (Chang, Schmid, Rixon, & Chiu, 2007; Xiang et al., 2006). Such portals not only break the icosahedral symmetry but also generally possess a symmetry mismatch with the underlying pentagonal vertex they are positioned over. Initial reconstructions are typically performed assuming full 60-fold symmetry, since this effectively provides 60 (or at least 55) times more information about each subunit, and can produce high-resolution structures with small numbers of particles. The 11 near-identical vertices dominate the average in such cases, and only residual hints of the unique vertex remain.

To gain information about the unique vertex in such situations, the particles could simply be treated as asymmetric objects in a standard refinement. However, there are several problems with this approach. When generating uniformly distributed projections on the unit sphere, the pattern of projections would be different within each pseudosymmetric unit. If the symmetry-breaking differences in the structure are subtle, the symmetry may potentially be broken by this angular inconsistency rather than by the true symmetry-breaking features in the object. Second, this method is far less efficient than it should be. Assuming an underlying near symmetry, particle orientations can be determined much more efficiently.

An option in *e2refine_easy*, called—breaksym is the preferred way to respond to this situation in EMAN2.1. When the pseudosymmetry is specified with—breaksym, projections are generated identically among the asymmetric units, maximizing the chance of detecting a true difference. The orientation determination strategy in EMAN2.1 always occurs in two stages for improved efficiency. In the first stage, coarse orientation is determined, among the least similar projections, and in the second stage the orientation is refined among a subset of similar particles. Projections are not grouped together for this first stage based on their relative orientations, but rather based on mutual similarity among projections. So, in the case of a pseudosymmetry, the symmetry-related copies are averaged together and the first-stage classification becomes orientation determination within one asymmetric unit, followed by the second-stage classification to determine which of the N symmetry-related orientations is the best match.

4.7 Subtomogram Averaging

The situations discussed so far have focused on discrete variability, or continuous variability largely localized to a specific region of a larger assembly. However, a significant fraction of large biomolecules and assemblies makes use of flexibility as a critical aspect of their function. In Section 4.2 we assumed that it was possible to unambiguously discriminate between changes of particle orientation and changes of conformation. For example, addition or subtraction of a ligand (Roseman, Chen, White, Braig, & Saibil, 1996) or large-scale discrete motions such as ribosome ratcheting (Frank & Agrawal, 2000) do not create an orientation ambiguity as long as the appropriate number of models are being refined. However, macromolecules, such as mammalian fatty acid synthase (Brignole, Smith, & Asturias, 2009; Brink et al., 2004), undergo multiple independent large-scale continuous motions. In such a situation, without a large set of accurate starting models, there can be ambiguity over whether a small difference observed in 2-D is due to a change in orientation or conformation. Even with human intuition involved in the process to eliminate physically unreasonable models, results are frequently still ambiguous. We have generally observed that once a set of accurate starting models for valid states of the system have been obtained, *e2refinemulti* will successfully refine normal single-particle data to produce better structures and accurately classify particles. However, this statement cannot be proven to be true for all conceivable situations, so caution is still required.

When the motions involved in a system are significant, perhaps ~15 Å, one viable approach is subtomogram averaging, also known as single-particle tomography (see chapter "Cryo-Electron Tomography and Subtomogram Averaging" by Briggs). This is a powerful technique with a range of applications, including in situ structural biology, but in the present context this method simply provides a tool for disambiguating complex macromolecular motion. The reason ambiguity exists in traditional single-particle analysis is each particle is imaged only in a single orientation. When comparing particles in two different orientations to look for self-consistency, these two projections share only a single common line of shared information in Fourier space, and this line simply may not possess sufficient information to discriminate among multiple functional states. If multiple putative orientations are considered simultaneously, the number of required simultaneous projections to resolve ambiguities rapidly becomes intractable, and even then, the necessary information may simply not be present in the data.

By collecting a tomographic tilt series on a traditional single-particle cryospecimen, the multiple views of each particle provide sufficient information to unambiguously disambiguate orientation vs conformational change. This is, of course, complicated by the presence of the missing wedge and the low dose in each tilt. Further, there are many intricacies involved in performing this analysis, which are too detailed for this manuscript. Several existing manuscripts document EMAN2's capabilities for subtomogram alignment, averaging, and classification (Galaz-Montoya et al., 2015; Schmid & Booth, 2008; Shahmoradian et al., 2013). The general convention is for these programs to be named *e2spt_**, and at present there are over 30 such programs covering a wide range of situations.

The goal of these methods in the present context is to determine multiple structures for a given system. This process always begins with tomographic reconstruction, followed by *e2spt_boxer* or *e2spt_autoboxer* to identify and extract particle volumes. All of the subsequent algorithms compensate for the missing wedge artifact present in all CryoET data. The stacks of subvolumes must then be aligned and classified. One approach is to use *e2spt_hac*, which performs all-vs-all comparison and hierarchical ascendant classification (Frank, Bretaudiere, Carazo, Verschoor, & Wagenknecht, 1988), in 3-D, to group together and average similar structures. Alternatively, *e2spt_refinemulti* is the tomographic equivalent of *e2refinemulti*, for situations where N putative starting models are available.

It is also possible to perform PCA-based classification of subtomograms using a sequence of programs. A detailed description of this workflow is available via the online tutorials, but briefly, *e2spt_align* is used to align all particles to a common reference, and *e2spt_average* is used to produce a single overall average volume for all of the particles. Next, this average volume is used to fill the missing wedge of the individual subtomograms using *e2spt_wedgefill*. This prevents the PCA analysis from focusing on the missing wedge instead of actual particle variations. *e2msa*, which works on both 2-D and 3-D data, is then used to perform PCA. This information is then combined using *e2basis*, and *e2classify* classifies the results. The classified particles can then be realigned and averaged using any of the available methods for this.

Once any of these methods have been applied to produce a set of characteristic maps, these can be used as initial models for *e2refinemulti*, targeting higher resolution from traditional single-particle refinement. While subtomogram averaging with CTF correction has successfully achieved subnanometer resolution (Bharat, Russo, Löwe, Passmore, & Scheres, 2015; Schur, Hagen, de Marco, & Briggs, 2013), this is a difficult and time-consuming process, and it is clear that high resolution is far easier to achieve using single-particle methods once the heterogeneity can be dealt with. Indeed, the presence of significant motion may remain the resolution-limiting factor in many cases. Consider a molecule undergoing a single mode of motion, moving ~20 Å. Achieving 4 Å resolution on this specimen would require subdividing the data into at least five subsets, requiring at least 5 × the number of particles as compared to a rigid structure to achieve comparable resolution. If an additional independent motion was present, the number of required subsets could easily rise to 50 or 100. Further, as discussed earlier, for this to be successful, it must be possible to at least partially discriminate among all of these possibilities for each individual particle. Again, the improved per-particle contrast provided by direct electron detectors has made many previously impossible problems feasible.

4.8 Tilt Validation

When performing single-particle reconstruction in a new system, initial reconstructions may be low resolution and largely provide information about quaternary or possibly tertiary structure. Lacking landmarks such as alpha helices or strands in beta sheets to clearly show a "protein-like" structure, it is often difficult to tell whether such structures are accurate interpretations of the single-particle data. As discussed earlier, roughly half of published single-particle reconstructions still fail to achieve the subnanometer resolution required to resolve alpha helices, so this problem is far from uncommon. If subtomogram averaging has been performed, then the use of tilt information implicit in this process already serves as a strong validation for structure. However, if only traditional single-particle analysis is being used, some additional validations may be required to demonstrate reliability of low-resolution maps. One strong validation is tilt validation (Rosenthal & Henderson, 2003) (and chapter "Testing the Validity of Single Particle Maps at Low and High Resolution" by Rosenthal). In this method a few tilt pairs are collected from the single-particle specimen, with one image having the specimen typically tilted to 10–20 degrees. Matching pairs of particles are selected, and the orientation of all of all tilted and untilted particles is determined using projections of a 3-D structure requiring validation. We can then ask whether the orientation relationship among these pairs of particles matches the known experimental tilt. If it matches consistently for a majority of the particles, then this demonstrates that the 3-D map is consistent with the data. EMAN2.1 implements this methodology through *e2RCTboxer* and *e2tiltvalidate* (Murray et al., 2013). There are a number of potential difficulties in this methodology, and even for very good specimens, the results are often not perfect. Nonetheless, it is generally possible to distinguish between valid and invalid quaternary structures using this methodology.

5. INTERACTIVE TOOLS

In addition to its workflows and utility programs, EMAN2.1 contains a range of graphical tools (Fig. 4) which may be useful as utilities regardless of what software is used to perform high-resolution reconstructions. Most, but not all of these tools are also available from the main *e2projectmanager* GUI, but all can also be launched directly from the command line. A few of these tools are summarized here.

e2display: This is a general-purpose CryoEM-aware file browser and display program. In addition to displaying data and metadata about all supported CryoEM image file types in the browser, this interface provides access to: 3-D display including multivolume isosurfaces, arbitrary slices, volume rendering, and annotations; 2-D single-image display including measurement and image manipulation capabilities; 2-D tiled image display for looking at image stacks all together, including capabilities for deleting images from stacks and forming new sets; X–Y plotting, including the ability to easily switch axes among multicolumn text files to explore multidimensional spaces.

e2filtertool: This versatile tool can process 3-D volumes, 2-D stacks, or individual 2-D images. This tool permits chains of image processing operations (filters, masks, transformations, etc.) to be constructed, and parameters to be adjusted interactively with real-time update of the image/volume being processed.

e2evalimage: A tool for assessment of whole micrographs. Provides a variety of tools for generating power spectra and other features, and performing automatic CTF fitting with manual adjustment. This includes keyboard shortcuts for rapidly traversing lists of thousands of micrographs and interactively marking those to in/exclude from a project.

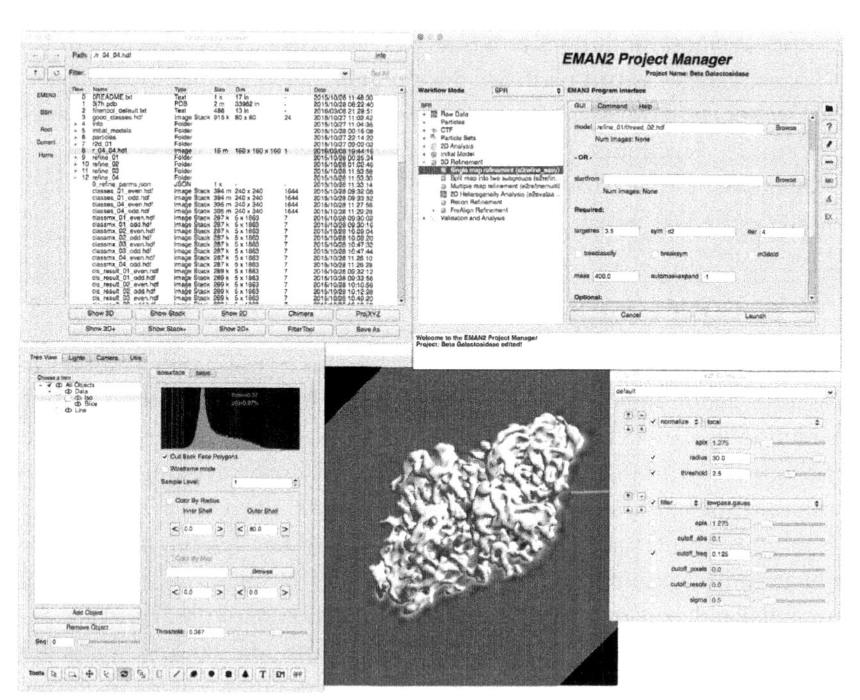

Fig. 4 A panel showing just a few of EMAN2.1's many graphical tools. The main project manager is shown in the *upper right*, with a portion of the dialog for a run of *e2refine_easy* shown. On the *upper left* is the file browser, note the metadata displayed in the columns of the browser, as well as the actions available at the *bottom* of the window specific to the selected file type. The *bottom* three windows show *e2filtertool* operating on a 3-D volume. The 3-D display in the *center* shows an isosurface with a slice through the *middle* and a 3-D *arrow* annotation. The control panel, which permits modifying this display, is shown on the *left*. On the *right* is the *e2filtertool* dialog, where a sequence of two image processing operations has been added: first, a "local normalization" operation, which helps compensate for low-resolution noise and ice gradients; second, a Gaussian low-pass filter set to 8 Å resolution. The parameters of these operations can be adjusted in real time with corresponding updates in the 3-D display. 2-D images and stacks can also be processed using this program. Any of over 200 image processing operations can be used from this interface, and command-line parameters are provided mimicking the final adjusted parameters. (See the color plate.)

e2boxer, *e2helixboxer*, *e2RCTboxer*, and *e2spt_boxer*: Tools for interactive or semiautomatic particle picking in 2-D and 3-D. Can be used with EMAN2.1 reconstructions or as independent extraction tools.

e2evalparticles: Primarily useful with EMAN2.1 results. This permits extraction of selected particles associated with user-selected class-averages. For instance, if *e2refine2d* is used to generate reference-free

class-averages, and the user wishes to mark particles associated with bad class-averages as bad particles, this program provides that capability.
e2eulerxplor: Primarily useful with EMAN2.1 refinements. This program permits visualization of the Euler angle distribution for a standard 3-D refinement. It also has some generic asymmetric unit visualization capabilities in 3-D which may be useful in other situations.

6. CONCLUSION

In this manuscript we have discussed the broad philosophy adopted by EMAN2.12 for single-particle reconstruction, targeting the highest possible resolution for each molecular system. We extensively discussed various methods for working with compositional and conformational variability both to extract information about the variability itself and to achieve higher resolutions. While many structures now being produced in this field are now achieving near-atomic resolution, it is still critical for the field to remain grounded and develop a strong culture of validation of results. EMAN2.12's interoperability capabilities and implementations of standard validation methods can play a very strong role in this process, and we strongly encourage other developers to embrace the interoperability concept. The algorithms described in this manuscript continue to evolve, and new methods in single-particle analysis, subtomogram averaging, and cellular tomography segmentation are being developed. We believe that significant room for improvement still exists in single-particle analysis software and reiterate that our aim is to produce the best structures possible from a given set of data.

EMAN2.12 remains a highly dynamic project, which has recently completed a transition to GitHub, becoming truly public open source, and we welcome direct contributions from other developers. A number of important developments are planned for EMAN2/SPARX in the near future. Anaconda is an open-source Python distribution, which incorporates many useful packages, including SciPy, an expansive library for mathematical computation in Python. In the near future, our binary distribution mechanism will switch to use Anaconda as its base. This will permit easy interoperability between EMAN2 and a range of scientific libraries available through this system. EMAN2 is also beginning the transition from Python2 to Python3, which represents a significant shift in the Python development community. In short, we are working to maintain EMAN2/SPARX as a cutting-edge development platform for the next generation of image processing scientists, while maintaining its easy-to-use and user-friendly nature.

ACKNOWLEDGMENTS

We would like to gratefully acknowledge the support of NIH Grants R01GM080139 and P41GM103832. We would like to thank Dr. Pawel Penczek and developers of the SPARX package whose contributions to the shared EMAN2/SPARX C++ library have helped make it into a flexible and robust image processing platform.

REFERENCES

Bell, J. M., Chen, M., Baldwin, P. R., & Ludtke, S. J. (2016). High resolution single particle refinement in EMAN2.1. *Methods*, *100*, 25–34.

Bharat, T. A., Russo, C. J., Löwe, J., Passmore, L. A., & Scheres, S. H. (2015). Advances in single-particle electron cryomicroscopy structure determination applied to sub-tomogram averaging. *Structure (London, England: 1993)*, *23*(9), 1743–1753.

Braig, K., Otwinowski, Z., Hegde, R., Boisvert, D. C., Joachimiak, A., Horwich, A. L., et al. (1994). The crystal structure of the bacterial chaperonin GroEL at 2.8 Å. *Nature*, *371*(6498), 578–586.

Brignole, E. J., Smith, S., & Asturias, F. J. (2009). Conformational flexibility of metazoan fatty acid synthase enables catalysis. *Nature Structural & Molecular Biology*, *16*(2), 190–197.

Brink, J., Ludtke, S. J., Kong, Y., Wakil, S. J., Ma, J., & Chiu, W. (2004). Experimental verification of conformational variation of human fatty acid synthase as predicted by normal mode analysis. *Structure (London, England: 1993)*, *12*(2), 185–191.

Chang, J. T., Schmid, M. F., Rixon, F. J., & Chiu, W. (2007). Electron cryotomography reveals the portal in the herpesvirus capsid. *Journal of Virology*, *81*(4), 2065–2068.

Chaudhry, C., Horwich, A. L., Brunger, A. T., & Adams, P. D. (2004). Exploring the structural dynamics of the *E. coli* chaperonin GroEL using translation-libration-screw crystallographic refinement of intermediate states. *Journal of Molecular Biology*, *342*(1), 229–245.

Chen, D. H., Luke, K., Zhang, J., Chiu, W., & Wittung-Stafshede, P. (2008). Location and flexibility of the unique C-terminal tail of Aquifex aeolicus co-chaperonin protein 10 as derived by cryo-electron microscopy and biophysical techniques. *Journal of Molecular Biology*, *381*(3), 707–717.

Chen, D. H., Song, J. L., Chuang, D. T., Chiu, W., & Ludtke, S. J. (2006). An expanded conformation of single-ring GroEL-GroES complex encapsulates an 86 kDa substrate. *Structure (London, England: 1993)*, *14*(11), 1711–1722.

Cong, Y., Baker, M. L., Jakana, J., Woolford, D., Miller, E. J., Reissmann, S., et al. (2010). 4.0-Å resolution cryo-EM structure of the mammalian chaperonin TRiC/CCT reveals its unique subunit arrangement. *Proceedings of the National Academy of Sciences of the United States of America*, *107*(11), 4967–4972.

de la Rosa-Trevín, J. M., Otón, J., Marabini, R., Zaldívar, A., Vargas, J., Carazo, J. M., et al. (2013). Xmipp 3.0: An improved software suite for image processing in electron microscopy. *Journal of Structural Biology*, *184*(2), 321–328.

DiMaio, F., Zhang, J., Chiu, W., & Baker, D. (2013). Cryo-EM model validation using independent map reconstructions. *Protein Science: A Publication of the Protein Society*, *22*(6), 865–868.

Frank, J., & Agrawal, R. K. (2000). A ratchet-like inter-subunit reorganization of the ribosome during translocation. *Nature*, *406*(6793), 318–322.

Frank, J., Bretaudiere, J. P., Carazo, J. M., Verschoor, A., & Wagenknecht, T. (1988). Classification of images of biomolecular assemblies: A study of ribosomes and ribosomal subunits of escherichia coli. *Journal of Microscopy*, *150*(Pt. 2), 99–115.

Frank, J., Radermacher, M., Penczek, P., Zhu, J., Li, Y., Ladjadj, M., et al. (1996). SPIDER and WEB: Processing and visualization of images in 3D electron microscopy and related fields. *Journal of Structural Biology*, *116*(1), 190–199.

Galaz-Montoya, J. G., Flanagan, J., Schmid, M. F., & Ludtke, S. J. (2015). Single particle tomography in EMAN2. *Journal of Structural Biology*, *190*(3), 279–290.

Glaeser, R. M., Han, B. G., Csencsits, R., Killilea, A., Pulk, A., & Cate, J. H. (2016). Factors that influence the formation and stability of thin, cryo-EM specimens. *Biophysical Journal, 110*(4), 749–755.

Goddard, T. D., Huang, C. C., & Ferrin, T. E. (2007). Visualizing density maps with UCSF chimera. *Journal of Structural Biology, 157*(1), 281–287.

Grigorieff, N. (2007). FREALIGN: High-resolution refinement of single particle structures. *Journal of Structural Biology, 157*(1), 117–125.

Henderson, R. (1995). The potential and limitations of neutrons, electrons and X-rays for atomic resolution microscopy of unstained biological molecules. *Quarterly Reviews of Biophysics, 28*(2), 171–193.

Henderson, R., Sali, A., Baker, M. L., Carragher, B., Devkota, B., Downing, K. H., et al. (2012). Outcome of the first electron microscopy validation task force meeting. *Structure (London, England: 1993), 20*(2), 205–214.

Hohn, M., Tang, G., Goodyear, G., Baldwin, P. R., Huang, Z., Penczek, P. A., et al. (2007). SPARX, a new environment for cryo-EM image processing. *Journal of Structural Biology, 157*(1), 47–55.

Lawson, C. L., Patwardhan, A., Baker, M. L., Hryc, C., Garcia, E. S., Hudson, B. P., et al. (2016). EMDataBank unified data resource for 3DEM. *Nucleic Acids Research, 44*(D1), D396–D403. http://dx.doi.org/10.1093/nar/gkv1126.

Leitner, A., Joachimiak, L. A., Bracher, A., Mönkemeyer, L., Walzthoeni, T., Chen, B., et al. (2012). The molecular architecture of the eukaryotic chaperonin TRiC/CCT. *Structure (London, England: 1993), 20*(5), 814–825.

Ludtke, S. J., Baldwin, P. R., & Chiu, W. (1999). EMAN: Semiautomated software for high-resolution single-particle reconstructions. *Journal of Structural Biology, 128*(1), 82–97.

Ludtke, S. J., Jakana, J., Song, J. L., Chuang, D. T., & Chiu, W. (2001). A 11.5 Å single particle reconstruction of GroEL using EMAN. *Journal of Molecular Biology, 314*(2), 253–262.

Marabini, R., Ludtke, S. J., Murray, S. C., Chiu, W., Jose, M., Patwardhan, A., et al. (2016). The electron microscopy exchange (EMX) initiative. *Journal of Structural Biology, 194*, 156–163.

Mindell, J. A., & Grigorieff, N. (2003). Accurate determination of local defocus and specimen tilt in electron microscopy. *Journal of Structural Biology, 142*(3), 334–347.

Murray, S. C., Flanagan, J., Popova, O. B., Chiu, W., Ludtke, S. J., & Serysheva, I. I. (2013). Validation of cryo-EM structure of IP_3R1 channel. *Structure (London, England: 1993), 21*(6), 900–909.

Nguyen, T. H., Galej, W. P., Bai, X. C., Savva, C. G., Newman, A. J., Scheres, S. H., et al. (2015). The architecture of the spliceosomal U4/U6.U5 tri-snRNP. *Nature, 523*(7558), 47–52.

Park, E., Ménétret, J. F., Gumbart, J. C., Ludtke, S. J., Li, W., Whynot, A., et al. (2014). Structure of the SecY channel during initiation of protein translocation. *Nature, 506*(7486), 102–106.

Penczek, P. A., Yang, C., Frank, J., & Spahn, C. M. (2006). Estimation of variance in single-particle reconstruction using the bootstrap technique. *Journal of Structural Biology, 154*(2), 168–183.

Rohou, A., & Grigorieff, N. (2015). CTFFIND4: Fast and accurate defocus estimation from electron micrographs. *Journal of Structural Biology, 192*(2), 216–221.

Roseman, A. M., Chen, S., White, H., Braig, K., & Saibil, H. R. (1996). The chaperonin ATPase cycle: Mechanism of allosteric switching and movements of substrate-binding domains in GroEL. *Cell, 87*(2), 241–251.

Rosenthal, P. B., & Henderson, R. (2003). Optimal determination of particle orientation, absolute hand, and contrast loss in single-particle electron cryomicroscopy. *Journal of Molecular Biology, 333*(4), 721–745.

Scheres, S. H. (2012). RELION: Implementation of a Bayesian approach to cryo-EM structure determination. *Journal of Structural Biology, 180*(3), 519–530.
Schmid, M. F., & Booth, C. R. (2008). Methods for aligning and for averaging 3D volumes with missing data. *Journal of Structural Biology, 161*(3), 243–248.
Schur, F. K., Hagen, W. J., de Marco, A., & Briggs, J. A. (2013). Determination of protein structure at 8.5 Å resolution using cryo-electron tomography and sub-tomogram averaging. *Journal of Structural Biology, 184*(3), 394–400.
Shahmoradian, S. H., Galaz-Montoya, J. G., Schmid, M. F., Cong, Y., Ma, B., Spiess, C., et al. (2013). TRiC's tricks inhibit huntingtin aggregation. *eLife, 2*, e00710.
Shaikh, T. R., Gao, H., Baxter, W. T., Asturias, F. J., Boisset, N., Leith, A., et al. (2008). SPIDER image processing for single-particle reconstruction of biological macromolecules from electron micrographs. *Nature Protocols, 3*(12), 1941–1974.
Stewart, A., & Grigorieff, N. (2004). Noise bias in the refinement of structures derived from single particles. *Ultramicroscopy, 102*(1), 67–84.
Tang, G., Peng, L., Baldwin, P. R., Mann, D. S., Jiang, W., Rees, I., et al. (2007). EMAN2: An extensible image processing suite for electron microscopy. *Journal of Structural Biology, 157*(1), 38–46.
van Heel, M., Harauz, G., Orlova, E. V., Schmidt, R., & Schatz, M. (1996). A new generation of the IMAGIC image processing system. *Journal of Structural Biology, 116*(1), 17–24.
Xiang, Y., Morais, M. C., Battisti, A. J., Grimes, S., Jardine, P. J., Anderson, D. L., et al. (2006). Structural changes of bacteriophage φ29 upon DNA packaging and release. *The EMBO Journal, 25*(21), 5229–5239.
Zhang, J., Baker, M. L., Schröder, G. F., Douglas, N. R., Reissmann, S., Jakana, J., et al. (2010). Mechanism of folding chamber closure in a group II chaperonin. *Nature, 463*(7279), 379–383.
Zhang, W., Kimmel, M., Spahn, C. M., & Penczek, P. A. (2008). Heterogeneity of large macromolecular complexes revealed by 3D cryo-EM variance analysis. *Structure (London, England: 1993), 16*(12), 1770–1776.
Zhang, J., Minary, P., & Levitt, M. (2012). Multiscale natural moves refine macromolecules using single-particle electron microscopy projection images. *Proceedings of the National Academy of Sciences of the United States of America, 109*(25), 9845–9850.

Frealign: An Exploratory Tool for Single-Particle Cryo-EM

N. Grigorieff[1]
Janelia Research Campus, Howard Hughes Medical Institute, Ashburn, VA, United States
[1]Corresponding author: e-mail address: niko@grigorieff.org

Contents

Abstract

Frealign is a software tool designed to process electron microscope images of single molecules and complexes to obtain reconstructions at the highest possible resolution. It provides a number of refinement parameters and options that allow users to tune their refinement to achieve specific goals, such as masking to classify selected regions within a particle, control over the refinement of specific alignment parameters to accommodate various data collection schemes, refinement of pseudosymmetric particles, and generation of initial maps. This chapter provides a general overview of Frealign functions and a more detailed guide to using Frealign in typical scenarios.

Methods in Enzymology, Volume 579
ISSN 0076-6879
http://dx.doi.org/10.1016/bs.mie.2016.04.013

1. INTRODUCTION AND PHILOSOPHY

Frealign (Grigorieff, 1998, 2007) is an image processing tool that can be used to calculate and refine three-dimensional (3D) structures of macromolecular assemblies that are calculated from images collected on an electron microscope. Its development began in 1996 at the MRC Laboratory of Molecular Biology (Cambridge, UK) with the aim to implement a fast and accurate projection matching algorithm, and to calculate 3D reconstructions that are fully corrected for the contrast transfer function (CTF) of the microscope. Besides the author, a number of other people have contributed to the development of Frealign in various ways, including Tim Grant, Richard Henderson, Dmitry Lyumkis, Alexis Rohou, Charles Sindelar, Alex Stewart, Douglas Theobald, Christine Villeneuve, and Matthias Wolf (Lyumkis, Brilot, Theobald, & Grigorieff, 2013; Sindelar & Grigorieff, 2012; Stewart & Grigorieff, 2004; Wolf, DeRosier, & Grigorieff, 2006), and a GPU-accelerated version was developed by Cheng and his group (Li, Grigorieff, & Cheng, 2010). Its primary application has remained the refinement of particle alignments and 3D reconstruction optimized to reveal details at the highest possible resolution. Other features were added over the years, including refinement of microscope defocus and magnification, correction for the Ewald sphere curvature (Wolf et al., 2006), processing of helical particles (Alushin et al., 2010), 3D classification (Lyumkis et al., 2013), and density masking. Furthermore, algorithms and run scripts were developed to take advantage of parallel computing environments to speed up processing. These developments have made Frealign one of the fastest, most versatile image processing tools for the refinement of single-particle structures, yielding some of the best-resolved reconstructions to date.

Frealign is freely available for download from the Grigorieff lab web page (http://grigorieffiab.janelia.org/frealign). Its primary purpose is to serve as a platform for the development of new image processing algorithms and to support projects in the Grigorieff lab and at the MRC Laboratory of Molecular Biology. Little attention has therefore been devoted to user friendliness and documentation, which require time and resources that were instead devoted to support the primary mission of Frealign. However, some help is provided by the Frealign user forum on the Grigorieff lab web page, which allows users to ask questions and read up on previously answered questions.

The more narrowly defined scope and purpose of Frealign distinguish it from software packages that are designed to offer a complete set of tools for single-particle image processing, including movie processing, CTF determination and correction, particle selection, initial map generation, 2D and 3D classification, refinement, and reconstruction. Using Frealign requires the use of other software to carry out many of the steps necessary to arrive at a 3D reconstruction. Tools for some of these steps have also been developed as stand-alone applications in the Grigorieff lab, such as Signature (Chen & Grigorieff, 2007), CTFFIND (Mindell & Grigorieff, 2003; Rohou & Grigorieff, 2015), Unblur/Summovie (Grant & Grigorieff, 2015b), and magnification distortion correction (Grant & Grigorieff, 2015a), which are also freely available to users. Some of these stand-alone applications will briefly be described at the end of this chapter. However, the main emphasis of the chapter will be on the use of Frealign, describing typical application scenarios and providing practical advice on how to achieve the highest possible resolution. This differs from an earlier paper on Frealign (Grigorieff, 2007) that focused more on algorithmic features.

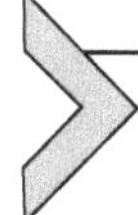

2. FREALIGN ELEMENTS AT A GLANCE

2.1 Running Frealign

The Frealign distribution is available from the Grigorieff lab web page (http://grigoriefflab.janelia.org/frealign) and contains compiled versions of the programs for 64-bit Linux and Mac OS systems. Installation requires unpacking of the archive and adding the path to the compiled programs and run scripts to the user environment.

Frealign has been developed to run on Linux and Mac OS workstations and is run from a command line inside a terminal. Many of the available commands can be listed by issuing the `frealign_help` command. These include `frealign_run_refine` and `frealign_calc_reconstructions`, commands that will start the refinement of a structure or calculate a 3D reconstruction using parameters from a previous Frealign run. The progress toward completion of a task can be followed by monitoring the file `frealign.log`. Apart from these high-level tasks (run scripts) there are also more primitive commands that call up one of the compiled programs that are part of the Frealign distributions, for example, `bfactor.exe`, which allows users to apply a low-pass filter and sharpen a 3D reconstruction using a specified B-factor (see below).

Structure refinement in Frealign is performed iteratively. Each iteration takes input files from the previous cycle and produces new output files that can serve as input for a new cycle. The user can specify the number of cycles to run and how many computing resources to dedicate to the job (see below).

2.2 Required Input

Most data used by Frealign is stored either in text files or image files. While the most extensively used and tested image format is the MRC format (Crowther, Henderson, & Smith, 1996), images stored using the Spider (Frank et al., 1996) and IMAGIC (van Heel, Harauz, Orlova, Schmidt, & Schatz, 1996) formats are also supported. A schematic overview of required and optional input and output files is shown in Fig. 1, together with some of the functional features discussed below. To run Frealign, the user has to set up a text file called `mparameters` (Fig. 2) that specifies

- the computing architecture, for example, cluster type and number of CPUs to use;
- the main Frealign control parameters, including the number of refinement cycles to run;
- data-specific parameters, such as input parameters and microscope settings;
- optional parameters to tune refinement; and
- masking parameters.

Each input line in `mparameters` is annotated to help users choose appropriate settings. Many of the settings involve flags that can be set to "T" (for "true") or "F" (for "false") to turn a feature on or off. Besides `mparameters` the user also has to supply a particle image stack and a particle parameter file—a text file providing some information about each image in the stack. The images in the stack should be uniformly scaled, ie, the average background (solvent) density should be set to zero and its variance to a specified value. In practice, it is sufficient to set the average density of each particle image to zero and its variance to a specified value. This scaling procedure includes the particle density and will therefore be less accurate than a procedure that only considered the background in the calculation. However, in tests comparing results with particles scaled with one of these two procedures, no noticeable differences were observed.

The particle parameter file contains one line of text for each image in the stack. It is therefore important to make sure that the number of lines (excluding comment lines) in a parameter file matches the number of images in the stack. Each line of text normally contains the following information:

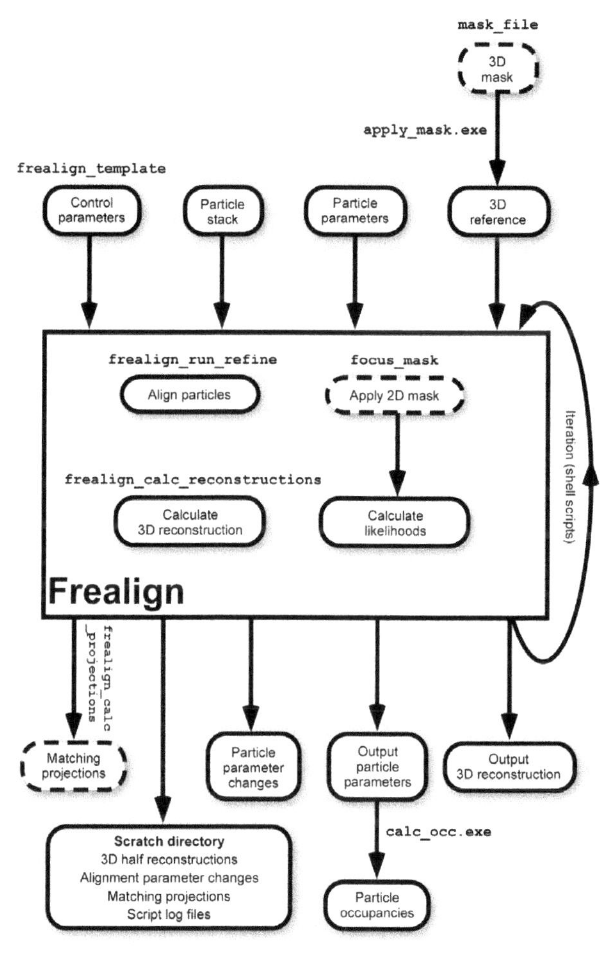

Fig. 1 Schematic overview of Frealign functions, input and output data. Some of the Frealign commands mentioned in the text are shown in Courier font above the functions or files they relate to. Features and files that are optional and not always used are shown with *dashed* borders. Frealign is run by a number of shell scripts that prepare input and output data, manage parallel execution, and perform iterations when more than one cycle is run.

- Euler angles (in degrees) and x,y translations (in Å) describing the alignment of the particle;
- micrograph identifier, image magnification, and defocus;
- a parameter describing the membership of the particle to a class (occupancy);
- relative log likelihood for the particle given the map and alignment parameters;

```
Control parameter file to run Frealign
======================================

This file must me kept in the project working directory from which the refinement scripts are launched.

Note: Please make sure that project and scratch directories (if specified) are accessible by all sub-processes that are run on cluster nodes.

# Computer-specific setting
cluster_type          none      ! Set to "sge", "lsf", "slurm", "stampede", "pbs" or "condor" when running on a cluster, otherwise set to "none".
nprocessor_ref        16        ! Number of CPUs to use during refinement.
nprocessor_rec        16        ! Number of CPUs to use during reconstruction.
mem_per_cpu         2048        ! Memory available per CPU (in MB).

# Refinement-specific parameters
MODE                   1        ! 1, 2, 3 or 4. Refinement mode, normally 1. Set to 2 for additional search.
start_process          2        ! First cycle to execute. Output files from previous cycle (n-1) required.
end_process            2        ! Last cycle to execute.
res_high_refinement    6.0      ! High-resolution limit for particle alignment.
res_high_class         8.0      ! High-resolution limit to calculate class membership (OCC).
nclasses               1        ! Number of classes to use.
DANG                   0.0      ! Mode 3 and 4: Angular step for orientational search. Will be set automatically if set to 0.
ITMAX                200        ! Mode 2 and 4: Number of repetitions of grid search with random starting angles.

# Dataset-specific parameters
data_input            particle  ! Root name for parameter and map files.
raw_images            /path/particle_stack
image_contrast         N        ! N or P. Set to N if particles are dark on bright background, otherwise set to P.
outer_radius         170.0      ! Outer radius of spherical particle mask in Angstrom.
inner_radius           0.0      ! Inner radius of spherical particle mask in Angstrom.
mol_mass            2500.0      ! Molecular mass in kDa of particle or helical segment.
Symmetry               C1       ! Symmetry of particle.
pix_size               1.237    ! Pixel size of particle in Angstrom.
dstep                  5.0      ! Pixel size of detector in micrometer.
Aberration             2.0      ! Sherical aberration coefficient in millimeter.
Voltage              300.0      ! Beam accelleration voltage in kilovolt.
Amp_contrast           0.07     ! Amplitude contrast.

# Expert parameters (for expert users)
XSTD                   0.0      ! Tighter masking of 3D map (XSTD > 0) or particles (XSTD < 0).
PBC                    2.0      ! Discriminate particles with different scores during reconstruction. Small values (5 - 10) discriminate more than large values (50 - 100).
parameter_mask    "1 1 1 1 1"   ! Five flags of 0 or 1 e.g. 1 1 1 1 1). Determines which parameters are refined (PSI, THETA, PHI, SHX, SHY).
refineangleinc         4        ! When larger than 1: Alternate between refinement of OCC and OCC + angles.
refineshiftinc         4        ! When larger than 1: Alternate between refinement of OCC and OCC + angles + shifts.
res_reconstruction     0.0      ! High-resolution limit of reconstruction. Normally set to Nyquist limit.
res_low_refinement     0.0      ! Low-resolution limit for particle alignment. Set to particle dimention or larger.
thresh_reconst         0.0      ! Particles with scores below this value will not be included in the reconstruction.
thresh_refine         50.0      ! Mode 4: Score threshold above which search will not be performed.
nbootstrap          1000        ! Number of bootstrap volumes to calculate real-space variance map.
FMAG                   F        ! T or F. Set to T to refine particle magnification. Not recommended in most cases.
FDEF                   F        ! T or F. Set to T to refine defocus per micrograph. Not recommended in most cases.
FASTIG                 F        ! T or F. Set to T to refine astigmatism. Not recommended in most cases.
FPART                  F        ! T or F. Set to T to refine defocus for each particle. Not recommended in most cases.
FFILT                  T        ! T or F. Set to T to apply optimal filter to reconstruction. Recommended in most cases.
FMATCH                 F        ! T or F. Set to T to output matching projections. Only needed for diagnostics.
FBEAUT                 F        ! T or F. Set to T to apply symmetry also in real space. Not needed in most cases.
FBOOST                 T        ! T or F. Set to F if potential overfitting during refinement is a concern.
RBfactor               0.0      ! B-factor sharpening (when < 0) applied during refinement. Not recommended in most cases.
beam_tilt_x            0.0      ! Beam tilt in mrad along X-axis.
beam_tilt_y            0.0      ! Beam tilt in mrad along y-axis.
mp_cpus                1        ! Number of CPUs to use for each reconstruction job.
restart_after_crash    F        ! T or F. Set to T to restart job after a crash.
delete_scratch         T        ! Delete intermediate files in scratch directory.
qsub_string_ref  ""             ! String to add to cluster jobs submitted for refinement (only for sge, lsf, slurm and pbs clusters).
qsub_string_rec  ""             ! String to add to cluster jobs submitted for reconstruction (only for sge, lsf, slurm and pbs clusters).
first_particle
last_particle
frealign_bin_dir
scratch_dir

# Masking parameters (for expert users)
mask_file
mask_edge              5        ! Width of cosine edge in pixels to add around mask. Set to 0 to leave mask unchanged.
mask_outside_weight    0.0      ! Factor to downweight density outside of mask (normally 0.0 - 1.0).
mask_filt_res          0.0      ! Filter radius (in A) to low-pass filter density outside density. Set to 0.0 to skip filtering.
mask_filt_edge         5        ! Width of cosine edge in reciprocal pixels to add to filter function.

focus_mask       ""             ! Four numbers (in Angstroms) describing a spherical mask (X, Y, Z for mask center and R for mask radius).
```

Fig. 2 Template for the `mparameters` file containing all the control parameters required to run Frealign. Each keyword has an assigned value or string (text) and, in most cases, a comment line explaining how the parameter should be set. The control parameters are divided into different sections to help users customize the parameters according to their environment and project. Parameters listed in the expert section usually do not need to be changed until refinement and classification have converged. These parameters allow users to tune their refinement

- standard deviation of the estimated background noise; and
- particle score.

At the beginning of a new project, when alignment parameters are not known, the user can supply an abbreviated particle parameter file that will contain only four numbers per line:

- an integer to identify the micrograph the particle came from and
- two defocus values and an astigmatic angle (in degrees) needed to describe the microscope CTF. These values are written out by the program CTFFIND (Mindell & Grigorieff, 2003; Rohou & Grigorieff, 2015), for example.

Frealign will convert this file into a full parameter file by adding random Euler angles, setting the particle translations to zero and filling in the rest of the parameters with nominal values that can then be updated in later refinement cycles.

2.3 Optional Input

In most cases, users will also supply a 3D reconstruction (or several reconstructions for multireference refinement and classification, see below) on input that was obtained either from a previous Frealign run or through other means. The 3D reconstruction is used by Frealign as a reference to refine the particle parameters (see later). If no 3D reconstruction is supplied, Frealign will calculate one from the input particle parameters. If these parameters are given in the abbreviated format, ie, no Euler angles and *x*,*y* translations are available, Frealign will calculate a 3D reconstruction using randomly assigned angles after converting the abbreviated parameter file to a full parameter file, thus creating a random startup structure (see below).

Finally, a 3D mask file can be specified in `mparameters` that Frealign will use to mask the input 3D reconstruction, allowing the user to focus on specific regions of the molecule during refinement (see below).

2.4 Output

Each refinement cycle performed by Frealign will generate a new particle parameter file and 3D reconstruction (if multiple references are used, parameter files and reconstructions are generated for each). A scratch directory used by Frealign to hold temporary files also contains 3D reconstructions calculated from half the data each. These are used by Frealign to calculate a Fourier shell correlation (FSC) curve to estimate the resolution of the final

3D structure (Harauz & van Heel, 1986). These half reconstructions can be ignored in a typical Frealign run but are sometimes useful if users would like to perform their own resolution estimation, for example, using ResMap (Kucukelbir, Sigworth, & Tagare, 2014). The FSC curve for each reconstruction can be found in a resolution table at the end of the particle parameter file belonging to the reconstruction, and plots can be generated using the command `frealign_plot_fsc`. The table also contains other information, for example, an adjusted FSC curve called `Part_FSC`, which estimates the resolution of the reconstruction (usually at the threshold value of 0.143, Rosenthal & Henderson, 2003) after removing all solvent noise. This adjusted curve should therefore be taken as the curve shown in publications to support resolution claims. While resolution estimates for reconstructions after removing solvent noise are often obtained by applying tight masks, Frealign obtains this estimate by considering the volume occupied by the particle (Sindelar & Grigorieff, 2012). The volume is calculated from the molecular mass of the particle provided by the user in `mparameters` using a conversion factor of 810 Da/nm^3, equivalent to 1.25 Å^3/Da (Matthews, 1968). The volume is also used to calculate the spectral signal-to-noise ratio (SSNR) present inside the reconstructed particle density (called `Rec_SSNR` in the final resolution table). The particle SSNR (called `Part_SSNR` in the final resolution table) is used to apply an optimal filter to the final reconstruction (Sindelar & Grigorieff, 2012) if `FFILT` is set in `mparameters`. In addition to the SSNR filter, it is recommended to apply a negative B-factor to the final reconstruction to sharpen the density, and a low-pass filter with a cosine-edged cutoff to set terms beyond the resolution limit to 0. Both operations can be accomplished with `bfactor.exe` (see below).

2.5 Naming Conventions

Frealign uses the following naming scheme for alignment parameter files and 3D reconstructions: file names consist of a seed that should identify the particle (for example, `70S` for the 70S ribosome), followed by "`_M_rN`" where `M` and `N` are integers that signify the refinement cycle number and reference, respectively. Typically, when initializing a Frealign run, only one reference is used (see below) and the refinement starts with cycle 1. Therefore `M` should be set to 0 and `N` to 1 (eg, "`70S_0_r1`"). The parameter files are expected to have the extension "`.par`" while particle image stacks and 3D reconstructions have extensions that depend on the file format. MRC/CCP4 files have the ending "`.mrc`," Spider "`.spi`," and IMAGIC "`.hed`" and "`.img`."

3. ALGORITHMS

A description of the algorithms employed by Frealign to refine alignment parameters and perform classification is given in Grigorieff (2007) and Lyumkis et al. (2013). Briefly, Frealign performs projection matching to determine more accurate alignment parameters. Projections are calculated using the reference map provided on input and alignment parameters for each particle are updated according to the projection that generates the highest correlation coefficient. The user has a choice to search for projections with parameters close to those previously found (local search) by setting `MODE` to 1 in `mparameters`, or to perform a global parameter search with randomly generated parameters (`MODE` set to 2 and `ITMAX` set to the desired number of trials, eg, 100) or with parameters systematically chosen according to a search grid (`MODE` set to 3 and `DANG` set to the desired angular step in degrees, eg, 10). If `MODE` is set to 3, the user can also set `DANG` to 0 to let Frealign calculate an appropriate angular step, based on the particle radius and refinement resolution limit (see below) set by the user. The correlation coefficient calculated by Frealign is weighted according to the SSNR present in the particle images. The average particle SSNR is contained in the final resolution table (`Part_SSNR`, see above), together with the FSC curve and other statistics. It is important not to delete this table from the end of a parameter file because it is read by Frealign in the next refinement cycle.

3D classification is done using a maximum likelihood approach. Frealign calculates relative likelihoods (ie, likelihood values that are missing constants necessary to put the likelihoods on an absolute scale) of each particle given a 3D reference map and a set of alignment parameters. The logarithms of these values are listed in the particle parameter files (column `LogP`). For classification, multiple reference structures are used and likelihood values are calculated for each. At the end of a refinement cycle, these likelihood values are converted by the program `calc_occ.exe` (part of the Frealign distribution and executed automatically by the run script) into weights, so-called occupancies (column `OCC`) that determine the partitioning of each particle into the different classes. To speed up classification and improve convergence, the user can specify if Frealign should refine alignment parameters with every classification cycle, or if alignment parameter refinement should only be done every *N*th cycle. Typically, alignment parameter refinement should only be done every third or fourth cycle by setting the values for both `refineangleinc` and `refineshiftinc` to 3 or 4 in `mparameters`. During cycles where alignment parameters are not refined, these parameters remain unchanged and the cycles complete more quickly. The user can also specify different numbers for `refineangleinc` and `refineshiftinc` to refine angles and shifts on different schedules but this is usually not necessary.

Frealign has been designed to reduce or entirely avoid overfitting of parameters. Overfitting is a well-known problem in single-particle work (Grigorieff, 2000; Stewart & Grigorieff, 2004). It manifests as features in the reconstruction that result from the alignment of noise rather than signal. If all particles in a dataset are aligned against the same reference, overfitting can also lead to inflation of the FSC and unrealistically high-resolution estimates. The latter problem can be mediated by calculating FSC curves from reconstructions that were refined entirely separately, thus not sharing a common reference (Grigorieff, 2002; Henderson et al., 2012). While this reduces the chances of generating inflated FSC values, it does not directly address overfitting. Frealign provides several ways to counter overfitting and with it, inflated FSC values:

(a) *Weighting of the correlation coefficient.* The SSNR-weighted correlation coefficient used during projection matching (see above) aims at giving data with a stronger signal a higher weight than noisier data. This helps the signal drive the alignments and reduces the impact of the noise. A potential weakness of this approach is an incorrect measurement of the SSNR, which might be biased toward higher values like the FSC.

(b) *Maximizing the absolute value of the correlation coefficient.* The user has the option to maximize an unsigned version of the weighted correlation coefficient (Stewart & Grigorieff, 2004) instead of the signed version by setting `FBOOST` to "F" in `mparameters`. When switching to the unsigned correlation coefficient, alignment of the strong low-resolution signal is usually unaffected (at very low resolution, below 30 Å, Frealign will always use a signed correlation to ensure proper centering of particles). However, as the SSNR decreases toward higher resolution, alignment may be driven more by the noise and, using the unsigned correlation coefficient, may end up being aligned in-phase (positive correlation) or out-of-phase (negative correlation). Subsequent averaging of images during 3D reconstruction will lead to strong attenuation of the incoherently aligned noise. Setting `FBOOST` to "T" may help the alignment in some cases and the user is encouraged to try different settings to get the best results (see below). However, setting

FBOOST to "T" also means that the chance of overfitting is increased. Careful validation of additional features appearing in the map is therefore necessary.

(c) *Setting a refinement resolution limit.* In Frealign, the data used during refinement are bandpass filtered. The low- and high-resolution limits can be set in mparameters but users usually only change the latter while the former is left at 0 to let Frealign set the value automatically. It is good practice to monitor the progress of refinement and limit the high-resolution limit to a value well below the current resolution limit. For example, if the Part_FSC curve suggests a resolution of 8 Å, the resolution limit should probably not exceed 10 Å. Limiting the resolution during refinement (and classification) means that the FSC values at higher resolution will show little or no bias from noise overfitting. If, on the other hand, the FSC curves suggest that the resolution of the reconstruction increases more or less in parallel with the resolution limit set by the user, this is usually a strong sign that the refinement is unsuccessful and does not produce reliable structural details.

4. TYPICAL APPLICATION SCENARIOS

In this section, a few typical application scenarios are described to help users get started with Frealign. The processing steps for cases not described here may be derived from the scenarios below, giving users the flexibility to adapt to their own situations.

4.1 Refinement of a Structure Generated with Different Software

This will be the most common scenario in which the following input data exist:

- a 3D reconstruction,
- a particle image stack,
- a list of defocus parameters (two defocus values and an astigmatic angle) for each particle,
- a list of micrograph numbers detailing where each particle came from (optional), and
- a list of alignment parameters (Euler angles and x,y translations) for each particle (optional).

If particle alignment parameters are not available, the user will have to generate a startup parameter file with a list that contains four numbers per line

```
1  21241.2  21446.7   32.56
1  21241.2  21446.7   32.56
1  21241.2  21446.7   32.56
1  21241.2  21446.7   32.56
1  21241.2  21446.7   32.56
1  21241.2  21446.7   32.56
1  21241.2  21446.7   32.56
1  21241.2  21446.7   32.56
1  21241.2  21446.7   32.56
1  21241.2  21446.7   32.56
2  19113.7  19301.8   41.41
2  19113.7  19301.8   41.41
2  19113.7  19301.8   41.41
2  19113.7  19301.8   41.41
2  19113.7  19301.8   41.41
2  19113.7  19301.8   41.41
2  19113.7  19301.8   41.41
2  19113.7  19301.8   41.41
2  19113.7  19301.8   41.41
2  19113.7  19301.8   41.41
```

Fig. 3 Example of a startup parameter file containing the required data for 20 particles. Each line lists a micrograph number that the particle originates from, as well as defocuses values and astigmatic angle determined for this micrograph. The defocus information can vary from particle to particle if more accurate information is available, for example, by using CTFTILT (Mindell & Grigorieff, 2003). If micrograph numbers are not known, they can be set to a constant number larger than 0 or to the particle number.

and one line per particle image (Fig. 3). In each line, the first number identifies the micrograph that the corresponding particle originates from and the last three numbers provide the defocus information (following CTFFIND conventions (Mindell & Grigorieff, 2003; Rohou & Grigorieff, 2015)). Numbers can be separated by commas or spaces; no other formatting is required. If micrograph numbers are not available, the user can set the identifiers for all particles to a constant number, for example, 1. The startup parameter file will be renamed by Frealign and replaced with a full Frealign-style parameter file (Fig. 4) that contains random Euler angles and 0,0 for the x,y translations.

If particle alignment parameters are also known, a Frealign-style parameter file should be generated from the file originating from the other software. Conversion scripts are available on the Frealign download page for different software. Users are cautioned, however, to make sure that the conversion worked correctly by checking the results for a few particles. The format of the parameter files originating from other software may change and these changes are usually not immediately accommodated in the conversion

C	PSI	THETA	PHI	SHX	SHY	MAG	FILM	DF1	DF2	ANGAST	OCC	LogP	SIGMA	SCORE	CHANGE
1	224.68	301.05	104.64	-0.99	-3.26	104011	1	21241.2	21446.7	32.56	100.00	-6342	0.8298	24.03	0.63
2	222.50	341.41	357.66	1.90	1.63	104011	1	21241.2	21446.7	32.56	100.00	-6444	0.8403	23.08	-0.06
3	192.03	97.21	178.83	-1.00	-0.38	104011	1	21241.2	21446.7	32.56	100.00	-7760	0.8809	17.92	2.41
4	127.34	143.60	123.36	1.25	2.54	104011	1	21241.2	21446.7	32.56	100.00	-7275	0.8725	18.48	-0.56
5	138.80	336.29	113.17	-1.37	-0.78	104011	1	21241.2	21446.7	32.56	100.00	-7426	0.8704	19.98	1.09
6	99.27	24.49	348.32	-0.52	-0.31	104011	1	21241.2	21446.7	32.56	100.00	-6681	0.8552	22.98	-1.34
7	317.37	195.53	124.72	0.30	-2.47	104011	1	21241.2	21446.7	32.56	100.00	-7145	0.8604	21.21	0.98
8	101.06	347.28	62.85	-1.95	-3.25	104011	1	21241.2	21446.7	32.56	100.00	-6974	0.8624	24.71	0.60
9	21.36	358.81	132.15	1.48	-1.59	104011	1	21241.2	21446.7	32.56	100.00	-8992	0.9527	20.33	1.89
10	334.18	264.77	261.20	-1.72	-1.30	104011	1	21241.2	21446.7	32.56	100.00	-7033	0.8573	22.35	-0.27
11	138.79	268.49	294.01	-1.09	-0.61	104011	2	19113.7	19301.8	41.41	100.00	-7000	0.8600	24.16	0.17
12	93.39	148.69	315.88	-3.19	-1.31	104011	2	19113.7	19301.8	41.41	100.00	-7273	0.8830	21.08	-0.98
13	132.31	228.06	266.01	-2.13	-0.40	104011	2	19113.7	19301.8	41.41	100.00	-7270	0.8675	23.83	1.77
14	178.55	200.04	86.08	-0.80	-0.83	104011	2	19113.7	19301.8	41.41	100.00	-7404	0.8740	20.65	0.16
15	22.43	177.64	118.28	1.60	-0.27	104011	2	19113.7	19301.8	41.41	100.00	-7102	0.8609	26.86	0.41
16	244.62	284.19	201.33	1.48	1.09	104011	2	19113.7	19301.8	41.41	100.00	-7206	0.8708	19.26	0.65
17	276.29	189.05	153.91	2.17	0.33	104011	2	19113.7	19301.8	41.41	100.00	-7605	0.8862	19.81	-0.43
18	182.05	332.75	208.28	0.25	2.47	104011	2	19113.7	19301.8	41.41	100.00	-7851	0.8958	22.53	-0.24
19	325.00	98.83	130.36	-0.41	-0.09	104011	2	19113.7	19301.8	41.41	100.00	-7463	0.8792	21.37	0.57
20	276.20	154.32	148.45	1.38	-1.24	104011	2	19113.7	19301.8	41.41	100.00	-7060	0.8579	17.51	0.97

Fig. 4 Example of a full Frealign alignment parameter file, after running Frealign with the startup file in Fig. 3, containing Euler angles (PSI, THETA, PHI) and *x,y* translations (SHX, SHY), as well as micrograph numbers (FILM), magnification (MAG) and defocus (DF1, DF2, ANGAST) information, occupancies (OCC), log likelihoods (LogP), and scores (SCORE). The SIGMA column lists estimates of the standard deviation of the noise present in the particle images while CHANGE lists the change in the score compared with the previous refinement cycle.

scripts. It is recommended that users familiarize themselves with a scripting language (eg, shell script or Python) and then adapt the available conversion scripts to their own needs.

The particle image stack must contain uniformly scaled images (see above) with an even box size. For both ice-embedded and negatively stained particles and underfocused images, the particles should be dark on lighter background (there is also an option to use the opposite contrast, see below). Finally, some parameters have to be set or adjusted in the `mparameters` file. A fresh `mparameters` file can be generated using the Frealign command `frealign_template`. Using a text editor, the following settings should be adjusted:

- `cluster_type`: should be set to the computing infrastructure used. If computations are done on a local workstation, this should be set to "`none`."
- `nprocessor_ref`, `nprocessor_rec`: should be set to the number of CPUs to be used for parallelized computation during parameter refinement and reconstruction, respectively. While there is relatively little overhead when adding CPUs for refinement (using 1000–2000 CPUs should work without problems), the speedup from additional CPUs used for reconstruction will depend on the speed of the disk storage. Values for `nprocessor_rec` should probably not exceed 100 and are more typically 30–50. It is recommended to set the values for both parameters to a multiple of the number of classes used for classification (see below).
- `MODE`: should be set to 1 if valid particle alignment parameters are available on input, or 3 if only defocus values and micrograph information are provided. Setting `MODE` to 1 will perform a local search for improved alignments while 3 will perform a global search that will take significantly more time and should therefore only be run when necessary. There are other run modes available (2 and 4) that are less commonly used and not discussed here. If the user decides to run with `MODE` set to 3, it is recommended to work with binned data. Using Frealign's tool `resample.exe` (or `resample_mp.exe` for multi-CPU environments), the pixel size of the particle image stack and 3D reference reconstruction can be changed to speed up processing. For example, if the native pixel size of a dataset is 1.5 Å but the global search with `MODE` set to 3 is performed at lower resolution, for example, at 20 Å (see below), the stack and 3D reference can be resampled to a pixel size of 9 Å, giving a Nyquist frequency limit of 18 Å, ie, just a little higher than the chosen resolution limit of 20 Å. At a later stage, processing can be switched back to a

smaller pixel size to enable higher resolution refinement. When changing the pixel size, it is important to also change the settings for `pix_size` and `dstep` in `mparameters` (see below). Furthermore, the 3D reconstruction from the previous cycle has to be recalculated using the new pixel size, or processed with `resample.exe` to generate a volume with the correct pixel size. Frealign recalculates the reconstruction automatically if the previous reconstruction is deleted (or moved to a different directory). Additional speedup can be obtained by reducing the margins around the particles (if possible) using the CROP tool available for download on the Grigorieff lab web page (see below). Images should not be cropped when refining at high resolution as this may lead to CTF aliasing loss of high-resolution signal (Rohou & Grigorieff, 2015).

- `start_process`, `end_process`: should be set to the first and last refinement cycle to be run. For example, if the initial parameter file carries cycle number 0 and 10 cycles should be run, the values for `start_process` and `end_process` should be set to 1 and 10, respectively.
- `res_high_refinement`: should be set to the desired resolution limit used during refinement. This will usually depend on the estimated resolution of the input 3D reference (see above). `mparameters` contains a second resolution limit, `res_high_class`, which determines the resolution used for classification. Users can keep this value at the default of 8 Å as Frealign will always adjust it internally to the value for `res_high_refinement` if that indicates a lower resolution than `res_high_class`.
- `nclasses`: determines the number of classes to be refined. This should be set to 1 when working with a parameter file that does not contain particle alignment parameters, or if the alignment parameters are not very accurate. Classification of particles into multiple classes is only recommended at a later stage of the refinement, when refinement with a single class does not improve the resolution further. To switch on classification, the user simply sets `nclasses` to a value larger than 1.
- `DANG`: determines the angular step size used in a global search (`MODE` set to 3). This should be set to the default of 0 to enable automatic step sizing by Frealign, based on the specified resolution limit (see above) and particle radius (see below). The user can specify a fixed value by entering a value larger than 0. A second parameter, `ITMAX`, is not used for `MODE` 1 and 3 and does not need to be changed by the user.
- `data_input`: the text string defining the seed used to generate the file names for alignment parameters files and reconstructions (see above).

- `raw_images`: name of the particle image stack. The path can either be relative to the working directory or absolute.
- `image_contrast`: normally set to N to indicate that particles appear dark against light background (see above). Users can also set this to P if the particle images have opposite contrast.
- `outer_radius`: determines the outer radius of the spherical mask to be applied to the final reconstruction, as well as the radius of a circular mask applied to the particle images during refinement. If this is set to a negative value no mask will be applied.
- `inner_radius`: determines the inner radius of the spherical mask to be applied to the final reconstruction. It is not used to mask the particle images. This parameter is normally set to 0 but if the particle is hollow or contains disordered density (eg, a clathrin coat or an icosahedral virus capsid), setting the inner radius to an appropriate value allows users to mask the inside of the particle and reduce noise.
- `mol_mass`: should be set to the total molecular mass of the particle. The value is given in kDa and determines how the FSC curve is scaled to calculate `Part_FSC`, which provides a more accurate resolution estimate of the particle density (see above).
- `Symmetry`: should be set to the assumed symmetry of the particle. The default is C1, ie, no symmetry.
- `pix_size`: should be set to the desired pixel size of the output reconstruction in Angstroms. Usually this is the same as the pixel size of the input particle images.
- `dstep`: should be set to the effective pixel size of the detector in micrometers. This is usually the physical pixel size of the detector, for example, 5 μm for the K2 detector (Gatan, Pleasanton, CA). However, if the particle images are binned, both the image pixel size and the effective detector pixel size change. For example, 2×2 pixel binning of the images doubles their pixel size (`pix_size`) and the effective detector pixel size (`dstep`). Users can check that they have set `pix_size` and `dstep` correctly by dividing `dstep` by `pix_size`. The result should be the particle magnification indicated in the alignment parameter file.
- `Aberration`, `Voltage`, `Amp_contrast`: should be set to the appropriate microscope parameters. `Aberration` and `Voltage` are given in millimeters and kilovolts, respectively. For example, for the FEI Titan Krios microscope, typical values for `Aberration`, `Voltage`, and `Amp_contrast` are 2.7, 300, and 0.07. If `Amp_contrast` is set to a negative number, CTF

correction is turned off in Frealign. However, this only works when image_contrast is set to *N* and the particle images have been pretreated to correct for the CTF (eg, phase flipping), yielding particles that appear light against a dark background.

The final list of files in the working directory needed to run Frealign includes mparameters, the particle parameter file (either with four values per line or a full Frealign parameter file with Euler angles, *x*,*y* translations and additional columns with occupancies, likelihood values, and scores), a 3D reference map and a particle image stack. For example, if the seed for the file names is 70S, and the stack is named particle_stack.mrc, the list of files includes: mparameters, 70S_0_r1.par, 70S_0_r1.mrc, particle_stack.mrc (assuming MRC file format). Issuing the frealign_run_refine command will then start the refinement. New parameter files and 3D reconstructions should appear in the working directory as refinement cycles are completed. Users can follow the status of the refinement by inspecting the file frealign.log (see above), as well as temporary files generated in the scratch directory, which is created in the working directory unless specified differently in mparameters.

Refinement progress can be monitored in several ways. Users are encouraged to check the resolution statistics appended to the end of the parameter files and verify that the resolution improves from cycle to cycle while observing the steps discussed above to avoid inflated resolution estimates. If there is no noticeable resolution improvement for a number of cycles (eg, 5), the refinement may have converged. Users can try increasing the refinement resolution limit if the resolution of the reconstruction is significantly higher than this limit (see above) and run a few more cycles to see if this leads to further improvement. To continue refinement, the numbers for start_process and end_process must be updated before issuing the frealign_run_refine command. Users can also check if the particle alignment parameters are still changing significantly between cycles. In the scratch directory, text files containing .shft_ in their names list the changes in the parameters of the current cycle relative to the previous cycle. These files are normally deleted at the end of a cycle, so users have to check them while a cycle is running. Finally, the command frealign_calc_stats will display the average score, relative log likelihood per particle (and occupancy, see below) for a specified round. Both scores and likelihood values should increase during refinement until convergence has been reached. However, if the resolution limit is changed during refinement, scores and likelihood values will be affected. Therefore changes in these values are only meaningful between cycles that use the same resolution limits.

4.2 3D Reconstruction Using Parameters from a Previous Frealign Run

To calculate (or recalculate) a 3D reconstruction using an existing particle parameter file containing Euler angles and *x*,*y* translations (ie, a full Frealign parameter file), the user has to set the following values in `mparameters`: `cluster_type`, `nprocessor_rec`, `nclasses`, `data_input`, `raw_images`, `image_contrast`, `outer_radius`, `inner_radius`, `mol_mass`, `Symmetry`, `pix_size`, `dstep`, `Aberration`, `Voltage`, and `Amp_contrast`. Details of how these values should be set are provided in the previous section. If the previous run used more than one reference/class, `nclasses` will have to be set accordingly. The command `frealign_calc_reconstructions` starts the calculation and a new reconstruction (or reconstructions if `nclasses` is larger than 1) will appear in the working directory when Frealign has finished. The parameter file (or files if `nclasses` is larger than 1) will be appended with the relevant resolution statistics. Therefore recalculating reconstructions multiple times will append multiple tables at the end of the parameter files. Only the last table in a parameter file will be used by Frealign in the next refinement cycle.

4.3 3D Classification

3D classification is typically done only after refinement with a single reference has converged, ie, there is no more improvement in resolution and particle parameters do not change much anymore from cycle to cycle (see above). To turn on classification, the value for `nclasses` (in `mparameters`) must be changed from 1 to the desired number of classes. After updating `start_process` and `end_process` to continue from the previous last cycle, refinement with classification is performed by issuing the `frealign_run_refine` command. Frealign will rename the particle parameter file and 3D reconstruction from the previous cycle and replace them with multiple parameter files and 3D reconstructions that result from randomly assigned particle occupancy values (`OCC` column in the parameter files). These occupancy values will be refined in subsequent cycles and, if successful, will indicate the membership of each particle to a class. Often, after refinement, occupancy values will be either 100 or 0, indicating that a particle is, or is not a member of a class, respectively. However, intermediate values are also possible, indicating that the assignment to a class is not unique. For each particle, the sum of the occupancy values from all classes always adds up to 100.

Typically, several tens of refinement/classification cycles (eg, 50) have to be run before convergence is reached. As before, progress can be monitored by inspecting the resolution statistics at the end of the particle parameter files, changes in the `.shft_` files inside the scratch directory and by running the `frealign_calc_stats` command. This command will also calculate the average particle occupancy for each class. Upon successful classification, significant differences should emerge in the density maps representing the different classes. Users should inspect and compare density maps using display programs such as UCSF Chimera (Pettersen et al., 2004) and TIGRIS (http://tigris.sourceforge.net).

If classes represented by different reconstructions are available from a previous Frealign run or another source, classification can also be initiated using these reconstructions. This can be done by following these steps:

- save the particle parameter file and reconstruction from the single-class refinement in a safe place (or rename them),
- copy the saved particle parameter file multiple times to generate new parameter files for each of the classes. For example, if classification should proceed with three classes starting at cycle 101 and the seed for the file names is `70S`, copy the original parameters file into `70S_100_r1.par`, `70S_100_r2.par`, and `70S_100_r3.par`. Similarly, copy the available reconstructions representing the previously obtained classes into `70S_100_r1.mrc`, `70S_100_r2.mrc`, and `70S_100_r3.mrc` (assuming MRC file format). Finally, set `nclasses` (in `mparameters`) to the number of classes used (here 3) and issue the command `frealign_run_refine`.

As before, the refinement should be run with `MODE` set to 1 unless previous alignment parameters are not available, in which case `MODE` should be set to 3 for the initial one or two cycles. It is important to remember that when more than one reference is used (`nclasses` set to a value larger than 1) Frealign will only refine the alignment parameters every *N*th round where *N* is specified by `refineangleinc` and `refineshiftinc` (see above). When initiating a refinement with `MODE` set to 3, `refineangleinc` and `refineshiftinc` should be temporarily set to 1 until the alignment parameters are deemed roughly correct and `MODE` is set to 1.

As a note of caution, users should be aware that differences in the densities that appear after several cycles of classification may reflect noise and not real structural differences in the particles. Emerging features should always be evaluated on the basis of plausibility and what else is known about the sample. If the classification results are mainly driven by noise the average occupancy is often very similar for each class. Users must therefore be suspicious of classification results that suggest similar average occupancies of all classes.

4.4 Selecting or Merging Particles from Different Classes

When classification of a dataset has converged, it is often useful to continue refinement and/or classification using a subset of the data. For example, if classification yielded five classes and, after inspection of the densities, two of the classes are so similar that they are considered the same structure, particles from these two classes can be merged into one class. The reconstruction representing the combined class should then reach higher resolution since the total number of particles will be larger compared with the two original classes. To merge particles from several classes, Frealign comes with a tool called `merge_classes.exe`. Input prompts request the file names of the particle parameter files belonging to the classes to be merged, the particle image stack, and criteria for including a particle in the merged output. The criteria include minimum values for occupancy and score. Normally, a minimum occupancy of 50 and score of 0 should be used but users can include or exclude more particles by changing these numbers. `merge_classes.exe` will then generate a new particle parameter file and image stack (using file names specified by the user) that contain only the selected particles. These can then be used in additional refinement and classification cycles using Frealign.

A different tool, `select_classes.exe`, allows the user to select particles from a subset of classes for further refinement. Unlike `merge_classes.exe`, `select_classes.exe` produces several particle parameter files on output that are related to the parameter files provided on input. `select_classes.exe` will remove particles from these parameter files and image stack that do not belong to any of the selected classes. The new parameter files and image stack can then be used for further refinement and classification by Frealign. `select_classes.exe` can therefore be used to remove unwanted particles, for example, because they belong to a "junk" class or a good class that represents a state that should not be refined or classified further. The latter situation may occur when a sample containing many different conformations is classified. To isolate all the different conformations, one strategy would be to allow a larger number of classes during classification. However, this increases the need for computational resources and may reduce the chance of finding smaller classes because particles may be misclassified to belong to a bigger class simply due to the better signal represented in this class (Yang, Fang, Chittuluru, Asturias, & Penczek, 2012). Therefore removing particles belonging to some of the bigger classes, or selecting one of the bigger classes and classifying it further into a smaller number of subclasses may result in new classes to emerge and will reduce the need for computational resources. When looking for new conformations that are not well represented in a dataset, it is often also useful to employ masking (see below).

4.5 Generating an Initial Map

Although there is no dedicated algorithm implemented in Frealign to generate an initial map, the following scheme can be applied. The user has to supply a startup parameter file with micrograph identifiers and defocus values for each particle (see above) and a particle image stack. It is recommended to limit the parameter list and stack to a subset of about 10,000 particles that have been selected from a larger stack based on defocus and some other "quality" criteria. The selected defocus range should be at the high end of the range used for the entire dataset. Particle quality can be ascertained either by manual picking of the particles by an experienced user, or by selecting particles based on 2D classification, for example, using ISAC (Yang et al., 2012). Furthermore, it is recommended to use the `resample.exe` tool (see above) to change to pixel size of the image stack to a value between 4 and 5 Å to speed up computation. The size of the particle images can be further reduced by using the CROP tool (see below) that is available for download on the Grigorieff lab web page and allows trimming of the margin around particles (users must make sure that the trimming does not cut into the particles). Finally, the refinement resolution limit should be set to a value between 30 and 40 Å (`res_high_refinement`), `nclasses` should be set to 1, `MODE` should be set to 3, the assumed particle symmetry should be specified (`Symmetry`), `refineangleinc` and `refineshiftinc` should be left unchanged (defaults are 4 and 4), and `FBOOST` should be set to T. Using the startup parameter file and stack (named according to Frealign's naming scheme, see above), as well as an appropriately set `mparameters` file, the user should run one "refinement" cycle by issuing the `frealign_run_refine` command. Using the previous example, the needed files include `mparameters`, `70S_0_r1.par`, `particle_stack.mrc`, and `start_process` and `end_process` should both be set to 1. Frealign will rename the startup parameter file `70S_0_r1.par` and replace it with a full Frealign parameter file with randomly set Euler angles and translations set to 0,0. The initial reconstruction generated by Frealign will therefore approximate a featureless sphere. When the initial refinement cycle has finished, between three and six classes should be specified by changing `nclasses` to the appropriate number, and seven more cycles should be run (`start_process` and `end_process` should be set to 2 and 8, respectively). As more refinement and classification cycles are executed, the spheres will gain features. When the specified number of cycles has been run, another eight cycles should be run but with a somewhat increased resolution limit, for example, increasing it from 40 to 30 Å. This should be repeated while increasing the resolution limit every time until a resolution of about 10 Å is reached. A possible schedule would therefore include

- cycle 1, resolution limit set to 40 Å, 1 class;
- cycles 2–8, resolution limit set to 40 Å, 3–6 classes;
- cycles 9–16, resolution limit set to 30 Å, 3–6 classes;
- cycles 17–24, resolution limit set to 20 Å, 3–6 classes;
- cycles 25–32, resolution limit set to 15 Å, 3–6 classes;
- cycles 33–40, resolution limit set to 10 Å, 3–6 classes.

When a resolution limit of about 10 Å has been reached, the class with the highest overall FSC should be selected and taken as a starting structure for another round of 40 refinement cycles that follow the same schedule. However, for this new round, the reconstruction representing the selected class with the highest FSC should be provided alongside the startup parameter file. Following the `70S` example, for the next 40 cycles, refinement could start with cycle number 101, `70S_100_r1.par` should contain the startup parameters (this could simply be copied from `70S_0.par`, which should be available from the previous round of refinement cycles) and `70S_100_r1.mrc` should be a copy of (or symbolic link to) the selected best reconstruction. The schedule of classification with successively increasing resolution should be repeated until one of the classes shows an FSC curve indicating a resolution that extends significantly beyond 10 Å—an indication that the corresponding reconstruction contains reliable signal beyond 10 Å and therefore, that the structure is likely correct. Users can vary the number of cycles, classes, and resolution thresholds used in every new round. A larger number of cycles and classes may increase the chances of finding classes that represent the correct structure(s) but the computational cost increases and it may be necessary to use more than 10,000 particles for the trials to make sure there is a sufficient number of particles in each class (a minimum of about 2000 particles is recommended). An example of generating an initial model from an initial reconstruction calculated with randomly assigned Euler angles is shown in Fig. 5 for a cryo-EM dataset of L protein of vesicular stomatitis virus (VSV-L) that was published previously (Liang et al., 2015).

4.6 Using Masks

Frealign can employ two types of masks. The first involves a 3D volume stored in the same format as the input references. The mask file should contain positive and negative numbers outlining regions of the reference to be included or excluded by the mask, respectively. The mask file must be specified in `mparameters` using the `mask_file` key (if no filename is provided, Frealign assumes that a 3D mask is not used). Frealign will apply (ie, multiply) the mask to the input references before starting refinement after making the following modifications to the mask file:

- all positive mask densities are reset to 1;
- all negative mask densities are reset to 0 or another value specified in `mparameters`, keyword `mask_outside_weight`;
- a soft edge is added to the mask (ie, the region now containing voxels set to 1) using a width specified in pixels in `mparameters`, keyword `mask_edge` (usually set to 5);
- if specifying a value for `mask_outside_weight` larger than 0, the user can also specify if the density outside the mask should be low-pass filtered by setting keyword `mask_filt_res` to specify the filter resolution (a value of 0 turns the filter off) and `mask_filt_edge` to specify the width of a smooth edge (in units of Fourier voxels, usually set to 5) in `mparameters`.

This set of parameters provides the user with different masking strategies. Usually the goal of masking will be to set all densities outside the mask to 0 and retain unmodified density inside the mask so that refinement and/or classification is driven solely by the density inside the mask. However, depending on the size of the particle and mask, the fraction of density left after masking may be too small to achieve reliable particle alignment. The result may be larger alignment errors and loss of resolution in the reconstructions. Using different combinations of `mask_outside_weight` and `mask_filt_res`, it is possible to retain density outside the mask that is downweighted and/or low-pass filtered to prevent significant particle misalignment against the masked references. Depending on the filtering and weighting, the alignment is still driven by the density inside the mask at high resolution while particles are prevented from major misalignment by the low-resolution signal retained both inside and outside the mask. Users can experiment with different filtering and weighting applied to the density

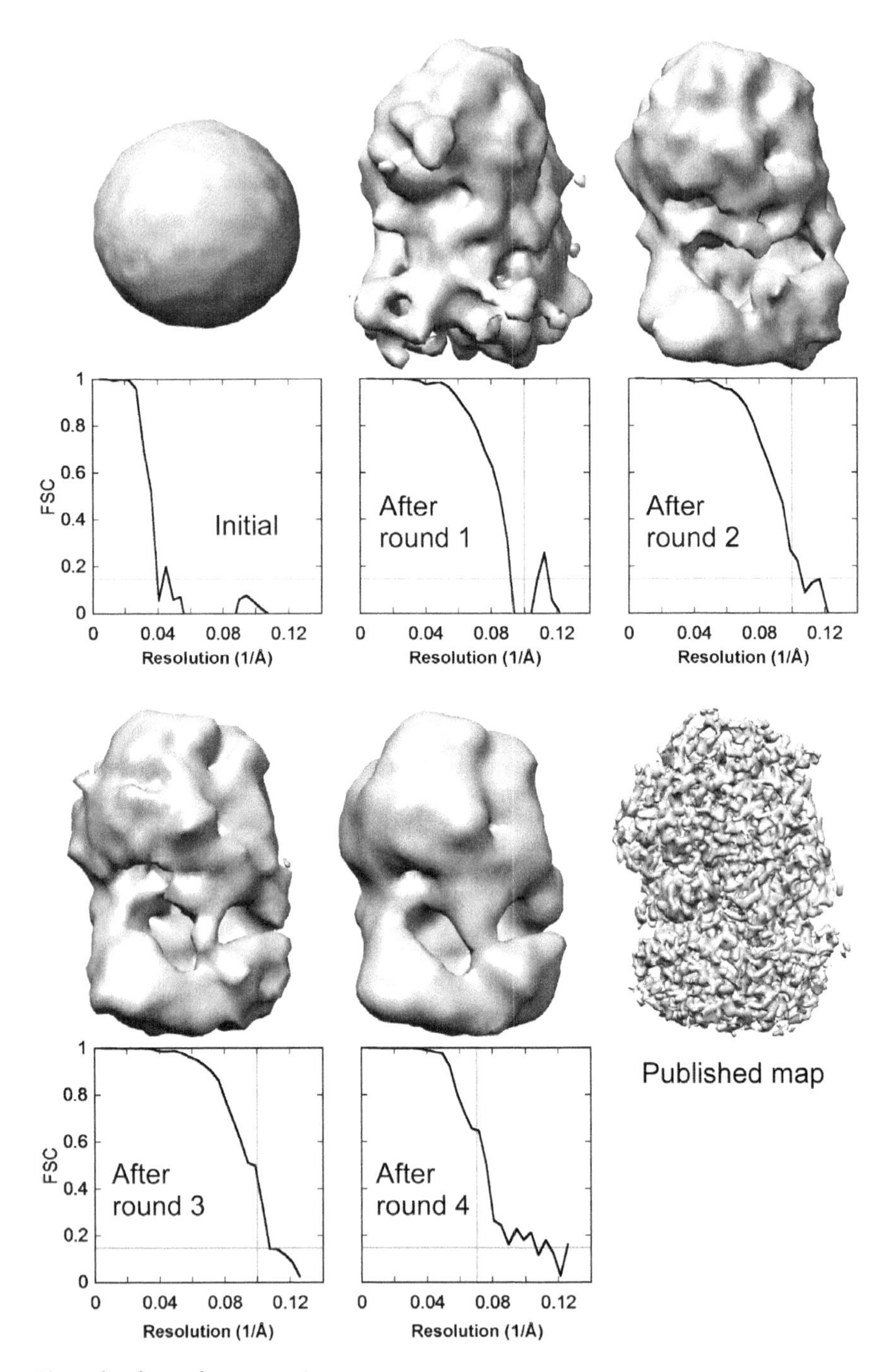

Fig. 5 See legend on opposite page.

outside the mask to achieve the best results. Preventing misalignment when using masks can also be accomplished by turning off the refinement of some of the alignment parameters (see below).

The second type of masking does not require a 3D volume on input. Instead, using the keyword `focus_mask` in `mparameters`, the user specifies the coordinates and radius of a sphere that is then used by Frealign to mask regions in the particle images. Therefore this second type of masking is applied to 2D images, not 3D volumes. For each particle image, the region inside the mask will be a disk that results from the thresholded projection of the specified sphere, in the direction of the view presented by the particle. This masking therefore depends on the correct alignment of the particles and should only be used once the alignments are reasonably accurate. The masked particle images will then be used to derive relative likelihood values that are used for classification (see above). The score function that is used for particle alignment is not affected by this type of masking and, therefore, the `focus_mask` option is only used for classification. Therefore unlike with 3D masking, the alignment accuracy should remain as high as without masking.

By defining a sphere around a region of interest, only structural variability in this region will be used to classify particles. Since the mask is applied in

Fig. 5 Example of an initial map generated from a reconstruction calculated using randomly assigned Euler angles. The dataset contained images of L protein of vesicular stomatitis virus (VSV-L) and led to a 3.8-Å reconstruction of this multienzyme (Liang et al., 2015). To initiate the startup procedure, the particle images were binned threefold to an effective pixel size of 3.711 Å and cropped to generate particle images of 60 × 60 pixels. A subset of 11,671 particles with an underfocus ranging between 1.7 and 2.5 μm were selected from the complete dataset (356,211 particles) and 40 rounds of multiresolution search and refinement were performed according to the scheme described in Section 4.5, using 5 classes and starting with a resolution limit of 40 Å that was gradually increased to 10 Å resolution. `FBOOST` was set to T for this startup procedure. Reconstructions at each stage are shown together with the calculated FSC curves, starting with the initial map obtained with randomly assigned angles (labeled "Initial"). The reconstruction with the best FSC curve from each round was selected to seed the next round of refinement and classification. The FSC curve gradually improved from round to round until the final round 4 in which only 25 refinement cycles were run and the resolution was limited to 14 Å. The FSC curve for the final round indicates a resolution of about 9.5 Å according to the 0.143 criterion (Rosenthal & Henderson, 2003) indicated by the horizontal gray line, thus significantly exceeding the resolution limit used in the refinement (indicated by the vertical gray line) and therefore reflecting an unbiased resolution estimate. The FSC curves in earlier rounds likely reflect some bias as the resolution limit during refinement exceeded to resolution indicated by the FSC. The final published map is shown in the last panel for comparison.

2D, parts of the volume overlapping with the region of interest in the views determined for the particles will be included inside the mask both in the particle image and projections of the reference structures. Differences between images and projections that drive classification will therefore result solely from true structural differences between the particles and references. This differs from 3D masking where density outside the masked region will also be missing in the projections of the masked reference structures. 2D masking may therefore increase the accuracy of classification, compared with 3D masking, an improvement that can also be achieved by subtracting constant parts of the 3D references from the particle images (Bai, Rajendra, Yang, Shi, & Scheres, 2016; Morais et al., 2003; Park et al., 2014) before determining their class memberships. However, unlike the approaches previously described, no shaped mask and no density subtraction in 2D or 3D is required. Masking in 2D offers the additional benefit of excluding noisy areas of the images that do not contain features to be classified, improving classification accuracy further. An example employing both 2D and 3D masking is shown in Fig. 6.

The `focus_mask` option can be used to "explore" different regions of a particle and obtain different classification results depending on which region the sphere is placed in. This opens up the possibility of classification based on different particle regions that display uncorrelated heterogeneity. Each classification task focuses on only one of the affected regions and separates particles based on variability in this region. If variability in one region is correlated with that in another region, classification based on the former will also separate variable features in the latter. The `focus_mask` option can therefore be used to simplify a classification problem and to test if structural variability in different regions of the complex are correlated.

4.7 Asymmetric Refinement

Symmetry present in a particle offers the advantage of additional averaging and the potential to increase the final resolution of a reconstruction. However, in many cases the individual particles will exhibit deviations from the nominal symmetry due to small distortions, disorder, conformational variability or the presence of pseudosymmetry that only becomes apparent at higher resolution. These deviations may limit the attainable resolution of the fully symmetrized reconstruction. One way to overcome this limitation is to treat the asymmetric subunits of each particle as separate entities (Ilca et al., 2015). Frealign provides an option to perform "asymmetric" refinement and reconstruction to explore if the alignment and classification of

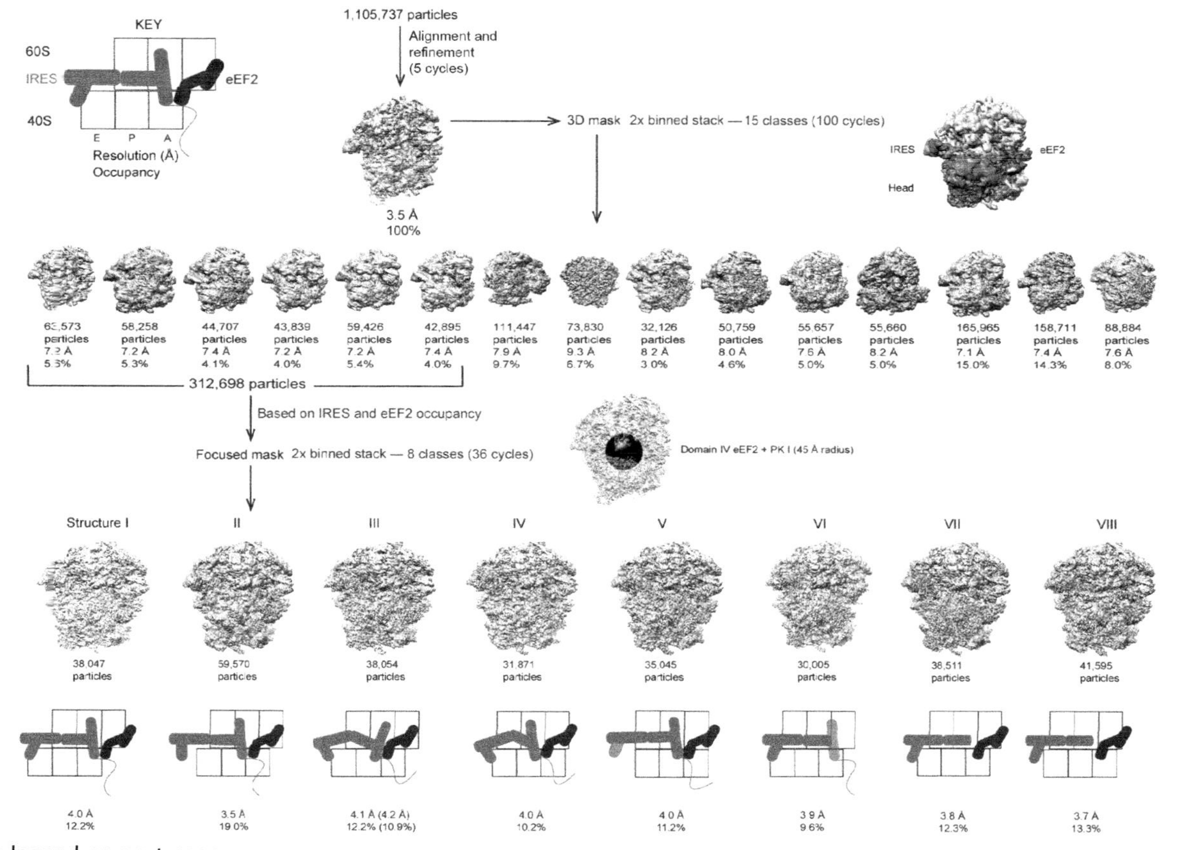

Fig. 6 See legend on next page.

asymmetric units improves the density. To switch to the asymmetric reconstruction mode, users can change the letter of the symmetry symbol from upper case to lower case. For example, to carry out asymmetric refinement of a structure with nominal C2 symmetry, users must specify "c2" (keyword `Symmetry` in `mparameters`). In subsequent refinement cycles, additional lines will appear in the alignment parameter file with alignment results for each symmetry-related orientation of the particle. In the c2 example, there would be two lines of parameters for each particle, thus doubling the number of lines in the parameter file. In the reconstruction step, Frealign inserts particle images into the 3D volume according to each line in the parameter file, without applying additional symmetry. Therefore if the alignment parameters on different lines belonging to the same particle differ, this will break the symmetry of the reconstruction. It is therefore possible to obtain a reconstruction that is not perfectly symmetrical when using asymmetric refinement.

The asymmetric mode can be used in two ways. In the first, particles with nominal symmetry experience distortion or disorder that displaces otherwise rigid subunits (or groups of subunits) from their symmetric positions. In this situation Frealign can align the subunits individually using a reference with an applied 3D mask (option `mask_file`, see above). For example, if a particle has a nominal C2 symmetry, a suitably designed 3D mask could be used to downweight or completely remove half of the reference density and align

Fig. 6 Example of a classification scheme using 2D and 3D masking. The dataset consisted of 80S ribosomes prepared with the Taura syndrome virus internal ribosome entry site (TSV IRES) and elongation factor 2 (eEF2) (Abeyrathne, Koh, Grant, Grigorieff, & Korostelev, 2016). The complete dataset of 1,105,737 images of 80S–IRES–eEF2 complex was initially aligned against a density map calculated from the atomic model of the non-rotated 80S ribosome bound with 2 tRNAs (PDB: 3J78, Svidritskiy, Brilot, Koh, Grigorieff, & Korostelev, 2014). This initial alignment was performed on data with a pixel size of 1.64 Å and limited to 20 Å resolution, resulting in a 3.5-Å resolution reconstruction. After 5 cycles of refinement the data were 2 × binned (by Fourier cropping using the `resample.exe` tool, new pixel size = 3.28 Å) and subjected to classification into 15 classes using a 3D mask that contained the IRES, eEF2, and head domain of the small subunit. Six of the resulting classes (312,698 particle images) contained density for the IRES and eEF2 and were further classified into eight classes. For this classification, a 2D mask was applied around the ribosomal A site to include IRES pseudoknot I and eEF2 domain IV. The figure shows this mask as a sphere which, when projected according to the orientation of a particle, results in a 2D mask correctly placed on the region of interest. In the case of the 80S–IRES–eEF2 complex, this focused classification resulted in the separation of different translocation states of the IRES, catalyzed by eEF2, as shown schematically below each reconstruction. The structures I–V containing clear density for the IRES and eEF2 are *highlighted in color*. (See the color plate.)

the particles against the remaining half. The particle images still contain both halves and Frealign will attempt to align each to the masked reference after applying symmetry-derived rotation matrices to transform each half into the correct frame of reference with respect to the masked reference. The resulting reconstruction should display improved density in the half of the reconstruction that survived the masking while the other half should display degraded density, depending on the amount of distortion/disorder present in the particles. It is important to note that the masking can also lead to increased alignment errors, degrading both halves of the density. Users should therefore carefully evaluate the asymmetric reconstructions and explore different masking options (see above).

The asymmetric refinement mode can also be used to classify symmetry-related regions (eg, subunits) in a particle if these differ in their conformation or composition (eg, ligand binding). In this case, classification (`nclasses` > 1, see above) has to be used together with asymmetric refinement. Using masking in 2D or 3D (see above), one of the symmetry-related regions has to be selected by the mask. Frealign will then assign each of the symmetry-related views of a particle to one of the classes using different occupancies in the alignment parameter file. In the resulting reconstructions, the region corresponding to the density inside the mask will contain the classification results while density outside this region will only show class-specific features if the variability inside the mask is correlated with that outside the mask.

Frealign currently does not offer a way to switch from asymmetric refinement back to regular refinement. Using their own scripts, users can reduce the "multiline" parameter files to "single-line" parameter files to continue with regular refinement. As a technical note, the multiple alignment parameters for each particle do not include the additional symmetry-derived transforms that generate each of the symmetry-related views. This means that the alignment parameters in multiline parameter files usually remain fairly similar to each other as differences only indicate the (typically small) differences in the alignments of the symmetry-related views.

4.8 Processing Images of Helical Structures

Frealign includes an option to impose helical symmetry on reconstructions calculated from segments of helical structures and filaments (see chapter "Cryo-EM Structure Determination Using Segmented Helical Image Reconstruction" by Sachse). The helical symmetry can be selected using symmetry symbol "H" or "HP." When "H" is specified, Frealign resets the alignment parameters at the beginning of a new refinement cycle to center all helical segments to be within a single "asymmetric unit" of the helical lattice. This will affect the Phi Euler angle, ie, the third angle listed in the alignment parameter file that determines the rotation of each segment around the helical axis, and the *x*,*y* translations. While this is useful in most cases, it prevents the refinement and reconstruction of pseudohelical structures, such as microtubules that have a seam. When symmetry "HP" is specified, the parameters are not reset and seam-sensitive alignments are preserved.

Frealign expects the helical axis to be aligned with the *z*-axis. The second Euler angle Theta, which describes the out-of-plane alignment of a helical segment, is therefore usually close to 90 degree. Most helical structures are also characterized by a persistence length that indicates how easy it is to bend them. The variability of the first Euler angle Psi, which describes the in-plane rotational alignment of a segment, is therefore limited depending on the value of the persistence length. While Frealign does not use persistence length as a parameter to restrict the angular alignment of segments, the user can define a `STIFFNESS` parameter as part of the helical symmetry section in `mparameters`. A large value for `STIFFNESS` will force the Psi and Theta angles to deviate less from the average values for a filament while a small value allows more variability. For this to work, it is important that segments belonging to the same filament are arranged consecutively in the alignment parameter file and image stack, and that the micrograph identifier ("Film" column in the parameter file) is the same for these segments. Other parameters that must be set when specifying helical symmetry are `ALPHA` (the rotation angle involved when going from one helical subunit to the next), `RISE` (the translation along the helical axis from one subunit to the next), `NSUBUNITS` (the number of unique subunits per segment; for overlapping segments that should only include the number of subunits in the non-overlapping parts), and `NSTARTS` (the number of helical starts present in the structure). Frealign provides a special `mparameters` template for helical image processing that includes these additional parameter keys. Users can access this template using the command `frealign_helical_template`. There is currently no option in Frealign to refine the helical symmetry parameters and users must therefore make sure that the `ALPHA` and `RISE` parameters are correct.

5. TUNING OPTIONS

Frealign offers a number of tuning options that allows users to optimize refinement and classification. These options are typically used only after refinement with the standard (default) parameters has converged, ie, no further improvement in the reconstructed density is observed when running additional refinement cycles. The following options, which are listed in the expert section in `mparameters`, are available:

- `XSTD`: determines if the input 3D reference should be masked (values larger than 0) or the 3D reference should be used to generate 2D masks (values smaller than 0) to be applied to the particle images before alignment (the masks are not used when calculating reconstructions). The default is 0, which disables this feature. Frealign interprets the provided value as a multiple of the standard deviation of the input 3D reference and values between 2 and 5 are usually appropriate. Higher values will tighten the masks while lower values will loosen them, both in 2D and 3D. Masking can reduce the noise present in the 3D reference or particle images but users must be careful not to overtighten the masks and cut into the 3D reference density or 2D particle density. 3D masking can also be achieved using the `mask_file` option (see above), which is preferred over the XSTD option because it provides the user with more control over different aspects of the masking.
- `PBC`: determines the weighting of individual particle images using a B-factor during reconstruction. The applied B-factor is calculated as $B = 4 \times \text{score}/$ `PBC` (in $Å^2$). Reconstructions will be corrected for the average applied weights and therefore, only relative weight differences between particles are significant. A large value for `PBC` (eg, 100) will effectively remove particle weighting while a small value (10 or smaller) will apply weighting. Users can recalculate reconstructions (using command `frealign_calc_reconstructions`) with different `PBC` values to see which value produces the best results.
- `parameter_mask`: determines which of the five alignment parameters (Psi, Theta, and Phi Euler angles, and x,y translations) will be refined. Normally, all parameters are refined and `parameter_mask` is set to "1 1 1 1 1." However, any of these flags can be set to "0," thereby forcing the corresponding parameter to remain constant during refinement. This allows users to reduce the degrees of freedom during refinement. For example, when images are collected as movies, it is possible to calculate movie sums with different numbers of frames, one with all frames to boost contrast and one with only the early frames to boost high resolution (Campbell et al., 2012). Frealign allows the use of different particle stacks for refinement and reconstruction by using the keywords `raw_images_ref` and `raw_images_rec` instead of `raw_images`. Using this simplified version of an exposure filter (Grant & Grigorieff, 2015b), it is possible to obtain

higher resolution. However, if the movie frames are not accurately aligned there might be small differences in the translational alignments of the particles between the refinement and reconstruction stacks (rotational differences are usually very small and can be ignored). Once the best possible alignments have been obtained using the stack corresponding to a higher exposure, one or two additional refinement cycles using the low-exposure stack can increase the resolution even further. Since only translational alignment is needed, `parameter_mask` should be set to "0 0 0 1 1" to avoid increased errors in the Euler angles due to the lower image SNR of the low-exposure stack. Keeping some of the parameters constant may also be useful in other situations, for example, when some of the alignment parameters are known from the experimental setup, such as in the random conical tilt (Radermacher, Wagenknecht, Verschoor, & Frank, 1987) or orthogonal tilt reconstruction (Leschziner & Nogales, 2006) methods.

- `thresh_reconst`: determines the particle score value below which particles will be excluded from the reconstruction. Normally this value is set to 0 to include all particles. However, if a large fraction of particles are damaged or otherwise compromised and do not contribute high-resolution signal, reconstructions can be improved by excluding them. These particle images tend to receive lower scores than particles that contribute the strongest signal. Users can therefore tune their reconstructions by testing different values for `thresh_reconst` and recalculating the reconstruction (using command `frealign_calc_reconstructions`).
- `FMATCH`: specifies that Frealign should also output matching projections after each refinement cycle. The matching projections are stored inside Frealign's scratch directory and will have names that include the pattern `_reproject_`. It is recommended to keep `FMATCH` set to "F" and use the command `frealign_calc_projections` instead to generate match projections, after refinement is completed. Inspecting matching projections may be useful to verify that particle alignment was successful, especially when starting without a set of alignment parameters and `MODE` set to 3 (see above).
- `FBEAUT`: if set to "T," Frealign will apply the specified particle symmetry also in real space. This will not improve the results of the refinement but may improve the appearance of symmetry in a reconstruction if some of the symmetry operators require interpolation to be represented in the orthogonal coordinate system used by Frealign. The feature is therefore usually only used when making figures for presentations and publications.

- FBOOST: determines if a signed (when set to "T") or an unsigned (when set to "F") correlation coefficient is maximized during particle parameter refinement. As explained earlier, refining with the signed correlation coefficient may improve alignments but also increases the chance of overfitting. Users should initially refine with FBOOST set to "F" and, after convergence, test if additional refinement cycles with FBOOST set to "T" improve the reconstruction. It is important to limit the resolution used during refinement and look for a clear improvement of the FSC curve beyond the resolution limit. Also, an improvement of the FSC should be accompanied by an improvement in the density that can be correlated with known structural features. Users may have to sharpen the reconstructed density using bfactor.exe (see below) to observe high-resolution features.
- beam_tilt_x, beam_tilt_y: allows users to specify a beam tilt that Frealign will include during CTF correction. The values must be given in units of milliradians. Frealign cannot refine these values. However, if the presence of beam tilt is suspected, users can try different values and recalculate the reconstructions (using command frealign_calc_reconstructions) to find the best values.
- FMAG, FDEF, FASTIG, FPART: setting any of these flags to "T" will enable refinement of the magnification, defocus, and astigmatism, optionally for individual particles (FPART set to "T"). Refinement of these parameters is usually not warranted, and users should not use these options in most cases.
- RBfactor: alters the correlation function used during refinement. This is not recommended and users should not use this option in most cases.

6. RELATED SOFTWARE

Besides Frealign, there are several other image processing tools that have been developed in the Grigorieff lab and that are freely available for download from the lab web page. Since some of these may be useful in combination with Frealign, they are briefly listed here for reference.

- *Signature*: software to display micrographs and select particles (Chen & Grigorieff, 2007). A semiautomatic mode is available that uses templates to identify particles using an algorithm first developed for FindEM (Roseman, 2004).
- *CTFFIND3/CTFTILT*: software used to determine accurate image defocus present in a micrograph (Mindell & Grigorieff, 2003) of untilted

and tilted samples. A recent update, CTFFIND4 (Rohou & Grigorieff, 2015), offers significant speedups over CTFFIND3.

- *BFACTOR*: a tool to low-pass filter images and 3D density maps, and to estimate and apply a B-factor to bring out high-resolution features in a map. This tool is also included with the Frealign distribution.
- *CROP*: a tool to cut out a region of density from a 2D image or 3D volume.
- *DIFFMAP*: a tool to perform amplitude scaling of one 3D map against another and to write out a difference map. The tool can be used for scaling of 3D maps even if these are not aligned with each other. While the difference map is not meaningful in this case, the scaled maps should have similar filtering and B-factor sharpening, making it easier to compare them in terms of quality and high-resolution features.
- *Unblur/Summovie*: software to align frames of movies collected on an electron microscope, and to apply exposure-dependent filtering to the frames to enhance high-resolution signal (Grant & Grigorieff, 2015b).
- *mag_distortion_estimate/correct*: a tool to measure and correct for magnification distortions present in electron microscope images (Grant & Grigorieff, 2015a).

REFERENCES

Abeyrathne, P., Koh, C. S., Grant, T., Grigorieff, N., & Korostelev, A. A. (2016). Ensemble cryo-EM uncovers inchworm-like translocation of a viral IRES through the ribosome. *eLife*. in press. http://dx.doi.org/10.7554/eLife.14874.

Alushin, G. M., Ramey, V. H., Pasqualato, S., Ball, D. A., Grigorieff, N., Musacchio, A., et al. (2010). The Ndc80 kinetochore complex forms oligomeric arrays along microtubules. *Nature*, *467*, 805–810.

Bai, X. C., Rajendra, E., Yang, G., Shi, Y., & Scheres, S. H. W. (2016). Sampling the conformational space of the catalytic subunit of human gamma-secretase. *eLife*, *4*, e11182.

Campbell, M. G., Cheng, A., Brilot, A. F., Moeller, A., Lyumkis, D., Veesler, D., et al. (2012). Movies of ice-embedded particles enhance resolution in electron cryomicroscopy. *Structure*, *20*, 1823–1828.

Chen, J. Z., & Grigorieff, N. (2007). SIGNATURE: A single-particle selection system for molecular electron microscopy. *Journal of Structural Biology*, *157*, 168–173.

Crowther, R. A., Henderson, R., & Smith, J. M. (1996). MRC image processing programs. *Journal of Structural Biology*, *116*, 9–16.

Frank, J., Radermacher, M., Penczek, P., Zhu, J., Li, Y., Ladjadj, M., et al. (1996). SPIDER and WEB: Processing and visualization of images in 3D electron microscopy and related fields. *Journal of Structural Biology*, *116*, 190–199.

Grant, T., & Grigorieff, N. (2015a). Automatic estimation and correction of anisotropic magnification distortion in electron microscopes. *Journal of Structural Biology*, *192*, 204–208.

Grant, T., & Grigorieff, N. (2015b). Measuring the optimal exposure for single particle cryo-EM using a 2.6 A reconstruction of rotavirus VP6. *eLife*, *4*, e06980.

Grigorieff, N. (1998). Three-dimensional structure of bovine NADH:ubiquinone oxidoreductase (complex I) at 22 A in ice. *Journal of Molecular Biology*, *277*, 1033–1046.

Grigorieff, N. (2000). Resolution measurement in structures derived from single particles. *Acta Crystallographica. Section D, Biological Crystallography, 56*, 1270–1277.

Grigorieff, N. (2002). Single particles always fit the mold. In *Paper presented at Frontiers in structural cell biology: How can we determine the structures of large subcellular machines at atomic resolution?* Asilomar, CA: The Biophysical Society.

Grigorieff, N. (2007). FREALIGN: High-resolution refinement of single particle structures. *Journal of Structural Biology, 157*, 117–125.

Harauz, G., & van Heel, M. (1986). Exact filters for general geometry 3-dimensional reconstruction. *Optik, 73*, 146–156.

Henderson, R., Sali, A., Baker, M. L., Carragher, B., Devkota, B., Downing, K. H., et al. (2012). Outcome of the first electron microscopy validation task force meeting. *Structure, 20*, 205–214.

Ilca, S. L., Kotecha, A., Sun, X., Poranen, M. M., Stuart, D. I., & Huiskonen, J. T. (2015). Localized reconstruction of subunits from electron cryomicroscopy images of macromolecular complexes. *Nature Communications, 6*, 8843.

Kucukelbir, A., Sigworth, F. J., & Tagare, H. D. (2014). Quantifying the local resolution of cryo-EM density maps. *Nature Methods, 11*, 63–65.

Leschziner, A. E., & Nogales, E. (2006). The orthogonal tilt reconstruction method: An approach to generating single-class volumes with no missing cone for ab initio reconstruction of asymmetric particles. *Journal of Structural Biology, 153*, 284–299.

Li, X., Grigorieff, N., & Cheng, Y. (2010). GPU-enabled FREALIGN: Accelerating single particle 3D reconstruction and refinement in Fourier space on graphics processors. *Journal of Structural Biology, 172*, 407–412.

Liang, B., Li, Z., Jenni, S., Rameh, A. A., Morin, B. M., Grant, T., et al. (2015). Structure of the L-protein of vesicular stomatitis virus from electron cryomicroscopy. *Cell, 162*, 314–327.

Lyumkis, D., Brilot, A. F., Theobald, D. L., & Grigorieff, N. (2013). Likelihood-based classification of cryo-EM images using FREALIGN. *Journal of Structural Biology, 183*, 377–388.

Matthews, B. W. (1968). Solvent content of protein crystals. *Journal of Molecular Biology, 33*, 491–497.

Mindell, J. A., & Grigorieff, N. (2003). Accurate determination of local defocus and specimen tilt in electron microscopy. *Journal of Structural Biology, 142*, 334–347.

Morais, M. C., Kanamaru, S., Badasso, M. O., Koti, J. S., Owen, B. A., McMurray, C. T., et al. (2003). Bacteriophage phi29 scaffolding protein gp7 before and after prohead assembly. *Nature Structural Biology, 10*, 572–576.

Park, E., Menetret, J. F., Gumbart, J. C., Ludtke, S. J., Li, W., Whynot, A., et al. (2014). Structure of the SecY channel during initiation of protein translocation. *Nature, 506*, 102–106.

Pettersen, E. F., Goddard, T. D., Huang, C. C., Couch, G. S., Greenblatt, D. M., Meng, E. C., et al. (2004). UCSF Chimera—A visualization system for exploratory research and analysis. *Journal of Computational Chemistry, 25*, 1605–1612.

Radermacher, M., Wagenknecht, T., Verschoor, A., & Frank, J. (1987). Three-dimensional reconstruction from a single-exposure, random conical tilt series applied to the 50S ribosomal subunit of Escherichia coli. *Journal of Microscopy, 146*, 113–136.

Rohou, A., & Grigorieff, N. (2015). CTFFIND4: Fast and accurate defocus estimation from electron micrographs. *Journal of Structural Biology, 192*, 216–221

Roseman, A. M. (2004). FindEM—A fast, efficient program for automatic selection of particles from electron micrographs. *Journal of Structural Biology, 145*, 91–99.

Rosenthal, P. B., & Henderson, R. (2003). Optimal determination of particle orientation, absolute hand, and contrast loss in single-particle electron cryomicroscopy. *Journal of Molecular Biology, 333*, 721–745.

Sindelar, C. V., & Grigorieff, N. (2012). Optimal noise reduction in 3D reconstructions of single particles using a volume-normalized filter. *Journal of Structural Biology*, *180*, 26–38.

Stewart, A., & Grigorieff, N. (2004). Noise bias in the refinement of structures derived from single particles. *Ultramicroscopy*, *102*, 67–84.

Svidritskiy, E., Brilot, A. F., Koh, C. S., Grigorieff, N., & Korostelev, A. A. (2014). Structures of yeast 80S ribosome-tRNA complexes in the rotated and nonrotated conformations. *Structure*, *22*, 1210–1218.

van Heel, M., Harauz, G., Orlova, E. V., Schmidt, R., & Schatz, M. (1996). A new generation of the IMAGIC image processing system. *Journal of Structural Biology*, *116*, 17–24.

Wolf, M., DeRosier, D. J., & Grigorieff, N. (2006). Ewald sphere correction for single-particle electron microscopy. *Ultramicroscopy*, *106*, 376–382.

Yang, Z., Fang, J., Chittuluru, J., Asturias, F. J., & Penczek, P. A. (2012). Iterative stable alignment and clustering of 2D transmission electron microscope images. *Structure*, *20*, 237–247.

CHAPTER NINE

Testing the Validity of Single-Particle Maps at Low and High Resolution

P.B. Rosenthal[1]
Francis Crick Institute, Mill Hill Laboratory, London, United Kingdom
[1]Corresponding author: e-mail address: peter.rosenthal@crick.ac.uk

Contents

Abstract

Single-particle electron cryomicroscopy may be used to determine the structure of biological assemblies by aligning and averaging low-contrast projection images recorded in the electron microscope. Recent progress in both experimental and computational methods has led to higher resolution three dimensional maps, including for more challenging low molecular weight proteins, and this has highlighted the problems of model bias and over-fitting during iterative refinement that can potentially lead to incorrect map features at low or high resolution. This chapter discusses the principles and practice of specific validation tests that demonstrate the consistency of a 3D map with

Methods in Enzymology, Volume 579
ISSN 0076-6879
http://dx.doi.org/10.1016/bs.mie.2016.06.004

projection images. In addition, the chapter describes tests that detect over-fitting during refinement and lead to more robust assessment of both global and local map resolution. Application of several of these tests together demonstrates the reliability of single-particle maps that underpins their correct biological interpretation.

1. INTRODUCTION

Single-particle analysis of frozen-hydrated specimens may be used to determine the structure of biological assemblies. Recent success in determining high-resolution structures of small, asymmetric proteins has brought more widespread interest in electron cryomicroscopy (cryoEM) as a tool for structural biology. As with structures obtained by other macromolecular techniques such as X-ray crystallography and nuclear magnetic resonance spectroscopy, single-particle maps and their interpretation by atomic coordinate models must be consistent with experimental image data while obeying chemical constraints on bonding and stereochemistry. Validation of all the steps en route to a structure is critical for success, particularly as the method is applied to more challenging low molecular weight and heterogeneous specimens.

In principle only a small number of projection views $\pi D/d$ are required to determine the structure of a particle of diameter D to a resolution d (Crowther, DeRosier, & Klug, 1970). However, the radiation sensitivity of frozen-hydrated biological specimens requires that images should be recorded with minimum dose to preserve high-resolution features. As a result the images have a low signal-to-noise ratio (SNR). High-resolution structures result from coherent averaging of many low SNR image projections of frozen-hydrated single particles.

Iterative, map-based refinement of the orientation of the individual particle images may be used to bring an initial low-resolution map of the structure into closer agreement with image data and to extend its resolution (Fig. 1). The main computational problem in single-particle analysis is therefore to assign the correct relative orientation and position of the particle in each image. Typically, the map is used to assign three particle orientation parameters and two position parameters by matching calculated projections of the map with experimental images according to a scoring criterion. Microscope parameters such as magnification and defocus are also required to calculate projections that match experimental images.

In such a procedure, problems may arise when map projections incorrectly match signal in the image or when they match noise resulting in

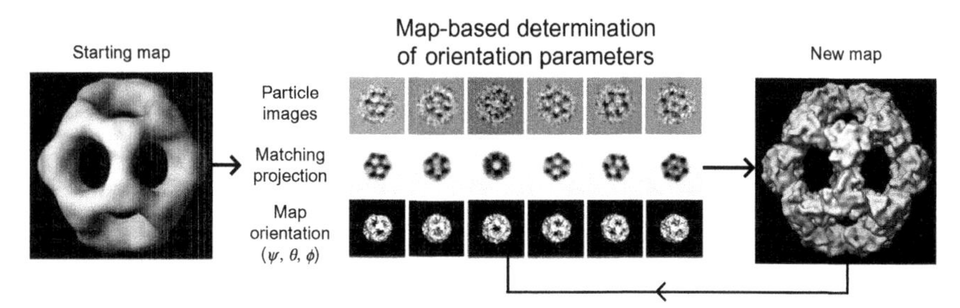

Fig. 1 Iterative refinement of single-particle orientation parameters. A low-resolution starting map is used to assign orientations to a particle image by finding the best matching projection. A new map is calculated from the new orientations and the procedure is repeated. *Based on the data described in Rosenthal, P. B., & Henderson, R. (2003). Optimal determination of particle orientation, absolute hand, and contrast loss in single-particle electron cryomicroscopy.* Journal of Molecular Biology, 333, *721–745.*

incorrect orientation assignment. One consequence of incorrect orientation assignment may be a degradation of map quality. A more serious problem occurs when an incorrect starting map fails to match signal in the images but nevertheless the incorrect map persists through refinement. The inability of an iterative refinement scheme to escape features of the initial model is called model bias. In the complete absence of structural signal in the images, a starting map may align to noise, as dramatically demonstrated by the appearance of "Einstein from noise" when pure noise images are aligned by a template (Henderson, 2013; Penczek, Radermacher, & Frank, 1992; Sigworth, 1998; van Heel, 2013). Another problem called over-fitting occurs when a map correctly matches structural features in images at low resolution, but aligns to noise at higher resolution, which is then reinforced during iterative refinement (Chen et al., 2013; Grigorieff, 2000; Stewart & Grigorieff, 2004). The reinforced noise may even be erroneously interpreted as high-resolution structural features (Scheres & Chen, 2012).

The introduction of direct detector devices (DDD) has improved the quality of images through better detective quantum efficiency (McMullan, Faruqi, Clare, & Henderson, 2014) and the correction of specimen motions through movie alignment (Bai, Fernandez, McMullan, & Scheres, 2013; Brilot et al., 2012; Li et al., 2013), making it possible to align and average smaller particles and to reach atomic resolution with fewer images. Progress in image analysis includes refinement protocols that use maximum likelihood targets that are less sensitive to model bias and over-fitting than methods based on cross-correlation (Lyumkis, Brilot, Theobald, & Grigorieff, 2013; Scheres, 2012; Scheres & Carazo, 2009; Sigworth, 1998).

Along with these recent experimental and computational developments in cryomicroscopy, there is also increasing consensus on how to assess maps and demonstrate their validity during the course of structure determination, publication, and when depositing maps in public databases (Henderson, 2013; Henderson et al., 2012; Rosenthal & Rubinstein, 2015). This chapter describes specific validation tests that build confidence in a single-particle map determination whether at low or high resolution. Several of these tests exploit the concept of cross-validation. Tilt-pair analysis (Henderson et al., 2011; Rosenthal & Henderson, 2003) validates map quality at low resolution by showing that it is consistent with structural information in images and shows that particle orientations are correctly assigned. It can also be applied to optimizing orientation determination, which is important for obtaining high-resolution maps.

Once a map is validated at low resolution, it may be further refined against many particle images with the expectation that resolution-dependent stereochemical features from protein or nucleic acid will appear that validate the structure determination at high-resolution. A statistical assessment of the map such as the Fourier shell correlation (FSC) is used to assign resolution. At high-resolution, the problem of over-fitting may lead to overestimation of resolution that is problematic during the course of refinement and in assessing the information content of a map. Refinement strategies that allow monitoring of true high-resolution signal (frequency-limited refinement, gold-standard refinement) are described. High-resolution noise substitution is a specific validation test that provides a quantitative measure of the true signal in the FSC at high resolution (Chen et al., 2013).

While this chapter emphasizes the internal consistency of map and data, a map may be validated by independently determined X-ray or NMR models, and the overall resolution of a map may be assessed by comparison to an atomic model. Lastly, maps may have nonuniform resolution that may be a feature of structure determination or a consequence of molecular properties related to biological function. Assessment of local map resolution may be important to map interpretation.

2. VALIDATION AT LOW RESOLUTION: TILT-PAIR ANALYSIS AND ORIENTATION DETERMINATION

Structural analysis begins with picking and analysis of the particles in two dimensions. Classification, alignment, and averaging of particle images in two dimensions are performed to obtain high SNR views (class averages) of the structure in projection. An initial low-resolution 3D map may be calculated following assignment of the correct relative orientation of the class averages.

The 3D map must be consistent with the particle image data. When an initial 3D map is proposed and subsequently refined, projections of the 3D structure should agree with both class averages and raw images. There should be evidence for the symmetry of the particle, and the orientations of all the individually aligned particles should include all the views required to make a 3D map for that symmetry.

Given that a map can align to noise images during refinement, at each stage of the analysis, methods should be employed that avoid template or model bias. This type of bias is a concern when using an automatic template matching procedure to pick particles. One recommendation is to record a few high-defocus, high-exposure images to confirm visually that the expected particles are present in the images (Henderson, 2013). During initial classification and averaging of particles, reference-free procedures, such as alignment to a rotationally averaged sum of particles, should be used (Penczek et al., 1992; Schatz & van Heel, 1990). The individual particle images that contribute to a class should look like the class average. The class averages may be validated by inspection to show that they do not arise from incoherent averaging of images, such as the averaging of mirror images of views of the particle (Rubinstein, Walker, & Henderson, 2003). These criteria are necessary but not sufficient to guarantee a correct map. A low-resolution map may be validated by rigid-body fitting with an atomic coordinate model, which may also determine the absolute hand of the map, though such a model is not always available.

The problem in assigning orientations to individual particles in low SNR images is that the map may align to the wrong features in the image, despite giving a reasonable score for the match. However, if the agreement is due to true structural features in the image, then after tilting the specimen holder, the map should match the new view after applying the same tilt transformation to the map. The tilt-pair test is thus a cross-validation test for orientation determination with applications to map validation, hand discrimination, and optimization of orientation determination (Henderson et al., 2011; Rosenthal & Henderson, 2003). Implementations of tilt-pair analysis are available in several image analysis packages and as a stand-alone web service (http://www.ebi.ac.uk/pdbe/emdb/validation/tiltpair/; Wasilewski & Rosenthal, 2014).

2.1 The Tilt-Pair Experiment

A tilt pair is a pair of images of the same field of particles recorded at two angles of the microscope goniometer, shown as $-\alpha$ and $+\alpha$ for the experiment in Fig. 2A. The experiment is easy to perform on modern microscopes. Imaging conditions may be similar to those for imaging the full dataset used for structure determination. However, this need not be the case, and the main additional considerations are the choice of tilt angle, magnification, and how to distribute radiation exposure over two images.

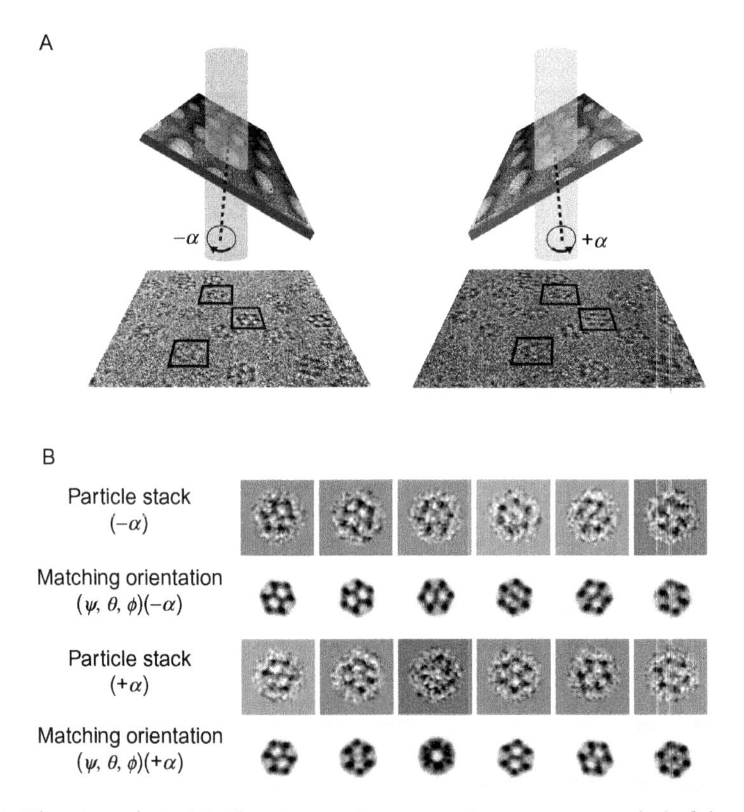

Fig. 2 Tilt-pair analysis. (A) Tilt-pair experiment: two images are recorded of the same position on the specimen at different angles of the microscope goniometer ($-\alpha$ and $+\alpha$). (B) Orientation determination for particles in the tilt-pair images. Matching orientations of a 3D map given by Euler angles (ψ, θ, ϕ) are assigned to particle image stack ($-\alpha$) and particle image stack ($+\alpha$). *Based on the data described in Rosenthal, P. B., & Henderson, R. (2003). Optimal determination of particle orientation, absolute hand, and contrast loss in single-particle electron cryomicroscopy.* Journal of Molecular Biology, 333, 721–745.

The first step in the analysis is to identify and box out the particle images and store them in two image stacks so that the two views of the particle appear with the same index in the stacks for $-\alpha$ and $+\alpha$. Programs such as Ximdisp (Smith, 1999), Tiltpicker (Voss, Yoshioka, Radermacher, Potter, & Carragher, 2009), Boxer (Cong & Ludtke, 2010), Relion (Scheres, 2012), or Web (Frank et al., 1996) have graphical interfaces for selecting particles.

Next, any protocol preferred by the investigator may be used to determine orientation parameters for the particle images in the two stacks using a 3D map, which are then stored as two separate parameters files. These parameters may be three Euler angle parameters and two translation coordinates (ψ, θ, ϕ, x, and y) (Frank, 2006) and the parameter file lists these in the same order as the particles in the stack. There are two slightly different forms of analysis of the tilt-pair experiment as described in Sections 2.2 and 2.3.

2.2 Tilt-Pair Parameter Plot

One form of tilt-pair analysis requires only the orientation parameters determined for each pair of images and identifies the tilt axis and tilt angle required to rotate the first orientation to the second. This may be performed by the program Tiltdiff (Henderson et al., 2011; Rosenthal & Henderson, 2003), which identifies the best tilt axis and tilt angle that relates the orientation at $-\alpha$ to $+\alpha$, assuming that the tilt axis lies in a plane perpendicular to the electron beam. This assumption is true of most microscope goniometers. The calculation also requires knowledge of the symmetry of the particle. A plot of the tilt axis and tilt angle for all the particles in the stack is called a tilt-pair parameter plot (TPPP), as shown in Fig. 3. If the 3D map exactly represents the structure in the image and there are no errors in orientation determination in both image stacks, then all the points should be located at the experimental tilt axis and tilt angle used in the experiment, knowledge of which is not included in the calculation of the TPPP. In fact, such a plot may show a wide scattering of tilt transformations (Fig. 3A), which may be due to errors in determining particle parameters or due to an incorrect map or one that is at too low resolution to differentiate orientations. In Fig. 3B, following an optimized procedure that produced more accurate orientation parameters with the same map (Rosenthal & Henderson, 2003), most of the points are clustered near the experimental tilt axis and tilt angle (10°), proving the consistency of the 3D map with the image pairs.

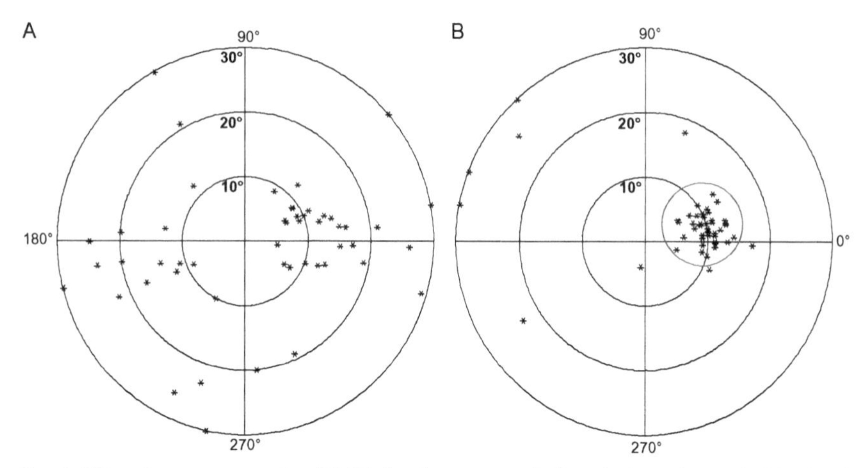

Fig. 3 Tilt-pair parameter plot (TPPP) for the map and tilt-pair data shown in Fig. 2 calculated using the program Tiltdiff (A) before and (B) after optimizing orientations, where the small *red* (*gray* in the print version) *circle* identifies the cluster of orientations near the tilt axis. Tilt angle is plotted in *bold* along the vertical axis. Tilt-axis direction is indicated by *angles* outside the last circle. *Based on fig. 6 of Rosenthal, P. B., & Henderson, R. (2003). Optimal determination of particle orientation, absolute hand, and contrast loss in single-particle electron cryomicroscopy.* Journal of Molecular Biology, 333, *721–745.*

2.3 Tilt-Pair Phase Residual Plot

A different form of tilt-pair analysis requires the two image stacks, but only requires orientation parameters for the particles in the first stack (Rosenthal & Henderson, 2003; Wasilewski & Rosenthal, 2014) (Fig. 4). For orientation parameters determined for the particles at $-\alpha$, all in-plane tilt transformations (tilt axis and tilt angle) are applied, and the predicted orientations are scored for agreement (eg, amplitude-weighted phase residual minimum or cross-correlation coefficient maximum), with the images in the second stack, following a search for the best translational parameters. The plot of the scores for all possible tilt axes and tilt angles, called a tilt-pair phase residual plot, when summed for all the particles, should show a pronounced phase residual minimum at the experimental tilt axis and tilt angle used in the experiment (Fig. 4A and B left side). The range of tilt angles tested must be large enough to include the experimental tilt angle. The resolution range for calculating the agreement of map projections with the second ($+\alpha$) stack, typically 100–25 Å, must be chosen by the investigator, though the results are relatively insensitive to this choice. A plot of the minimum position for each particle also yields a TPPP (Fig. 4A and B right side). An advantage of the second method of tilt-pair analysis is that

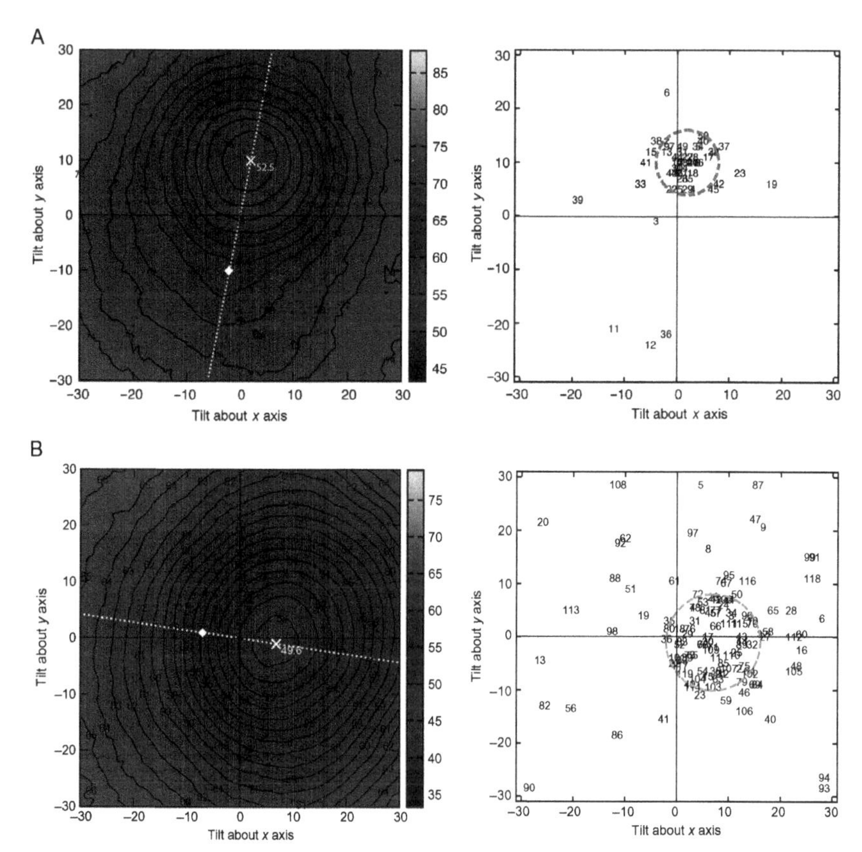

Fig. 4 Tilt-pair phase residual plot (*left*) and tilt-pair parameter plot (TPPP, *right*) for (A) 50 particles of the pyruvate dehydrogenase E2 core (1.5 MDa) and (B) 119 particles of β-galactosidase (450 kDa). For both cases, the phase residual plots show a minimum at the tilt angle (10°) and tilt axis of the experiment. Note that the orientation of the tilt axis is different for the two cases. The "free hand difference" for hand discrimination is calculated from the difference between the phase residual at the minimum (×) and the negative tilt angle (♦). The width of the cluster of orientations at the tilt axis and tilt angle for β-galactosidase is approximately twice that for pyruvate dehydrogenase. *Based on figs. 4 and 5 of Wasilewski, S., & Rosenthal, P. B. (2014). Web server for tilt-pair validation of single-particle maps from electron cryomicroscopy.* Journal of Structural Biology, 186, *122–131 and data in Henderson, R., Chen, S., Chen, J. Z., Grigorieff, N., Passmore, L. A., Ciccarelli, L., et al. (2011). Tilt-pair analysis of images from a range of different specimens in single-particle electron cryomicroscopy.* Journal of Molecular Biology, 413, *1028–1046.*

a full search of orientations is only required for the first image, which is of better quality and contains higher resolution information than the second image due to radiation damage, but both forms of analysis will reach similar conclusions.

Fig. 4 shows TPPPs and tilt-pair phase residual plots for icosahedral pyruvate dehydrogenase E2 core (1.5 MDa) and D2-symmetric β-galactosidase (450 kDa). The residual plots show a minimum near the tilt axis and tilt angle used in the experiment. The TPPPs show that the majority of the particles are clustered near the correct tilt axis and tilt angle indicated by the small circle. The angular width of the cluster is smaller in the case of pyruvate dehydrogenase, indicating more accurate orientation determination for the larger particle. The width of the cluster on the TPPP is a measure of the accuracy of orientation determination. Comparison of TPPPs from specimens with a wide range of sizes and shapes shows that higher molecular mass specimens can be oriented more accurately (Henderson et al., 2011).

The test requires approximately 100 particles, though more or less may be required depending on the accuracy of the orientations. If there are relatively few particles per image, then particle pairs from different images may be combined into single stacks, provided that they have been recorded at conditions where the tilt axis and tilt angle are the same.

Several choices of tilt angle should be tested during the experiment. A small absolute tilt angle has the advantage of not producing a large defocus gradient, though this does not cause major problems when matching particle projections at low resolution on the second particle image. A larger tilt angle may be useful to obtain views with a different appearance and discrimination of absolute hand. The optimal tilt angle may depend on the size and shape features of the particle. We can see in the TPPPs in Fig. 4 that the tilt angle of the experiment should be large enough to distinguish the mean value of the cluster of orientations from the plot origin, where the cluster may be located when the orientations are random.

2.4 Map Validation

The tilt-pair test shows whether particle orientations are accurately determined indicating that the map represents the structural information in the images. Tilt-pair analysis may similarly be used to evaluate the quality of different maps or monitor map improvement. A wide distribution of views should be used in the tilt-pair test. Selection and presentation of only the subset of tilt-pair data that matches the known tilt axis and tilt angle contradicts the use of the test for validation purposes, which ideally would include all the particles in the image field. If a small number of tilt-pair image stacks are deposited along with the 3D map, the TPPP may be easily verified by independent users.

A valid map should have a well-defined minimum on the residual plot and a clear cluster on the TPPP. Based on examples studied so far, a simple majority criterion, such as 60% of particle tilt-pair parameters clustered at the known tilt axis and tilt angle, is a strong support for the structure determination (Henderson et al., 2011). If this is not the case, then the experiment and data analysis should be further evaluated to assess why significant numbers of particles lie outside the cluster. Possible explanations include image quality and specimen size or heterogeneity.

The program TiltStats (Russo & Passmore, 2014) may be used to obtain a quantitative analysis of the clustering on the TPPP (from output of the Tiltdiff program) using the concentration parameter (κ) of the Fisher distribution. If $\kappa < 10$, then there is no evidence of clustering, whereas the specimens shown in Fig. 4 have $\kappa = 175$ and $\kappa = 93$ (Russo & Passmore, 2014). The tilt-pair experiment may also be analyzed by the tilt-pair alignment test (Baker, Watt, Runswick, Walker, & Rubinstein, 2012) which shows the percentage of particles with specified angular error and may be assessed against the expected result for random orientations.

Though in principle maps studied by single-particle analysis in ice or in negative stain may be validated by tilt-pair analysis, extensive flattening of the particles in negative stain may change the expected relationship between untilted and tilted views, making the test less informative.

2.5 Hand Discrimination

A map of either absolute hand will match 2D projection images, and therefore experimental determination of absolute hand requires the comparison of images recorded at more than one tilt angle. If tilt-pair analysis performed with a map indicates a minimum at the expected relative tilt angle 2α, then the map of opposite hand will indicate a minimum at a tilt angle of -2α. Thus, the difference in the value (eg, phase residual) between the minimum at -2α and $+2\alpha$ (cross $\times$ and diamond ◆, respectively, on the plots in Fig. 4), called the "free hand difference," measures the discrimination of absolute hand (Rosenthal & Henderson, 2003). The free hand difference will be greater when a larger relative tilt angle is used in the experiment. While handedness is always a feature of biological assembles, a map without handedness will not discriminate these, and not produce a unique validation result in this test.

The true sign of the rotation angle, recorded during the experiment, may be used to verify the absolute hand of the map though care must be taken in all image-processing steps that could potentially invert the hand determination (Rosenthal & Henderson, 2003). The location in the tilt-pair residual plot where the minimum is expected depends on the direction of the tilt axis in the experiment and often depends on the magnification setting of the microscope. Recording a tilt-pair under identical conditions for a specimen of known handedness may be used to indicate where the minimum is expected for a specimen of unknown absolute hand. When high-resolution features such as the pitch of α-helices are observed in a map, it is possible to assess from the map alone whether the hand is correct. At lower resolution, experimental determination of absolute hand determination can validate the rigid-body fitting of an atomic coordinate model to the map.

2.6 Optimization of Orientation Determination and Alignment Accuracy

Both types of tilt-pair analysis may be applied to the optimization of single-particle analysis by testing refinement parameters that influence orientation determination. Some optimization studies may benefit from having the first particle image stack recorded at 0° and at the same magnification as for the main dataset. Parameters may be varied in refining orientations of particle images in the first stack and then the accuracy scored at constant computational conditions against images on the second stack usually at low resolution due to the prior radiation exposure. The tilt-pair phase residual plot may show a weak minimum from only a small fraction of particles being assigned correct orientations, which may be a starting point for successful optimization. As an important example, the resolution range used in refining orientations may be varied on the first stack. Evaluation of the width of the cluster on the TPPP or the depth of the minimum on the phase residual plot shows which resolution ranges produce the most accurate orientations. While refinement with a high-resolution limit of 30 Å may produce less accurate orientations than using data to 15 Å, inclusion of the highest resolution data beyond 10 Å may not improve the accuracy of orientation determination (Henderson et al., 2011; Rosenthal & Henderson, 2003). Excluding high-resolution information from refinement may be important for resolution assessment as discussed in the next section. Fig. 3 shows the difference in success in orientation determination before and after a systematic optimization of several parameters used in particle refinement.

The Tiltdiffmulti version of the program Tiltdiff facilitates optimization by taking as input several different sets of orientation per particle, each obtained using slightly different refinement parameters, and then finds the best tilt transformation that relates the two particle orientations (Henderson et al., 2011).

In summary, tilt-pair analysis shows that orientations are accurately determined, validating map determination. High-resolution map features may also validate a map, but the investigator may apply tilt-pair analysis in order to know that the experimental and computational procedures have been optimized. The width of the clusters on a TPPP may be connected to the contrast loss at high resolution due to errors in computations, described by a temperature factor $B_{\text{computation}}$ (Henderson et al., 2011; Rosenthal & Henderson, 2003). This gives an indication whether orientations are accurate enough to reach a specified resolution with a specified number of particles, an important concept for validation.

2.7 Alignment Reliability Test

Another measure of whether maps are consistent with data is the alignment reliability test (Vargas, Oton, Marabini, Carazo, & Sorzano, 2016). For the map to be tested, the distribution of orientations that best agree with a given image are identified. If these are clustered at a particular orientation rather than randomly distributed, this provides evidence that the image is alignable by the map. The test addresses precision rather than accuracy of the orientations and is thus complementary to tilt-pair analysis. Ideally this test can be performed with a subset of the images used to calculate the final map, and it can be performed post hoc on any dataset without requiring additional experiments. Like the tilt-pair test it does not directly validate map resolution, because the alignment does not require the highest resolution information.

3. VALIDATION OF MAP RESOLUTION

The tools described in the last section cannot validate map features at high resolution. Resolution may be assessed by the interpretability of molecular features typical of protein or nucleic acid. For proteins, resolution milestones are encountered when α-helices become resolved below 10 Å, β-sheets below 5 Å, and side chains at better than 3.5 Å. Even where high-resolution features are present, their significance needs to be assessed and there remains the concern that these features may have been introduced by refinement of an initial map that already possessed them. In addition, statistical measures of map consistency are necessary for monitoring the progress of refinement and for communicating results.

The resolution of a three-dimensional density map is commonly estimated by the FSC which is the correlation of structure factors (F_1,F_2) of two maps, each calculated from the half the particle images:

$$\mathrm{FSC} = \frac{\sum F_1 \cdot F_2^*}{\sqrt{\sum |F_1|^2 \sum |F_2|^2}},$$

such that sums run over reciprocal space shells (Saxton & Baumeister, 1982; van Heel & Harauz, 1986; van Heel & Stoffler-Meilicke, 1985). The structure factors are calculated from the 3D Fourier transform of the map and each has an amplitude ($|F|$) and phase (ϕ) component. The FSC is reported by the major software packages, and the FSC validation web service at the EMDB will calculate the FSC from half-maps (http://www.ebi.ac.uk/pdbe/emdb/validation/fsc/).

At low resolution, where the value of the FSC curve is near 1.0, the map is nearly perfect and has little noise, but the map contains increasing amounts of noise toward its resolution limit. Typically a threshold value for the FSC is chosen to define the resolution limit of the map, such as 0.143 (Rosenthal & Henderson, 2003). The FSC curve should be reported in all resolution shells to the Nyquist frequency, which may then identify resolution ranges with poor signal, such as when there is missing information due to zeroes of the contrast transfer function. The FSC curve may reveal other problems, such as a precipitous decline in the FSC precisely at the resolution limit of the data used in refinement (Penczek, 2010). This suggests that the map has matched noise features for the data included in the refinement.

Thus a potential problem with the FSC is false correlation at high resolution due to over-fitting. Over-fitting is particularly a concern when a single reference map is used to align both halves of the image data that are used in the FSC calculation, in which case the same noise features on the map will be reinforced through alignment of both data half-sets (Grigorieff, 2000). In principle, refinement may be cross-validated by leaving out some of the data which may be used to monitor map improvement and detect over-fitting (Shaikh, Hegerl, & Frank, 2003). This approach is conceptually similar to the cross-validation protocol in X-ray crystallography in which a subset of reflection data (the free set) are excluded from refinement to calculate the free R factor that measures true agreement between model and data and not over-fitting (Brunger, 1992, 1997). There may be, however, concerns about leaving out data that is essential for the accurate alignment of particles.

Two refinement protocols are widely used that can detect and therefore hopefully reduce over-fitting during iterative map refinement. One approach is called resolution-limited or frequency-limited refinement, in which the data in the highest resolution shells are not included in orientation determination and therefore cannot be over-fit. Thus during refinement, an increasing value for the FSC in the "free shells" is an indication that the map is really improving. The highest resolution shells are precisely the data that need to be assessed to define the resolution of the map. Furthermore, tilt-pair analysis has demonstrated that these highest resolution shells, eg, higher than 10 Å, may not contribute to orientation determination (Henderson et al., 2011; Rosenthal & Henderson, 2003). Under these conditions, and where there is strong structural signal in the data at low resolution, it may be sufficient simply to separate the data into two halves at the final iteration of map calculation for FSC calculation.

A second approach to the over-fitting problem is now in common use and is sometimes referred to as "gold-standard refinement," though there is less agreement on labeling any procedure as a "gold standard." This procedure requires that the two halves of the image dataset be kept fully independent during refinement and that independent reference maps should be used for assigning particle orientation parameters (Grigorieff, 2000; Henderson et al., 2012; Scheres & Chen, 2012). When half-maps are kept entirely independent, an additional procedure is required to align the independent maps for FSC calculation. Consistent with frequency-limited refinement and tilt-pair analysis, the use of half-maps for refinement, which have lower resolution than the full map, does not produce worse alignments and indeed reduces over-fitting (Scheres & Chen, 2012). Frequency-limited refinement may be implemented by using the FSC between independent half-maps as an unbiased weight of the data used during each round of refinement to limit resolution so that only the most reliable map signal and the least noise are used for particle alignment. As the map resolution improves, the FSC reveals more signal at high resolution (eg, up to the 0.143 threshold) which may then be automatically introduced into refinement.

Even with protocols that detect or prevent over-fitting, caution is required in using an initial model that possesses high-resolution feature for refinement, and the starting map should be low-pass filtered to an extremely low resolution.

3.1 The High-Resolution Noise-Substitution Test

While both frequency-limited refinement and independent half-maps are recognized as good practice, there always remains a concern about over-fitting in any refinement. Chen et al. (2013) have devised a test to quantitatively measure over-fitting by making use of images where the phases (ϕ) of high-resolution structure factors have been substituted by the structure factors of noise images.

The procedure is to refine the maps against images with and without high-resolution noise substituted:

1. Refine the 3D map against the original images using any protocol to be tested by the investigator and calculate the FSC (called FSCt).
2. A high-resolution value is selected beyond which the Fourier components of the images are substituted by noise (program *make-stack_HRnoise*). The substitution may be done in one of two ways: (a) they may be obtained using images from the parts of the particle field that do not contain particles, ie, signal from only ice and noise (b) randomize the phase values for structure factors above the high-resolution value producing images that have the same power spectrum as the original images. The latter approach is easier in practice.
3. Refine the 3D map against the noise substituted images using the same protocol to be tested and calculate the FSC (call FSCn).
4. The FSCt curve is compared with the FSCn curve for evidence of over-fitting by plotting.
5. FSCtrue is then calculated according to the following formula, which assesses the true signal in the map and not spurious signal due to over-fitting: $\text{FSCtrue} = (\text{FSCt} - \text{FSCn})/(1 - \text{FSCn})$.

Fig. 5A shows examples of images, noise images, and images with high-resolution noise substituted above 10 Å resolution.

Fig. 5B shows the case in which high-resolution data have been given a large weight in particle alignment. The difference between FSCt and FSCn indicates a significant overestimate of resolution due to over-fitting. Fig. 5C shows the case of "gold-standard" refinement with FSC weighting (Scheres & Chen, 2012) showing no over-fitting beyond the high-resolution value chosen for noise substitution. Thus the analysis has led to a further demonstration of the principles stated in the last section. HR-noise substitution may be used to identify refinement protocols that reduce over-fitting.

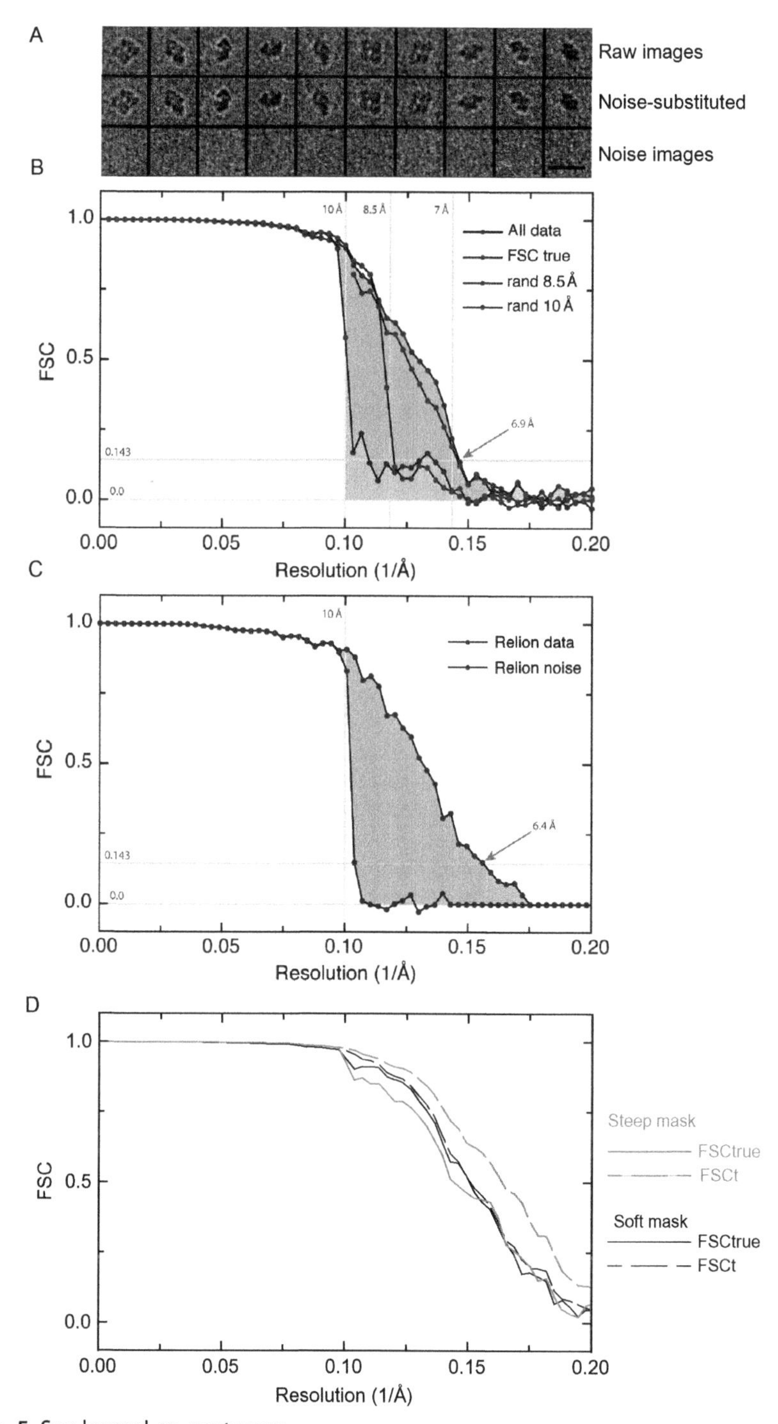

Fig. 5 See legend on next page.

3.2 The Effect of Masks

A mask may be applied to remove noise from the region outside a molecular envelope (solvent flattening) and therefore improve the FSC. However, correlations between identical masks applied to each half-map will themselves produce correlations to high-resolution independent of the structure of interest leading to a resolution overestimate (Penczek, 2010). A mask with a soft edge rather than a sharp edge will somewhat reduce the worst effects of mask correlation. Access to both the original map and masks may therefore be important to validating the resolution assessment (Henderson et al., 2012). The noise-substitution test in Fig. 5D shows that a sharp-edge mask produces a spurious high-resolution signal but that this is reduced for a soft-edge mask (Chen et al., 2013).

The program Relion employs the high-resolution noise-substitution procedure to measure the effect of a mask on the FSC curve as follows:

1. Calculate the FSC curve between masked maps.
2. Randomize phases of the two unfiltered maps for structure factors above 2 × the Nyquist frequency before masking.
3. Mask the maps with randomized phases.
4. Calculate the FSC.

Fig. 5 High-resolution noise-substitution test to validate resolution assessment by the Fourier shell correlation (FSC). (A) Single-particle images of β-galactosidase. *Top row*: raw images, *middle row*: images with phases randomized higher than 10 Å have a similar appearance, *bottom row*: noise images obtained by selecting regions of ice only from the same images where particle images are obtained. (B) Case where particle orientation refinement overweights high-resolution data leading to over-fitting. FSC is shown between half-maps for original image data (*red circles*) and image data with randomized phases beyond 10 Å (*blue circles*) or beyond 8.5 Å (*green circles*). Portion of FSC signal due to over-fitting is shown by the shaded *blue region* (FSCn), and the portion due to true structural signal is given by the shaded *pink region* or the curve FSCtrue (*purple circles*). (C) FSCt between half-maps refined using completely independent half-maps and FSC weighting (program *Relion*) for original image data (*red circles*) or image data with randomized phases beyond 10 Å (*blue circles*). FSCn (shaded *blue*) indicates that there is no over-fitting. (D) Comparison of FSCt and FSCn for cases where a sharp-edge mask or a soft-edge mask is used to calculate the FSC between half-maps. Exaggerated resolution results from a tight-edge mask but not from a soft-edge mask. *Panel (A) is from fig. 1 of, (B and C) are from fig. 3A and B of, and (D) shows replotted graphs from fig. 4A and C of Chen, S., McMullan, G., Faruqi, A. R., Murshudov, G. N., Short, J. M., Scheres, S. H., et al. (2013). High-resolution noise substitution to measure overfitting and validate resolution in 3D structure determination by single particle electron cryomicroscopy.* Ultramicroscopy, 135, *24–35.* (See the color plate.)

5. Compare FSCs with and without randomization (program *rlnFourierShellCorrelationCorrected*).

Different programs may apply masks differently when reporting the FSC between half-maps and this may affect comparisons.

3.3 Another Noise Test

Another simple test of the validity of a map has been proposed (Heymann, 2015) by assessing whether the resolution obtained by aligning a small subset of images to the map is better than aligning pure noise images. Real particles should produce higher resolution. For a given map and images, this test is able to assess whether there is any information in the images that is alignable by the map.

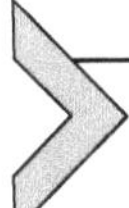

4. RESOLUTION ASSESSMENT OF A MAP WITH AN ATOMIC COORDINATE MODEL

An estimate for the correlation between a map calculated from the full dataset, which has improved resolution compared to a half-map due to two-fold averaging, and a perfect reference map without errors may be derived from the FSC curve between half-maps and is called C_{ref} (Rosenthal & Henderson, 2003):

$$C_{\text{ref}} = \sqrt{2\text{FSC}/(1 + \text{FSC})}$$

C_{ref} is equivalent to the crystallographic figure of merit (Blow & Crick, 1959). The resolution estimated by the FSC between half-maps at the threshold of 0.143 is the same resolution as estimated by C_{ref} at a threshold of 0.5 and is equivalent to a final resolution shell phase error of 60°.

Where a high-resolution atomic map is obtained from a coordinate model such as an X-ray structure that represents the structure in the EM map exactly, we can calculate C_{ref} between the map and the model and expect the $C_{\text{ref}}=0.5$ to indicate the resolution. C_{ref} may indicate lower resolution than expected from the FSC between half-maps when there are differences between map and model or where the map has variable resolution. On the other hand, C_{ref} may show false correlations due to over-fitting if the atomic coordinate model is refined against the map without safeguards well known from crystallographic refinement, such as restricting the degrees of freedom of the model to those justified by the number of structure factor observations.

To test over-fitting during refinement of coordinate models in single-particle maps, frequency-limited refinement has been used to monitor the improvement in signal in "free shells" at high resolution (Falkner & Schroder, 2013). Also, the atomic model may be refined against a half-map and the progress evaluated in the FSC between the coordinate model and the second half-map, assuming that the data for the two half-maps have been kept separate during map refinement (Brown et al., 2015; DiMaio, Zhang, Chiu, & Baker, 2013).

5. CONTRAST RESTORATION

A map with a valid FSC curve may not show the expected resolution features unless the relative weighting of low- and high-resolution structure factor amplitudes are corrected for the contrast loss at high resolution which otherwise will make the map look featureless. This contrast loss may be modeled by $e^{-(B_{\mathrm{overall}}/4d^2)}$, where B_{overall} is the overall positive temperature factor for the map, which includes the effects of contrast fall-off in images (B_{image}) and errors due to computation ($B_{\mathrm{computation}}$) (Bottcher, Wynne, & Crowther, 1997; Rosenthal & Henderson, 2003). B_{overall} may be measured by comparing figure-of-merit-weighted map amplitudes, C_{ref}F, with theoretical scattering criteria (Rosenthal & Henderson, 2003). Map features may be enhanced by a sharpening procedure in which C_{ref}F is multiplied by $e^{-(B_{\mathrm{restore}}/4d^2)}$, where the negative temperature factor B_{restore} increases the relative weight of high-resolution structure factor amplitudes, and the figure-of-merit weight (C_{ref}) prevents the amplification of noise (Fernandez, Luque, Caston, & Carrascosa, 2008; Rosenthal & Henderson, 2003).

The temperature factor obtained from the fall-off in the figure-of-merit-weighted map amplitudes may be used to calculate the number of particles required to reach a given resolution (Rosenthal & Henderson, 2003). Indeed, the recent progress in cryomicroscopy of single particles may be understood as a reduction of B_{image} by the use of DDDs which record better images, as well as a reduction in $B_{\mathrm{computation}}$ by improved approaches to image analysis described in this chapter. Therefore, the temperature factor is an important indicator of the overall quality of the data and computations in the structure determination. In conjunction with the other major indicators of global resolution and success in orientation determination, it indicates whether the signal in the map is consistent with the number of particles used or if the experiment and computations could be optimized further. Graphical plots of map resolution for subsets of the data containing different numbers of particles (Bottcher et al., 1997) have been advocated as a validation tool called ResLog plots (Stagg, Noble, Spilman, & Chapman, 2014).

6. LOCAL RESOLUTION ASSESSMENT

Global measures of resolution such as the FSC do not report variation of map quality at positions within the map and are unaffected by small changes in the maps. Errors in the reconstruction procedure may contribute to variable resolution, including errors in orientation assignment that have a greater effect on the periphery of the map (Kishchenko & Leith, 2014). The variation may also be due to flexibility, disorder, or compositional heterogeneity that may be important to understanding molecular function. Local resolution measures are of increased significance now that higher resolution structures are possible in solution by single-particle analysis.

In X-ray crystallography, where the building of an atomic model is an essential part of improving phases for the map, measures for assessing local map quality have been used for many years (Read et al., 2011). These include real space correlation for electron density at atomic positions in the model, and assignment of occupancy and temperature factors to atoms. These local measures are equally applicable when building and refining atomic models into single-particle maps from electron cryomicroscopy.

In single-particle maps, local variation in map resolution has been evaluated in radial shells in the context of icosahedral virus calculation (Harris et al., 2006; Huiskonen, Kivela, Bamford, & Butcher, 2004; Mancini, Clarke, Gowen, Rutten, & Fuller, 2000). Local evaluation of an atomic model against a map may be achieved by masking a subvolume of a map with a soft-edge mask (Brown et al., 2015). Local resolution may be calculated for the whole map using a windowed FSC (as implemented in the program *blocres* in the Bsoft package (Cardone, Heymann, & Steven, 2013; Heymann & Belnap, 2007) or *sxlocres* in SPARX (Hohn et al., 2007)). The FSC between half-maps is calculated within a small region at all points of the map, while taking care that the isolation of the subvolume does not introduce high-frequency artifacts. A similar approach can be used to study local variation in other resolution measures.

The calculation of local resolution by the program ResMap (Kucukelbir, Sigworth, & Tagare, 2014) and its graphical representation using the program UCSF Chimera (Pettersen et al., 2004) have become widely reported.

Using orientable basis functions, the local resolution at a point in the map is defined as the smallest scale at which a local signal feature is statistically detectable above noise (Kucukelbir, 2014). An advantage to the approach is that no resolution threshold choice is required. Ideally the map should not be low-pass filtered and the program provides an interactive map-whitening routine as a first step. In some cases the resolution value assigned to the full map by ResMap agrees with estimates by C_{ref} (Kucukelbir, 2014).

Fig. 6 shows ResMap analysis of two important single-particle structures where variable resolution has a biological interpretation. Fig. 6A shows the structure of mammalian mitochondrial complex I (EMD-2676), an integral membrane protein, where a detergent-phospholipid belt surrounding the transmembrane region has lower resolution than most of the protein regions of the map (Vinothkumar, Zhu, & Hirst, 2014). Fig. 6B shows the ATPase NSF (*N*-ethylmaleimide sensitive factor, EMD-6204), a molecular machine

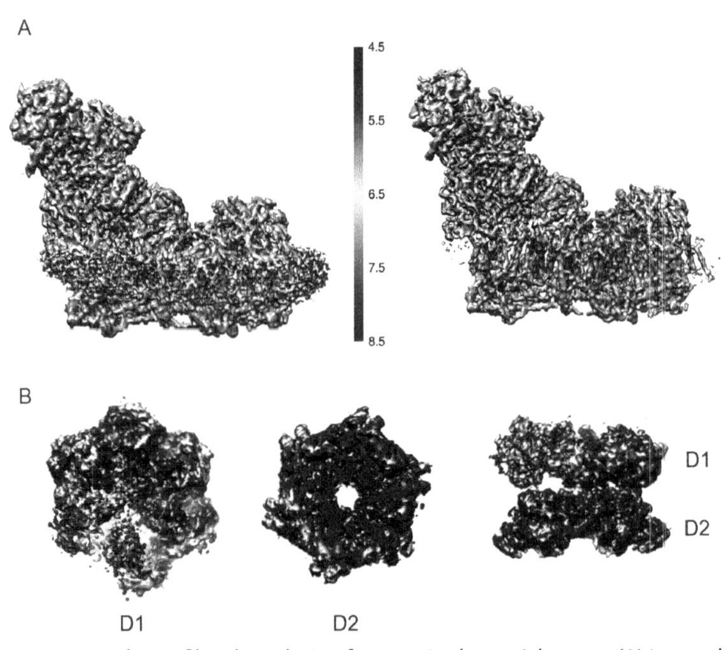

Fig. 6 ResMap analysis of local resolution for two single-particle maps (A) integral membrane protein mammalian mitochondrial complex I (EMD-2676) shows that the detergent-phospholipid belt surrounding the transmembrane region on map (*left*) is at lower resolution than other parts of the map (*right*). (B) NSF (*N*-ethylmaleimide sensitive factor, EMD-6204) ATPase shows uniform resolution for subunits of the D2 ring, but variable resolution for two subunits in the D1 ring. *Color legend* indicates resolution in Å units. (See the color plate.)

for disassembling SNARE proteins (Zhao et al., 2015). Variable resolution is related to dynamic structural changes related to function.

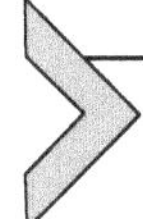

7. SUMMARY OF VALIDATION TESTS AND FUTURE PROSPECTS

Electron cryomicroscopy of single particles has entered a new era in which the atomic scale structure of proteins and dynamics may be studied outside the crystalline state. Computational tools that demonstrate the reliability of single-particle maps will increase the impact of these studies. The validation tests described in this chapter play the dual role of communicating the accuracy of the final structure and providing the investigator with tools that may be useful throughout structure determination.

Recognizing that an incorrect low-resolution starting model may bias subsequent refinement, demonstration of the consistency of a low-resolution map with image data is essential and should include the most stringent tests, such as tilt-pair analysis (Henderson et al., 2011; Rosenthal & Henderson, 2003). New tests evaluate how precisely a map can align particle images (Vargas et al., 2016) or whether a map can align and average particle images to a higher resolution than pure noise images (Heymann, 2015).

The effect of either over-fitting at high resolution or sharp-edge masks on the FSC may be measured through the high-resolution noise-substitution test (Chen et al., 2013). The test also assesses the amount of over-fitting in different refinement schemes. When an atomic coordinate model is available or may be built de novo, the presence of protein features in the map support the reliability. In addition, the FSC calculated between a map and an atomic coordinate model may support the overall structure determination and its interpretation by the model. For maps that are well resolved, local resolution measures will help interpretation.

Even where high-resolution features validate a structure, the validation tests guard against the worst types of model bias and may help assess whether the image analysis has been performed optimally and provides feedback on the imaging experiment. Reporting the temperature factor for a map indicates the prospects for extending the resolution further with more particle images.

Theoretical predictions suggest that atomic resolution structures may be obtained from significantly fewer particle images through experimental and computational improvements (Glaeser, 1999, 2016; Henderson, 1995; Rosenthal & Henderson, 2003) and that they may be achieved for smaller molecular weight specimens. The range of specimens that may be studied will expand and new analysis tools may be required, particularly for validating small differences in heterogeneous structures at high-resolution.

ACKNOWLEDGMENTS

P.B.R. is supported by the Francis Crick Institute, which receives its core funding from Cancer Research UK, the UK Medical Research Council (program code U117581334), and the Wellcome Trust. I thank Tony Crowther, Richard Henderson, and John Rubinstein for their discussions and comments on the manuscript. I thank Kutti R. Vinothkumar and Yifan Cheng for assistance in preparing Fig. 6.

REFERENCES

Bai, X. C., Fernandez, I. S., McMullan, G., & Scheres, S. H. (2013). Ribosome structures to near-atomic resolution from thirty thousand cryo-EM particles. *eLife*, *2*, e00461.

Baker, L. A., Watt, I. N., Runswick, M. J., Walker, J. E., & Rubinstein, J. L. (2012). Arrangement of subunits in intact mammalian mitochondrial ATP synthase determined by cryo-EM. *Proceedings of the National Academy of Sciences of the United States of America*, *109*, 11675–11680.

Blow, D. M., & Crick, F. H. C. (1959). The treatment of errors in the isomorphous replacement method. *Acta Crystallographica*, *12*, 794–802.

Bottcher, B., Wynne, S. A., & Crowther, R. A. (1997). Determination of the fold of the core protein of hepatitis B virus by electron cryomicroscopy. *Nature*, *386*, 88–91.

Brilot, A. F., Chen, J. Z., Cheng, A., Pan, J., Harrison, S. C., Potter, C. S., et al. (2012). Beam-induced motion of vitrified specimen on holey carbon film. *Journal of Structural Biology*, *177*, 630–637.

Brown, A., Long, F., Nicholls, R. A., Toots, J., Emsley, P., & Murshudov, G. (2015). Tools for macromolecular model building and refinement into electron cryo-microscopy reconstructions. *Acta Crystallographica. Section D, Biological Crystallography*, *71*, 136–153.

Brunger, A. T. (1992). Free R value: A novel statistical quantity for assessing the accuracy of crystal structures. *Nature*, *355*, 472–475.

Brunger, A. T. (1997). Free R value: Cross-validation in crystallography. *Methods in Enzymology*, *277*, 366–396.

Cardone, G., Heymann, J. B., & Steven, A. C. (2013). One number does not fit all: Mapping local variations in resolution in cryo-EM reconstructions. *Journal of Structural Biology*, *184*, 226–236.

Chen, S., McMullan, G., Faruqi, A. R., Murshudov, G. N., Short, J. M., Scheres, S. H., et al. (2013). High-resolution noise substitution to measure overfitting and validate resolution in 3D structure determination by single particle electron cryomicroscopy. *Ultramicroscopy*, *135*, 24–35.

Cong, Y., & Ludtke, S. J. (2010). Single particle analysis at high resolution. *Methods in Enzymology*, *482*, 211–235.

Crowther, R. A., DeRosier, D. J., & Klug, A. (1970). The reconstruction of a three-dimensional structure from projections and its application to electron microscopy. *Proceeding of the Royal Society of London*, *317*, 319–340.

DiMaio, F., Zhang, J., Chiu, W., & Baker, D. (2013). Cryo-EM model validation using independent map reconstructions. *Protein Science: A Publication of the Protein Society*, *22*, 865–868.

Falkner, B., & Schroder, G. F. (2013). Cross-validation in cryo-EM-based structural modeling. *Proceedings of the National Academy of Sciences of the United States of America, 110*, 8930–8935.
Fernandez, J. J., Luque, D., Caston, J. R., & Carrascosa, J. L. (2008). Sharpening high resolution information in single particle electron cryomicroscopy. *Journal of Structural Biology, 164*, 170–175.
Frank, J. (2006). *Three-dimensional electron microscopy of macromolecular assemblies: Visualization of biological molecules in their native state* (2nd ed.). Oxford/New York: Oxford University Press.
Frank, J., Radermacher, M., Penczek, P., Zhu, J., Li, Y., Ladjadj, M., et al. (1996). SPIDER and WEB: Processing and visualization of images in 3D electron microscopy and related fields. *Journal of Structural Biology, 116*, 190–199.
Glaeser, R. M. (1999). Review: Electron crystallography: Present excitement, a nod to the past, anticipating the future. *Journal of Structural Biology, 128*, 3–14.
Glaeser, R. M. (2016). How good can cryo-EM become? *Nature Methods, 13*, 28–32.
Grigorieff, N. (2000). Resolution measurement in structures derived from single particles. *Acta Crystallographica. Section D, Biological Crystallography, 56*, 1270–1277.
Harris, A., Belnap, D. M., Watts, N. R., Conway, J. F., Cheng, N., Stahl, S. J., et al. (2006). Epitope diversity of hepatitis B virus capsids: Quasi-equivalent variations in spike epitopes and binding of different antibodies to the same epitope. *Journal of Molecular Biology, 355*, 562–576.
Henderson, R. (1995). The potential and limitations of neutrons, electrons and x-rays for atomic-resolution microscopy of unstained biological molecules. *Quarterly Reviews of Biophysics, 28*, 171–193.
Henderson, R. (2013). Avoiding the pitfalls of single particle cryo-electron microscopy: Einstein from noise. *Proceedings of the National Academy of Sciences of the United States of America, 110*, 18037–18041.
Henderson, R., Chen, S., Chen, J. Z., Grigorieff, N., Passmore, L. A., Ciccarelli, L., et al. (2011). Tilt-pair analysis of images from a range of different specimens in single-particle electron cryomicroscopy. *Journal of Molecular Biology, 413*, 1028–1046.
Henderson, R., Sali, A., Baker, M. L., Carragher, B., Devkota, B., Downing, K. H., et al. (2012). Outcome of the first electron microscopy validation task force meeting. *Structure, 20*, 205–214.
Heymann, J. B. (2015). Validation of 3D EM reconstructions: The phantom in the noise. *AIMS Biophysics, 2*, 21–35.
Heymann, J. B., & Belnap, D. M. (2007). Bsoft: Image processing and molecular modeling for electron microscopy. *Journal of Structural Biology, 157*, 3–18.
Hohn, M., Tang, G., Goodyear, G., Baldwin, P. R., Huang, Z., Penczek, P. A., et al. (2007). SPARX, a new environment for cryo-EM image processing. *Journal of Structural Biology, 157*, 47–55.
Huiskonen, J. T., Kivela, H. M., Bamford, D. H., & Butcher, S. J. (2004). The PM2 virion has a novel organization with an internal membrane and pentameric receptor binding spikes. *Nature Structural & Molecular Biology, 11*, 850–856.
Kishchenko, G. P., & Leith, A. (2014). Spherical deconvolution improves quality of single particle reconstruction. *Journal of Structural Biology, 187*, 84–92.
Kucukelbir, A. (2014). *Sparse and steerable representations for 3D electron cryomicroscopy*. Ph.D. thesis, New Haven, CT: Yale University.
Kucukelbir, A., Sigworth, F. J., & Tagare, H. D. (2014). Quantifying the local resolution of cryo-EM density maps. *Nature Methods, 11*, 63–65.
Li, X., Mooney, P., Zheng, S., Booth, C. R., Braunfeld, M. B., Gubbens, S., et al. (2013). Electron counting and beam-induced motion correction enable near-atomic-resolution single-particle cryo-EM. *Nature Methods, 10*, 584–590.

Lyumkis, D., Brilot, A. F., Theobald, D. L., & Grigorieff, N. (2013). Likelihood-based classification of cryo-EM images using FREALIGN. *Journal of Structural Biology, 183*, 377–388.
Mancini, E. J., Clarke, M., Gowen, B. E., Rutten, T., & Fuller, S. D. (2000). Cryo-electron microscopy reveals the functional organization of an enveloped virus, Semliki Forest virus. *Molecular Cell, 5*, 255–266.
McMullan, G., Faruqi, A. R., Clare, D., & Henderson, R. (2014). Comparison of optimal performance at 300 keV of three direct electron detectors for use in low dose electron microscopy. *Ultramicroscopy, 147*, 156–163.
Penczek, P. A. (2010). Resolution measures in molecular electron microscopy. *Methods in Enzymology, 482*, 73–100.
Penczek, P., Radermacher, M., & Frank, J. (1992). Three-dimensional reconstruction of single particles embedded in ice. *Ultramicroscopy, 40*, 33–53.
Pettersen, E. F., Goddard, T. D., Huang, C. C., Couch, G. S., Greenblatt, D. M., Meng, E. C., et al. (2004). UCSF chimera—A visualization system for exploratory research and analysis. *Journal of Computational Chemistry, 25*, 1605–1612.
Read, R. J., Adams, P. D., Arendall, W. B., 3rd., Brunger, A. T., Emsley, P., Joosten, R. P., et al. (2011). A new generation of crystallographic validation tools for the protein data bank. *Structure, 19*, 1395–1412.
Rosenthal, P. B., & Henderson, R. (2003). Optimal determination of particle orientation, absolute hand, and contrast loss in single-particle electron cryomicroscopy. *Journal of Molecular Biology, 333*, 721–745.
Rosenthal, P. B., & Rubinstein, J. L. (2015). Validating maps from single particle electron cryomicroscopy. *Current Opinion in Structural Biology, 34*, 135–144.
Rubinstein, J. L., Walker, J. E., & Henderson, R. (2003). Structure of the mitochondrial ATP synthase by electron cryomicroscopy. *The EMBO Journal, 22*, 6182–6192.
Russo, C. J., & Passmore, L. A. (2014). Robust evaluation of 3D electron cryomicroscopy data using tilt-pairs. *Journal of Structural Biology, 187*, 112–118.
Saxton, W. O., & Baumeister, W. (1982). The correlation averaging of a regularly arranged bacterial cell envelope protein. *Journal of Microscopy, 127*, 127–138.
Schatz, M., & van Heel, M. (1990). Invariant classification of molecular views in electron micrographs. *Ultramicroscopy, 32*, 255–264.
Scheres, S. H. (2012). RELION: Implementation of a Bayesian approach to cryo-EM structure determination. *Journal of Structural Biology, 180*, 519–530.
Scheres, S. H., & Carazo, J. M. (2009). Introducing robustness to maximum-likelihood refinement of electron-microscopy data. *Acta Crystallographica. Section D, Biological Crystallography, 65*, 672–678.
Scheres, S. H., & Chen, S. (2012). Prevention of overfitting in cryo-EM structure determination. *Nature Methods, 9*, 853–854.
Shaikh, T. R., Hegerl, R., & Frank, J. (2003). An approach to examining model dependence in EM reconstructions using cross-validation. *Journal of Structural Biology, 142*, 301–310.
Sigworth, F. J. (1998). A maximum-likelihood approach to single-particle image refinement. *Journal of Structural Biology, 122*, 328–339.
Smith, J. M. (1999). Ximdisp—A visualization tool to aid structure determination from electron microscope images. *Journal of Structural Biology, 125*, 223–228.
Stagg, S. M., Noble, A. J., Spilman, M., & Chapman, M. S. (2014). ResLog plots as an empirical metric of the quality of cryo-EM reconstructions. *Journal of Structural Biology, 185*, 418–426.
Stewart, A., & Grigorieff, N. (2004). Noise bias in the refinement of structures derived from single particles. *Ultramicroscopy, 102*, 67–84.

van Heel, M. (2013). Finding trimeric HIV-1 envelope glycoproteins in random noise. *Proceedings of the National Academy of Sciences of the United States of America*, *110*, E4175–E4177.

van Heel, M., & Harauz, G. (1986). Resolution criteria for three dimensional reconstruction. *Optik*, *73*, 119–122.

van Heel, M., & Stoffler-Meilicke, M. (1985). Characteristic views of E. coli and B. stearothermophilus 30S ribosomal subunits in the electron microscope. *The EMBO Journal*, *4*, 2389–2395.

Vargas, J., Oton, J., Marabini, R., Carazo, J. M., & Sorzano, C. O. (2016). Particle alignment reliability in single particle electron cryomicroscopy: A general approach. *Scientific Reports*, *6*, 21626.

Vinothkumar, K. R., Zhu, J., & Hirst, J. (2014). Architecture of mammalian respiratory complex I. *Nature*, *515*, 80–84.

Voss, N. R., Yoshioka, C. K., Radermacher, M., Potter, C. S., & Carragher, B. (2009). DoG Picker and TiltPicker: Software tools to facilitate particle selection in single particle electron microscopy. *Journal of Structural Biology*, *166*, 205–213.

Wasilewski, S., & Rosenthal, P. B. (2014). Web server for tilt-pair validation of single particle maps from electron cryomicroscopy. *Journal of Structural Biology*, *186*, 122–131.

Zhao, M., Wu, S., Zhou, Q., Vivona, S., Cipriano, D. J., Cheng, Y., et al. (2015). Mechanistic insights into the recycling machine of the SNARE complex. *Nature*, *518*, 61–67.

CHAPTER TEN

Tools for Model Building and Optimization into Near-Atomic Resolution Electron Cryo-Microscopy Density Maps

F. DiMaio[*,†,1], **W. Chiu**[‡,1]
[*]University of Washington, Seattle, WA, United States
[†]Institute for Protein Design, University of Washington, Seattle, WA, United States
[‡]National Center for Macromolecular Imaging, Baylor College of Medicine, Houston, TX, United States
[1]Corresponding authors: e-mail address: dimaio@u.washington.edu; wah@bcm.edu

Contents

Abstract

Electron cryo-microscopy (cryoEM) has advanced dramatically to become a viable tool for high-resolution structural biology research. The ultimate outcome of a cryoEM study is an atomic model of a macromolecule or its complex with interacting partners. This chapter describes a variety of algorithms and software to build a de novo model based on the cryoEM 3D density map, to optimize the model with the best stereochemistry restraints and finally to validate the model with proper protocols. The full process of atomic structure determination from a cryoEM map is described. The tools outlined in this chapter should prove extremely valuable in revealing atomic interactions guided by cryoEM data.

1. INTRODUCTION

Recent advances in direct electron detectors as well as reconstruction algorithms for single particles have led to the structure determination of

Methods in Enzymology, Volume 579
ISSN 0076-6879
http://dx.doi.org/10.1016/bs.mie.2016.06.003

macromolecular complexes ranging from 2 to 5 Å resolution (Henderson, 2015; Kuhlbrandt, 2014). At these resolutions, also referred to as "near-atomic" resolution, it is possible to infer all-atom structures de novo. The ability to do this of course depends highly on the overall resolution of the data. Since map resolution can vary significantly, we describe here map resolution broadly in terms of map features (Fig. 1). At high resolution (roughly, 3.5 Å or better), sidechains are clearly visible in maps and individual rotamers may be distinguished. Generally, at this resolution the topology of the protein is unambiguous. At medium-high resolutions (3.5–4.5 Å) only some sidechains (generally aromatics) are visible. Beta strands are separated, but the topology may be ambiguous as the density corresponding to connecting loops may be difficult to resolve. At medium resolution

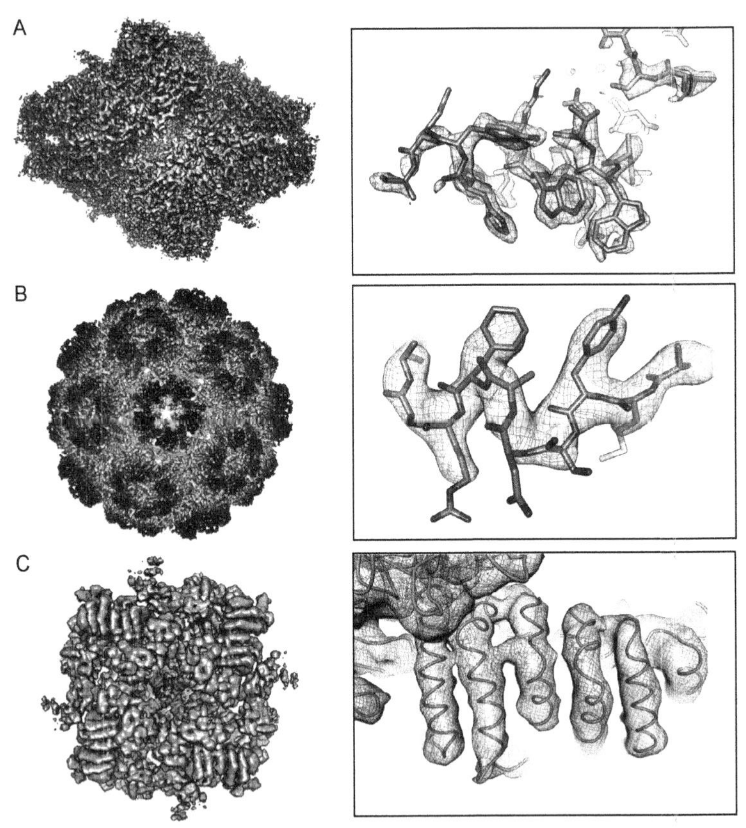

Fig. 1 Structure features of CryoEM maps determined at different resolutions. (A) Beta-galactosidase at 2.2 Å (EMDB2984, PDB 5A1A). (B) Brome mosaic virus at 3.8 Å (EMDB6000, PDB 3J7L). (C) IP3R1 at 4.7 Å (EMDB6369, PDB 3JAV). *These figures are kindly generated by Corey F. Hryc and Matthew L. Baker.*

(4.5–6 Å), helices are discernable, but individual beta strands are often no longer separated. The overall protein topology is often quite difficult to determine at this resolution. Generally speaking, at high-resolution, de novo map interpretation is relatively straightforward; at medium-high resolution it is possible but often difficult and error-prone, and at medium-resolution, it is generally not possible.

Most electron cryo-microscopy (cryoEM) maps tend to be determined to worse than 3.5 Å resolution, and even those at better resolutions often have nonisotropic resolution, which poses additional challenges in determination of accurate all-atom models from data. Consequently, a wide variety of approaches and strategies have been developed to deal with the limited resolution of the cryoEM data. A number of different tasks arise in the process of structure determination from a cryoEM reconstruction, including de novo model building, model optimization, and model validation. Broadly speaking, tools to address these tasks have been derived from two sources. The first source is tools developed initially for X-ray crystallographic model building and optimization that have been adapted for cryoEM reconstructions. The second source is tools that have been developed directly for cryoEM reconstructions and have been targeted for the near-atomic resolution maps that have become available. In both cases, there is a wide range of tools available for each of the three steps of structure determination.

There are also a number of other tools aimed at interpretation of moderate-resolution density (>6 Å). Due to the low information content of such maps, interpretation is generally limited to placement of existing high-resolution structures into the density maps of large macromolecular complexes. Such approaches typically draw off protein/protein docking, to identify the subunit arrangement with best agreement with the experimental map (Lasker, Topf, Sali, & Wolfson, 2009). Despite the valuable insights garnered by such approaches, this chapter does not cover the details of these methodologies, and instead is focused on the process of atomic-level structure determination from near-atomic resolution density maps.

The remainder of this chapter covers the workflow of structure determination from cryoEM density data, as illustrated in the schematic of Fig. 2. The chapter will cover the process of structure determination from a near-atomic resolution map at high and medium-high resolutions (2–4.5 Å), providing details on the various tools available to aid in structure determination at each step, as well as some biological examples where the corresponding approach was employed. Additionally, each section will also present some of the tools for gleaning structural insights at medium (4.5–6 Å) resolutions,

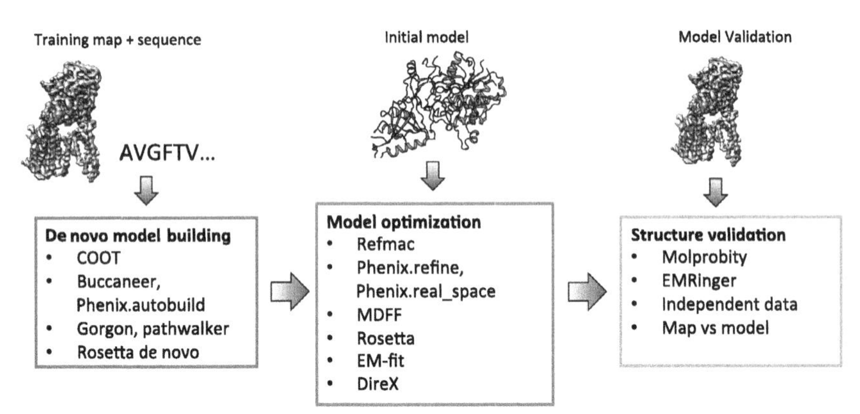

Fig. 2 An overview of three steps of atomic model determination from near-atomic resolution data. (*Left*) De novo building methods take primary sequence and map, and automatically produce a backbone model with sequence registered, identifying which regions in the map correspond to particular sequences. (*Center*) Model optimization takes an initial model—either produced from de novo building, or from a high-resolution homologue—and optimizes the coordinates to better agree with the map, as well as adopt more physically realistic geometry. (*Right*) Model validation aims to assess—both globally and locally—the accuracy of a model, given experimental data. Such tools are useful not only for assessing overall accuracy but also for tuning parameters of optimization.

though this resolution is generally beyond that for which a model may be determined to atomic-level accuracy. Each section will describe one of the three major steps in cryoEM structure determination, corresponding to the blocks in Fig. 2. The first step, de novo structure determination, describes how an initial model may be constructed given only a primary sequence and a reconstruction, when no other or limited structural information is known. In the second step, model optimization, we describe a broad class of methods to improve the fit of a model to data and improve the geometry of a model. Finally, we describe tools for model validation, which attempt to quantify the overall accuracy of a model given a reconstruction.

2. DE NOVO MODEL BUILDING

The first challenge in interpreting a map de novo is in segmentation. Generally, an electron microscopy single particle reconstruction will consist of many different subunits, both symmetrically and nonsymmetrically related to one another. Segmentation divides the map into submaps which each contain one polypeptide or one nucleic acid chain. There are several tools available for automatic, model-free segmentation, such as the program Segger (Pintilie, Zhang, Chiu, & Gossard, 2009; Pintilie, Zhang, Goddard, Chiu, & Gossard, 2010). However, such approaches perform inconsistently, with errors arising from either nonuniform resolution throughout the map, which makes segmentation difficult, or from tightly intertwined subunits, in which the protein–protein interfaces are indistinguishable from protein cores. For the remainder of this section, we will assume segmentation has been determined, or that we are performing "segmentation through interpretation," that is, using the building of a multichain model to provide segmentation of the cryoEM map. An interesting example of the latter arose in the epsilon 15 bacteriophage: model building of the major coat protein into the cryoEM map led to discovery (and corresponding segmentation) of a second, previously unknown, coat protein (Jiang et al., 2008).

Some de novo model building, particularly at the highest resolutions, directly uses tools originally developed for X-ray crystallography. One of the most widely used crystallographic tools for structure determination in cryoEM reconstructions is COOT (Emsley, Lohkamp, Scott, & Cowtan, 2010), a system for model building and real-space refinement into density maps. It displays maps and models and has a variety of tools—accessible through an interactive GUI interface—that allow for building of backbone into density, assignment of sequence to backbone, rotamer building, and real-space refinement (Brown et al., 2015). However, the process is largely manual, and consequently, determining atomic structure from cryoEM density can be labor-intensive.

In the highest-resolution cases, automated crystallographic structure determination tools, such as phenix autobuild (Adams et al., 2010) and Buccaneer (Cowtan, 2006), may be applied to problems in cryoEM. While these methods are widely used for crystallographic datasets of 3 Å or better, they have been used in crystallographic data as low as 3.8 Å, though the results at this resolution are inconsistent. The two programs differ in implementation details but both attempt to find "seed placements" of either helices or strands, extend these seeds guided by density, then finally place sequence on these seed placements. Since the majority of cryoEM maps have not reached sufficiently high enough resolution except a few exceptional cases, these tools have not been fully explored but definitely are options as cryoEM maps continue to improve in quality, or as methodological improvements permit their use at lower resolutions.

More recently, a de novo model-building tool based on Rosetta structure prediction has been developed (Wang et al., 2015). Unlike crystallographic model-building programs, this approach combines the steps of backbone placement and sequence determination, and makes use of predicted backbone conformation to enable sequence registration in maps where sidechain density is ambiguous. Seed fragments are placed—not based on ideal helices and strands—but rather based on predicted backbone conformations given local sequence. The correct placements are selected using Monte Carlo sampling with a score function that measures the consistency of a set of placements. The approach is then iterated—fixing fragments previously placed—until at least 70% of the structure is rebuilt. In a benchmark set of nine maps ranging from 3.1 to 4.8 Å resolution, six structures were completely interpreted using this approach. Fig. 3 shows one such example, the determination of the structure of the contractile sheath of the type VI secretion system (Kudryashev et al., 2015). This approach has also been used in the structure determination of several domains of the Coronavirus spike protein trimer (Walls et al., 2016).

Helixhunter and SSEhunter were a pair of tools originally intended for quantitative detection of alpha helices and beta sheets in early subnanometer

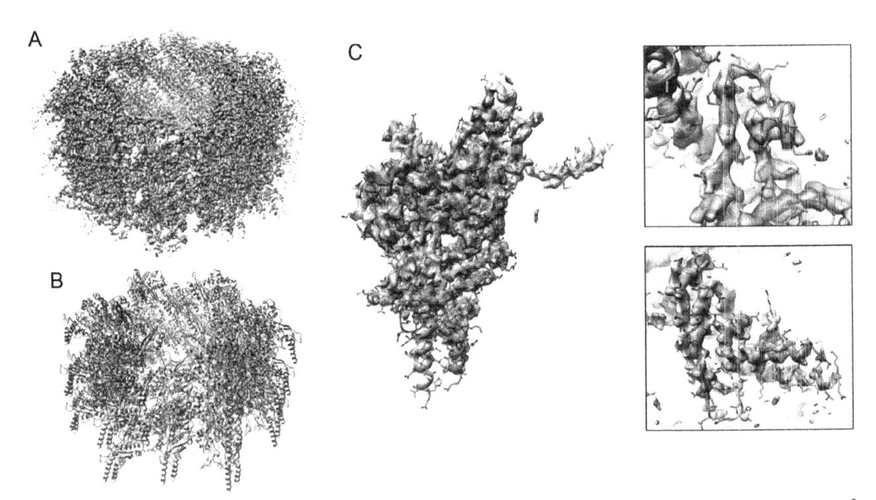

Fig. 3 Modeling a 3.5 Å cryoEM map of VipA/B (Kudryashev et al., 2015). (A) The 3.5 Å reconstruction (EMDB2699) of VipA/B, the contractile sheath of the type VI secretion system. (B) A model of the two protein components, built using Rosetta de novo building followed by optimization with RosettaCM (PDB 3J9G). (C) A close-up view of the asymmetric unit model, shown in density. The two panels on the *right* show regions of relatively low local resolution; Rosetta de novo allowed placement of the models in these regions.

resolution cryoEM maps (Baker, Ju, & Chiu, 2007; Jiang, Baker, Ludtke, & Chiu, 2001). Beyond secondary structure element identification, the skeletonization algorithm in SSEhunter provides secondary structure element connectivity. Subsequently, the Gorgon molecular modeling toolkit (Baker et al., 2011) was developed to utilize a graph matching approach to find the correspondence between the location, position, and connectivity of secondary structure elements found in the density map with those predicted in the sequence. This approach was useful for general protein topology determination in medium-high and medium resolution cryoEM density maps. While they are able to produce a gross protein topology, they are still from being "perfect." Relatedly, EM-fold (Lindert et al., 2009) identifies secondary structure elements in medium-resolution density maps; however, it attempts to register sequence to these elements using a combination of density fit and the Rosetta force field.

Pathwalking treats de novo modeling as a computational optimization problem (Baker, Baker, Hryc, Ju, & Chiu, 2012). It nearly automatically determines plausible topologies as an instance of the travelling salesman problem (TSP): "pseudoatoms" are placed into a map corresponding to regions of high density, then a TSP solver finds a minimal path through these pseudoatoms, where the cost function is related to the deviation from the ideal Cα–Cα distance. On a wide variety of cases—even as low as 6 Å resolution—this approach successfully determines the topology of a protein (Baker, Rees, Ludtke, Chiu, & Baker, 2012). Several benchmark examples of cryoEM maps drawn from EMDB have been used to demonstrate its applicability and relative accuracy for modeling their protein components (Baker et al., 2011). Fig. 4 shows the case of a rotavirus capsid protein map where the topology trace is very accurate and most errors in the Cα positions occur where the map is not as well resolved. Of note, this method does not directly use any sequence and/or protein topology information apart from the number of amino acids in the protein and the ideal Cα–Cα distance. Not even the cryoEM density is explicitly used to limit topological assignment. As such, subsequent optimization of an initial pathwalker model by COOT, Rosetta, or phenix is needed to improve overall model quality. Pathwalker, distributed with EMAN2, now offers improved de novo modeling performance with the incorporation of density filtering, geometry filtering, improved pseudoatom placement, automatic secondary structure element assignment, and iterative model refinement from a fully automated command-line utility (Chen, Baldwin, Ludtke, & Baker, 2016).

While many cryoEM structural models have been built solely using one of the above approaches, some of the more recent structures are large,

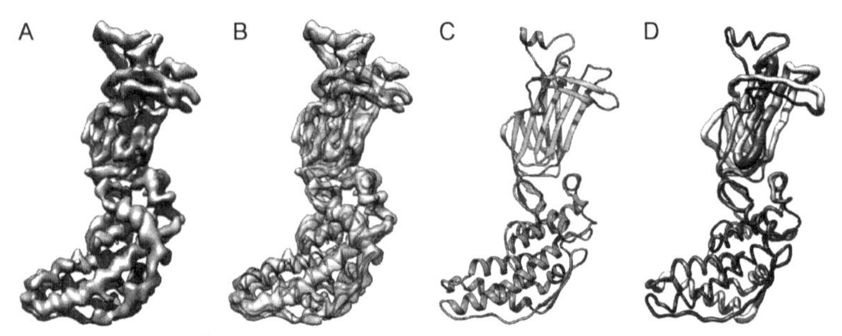

Fig. 4 Modeling a 3.8 Å cryoEM map of VP6 of rotavirus. (A) A segmented density map of a capsid protein subunit of rotavirus (VP6) determined at 3.8 Å (EMDB1460). (B) A de novo model built by pathwalker superimposed on the density map. (C) A crystal structure of the same protein (PDB 1QHD). (D) A Cα rms deviation between the cryoEM model and crystal structure with the most and least deviation in *red* (*gray* in the print version) and *blue* (*dark gray* in the print version), respectively. *These figures are kindly provided by Matthew L. Baker reproduced after Baker M.R. et al. (2012).*

complex, and have variable resolution throughout the map, necessitating the use of multiple modeling techniques. For instance, the inositol-1,4,5-trisphosphate receptor (IP3R1), a tetrameric cation channel, has 10 discrete structural domains arranged over 2750 amino acids per polypeptide chain (Fan et al., 2015). Among these domains, only ~600 residues have a corresponding high-resolution crystallographic structure. Building a model of such a large protein (Fig. 5) requires the use of a cocktail of modeling methods including sequence prediction (Cole, Barber, & Barton, 2008; Kelley & Sternberg, 2009), secondary structure element localization (Baker et al., 2007), homology modeling (Topf, Baker, Marti-Renom, Chiu, & Sali, 2006; Webb & Sali, 2014), rigid body and flexible fitting (Jiang et al., 2001; Pettersen et al., 2004; Wriggers & Chacon, 2001), and de novo modeling (Baker, Baker, et al., 2012; Baker, Rees, et al., 2012; Emsley et al., 2010). As such, modeling of these types of complexes requires a more complex strategy that integrates the various software packages and utilizes other structural data. Such integrative methods will likely prove commonplace as cryoEM is applied to large multisubunit molecular machines.

3. MODEL OPTIMIZATION

Once an initial model is constructed, or if a high-resolution homologous structure is already known, the next step in structure determination is model optimization. In this case, the task is to move atom positions to improve agreement to data, improve model geometry (eg, eliminating

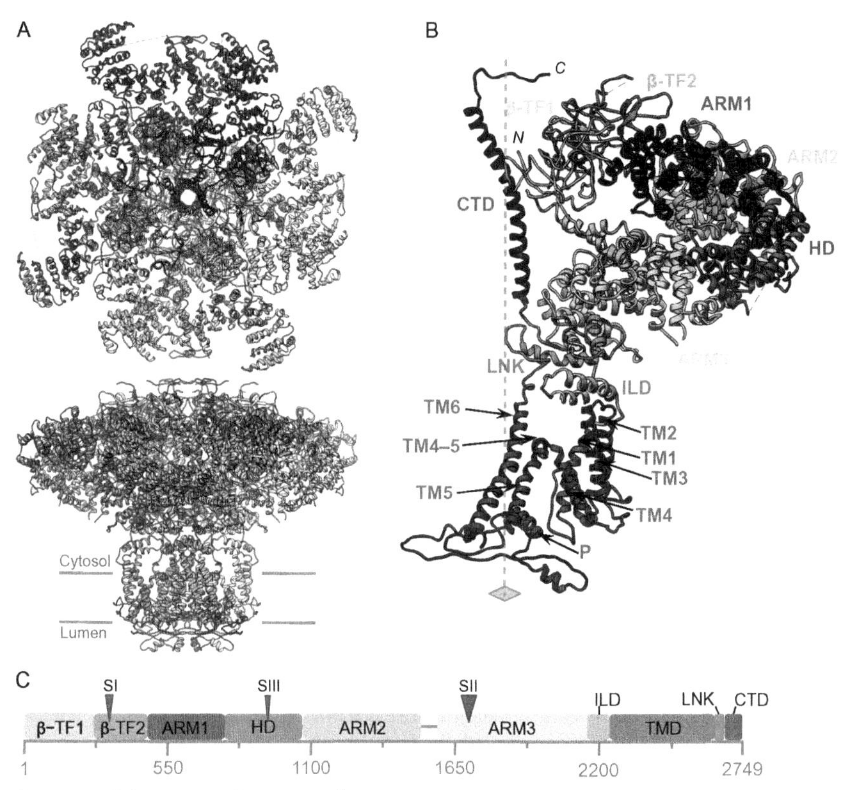

Fig. 5 Modeling IP3R1 from a 4.7 Å cryoEM map (EMDB6369) (Fan et al., 2015). (A) The model (PDB 3JAV) was built using a variety of modeling protocols, shown from two views. The model is of the entire tetramer with 85% chain connectivity per chain, partly due to the presence of isoforms at the SI, SII, and SIII sites causing specimen heterogeneity and partly due to the limited map resolution. (B) The annotation of the 10 structural domains of a single IP3R1 subunit with 2700 amino acids. (C) A schematic of the corresponding domains in the linear sequence. *Reproduced from Fan, G., Baker, M. L., Wang, Z., Baker, M. R., Sinyagovskiy, P. A., Chiu, W., et al. (2015). Gating machinery of InsP3R channels revealed by electron cryomicroscopy. Nature, 527, 336–341 and provided by Matthew L. Baker.* (See the color plate.)

clashes and unreasonable torsions), or some combination of both. Specifically, this step aims to optimize protein coordinates to minimize a target function $E = E_{\mathrm{geom}} + w \cdot E_{\mathrm{data}}$, where E_{geom} assesses the geometric goodness of a model, E_{data} assesses model-map agreement, and w is a weighing factor controlling the relative contributions of geometry and agreement to data in optimization. This section describes a number of different tools for model optimization, which vary in the functional form of E_{geom} and E_{data}, the parameter space describing protein motion, and the types of movements used to optimize E. Consequently, these methods may vary quite a bit in terms of recommended resolutions, magnitude of movements, and runtime of the corresponding approaches. This section gives an overview of each of these methods, when they may best be used and how to interpret the resulting output.

As with the de novo section above, several of the tools commonly used in model refinement are based on tools originally developed for X-ray crystallography. In particular, both Phenix.refine (Afonine et al., 2012) and Refmac have been commonly used (eg, phenix.refine was used to refine the first cryoEM structure of epsilon15 bacteriophage capsid (Baker et al., 2013)). In using these crystallographic tools, the data are first processed as if it were crystal data, assigning an artificial unit cell and symmetry to the data, and computing reciprocal space intensities and phases from the real-space density. Then, refinement is carried against a function that takes into account both Fourier intensities and phases of the data. The geometry function used by both is a relatively simple macromolecular energy function that takes into account stereochemistry and steric clashes, with optional support for torsional potentials or user-defined "restraints." Generally, function optimization consists of cycles of minimization, but may also include discrete rotamer optimization or cycles of simulated annealing. Consequently, these refinement methods are fast, but tend to make relatively small motions from the starting model, leading to a small "radius of convergence"; that is they are unable to correct errors in the starting model of large magnitude. However, they are quite widely used to improve model geometry and improve fit of models to data.

Much like the use of crystallographic tools for de novo model building, one weakness of these tools is that they may not perform particularly well when used at medium-high resolutions. However, there have been a number of recent advances for refinement against low-resolution data. One such approach employs secondary structure element restraints (Nicholls, Long, & Murshudov, 2012), where secondary structure elements in the initial model are identified, and harmonic constraints are used to maintain backbone hydrogen bond patterning throughout optimization, ensuring backbone hydrogen bond patterning stays intact even as refinement moves the structure far from the starting point. Alternately, some approaches make use of "reference-model restraints," where atom-pair distances or torsions from a related high-resolution structure are applied, rigidifying the structure during the course of optimization (Headd et al., 2012). In this way, refinement fills in ambiguity in the experimental data by enforcing agreement with high-resolution crystallographic data of a related structure. In both approaches, one can think of the restraints as adding additional restraints into E_{geom} in the equation above.

Several methods have been developed to perform model optimization in real space instead. One advantage of real-space optimization is "locality": if a map contains contaminants not present in the model (eg, detergents or amphipoles), when the data are converted to reciprocal space, these contaminants will affect the entirety of reciprocal space, but only the affected regions when refining in real space (Brunger & Rice, 1997), as local regions in real space contribute globally in reciprocal space. Similar advantages in real-space optimization arise when optimization is carried out with only partial models. This effect may be ameliorated by masking relevant regions before converting the data to reciprocal space, but this requires that a mask be defined a priori. Much like the crystallographic methods, these approaches also need methods to deal with the relatively low resolution of the data, typically using the same additional terms outlined earlier, improving the sensitivity of E_{geom}.

One such real-space optimization routine is based on the phenix crystallographic refinement software, called phenix.real_space_refine. This optimization protocol ensures optimal fit-to-density, while maintaining good stereochemistry and rotamer assignments. It is very efficient (generally taking minutes or less even for very large systems) but suffers from the same limitation (of having a small radius of convergence) as its crystallographic counterpart. One recent example where this approach was successfully employed was in structure determination of the brome mosaic virus: an all-atom model was optimized into a 3.8 Å cryoEM density map, resulting in outstanding MolProbity statistics (Chen et al., 2010) compared to the input model or the previously determined 3.4 Å X-ray crystal structure of the same virus particle (Wang et al., 2014) (Table 1).

Another real-space optimization tool makes use of Rosetta (DiMaio et al., 2015). By replacing the relatively coarse-grained crystallographic E_{geom} with a richer, physically realistic potential that accounts for the hydrophobic effect, hydrogen bonding, electrostatics, and torsional preferences, the number of effective degrees of freedom during optimization is dramatically limited. This allows for optimization to move the structure significantly while energetically favorable interactions are maintained. Rosetta-based optimization makes use of a combination of minimization and Monte Carlo sampling of both backbone and sidechain conformations, where extensive minimization after each Monte Carlo sampling step allows exploration of a relatively large portion of conformational space.

Table 1 MolProbity Statistics Comparing the cryoEM Map-Derived Models of Brome Mosaic Virus Before and After Real-Space Optimization (RSO) (PDB 3J7L) and the Corresponding X-ray Structure (PDB 1JS9) (Wang et al., 2014)

Asymmetric Unit	3 Subunits	CryoEM 477 Residues at 3.8 Å Resolution After RSO		CryoEM 477 Residues at 3.8 Å Resolution Before RSO		X-ray (PDB id: 1JS9) 503 Residues at 3.4 Å Resolution	
Density agreement	Correlation coefficient	0.84		0.76		0.68	
All-atom contacts	Clash score (all atoms)	13.35	97th Percentile	16.02	97th Percentile	31.77	78th Percentile
Protein geometry	Poor rotamers	0	0%	172	46%	181	49%
	Ramachandran Outliers	12	2.55%	48	10%	44	9%
	Ramachandran favored	434	92.14%	345	69%	351	71%
	MolProbity score	2.11	100th Percentile	3.82	46th Percentile	4.1	21st Percentile
	Cβ deviations	0	0%	0	0%	0	0%
	Bad backbone bonds	0	0%	1	0.05%	0	0%
	Bad backbone angles	0	0%	8	0.32%	5	0.2%

Percentile values in the table based on deposited structures at the reported resolution.
A complete asymmetric unit was analyzed, but the number of amino acids varies due to resolvability in the density map (EMDB6000). In addition, cross-correlation values were computed between the map and the model for the asymmetric unit. Percentiles were calculated based on the deposited structures at the reported resolution.

This sampling and energy function allow for larger conformational changes during optimization, however, it also comes at increased computational cost. One further advantage of Rosetta is the symmetric degrees of freedom are explicitly represented rather than restrained with noncrystallographic symmetry restraints. This further reduces degrees of freedom as well as improving performance on very highly symmetric systems (for example, icosahedral viral capsids). An example of the types of movement that may be obtained from this optimization protocol is shown in Fig. 6.

Another tool, DireX (Schroder, Levitt, & Brunger, 2010), uses ideas from crystallographic refinement methods as well, but instead is applied in real space, and—to address the poor data-to-parameters ratio of typical crystal refinement—makes use of reference model restraints. However, unlike typical crystallographic refinement programs, these restraints are allowed to change during the course of model optimization. The restraints very slowly adapt to the current model in the course of model optimization, with two metaparameters describing the "stiffness" and "speed" at which these constraints adapt. By exploring the results of refinement over various settings of these metaparameters, an optimal setting may be identified. A key advantage of

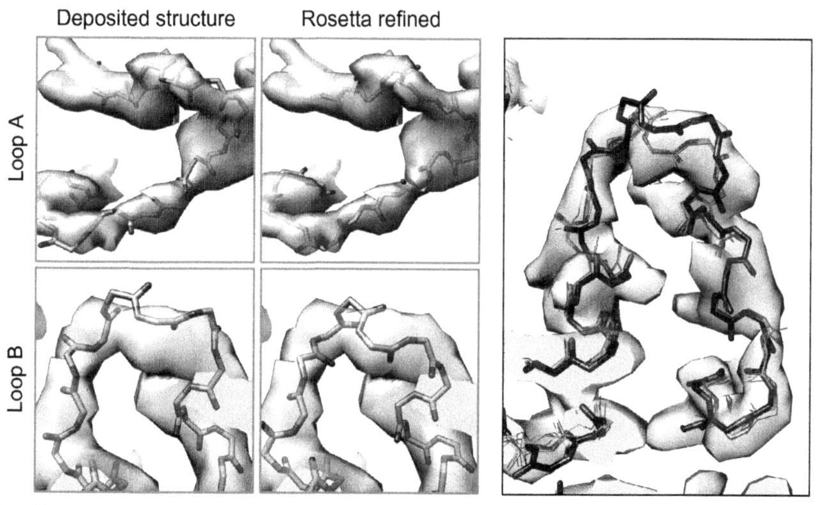

Fig. 6 The types of motion possible during Rosetta optimization. (*Left*) Two different regions of relatively low local resolution in the 3.4 Å resolution map of TRPV1; Rosetta refinement (*right two panels*) allows for significant conformational difference from the deposited structure (*left two panels*). (*Right*) Despite the significant backbone movement in the course of optimization, an ensemble of low energy models, resulting from independent trajectories, are well converged.

these adaptable constraints is that they identify a "minimally perturbed conformation," only violating constraints in the initial model if there is sufficient evidence from the data that these constraints should not hold. Consequently, the approach can allow for relatively large motions during optimization and often works reasonably well at low resolution if a corresponding high-resolution structure is known (Chen, Madan, et al., 2013). Fig. 7 illustrates an application of this approach in the determination of several different forms of F-actin refined against cryoEM data (Galkin et al., 2015).

Another class of tools is based on molecular dynamics guided by experimental data. The most commonly used tool is MDFF (Trabuco, Villa, Mitra, Frank, & Schulten, 2008), which combines the VMD molecular dynamics package with a score term assessing the agreement of a model to real-space density. Like Rosetta, the rich, physically realistic force field makes the approach well suited to modeling large conformational changes, as it maintains physically realistic geometry. One weakness of this approach is that it may be time consuming, particularly if explicit solvent molecules are used, due to the increased number of interactions introduced by explicit

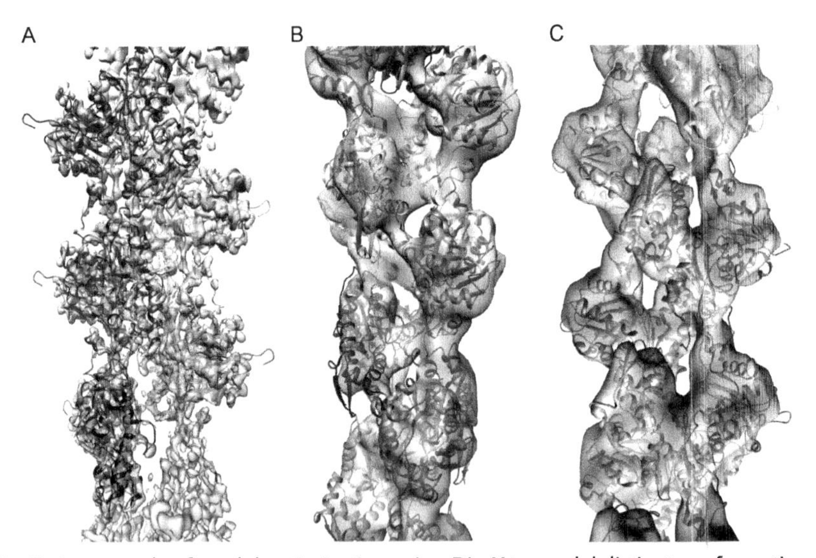

Fig. 7 An example of model optimization using DireX to model distinct conformational states of F-actin from a 4.8 Å cryoEM map (Galkin, Orlova, Vos, Schroder, & Egelman, 2015). (A) A 4.8 Å resolution reconstruction of F-actin (EMDB6179) into which a model has been built and optimized (PDB 3J8I). (B and C) Two alternate, low-occupancy conformations of actin, titled T1 and T2, into which the initial model has been refined. Even though the data are of relatively low resolution, DireX attempts to maintain as many contacts as possible during refinement. (See the color plate.)

solvent molecules and the long equilibration time consequently necessary. However, it may be parallelized to run in a reasonable amount of wall clock time. To date, it is often preferred for modeling large conformational changes subject to medium-resolution density maps, as in the ribosome (Trabuco et al., 2008) and HIV capsid (Zhao et al., 2013).

Finally, FlexEM is another method that approaches the problem hierarchically (Topf et al., 2008). A protein system is first broken into rigid bodies, which are refined, and then full flexibility is allowed. The score function used is similar to that of crystallographic force fields, however, the initial rigid-body refinement step has relatively few parameters, and thus allows for large motions of the system.

4. MODEL VALIDATION

Once after model optimization is completed, the final steps are model selection and model validation. Model validation attempts to assess the accuracy of the refined model. This is desirable for several reasons, and model validation may be used to address several different questions that arise during model building. One may want to estimate the *absolute accuracy* of a model fit to data. Alternately, one may want to compare models to find the most accurate, either to select models from stochastic refinement trajectories, or to tune parameters of model building, such as the weight on the experimental data in refinement. Finally, one may wish to see if the model is improved following optimization and identify when optimization can be stopped (if there is no more improvement).

This section is broadly divided into two parts: validation using model geometry and validation using model-map agreement. However, for both measures, a key to validating models is the use of *independent data*, that is, data that is not used in optimization—not optimized against—in order to evaluate accuracy of a model. While the agreement of model to data used in fitting is informative, it does not identify *overfitting*, fitting to noise in the original reconstruction. To identify overfitting, independent data are required (DiMaio, Zhang, Chiu, & Baker, 2013). This is critically important at near-atomic resolutions, as—at these resolutions—it is much easier to trace models incorrectly that fit the data well. Only by assessing the agreement to independent data can we be sure that we are improving the model.

MolProbity, a widely used model validation metric used in both crystallography and with NMR-derived models, assesses the geometry of a protein model (Chen et al., 2010). More specifically, it looks at certain geometric features of a model and compares them to the expected values of such features, as derived from very high-resolution crystal structures. Such features include bond-length and bond-angle distributions, backbone torsional distributions (specifically, counting the number of "allowed" and "favored" backbone dihedrals, which are seen in <2% and <0.05% of residues in high-resolution structures, respectively), sidechain rotamer distributions (counting the number of rotamers seen with <0.3% frequency in high-resolution structures), and atom pairs closer than the sum of their van der Waals radii. For each metric, a corresponding Z-score is computed, and linear regression is used to compute an aggregate score. This aggregate score is trained to predict the *resolution* of data at which a particular structure was solved using only the geometry of the model itself. The resulting "MolProbity score" can very loosely be thought of in terms of map resolution, where models scoring under 2 are of high quality, while models scoring higher than 4 are of relatively low quality. MolProbity reports (Chen et al., 2010) also describe the ranking of the examined structure relative to all the other structures in the PDB determined at the equivalent resolution (Table 1).

It is also important to point out that for many of the model optimization methods outlined in the previous section, the geometric data are not considered "independent data" for the purposes of validation. Both Rosetta and phenix.refine (when run in a certain mode) restrain sidechains to rotameric identity for example. This is not a weakness of these approaches; indeed, at low resolution such incorporation may be necessary, and given two otherwise equivalent models, the one with better geometry is more likely to be correct. However, for these approaches it is important to realize that these measures are not independent data for assessing overfitting or for parameter tuning.

One may also want to validate models based on fit between model and experimental data. This is often done by evaluating the Fourier shell correlation (FSC) between a model and the corresponding map, quantifying the fit by calculating the correlation between model and map in reciprocal space, in the complex plane. One advantage of this measure over something like real-space correlation is that the measure is independent of dampened intensities in high-resolution shells (real-space correlation is sensitive to this effect). This measure is often computationally corrected during the reconstruction process, and so an assessment measure that ignores this is preferred. In the near-atomic resolution regime, the FSC in high-resolution shells alone is most informative as to the accuracy of the high-resolution details of the model (DiMaio et al., 2015).

There are several different ways in which this measure is used to generate an independent validation measure. All compute a "free FSC" analogous to the R_{free} measure in X-ray crystallography: the agreement of model and map on a subset of data held aside during refinement. In crystallography, the R_{free} measure omits a subset (generally 5–10%) of reciprocal space intensities from refinement and evaluates their agreement to the model as refinement progresses. Since it is now a common practice to produce two independent maps from two independent sets of raw particle images to assess map resolution (Chen, McMullan, et al., 2013; Henderson et al., 2012), one can generate two independent models from the two independent maps. These two independent models can be compared against the two maps by either FSC or variance of the backbone between the two models (Rosenthal & Henderson, 2003). In addition, one can compute the FSC between the model derived from one map relative to the other independently determined map (Wang et al., 2014). It has been shown that FSC equal to 0.5 is a practical measure of the agreement between the model and map. If this measure exceeds the gold standard map resolution (Henderson et al., 2012; Scheres & Chen, 2012), it would imply that the model is overfitted. These multiple crosschecks should provide consistent results to assure that the model is not overfitted in each case. Fig. 8 shows an example application of this model validation metric in the 3.8 Å cryoEM structure of brome

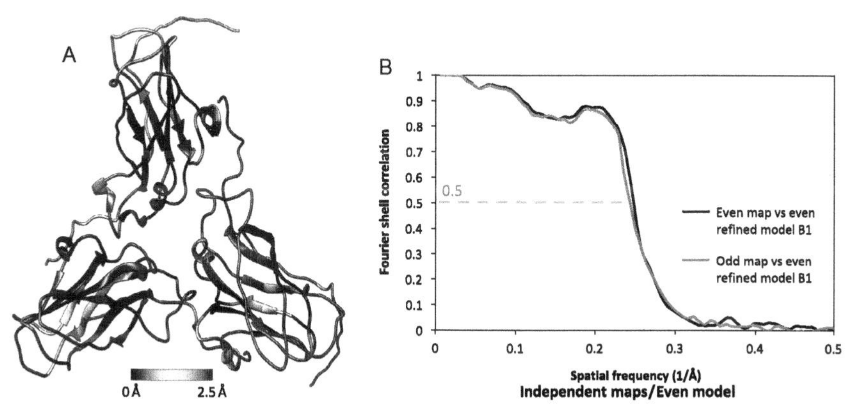

Fig. 8 Model validation of a 3.8 Å cryoEM map of brome mosaic virus (EMDB6000) (Wang et al., 2014) by (A) deviation between two independent models at the Cα level (PDB 3J7M and PDB 3J7N). (B) FSC between model and experimental map from two independent data sets. *These figures are reproduced from Wang, Z., Hryc, C. F., Bammes, B., Afonine, P. V., Jakana, J., Chen, D. H., et al. (2014). An atomic model of brome mosaic virus using direct electron detection and real-space optimization. Nature Communications, 5, 4808 and kindly provided by Corey F. Hryc.*

mosaic virus (Wang et al., 2014). This validation approach is additionally informative for tuning the relative weight of experimental data and geometric data during the optimization process. It is very much analogous to the R_{free} measure in X-ray crystallography: the agreement of model and map on a subset of data held aside during refinement. To deal with the somewhat reduced resolution, it is also proposed to perform additional refinement against the reconstruction of the entire data set (Brown et al., 2015), however, this final refinement is no longer independent with respect to the "independent map."

Alternatively, another approach (Falkner & Schroder, 2013), rather than using two different reconstructions, instead truncates the full reconstruction at some particular resolution (at the point where the FSC of two independent maps is 0.5). Fitting is carried out against this truncated reconstruction, while the truncated high-resolution information is used as an independent validation set. The advantage of the latter approach is that independent maps are not required. However, both techniques have been successfully employed in order to detect overfitting.

A final method for map validation, EMRinger (Barad et al., 2015), also comes from X-ray crystallography. This method, outlined in Fig. 9, identifies the fraction of amino acids with "rotameric sidechain density." While conceptually it may seem to be similar to rotamer probabilities reported by

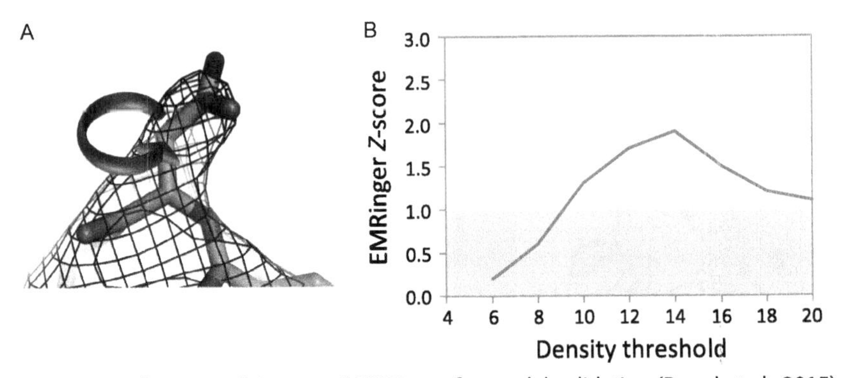

Fig. 9 (A) A schematic of the use of EMRinger for model validation (Barad et al., 2015). Given a backbone model and a density map, EMRinger considers all possible positions for a putative Cγ and identifies density peaks at a given threshold; the fractions of these peaks over the whole structure, which are rotameric are used to assess the quality of the model. (B) The results of EMRinger analysis on a sample system: the *x*-axis plots various density value cutoffs and the *y*-axis shows the EMRinger *Z*-score. Higher values are better, with *Z*-score of >2 indicating high-quality structures. *These figures are reproduced from Barad, B. A., Echols, N., Wang, R. Y., Cheng, Y., DiMaio, F., Adams, P. D., et al. (2015). EMRinger: Side chain-directed model and map validation for 3D cryo-electron microscopy. Nature Methods, 12, 943–946.*

MolProbity, it actually is quite different. It measures whether density is rotameric for a particular sidechain, that is, by looking along the Cα–Cβ vector of each residue, and identifying if the putative Cγ peak is rotameric. In doing so, it ignores whether the modeled sidechain is actually rotameric. Instead, it identifies backbone placements where sidechain density seems reasonable; incorrectly placed backbone will have nonrotameric sidechain density, even if the modeled sidechain is rotameric. Important to note is that this measure only depends upon the placement of backbone atoms, and it can be thought of as an orthogonal measure to those above.

5. DISCUSSION

This review provides a detailed view of the steps required in going from a cryoEM map to all-atom model. As illustrated, there are a wide variety of tools, each with tradeoffs in terms of most effective resolutions, run time, conformational state explored. However, these methods show that it is possible to obtain accurate, all-atom reconstructions from cryoEM density.

We have primarily focused on the determination of a single model that best explains the data. However, the reality is that—as an averaging method—a single cryoEM reconstruction may contain many different conformations of individual molecules. It might then make sense to consider fitting ensembles of models to cryoEM reconstructions. However, as X-ray crystallographic methods show us (Terwilliger et al., 2012), it is difficult to do this for two reasons: it introduces a significant number of parameters to optimization, and it is difficult to separate out the effects of uncertainty from conformational variability. An open challenge in future years remains how to represent and validate the various conformations possibly seen in different frequencies in a single molecular machine.

Finally, with the relatively low resolution of cryoEM reconstructions compared to those in crystallography, it is important to assess the accuracy of computed models. This is commonly done in several ways: by exploring the space of solutions consistent with data (DiMaio et al., 2015), by looking at consistency in models fit to independent datasets (DiMaio et al., 2013; Wang et al., 2014), and by explicitly fitting models that contain uncertainty (Pintilie, Chen, Haase-Pettingell, King, & Chiu, 2016), essentially putting error bars on generated models. The question is still an open one; however, as there is no consensus on the best way to compute or to represent uncertainty of a model given a reconstruction. So far, all the modeling is based on the assumption of the density map being correct and having an isotropic resolution. It has been shown in numerous cases that resolutions are nonuniform throughout the map. It will be very important in the future to explore this problem more rigorously. Accounting for the uncertainty of the model and the map is key to interpreting the biology of the corresponding system and planning next set of experiments based on the structures.

ACKNOWLEDGMENTS

Our research has been supported by NIH Grants (P41GM103832 and R01GM079429) and the Robert Welch Foundation (Q1242) to W.C. We thank Dr. Matthew L. Baker and Corey F. Hryc at Baylor College of Medicine for their comments and in preparation of some of the figures.

REFERENCES

Adams, P. D., Afonine, P. V., Bunkoczi, G., Chen, V. B., Davis, I. W., Echols, N., et al. (2010). PHENIX: A comprehensive Python-based system for macromolecular structure solution. *Acta Crystallographica. Section D, Biological Crystallography*, *66*, 213–221.

Afonine, P. V., Grosse-Kunstleve, R. W., Echols, N., Headd, J. J., Moriarty, N. W., Mustyakimov, M., et al. (2012). Towards automated crystallographic structure refinement with phenix.refine. *Acta Crystallographica. Section D, Biological Crystallography*, *68*, 352–367.

Baker, M. L., Abeysinghe, S. S., Schuh, S., Coleman, R. A., Abrams, A., Marsh, M. P., et al. (2011). Modeling protein structure at near atomic resolutions with Gorgon. *Journal of Structural Biology*, *174*, 360–373.

Baker, M. L., Baker, M. R., Hryc, C. F., Ju, T., & Chiu, W. (2012). Gorgon and pathwalking: Macromolecular modeling tools for subnanometer resolution density maps. *Biopolymers*, *97*, 655–668.

Baker, M. L., Hryc, C. F., Zhang, Q., Wu, W., Jakana, J., Haase-Pettingell, C., et al. (2013). Validated near-atomic resolution structure of bacteriophage epsilon15 derived from cryo-EM and modeling. *Proceedings of the National Academy of Sciences of the United States of America*, *110*, 12301–12306.

Baker, M. L., Ju, T., & Chiu, W. (2007). Identification of secondary structure elements in intermediate-resolution density maps. *Structure*, *15*, 7–19.

Baker, M. R., Rees, I., Ludtke, S. J., Chiu, W., & Baker, M. L. (2012). Constructing and validating initial Calpha models from subnanometer resolution density maps with pathwalking. *Structure*, *20*, 450–463.

Barad, B. A., Echols, N., Wang, R. Y., Cheng, Y., DiMaio, F., Adams, P. D., et al. (2015). EMRinger: Side chain-directed model and map validation for 3D cryo-electron microscopy. *Nature Methods*, *12*, 943–946.

Brown, A., Long, F., Nicholls, R. A., Toots, J., Emsley, P., & Murshudov, G. (2015). Tools for macromolecular model building and refinement into electron cryo-microscopy reconstructions. *Acta Crystallographica. Section D, Biological Crystallography*, *71*, 136–153.

Brunger, A. T., & Rice, L. M. (1997). Crystallographic refinement by simulated annealing: Methods and applications. *Methods in Enzymology*, *277*, 243–269.

Chen, V. B., Arendall, W. B., 3rd, Headd, J. J., Keedy, D. A., Immormino, R. M., Kapral, G. J., et al. (2010). MolProbity: All-atom structure validation for macromolecular crystallography. *Acta Crystallographica Section D, Biological Crystallography*, *66*, 12–21.

Chen M., Baldwin P. R., Ludtke S. J., & Baker M. L. (2016). De novo modeling of cryo-EM density maps with pathwalker, *Journal of Structural Biology*, (in press).

Chen, D. H., Madan, D., Weaver, J., Lin, Z., Schroder, G. F., Chiu, W., et al. (2013). Visualizing GroEL/ES in the act of encapsulating a folding protein. Cell, 153, 1354–1365.

Chen, S., McMullan, G., Faruqi, A. R., Murshudov, G. N., Short, J. M., Scheres, S. H., et al. (2013). High-resolution noise substitution to measure overfitting and validate resolution in 3D structure determination by single particle electron cryomicroscopy. *Ultramicroscopy, 135C*, 24–35.

Cole, C., Barber, J. D., & Barton, G. J. (2008). The Jpred 3 secondary structure prediction server. *Nucleic Acids Research, 36*, W197–W201.

Cowtan, K. (2006). The Buccaneer software for automated model building. 1. Tracing protein chains. *Acta Crystallographica. Section D, Biological Crystallography, 62*, 1002–1011.

DiMaio, F., Song, Y., Li, X., Brunner, M. J., Xu, C., Conticello, V., et al. (2015). Atomic-accuracy models from 4.5-A cryo-electron microscopy data with density-guided iterative local refinement. *Nature Methods, 12*, 361–365.

DiMaio, F., Zhang, J., Chiu, W., & Baker, D. (2013). Cryo-EM model validation using independent map reconstructions. *Protein Science: A Publication of the Protein Society, 22*, 865–868.

Emsley, P., Lohkamp, B., Scott, W. G., & Cowtan, K. (2010). Features and development of Coot. *Acta Crystallographica. Section D, Biological Crystallography, 66*, 486–501.

Falkner, B., & Schroder, G. F. (2013). Cross-validation in cryo-EM-based structural modeling. *Proceedings of the National Academy of Sciences of the United States of America, 110*, 8930–8935.

Fan, G., Baker, M. L., Wang, Z., Baker, M. R., Sinyagovskiy, P. A., Chiu, W., et al. (2015). Gating machinery of InsP3R channels revealed by electron cryomicroscopy. *Nature, 527*, 336–341.

Galkin, V. E., Orlova, A., Vos, M. R., Schroder, G. F., & Egelman, E. H. (2015). Near-atomic resolution for one state of F-actin. *Structure, 23*, 173–182.

Headd, J. J., Echols, N., Afonine, P. V., Grosse-Kunstleve, R. W., Chen, V. B., Moriarty, N. W., et al. (2012). Use of knowledge-based restraints in phenix.refine to improve macromolecular refinement at low resolution. *Acta Crystallographica. Section D, Biological Crystallography, 68*, 381–390.

Henderson, R. (2015). Overview and future of single particle electron cryomicroscopy. *Archives of Biochemistry and Biophysics, 581*, 19–24.

Henderson, R., Sali, A., Baker, M. L., Carragher, B., Devkota, B., Downing, K. H., et al. (2012). Outcome of the first electron microscopy validation task force meeting. *Structure, 20*, 205–214.

Jiang, W., Baker, M. L., Jakana, J., Weigele, P. R., King, J., & Chiu, W. (2008). Backbone structure of the infectious epsilon15 virus capsid revealed by electron cryomicroscopy. *Nature, 451*, 1130–1134.

Jiang, W., Baker, M. L., Ludtke, S. J., & Chiu, W. (2001). Bridging the information gap: Computational tools for intermediate resolution structure interpretation. *Journal of Molecular Biology, 308*, 1033–1044.

Kelley, L. A., & Sternberg, M. J. (2009). Protein structure prediction on the Web: A case study using the Phyre server. *Nature Protocols, 4*, 363–371.

Kudryashev, M., Wang, R. Y., Brackmann, M., Scherer, S., Maier, T., Baker, D., et al. (2015). Structure of the type VI secretion system contractile sheath. *Cell, 160*, 952–962.

Kuhlbrandt, W. (2014). Biochemistry. The resolution revolution. *Science, 343*, 1443–1444.

Lasker, K., Topf, M., Sali, A., & Wolfson, H. J. (2009). Inferential optimization for simultaneous fitting of multiple components into a CryoEM map of their assembly. *Journal of Molecular Biology, 388*, 180–194.

Lindert, S., Staritzbichler, R., Wotzel, N., Karakas, M., Stewart, P. L., & Meiler, J. (2009). EM-fold: De novo folding of alpha-helical proteins guided by intermediate-resolution electron microscopy density maps. *Structure, 17*, 990–1003.

Nicholls, R. A., Long, F., & Murshudov, G. N. (2012). Low-resolution refinement tools in REFMAC5. *Acta Crystallographica. Section D, Biological Crystallography, 68*, 404–417.

Pettersen, E. F., Goddard, T. D., Huang, C. C., Couch, G. S., Greenblatt, D. M., Meng, E. C., et al. (2004). UCSF Chimera—A visualization system for exploratory research and analysis. *Journal of Computational Chemistry, 25*, 1605–1612.

Pintilie, G. D., Chen, D. H., Haase-Pettingell, C. A., King, J. A., & Chiu, W. (2016). Resolution and probabilistic models of components in CryoEM maps of mature P22 bacteriophage. *Biophysical Journal, 110*, 827–839.

Pintilie, G., Zhang, J., Chiu, W., & Gossard, D. (2009). Identifying components in 3D density maps of protein nanomachines by multi-scale segmentation. *IEEE/NIH Life Science Systems and Applications Workshop, 2009*, 44–47.

Pintilie, G. D., Zhang, J., Goddard, T. D., Chiu, W., & Gossard, D. C. (2010). Quantitative analysis of cryo-EM density map segmentation by watershed and scale-space filtering, and fitting of structures by alignment to regions. *Journal of Structural Biology, 170*, 427–438.

Rosenthal, P. B., & Henderson, R. (2003). Optimal determination of particle orientation, absolute hand, and contrast loss in single-particle electron cryomicroscopy. *Journal of Molecular Biology, 333*, 721–745.

Scheres, S. H., & Chen, S. (2012). Prevention of overfitting in cryo-EM structure determination. *Nature Methods, 9*, 853–854.

Schroder, G. F., Levitt, M., & Brunger, A. T. (2010). Super-resolution biomolecular crystallography with low-resolution data. *Nature, 464*, 1218–1222.

Terwilliger, T. C., Read, R. J., Adams, P. D., Brunger, A. T., Afonine, P. V., Grosse-Kunstleve, R. W., et al. (2012). Improved crystallographic models through iterated local density-guided model deformation and reciprocal-space refinement. *Acta Crystallographica. Section D, Biological Crystallography, 68*, 861–870.

Topf, M., Baker, M. L., Marti-Renom, M. A., Chiu, W., & Sali, A. (2006). Refinement of protein structures by Iterative comparative modeling and cryoEM density fitting. *Journal of Molecular Biology, 357*, 1655–1668.

Topf, M., Lasker, K., Webb, B., Wolfson, H., Chiu, W., & Sali, A. (2008). Protein structure fitting and refinement guided by Cryo-EM density. *Structure, 16*, 295–307.

Trabuco, L. G., Villa, E., Mitra, K., Frank, J., & Schulten, K. (2008). Flexible fitting of atomic structures into electron microscopy maps using molecular dynamics. *Structure, 16*, 673–683.

Walls, A. C., Tortorici, M. A., Bosch, B. J., Frenz, B., Rottier, P. J., DiMaio, F., et al. (2016). Cryo-electron microscopy structure of a coronavirus spike glycoprotein trimer. *Nature, 531*, 114–117.

Wang, Z., Hryc, C. F., Bammes, B., Afonine, P. V., Jakana, J., Chen, D. H., et al. (2014). An atomic model of brome mosaic virus using direct electron detection and real-space optimization. *Nature Communications, 5*, 4808.

Wang, R. Y., Kudryashev, M., Li, X., Egelman, E. H., Basler, M., Cheng, Y., et al. (2015). De novo protein structure determination from near-atomic-resolution cryo-EM maps. *Nature Methods, 12*, 335–338.

Webb, B., & Sali, A. (2014). Comparative protein structure modeling using MODELLER. *Current Protocols in Bioinformatics/Editoral Board, Andreas D Baxevanis [et al], 47*, 5.6.1–5.6.32.

Wriggers, W., & Chacon, P. (2001). Modeling tricks and fitting techniques for multiresolution structures. *Structure, 9*, 779–788.

Zhao, G., Perilla, J. R., Yufenyuy, E. L., Meng, X., Chen, B., Ning, J., et al. (2013). Mature HIV-1 capsid structure by cryo-electron microscopy and all-atom molecular dynamics. *Nature, 497*, 643–646.

CHAPTER ELEVEN

Refinement of Atomic Structures Against cryo-EM Maps

G.N. Murshudov[1]
MRC Laboratory of Molecular Biology, Cambridge, United Kingdom
[1]Corresponding author: e-mail address: garib@mrc-lmb.cam.ac.uk

Contents

Abstract

This review describes some of the methods for atomic structure refinement (fitting) against medium/high-resolution single-particle cryo-EM reconstructed maps. Some of the tools developed for macromolecular X-ray crystal structure analysis, especially those encapsulating prior chemical and structural information can be transferred directly for fitting into cryo-EM maps. However, despite the similarities, there are significant differences between data produced by these two techniques; therefore, different likelihood functions linking the data and model must be used in cryo-EM and crystallographic refinement. Although tools described in this review are mostly designed for medium/high-resolution maps, if maps have sufficiently good quality, then these tools can also be used at moderately low resolution, as shown in one example. In addition, the use of several popular crystallographic methods is strongly discouraged in cryo-EM refinement, such as $2F_o-F_c$ maps, solvent flattening, and feature-enhanced maps (FEMs) for visualization and model (re)building. Two problems in the cryo-EM field

Methods in Enzymology, Volume 579
ISSN 0076-6879
http://dx.doi.org/10.1016/bs.mie.2016.05.033

are overclaiming resolution and severe map oversharpening. Both of these should be avoided; if data of higher resolution than the signal are used, then overfitting of model parameters into the noise is unavoidable, and if maps are oversharpened, then at least parts of the maps might become very noisy and ultimately uninterpretable. Both of these may result in suboptimal and even misleading atomic models.

1. INTRODUCTION

The results of single-particle cryo-EM reconstructions are three-dimensional maps representing macromolecules. Subsequent modeling of these maps as atomic structures is convenient, as the interpretation and extraction of biologically relevant information from atomic models are intuitive and well established. However, care must be exercised when using such models, as their reliability can never be better than the experimental data from which they were derived. Moreover, it is likely that atomic models represent only a subset of the structure of the molecule under study. Over-interpretation of atomic models, especially those derived at medium and low resolution, can result in wrong or misleading conclusions; any conclusions drawn from such models must refer to the original map and should also be supported by complementary experiments.

Before the resolution revolution (Cheng, 2015; Cheng, Grigorieff, Penczek, & Walz, 2015; Kühlbrandt, 2014; Liao, Cao, Julius, & Cheng, 2013; Smith & Rubinstein, 2014) in cryo-EM that started a few years ago, atomic models of different components of large macromolecular complexes would usually be derived using medium/high-resolution crystal structure analysis, and then these structures would be fitted into relatively low-resolution EM maps, accounting for flexibility where possible. The procedure was similar to solving a puzzle with imperfect information, sometimes resulting in ambiguous results. Many high-quality software tools like flexEM (Topf et al., 2008), direX (Shröder, Brünger, & Levitt, 2010), Rosetta (DiMiao, Zhang, Chiu, & Baker, 2013), MDFF (Trabuco, Viia, Mitra, Frank, & Schulten, 2008), and Segger (Printilie & Chiu, 2012) have been developed to tackle this problem from various angles. Chimera (Pettersen et al., 2004)—popular among cryo-EM researchers—offers many visualization, map interpretation, and map/model analysis tools. Software tools developed for low-resolution EM map interpretation have been and continue to be applied successfully to study many biologically relevant macromolecules (Berman et al., 2002; Lawson et al., 2011). When optimizing the fit of atomic models into maps, these software tools would often maximize the real space correlation between "observed" and calculated maps. This is a perfectly valid procedure, especially when local information or local interpretation is needed. However, there are better statistical tools for information transfer from experimental/observational data to the model, one of which is a widely used Bayesian technique—maximum a posteriori (MAP) estimation (O'Hagan, 1994). If accurately implemented and used with care, this technique can allow optimal information transfer from the data to the model, while ensuring consistency of the derived models with our current state of knowledge. It must be remembered that any Bayesian or related technique is intrinsically biased toward prior information: the trickier question is how to bias the model toward the correct knowledge while reducing "bad" bias toward the model errors (see, for example, Kay, 2010).

Rapid advances in single-particle cryo-EM reconstruction techniques over the past few years (Li et al., 2013; Lyumkis, Brilot, Thebald, & Grigoreff, 2013, Scheres, 2014) driven by the developments of electron microscopes and direct electron counting detectors (Faruqi & McMullan, 2011), as predicted by Henderson (Henderson, 1995), have changed the rules of the game in structural biology. The derivation of high-resolution cryo-EM maps comparable to medium/high-resolution crystallographic data is now routine, and de novo model building of atomic structures using high-resolution, high-quality experimental maps is becoming standard (Amunts et al., 2014; Bharat, Murshudov, Sachse, & Löwe, 2015; Fernandez, Bai, Murshudov, Scheres, & Ramakrishnan, 2014; Liao et al., 2013). Moreover, it is often possible to capture several conformations of the same molecule (Fernandez et al., 2013) in one experimental setup, shedding light on their mechanism. In response to demands created by these sudden advances, some of the existing X-ray crystallographic tools such as COOT (Emsely & Cowtan, 2004; Emsley, Lohkamp, & Cowtan, 2010), Phenix.refine (Adams et al., 2010), REFMAC5 (Murshudov, Vagin, & Dodson, 1997) had to be adapted rapidly. Although these software tools work reasonably well in most cases, it should be remembered that they were originally designed for crystal structure solution, model building, and refinement. As a result, atomic models derived using these programs could be suboptimal, and in future a project similar to PDB-redo (Joosten, Joosten, Murshudov, & Perrakis, 2012) may need to be started in order to sort out any problems that may have been introduced due to the deficiencies of current software.

Since the same software is used for the refinement of atomic structures in cryo-EM as in crystallography, it is tempting to use popular crystallographic procedures such as $2F_o - F_c$ maps (Main, 1979; Read, 1986), solvent flattening (Cowtan & Main, 1998; Terwilliger, 2003), and FEMs (Afonine et al., 2015). One should remember that crystallographic measurements correspond to intensities or amplitudes of Fourier coefficients; there is no way of measuring phases directly. The whole field of X-ray crystallography evolved around recovering the lost phases, the well-known "phase problem." In contrast, in cryo-EM, phases and amplitudes are recovered at the same time; there are even claims that the measured phases are more accurate than the amplitudes. Although both phases and amplitudes are available, their reliability as resolution increases falls rapidly, reducing the signal-to-noise ratio dramatically. In cryo-EM, care should be taken to utilize the signal in the data while avoiding fitting into the noise, which inevitably exists in the data. The signal-to-noise ratio for higher-resolution data can be diminishingly small, approaching zero. Exploiting such noisy data requires a robust statistical framework. This chapter attempts to describe some of the necessary tools for such a framework.

The main purpose of this review is to describe some of the methods used for atomic structure refinement, as well as to highlight some of the current problems encountered during the refinement of atomic models using cryo-EM maps. If these problems are not dealt with soon, then persistent errors may be introduced into models derived using these methods, the removal of which may become a challenging problem.

Methods and techniques described in this chapter have been, or will be, implemented in the program REFMAC5 (Murshudov et al., 2011) and associated programs that are available from the CCP4 (Winn et al., 2011) and/or CCP-EM (Wood et al., 2015) suites.

1.1 Notation

In the following sections, we slightly abuse notation and use the same symbols for random variables as well as for their realizations.

$\langle \boldsymbol{X} \rangle$ denotes the expectation value of a random variable X

$\boldsymbol{s}$ a vector in Fourier space (reciprocal space), which has length $|s|$

$\boldsymbol{F(s) = (A(s), B(s))}$ Fourier coefficients of a map with real and imaginary parts. Fourier coefficients are used interchangeably as complex numbers and two-dimensional vectors

$\boldsymbol{F_c(s)}$ Fourier coefficients calculated from the current atomic model

$\boldsymbol{F_o(s)}$ Fourier coefficients of the observed maps

$\boldsymbol{F_t(s)}$ Fourier coefficients of the "true" map

Δx Random error in atomic positions
ΔB Random error in atomic temperature factors—*B*-values
$D = \left\langle \cos(2\pi s\Delta x)e^{-\Delta B|s|^2/4} \right\rangle$ The effect of positional and *B*-value errors on Fourier coefficients; Luzzati's scale parameter (Luzzati, 1952)
$\Sigma(s) = \left\langle |F_t(s) - DF_c(s)|^2 \right\rangle$ The variance of unexplained signal, or variance of the Luzzati distribution for Fourier coefficients
$\Sigma_0(s) = \left\langle |F_t(s)|^2 \right\rangle$ The variance of total signal, or variance of the Wilson distribution for Fourier coefficients (Wilson, 1949)
$\sigma^2, \sigma^2_{1/2}$ The variances of noise corresponding to the full and half-data reconstructions, respectively. In general, these depend on individual vector *s*; in practice, they are calculated in resolution bins and therefore depend on $|s|$
$\mathrm{cov}(X,Y) = \langle (X - \langle X \rangle)(Y - \langle Y \rangle) \rangle = \langle XY \rangle - \langle X \rangle \langle Y \rangle$ covariance between random variables *X* and *Y*
$\mathrm{var}(X) = \mathrm{cov}(X,X)$ variance of the random variable *X*
$\mathrm{FSC}(s) = \dfrac{\mathrm{cov}(F_1(s), F_2(s))}{\sqrt{\mathrm{var}(F_1(s))\mathrm{var}(F_2(s))}}$ Fourier shell correlation between two Fourier coefficients, covariance and variances, are calculated in narrow resolution shells
$N_k(\mu, A) = \dfrac{1}{(2\pi)^{k/2}|A|^{1/2}} e^{-1/2(x-\mu)^T A^{-1}(x-\mu)}$ is a *k*-dimensional Gaussian (normal) distribution with a mean vector μ and covariance matrix *A*. The covariance matrix is assumed to be symmetric and nonsingular
$P(X;Y) = \dfrac{P(X,Y)}{P(Y)}$ Conditional probability distribution of the random variables *X* given *Y*, both of which can be vectors of different sizes

Relationship and formulas related to the multivariate normal distribution used in this review can be found in many standard textbooks on multivariate statistics. For example, see Eaton (2007).

2. TARGET FUNCTION

Fitting of an atomic model into a cryo-EM map can be viewed as a standard statistical modeling problem, where some knowledge about the internal structure of the system under study is available. Observations are typically incomplete (medium/low resolution) and noisy (signal-to-noise ratio is small), thus they alone are not sufficient to build and optimize a physically sensible model. There are many models that are consistent with the same set of observations, and we are looking for a model that is consistent with both our current state of knowledge as well as the data. One technique for deriving models that are equally well fitted to experimental data and prior information is maximizing a posterior probability (MAP). As the name suggests, this technique maximizes the posterior conditional probability distribution of the model parameters given the current observations. According to the Bayes' theorem (see, for example, O'Hagan, 1994), the probability distribution of a model given observations has the form:

$$P(m;o) = P(m)\frac{P(o;m)}{P(o)} \quad (1)$$

where $P(m;o)$ is the conditional probability distribution of model parameters when observations are known—posterior probability distribution, $P(m)$ is the prior probability distribution of model parameters, $P(o;m)$ is the conditional probability distribution of observations given current model parameters—likelihood function, and $P(o)$ is the probability distribution of observations—normalization factor. m denotes model parameters and o denotes observations. It is often convenient to minimize the negative log posterior probability distribution:

$$\begin{aligned} f(m;o) &= -1/2\log(P(m)) - 1/2\log(P(o;m)) + 1/2\log(P(o)) \\ &= f_G(m) + f_D(m;o) + K \end{aligned}$$

Since the last term $K=1/2\log(P(o))$ does not depend on model parameters, it is often dropped. Without further justification, we will interchangeably denote model parameters by m in coordinate space and F_c in Fourier space, and o will correspond to Fourier coefficients of observed maps F_o. All atomic parameters, including positional and thermal parameters, affecting the likelihood function are assumed to be inside F_c.

Minimization of the function $f(m;o)$ ensures that as much information is transferred from the observed data to the model as possible (via the likelihood function), and the model is kept consistent with our current state of knowledge (via the prior probability distribution). Obviously, it is impossible and unnecessary to use all knowledge about the molecules in designing such a prior probability distribution. It is necessary to analyze the information contained within the data and use complementary prior knowledge. Both crystallography and cryo-EM techniques produce data pertaining to the shape and long-range interactions of atoms in the molecule. As resolution of the data increases, shorter and shorter-range information becomes available. MX and EM rarely contain information about bond lengths or angles; only data beyond 1.2 Å can contain enough information to allow accurate estimation of bond lengths between well-defined atoms. The prior probability distribution must contain information about bond lengths, angles. As resolution decreases, longer and longer-range information such as torsion angles and information relating to secondary structures and domains might be needed.

2.1 Prior Knowledge

The negative-log prior probability distribution used in crystallographic refinement and fitting into cryo-EM maps has the form:

$$f_G(m) = \sum_{\text{class}} \sum_{\text{list}} \frac{1}{\sigma^2} w(t_m, t_i)(t_m - t_i)^2 + \sum_{\text{list}} \frac{1}{\sigma^2} (d_c - d_m)^2 \quad (2)$$

where *class* is a restraint class—bonds, angles, torsion angles, and others, *list* is the list of the restraints in this class, subscript m denotes values calculated from the current model, and i denotes ideal values tabulated in the dictionary or calculated using reference homologous structures, $w(t_m,t_i)$ is an additional weighting factor that may depend on the current value calculated from the model. Usually, $w(t_m,t_i)=1$ except when robust estimators are used, as well as for nonbonding interactions. In REFMAC5, robust estimator weights are used for reference structure and local NCS restraints (Nicholls, Long, & Murshudov, 2012). For nonbonding interactions, $w(t_m,t_i)=1$ when $t_m < t_i$ and 0 otherwise. σ is a standard deviation that is usually listed alongside the "ideal" values in the pretabulated dictionary (Vagin et al., 2004).

The last term in Eq. (2) represents so-called jelly-body restraints, where at every cycle of refinement $d_c = d_m$, ie, current values of the interatomic distances are taken as the ideal values. This means that both the value and the first derivative of this term with respect to d_m are 0, therefore this term does not change the value and gradient of the target function. Since the second derivative of this term is not zero, it does change the curvature of the parameter space affecting how parameters change during refinement. This term is similar to standard regularization terms that are used in Tikhonov regularization of ill-posed problems (Tikhonov & Arsenin, 1977) or ridge regression (Stuart, Ord, & Arnold, 2009) in statistics, with the one difference that the regularization term in Eq. (2) is applied in distance space instead of parameter space. The purpose of this term is to help keep the local conformation of the molecule intact while moving atoms (secondary structures, domains) in a concerted fashion, thus avoiding local minima and increasing the radius of convergence. If all pairs of distances were restrained with large weights, then the only allowed movement would be a rigid body movement with six parameters for each domain, ie, refinement would reduce to rigid body fitting. In practice, restraining interatomic distances up to 4.2 Å with

$\sigma = 0.01$ or 0.02 works sufficiently well for a large class of problems. It should be noted that despite apparent similarities to Deformable Elastic Network (Shröder et al., 2010) it is a different technique. The regularization term in Eq. (2) can only be used if the second derivative of the target function or its approximation (Steiner, Lebedev, & Murshudov, 2003) is used for minimization. If only first derivative methods, such as conjugate gradient, are used as an optimization technique, then the minimization iteration must be restarted after a few cycles to account for the updated interatomic distances, making convergence very slow.

Since the models in both crystallography and cryo-EM correspond to macromolecules, the prior knowledge used in both of these techniques is essentially same. However, one difference is that crystals belong to one of the space groups, with corresponding symmetry operators, and therefore when interactions between atoms are calculated symmetry must be accounted for. In contrast, cryo-EM maps are put in a box large enough to avoid interactions between molecules in neighboring boxes; cryo-EM boxes are not unit cells of a crystal, they are used for speed and convenience of calculations only.

2.2 Likelihood

This part of the target function is a tool that facilitates information transfer from the data to an atomic model. Having an accurate likelihood function ensures that the signal from the data is transferred to the model; the probability distribution must ensure that the model is not fitted into the noise. Let us assume that the observed Fourier coefficients of the map F_o are made for the "true" map with corresponding Fourier coefficients F_t, and the probability distribution of each observation given the true map is a two-dimensional normal distribution—$P(F_o;F_t) = N_2(F_t;\sigma^2)$. Let us also assume that we have a model that needs to be optimized using current observations. The model is incomplete and has some errors. Then, according to Luzzati (1952), the distribution of true Fourier coefficients given model parameters is $P(F_t;F_c) = N_2(DF_c,\Sigma)$, where D is a scale factor that accounts for errors in the model, and Σ is the variance that accounts for incompleteness as well as errors in the model—variance of unexplained signal. If we know the true Fourier coefficients, then the model Fourier coefficients would not tell us anything about the observations, ie, F_o and F_c are conditionally independent under the condition that we know F_t, in other words:

$$P(F_o;F_c,F_t) = P(F_o;F_t)$$

Now we can use the following chain for the joint conditional probability distributions of true and observed Fourier coefficients:

$$P(F_o, F_t; F_c) = P(F_o; F_t, F_c)P(F_t; F_c) = P(F_o; F_t)P(F_t; F_c) \quad (3a)$$

Integration of the right hand side over F_t, ie, marginalization over a so-called nuisance parameter that we do not know anything about, we get:

$$P(F_o; F_c) = N_2(DF_c, \Sigma + \sigma^2) \quad (3b)$$

Essentially, this means that the observed Fourier coefficients have a Gaussian distribution centered on a scaled version of calculated Fourier coefficients with inflated variance (total variance is the sum of the variances due to noise in the data and errors in the model). Eq. (3b) is the likelihood function for a single Fourier coefficient. If we assume that noise in different Fourier coefficients is independent (it must be said that it is a very strong assumption and it is not necessarily true for cryo-EM reconstructions), then the total likelihood function can be written as a product of the individual probability distributions. Thus, the negative-log likelihood function that is the part of the total target function has the form:

$$f_o = \sum -\log\left(N_2\left(DF_c, \Sigma + \sigma^2\right)\right) = \frac{1}{2}\sum \frac{|F_o - DF_c|^2}{\Sigma + \sigma^2} + \log\left(\Sigma + \sigma^2\right) \quad (3c)$$

The formula (3c) says that the conditional variances of the observed Fourier coefficients with known model parameters can never be less than the variances of noise in the observations.

It is natural to ask: what happens if observations have no relevance to the model we are building and refining at the moment? For example, what happens when observations are pure noise? This can happen if the resolution of the map is severely overestimated and is well beyond the signal. In this case, $P(F_o; F_t) = P(F_o)$, ie, the distribution of the observed data beyond the resolution of the data does not depend on the "true" Fourier coefficients, and therefore $P(F_o; F_c) = P(F_o)$, meaning that there is no information whatsoever in the data about the model we are refining. In this case, using the formula (3c) for the likelihood function would only result in the model parameters being fitted into the noise. Usually, optimization programs are very good at finding a local optimum of a function. If pure noise is used, then the impression of an "accurate" model can be created, and if self-critical care is not exercised, then a wrong model can be derived. Therefore, it is important to identify the resolution limit of the signal as accurately as possible.

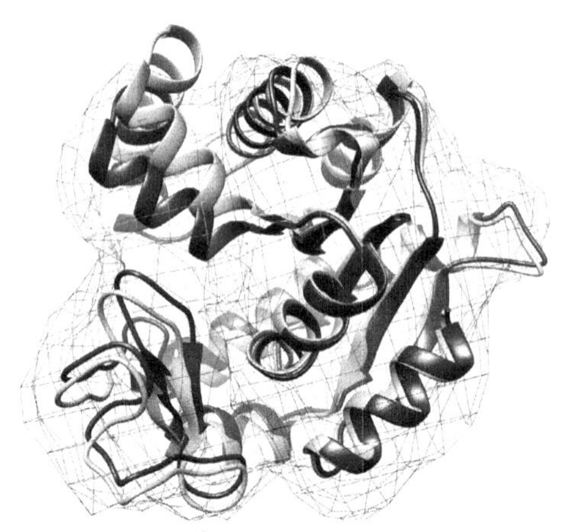

Fig. 1 Jelly-body refinement at 10 Å resolution against a calculated map. The map is calculated from coordinates and the model is distorted (see Topf et al., 2008). This figure demonstrates that if the map is of sufficiently good quality, jelly-body refinement can guide atoms into the map in a concerted fashion, keeping secondary structures intact. The model before refinement is shown in *gray*, and the model after 150 cycles of jelly-body refinement in *black*.

Clearly, there is always a trade-off; if resolution is overestimated, then fitting into the noise is unavoidable, if it is underestimated, then useful signal may not be used. In both cases, the quality of the model will suffer.

Eqs. (2) and (3c) have been used successfully for the refinement of number of macromolecular complexes (Amunts et al., 2014; Bharat et al., 2015; Brown et al., 2015; Fernandez et al., 2014). They do seem to work well even without the inclusion of the variances of experimental data. Although currently these procedures are recommended for refinement at medium and high-resolution (higher than 5 Å), they can also be applied for fitting into high-quality low-resolution maps. Fig. 1 shows the effect of jelly-body refinement applied to a 10 Å calculated map. At least for these types of map, jelly-body refinement allows secondary structures to move in a concerted fashion, and for a model to be fitted into the map. It must be stressed that, at the same resolution, the quality of experimental maps is much worse than of calculated maps. Therefore, the current implementation of this procedure in REFMAC5 should be used with extreme care at low resolutions.

3. VARIANCE OF OBSERVATIONAL NOISE

The likelihood function defined by the formulas (3a–3c) requires an estimation of the variance of noise in the data. Although current implementation of the formula (3c) in REFMAC5 (Brown et al., 2015) produces sufficiently good quality atomic models, it does not use the noise variance and therefore is prone to overfitting into the noise.

There are at least two ways to estimate the variance of noise in the data. One of them is well established, using a similar idea to that employed for the gold standard FSC calculation (Scheres & Chen, 2012) as implemented in the program RELION (Scheres, 2012). A slightly modified version of this procedure can be used for noise variance estimation.

3.1 Half-Data Reconstruction

Let us assume that we have two independent reconstructed maps with Fourier coefficients $F_{1o}(s)$ and $F_{2o}(s)$. In addition we assume that, in narrow resolution bins, noise in the Fourier coefficients for the half-maps have the same distribution. We also assume that noise is independent from each other in two half-maps, as well as being independent from the signal. We also assume that the variance of the signal is Σ_0. Then, using the algebra of covariances, we can write:

$$
\begin{aligned}
&F_{1o} = F_t + n_1, F_{2o} = F_t + n_2, F_o = \frac{F_{1o} + F_{2o}}{2}, G = F_{1o} - F_{2o} = n_1 - n_2, n = \frac{n_1 + n_2}{2} \\
&\mathrm{var}(G) = 2\,\mathrm{var}(n_1),\ \mathrm{var}(F_o) = \mathrm{var}(F_t) + \mathrm{var}(n) = \Sigma_0 + \sigma^2 \\
&\mathrm{var}(n) = \frac{\mathrm{var}(n_1)}{2} = \frac{\mathrm{var}(G)}{4},\ \mathrm{cov}(F_{1o}, F_{2o}) = \mathrm{var}(F_t) = \Sigma_0
\end{aligned}
\tag{4}
$$

So variance of the noise can simply be calculated using two half-reconstructed maps. However, it has to be remembered that the two reconstructions must be completely independent. Even then there will be some overfitting and associated noise. Assuming that overfitting in both maps is the same, the variance estimated with the earlier procedure will reflect random noise, and therefore will be underestimated, whereas the "signal" variance will reflect signal plus overfitting noise. Therefore, FSC will be overestimated and the estimated resolution will be higher than it really is. Chen et al. (2013) use a high-resolution noise substitution technique to solve some of the problems, giving a more accurate FSC calculation and thus resolution limit estimation.

3.2 Noise Reconstruction

Once the whole reconstruction procedure is complete, with all parameters refined to their optimal values, the reconstruction operation can be applied to the noise also. To do this, while picking particles, the corresponding noise "particles" around them (with exactly the same size) must also be picked. This will ensure that the systematic behavior of particles and "noise particles" will be the same. Assuming that the Fourier coefficients of the noise map are G and that of particle map are F_o, the variance of the noise and signal can be calculated using:

$$\mathrm{var}(G) = \sigma^2, \mathrm{var}(F_o) = \Sigma_0 + \sigma^2 \tag{5}$$

If this procedure is followed, then, before calculating signal and noise, any operation applied to the particle map must also be applied to the noise map. This can be done easily if operations are linear. For example, masking and symmetrization are linear operations. However, nonlinear operations, such as histogram matching or positivity constraints, would be much harder to apply equivalently to both noise and particle maps.

Note that the noise variance calculated using half-data reconstruction corresponds to the random noise only, and signal variance reflects both signal and overfitting noise. However, the noise variance calculated using the noise map seems to correspond to both systematic and random noise, and the signal variance is closer to the true signal variance. It is expected that FSC and resolution limits calculated using this procedure would be smaller, perhaps better reflecting the real situation. RELION does a good job of reducing overfitting by modification of the reconstruction procedure (Scheres & Chen, 2012). However, it is known that any regularization, while reducing overfitting noise, also reduces information extracted from the data. Finding optimal values for regularization is a tricky problem and sometimes might take a long time to achieve.

4. MAP CALCULATION

Now let us turn our attention to the "best" map calculation. To do this we need to have the conditional probability distribution of the Fourier coefficients of the "true" map given the observations and the current state of the atomic parameters. This distribution can be derived using the definition of the conditional probability distribution and has the form:

$$P(F_t; F_o, F_c) = \frac{P(F_o, F_t; F_c)}{P(F_o; F_c)} = N_2(w_1 F_o + w_2 D F_c, \Sigma_F)$$
$$w_1 = \frac{\Sigma}{\Sigma + \sigma^2}, w_2 = \frac{\sigma^2}{\Sigma + \sigma^2}, \Sigma_F = \sigma^2 \frac{\Sigma}{\Sigma + \sigma^2}$$

ie, the probability distribution of the "true" Fourier coefficients is a two-dimensional Gaussian with mean value equal to the weighted mean of observed and calculated Fourier coefficients, and variance equal to the weighted-down variance of the observations. Note that Σ is the variance of unexplained signal, and D reflects the errors in the atomic model parameters. Using the conditional distribution of Fourier coefficients of the "true" map, we can derive their expected values:

$$\langle F_t \rangle = \frac{\Sigma}{\Sigma + \sigma^2} F_o + \frac{\sigma^2}{\Sigma + \sigma^2} DF_c \tag{6}$$

This formula essentially means that Fourier coefficients for maps after refinement are a linear combination of the observed and calculated ones. If D is not estimated accurately, then there will be bias toward model errors due to the second term of Eq. (6). To exploit this formula, one needs to estimate D after refinement using estimated coordinate and B-value errors. This would allow maps to be calculated using higher frequency Fourier coefficients than those observed. Even if these types of map are calculated, one must always refer to the original observed maps, and only those parts of the map that have not seen the model should be inspected.

Considering various limiting cases might shed some light on the behavior of the map with Fourier coefficients defined by the formula (6).

Case 1. When there is no model, or model errors are too large: $D \rightarrow 0$ and $\Sigma \rightarrow \Sigma_0$. In this case:

$$\langle F_t \rangle = \frac{\Sigma_0}{\Sigma_0 + \sigma^2} F_o \tag{7}$$

ie, after the reconstruction, FSC-weighted maps should be used, as was suggested by Rosenthal and Henderson (2003). These weighted Fourier coefficients will reflect the signal in the map better and will damp the noise down, especially at high-resolution where the signal-to-noise ratio is very small.

Case 2. When observed data are very noisy: $\sigma^2 \rightarrow \infty$. In this case, the map coefficients become:

$$\langle F_t \rangle = DF_c$$

ie, map coefficients are a weighted-down version of the calculated Fourier coefficients. The weights must reflect the current errors in the model. In practice, variances of the observed noise are never infinite. The problem can also be considered from a different angle. If we assume that $P(F_o;F_t) = P(F_o)$, ie, the observed data have no information about the "true" Fourier coefficients, then $P(F_t;F_o,F_c) = P(F_t;F_c)$ and consequently $\langle F_t \rangle = DF_c$, ie, again the map coefficients are weighted-down versions of the calculated ones. Note that if $D \rightarrow 1$ then $\langle F_t \rangle = F_c$ independent of the observations. That is, if the model is a perfect reflection of the "true" structure, then observations are not needed. In practice this would never happen. It is more likely that D will be overestimated, causing model bias to obscure the signal to be interpreted.

Case 3. Experimental observations are exact, ie, $\sigma^2 \rightarrow 0$, then $\langle F_t \rangle = F_o$, ie, contributions from the calculated Fourier coefficients are not needed. In practice this case would never happen; there may be very accurate reconstructions with very small noise variance, but never perfect ones.

One of the popular maps used in crystallography is a difference map, which is meant to show peaks signifying differences between the "true" structure and the current model. In cryo-EM this procedure is not yet routine. One of the obvious maps that can be calculated easily with any program is a vector difference map—$F_o - DF_c$. However, this map is expected to have high variance due to the noise in the observed F_o maps. Another way of calculating difference maps is to subtract (weighted) calculated Fourier coefficients DF_c from those defined in Eq. (6) resulting in:

$$\Delta F = \frac{\Sigma}{\Sigma + \sigma^2}(F_o - DF_c) \tag{8}$$

These coefficients are just weighted vector difference map coefficients. In general, Fourier coefficients defined in Eqs. (6) and (8) can be combined to give various maps similar to weighted $nF_o - mF_c$ maps, where n and m should depend on the quality of the data, the variances, and the model. In future, we hope that these kind of difference maps may become more routine.

5. CROSSVALIDATION

Just like in macromolecular crystallography, cryo-EM maps have limited resolution, and the number of adjustable parameters is very large. Overfitting of model parameters to the noise is a real concern. One way of making sure that the fitted model describes signal in the density is always to check the model against the experimental map, even if modified "improved" maps are available. In macromolecular crystallography it is customary to use a set of reflections for validation purposes. Specifically, refinement statistics are calculated for Fourier coefficients included in refinement, eg, R-factors, as well as for those excluded from refinement, eg, free R-factors (Brünger, 1992). The implicit assumption in this procedure is that errors in different observed Fourier coefficients are independent. Unfortunately, in cryo-EM we cannot assume independence of noise in different Fourier coefficients. Increasing the number of particles used in reconstruction can reduce this dependence. However, removal of this dependence altogether is impossible since there are always adjacent pixel correlations. Therefore, free R-factors or associated statistics cannot be used in cryo-EM map interpretation and refinement. It is tempting to use thin resolution slices in Fourier space, although the width of these slices would have to depend on the correlation radius of errors in Fourier space. At the moment, there is no procedure for estimating this. Moreover, even if thin slices were to be used, coefficients near the borders of the slices would always be dependent on each other. Accurate crossvalidation would only be possible if performing refinement of atomic structures against raw images, but this procedure is not yet available.

For validation purposes, we can use half-maps as done in Fernandez et al. (2014):

(1) Build and refine the full model using the full-data map;

(2) Randomize coordinates as much as reasonably possible; in practice, uniform random sampling with maximum shift 0.5–1 Å seems to work reasonably well. Randomization in general may require sampling of the conformational space of the molecule. However, currently a simple randomization with maximum shift 0.5–1 Å of atoms independently seems to work reasonably well;

(3) Refine the randomized model against half-map 1 and half-map 2;

(4) Calculate FSC and all other statistics using half-map 2 and half-map 1, respectively;

(5) Calculate differences between models refined against map 1 and map 2.

Difference between the FSC after refinement using half-map 1 and that calculated using half-map 2 would indicate the degree of overfitting. It should be noted that this difference would always be underestimated.

Note that this procedure can also be used for weight optimization for each domain or subunit of a macromolecular complex, thus exploiting variation of the resolution throughout the map, deriving models with varying quality reflecting the local resolution (Kucukelbir, Sigworth, & Tagare, 2014).

6. EFFECT OF OVERSHARPENING

Since unfiltered maps produced by cryo-EM reconstruction programs are usually blurred, they are often sharpened before visualization. Recommendation for sharpening parameters can be found in Rosenthal and Henderson (2003). One should remember that sharpening or blurring does not change the information content of the map, but sharpening can highlight features that may be obscured in the original maps. Sharpening or blurring should only be used for visualization and model building purposes. If local resolution is estimated using ResMap (Kucukelbir et al., 2014), then it should be done directly on the map generated by the reconstruction program, ie, before any sharpening and/or filtering is applied. Although sharpening is a recommended procedure, which usually has a dramatic effect on map interpretability, selecting the "correct" sharpening parameter may be a tricky problem. If the map is under-sharpened, then features that are needed for interpretation of the molecule in a biological context could be missing. However, oversharpening can increase both the noise level and the series termination effect. This can cause interpretation of regions of the map, or even the whole map, to become impossible. Since map sharpening is intimately related with the resolution of the signal in the map, if it is overestimated then severe sharpening reflecting the claimed resolution might cause signal loss and noise amplification.

One of the examples where features of the map were obscured by oversharpening is described by Bartesaghi et al. (2015) with EMDB code emd-2984. The quality of this map is even better than described in the paper, and if the correctly sharpened map is used, then almost all features of the molecule, including ligands, are visible. However, the map deposited to the EMDB is oversharpened. Fig. 2A shows average $|F_o|$ vs resolution, in which the sudden jump at around 2.2 Å indicates clear oversharpening. Such behavior of average $|F_o|$ is a clear indication that a map has been severely oversharpened, and it is not surprising that noise and series termination effects have been amplified so much so that they obscure signal completely in some regions. Fig. 2B and C shows the original map from the EMDB and the map recalculated with blurring factor $B=60$, respectively. If the blurred map is used, then it is evident that the model (PDB code 5a1a) can be modified further. However, the original oversharpened map exhibits strong noise that makes interpretation impossible. With slight blurring it is possible to refine the model, resulting in a reasonable

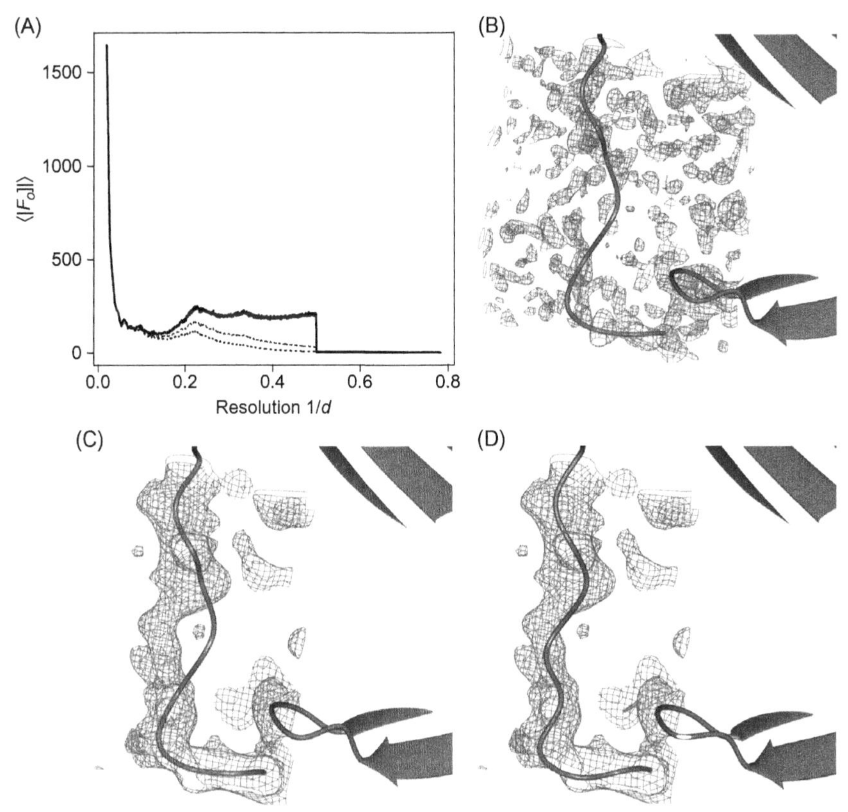

Fig. 2 Effect of oversharpening. This figure demonstrates that if a map is oversharpened, then the noise level can become large enough to obscure real features of the map. The map is taken from the EMDB (code emd-2984). (A) Average $|F_o|$ vs resolution. The *bold line* corresponds to the original map deposited in the EMDB, and the *dashed* and *dotted lines* correspond to the maps recalculated with blurring *B*-values 60 and 40 Å^{-2}, respectively. The *bold line* falls sharply at the high-resolution range, indicating the series termination effect. Using either of the blurred maps (*dashed* and *dotted lines*) results in much smaller series termination effects. (B) A portion of the map corresponding to the region around residue 733 of chain A. (C) Map corresponding to the same region as in subfigure (B) recalculated with blurring *B*-value 60 Å^{-2}. This map clearly shows nice connected density where the model should be. (D) Model after 100 cycles of refinement using the blurred map. The model moves toward the density, explaining that portion of the map. As it might be expected, equivalent refinement using the original map does not move the model into the map.

agreement between refined model and map. However, using the oversharpened map it is impossible to refine the model, *B*-values would try to become negative and convergence of the refinement procedure would never be achieved.

7. $2F_o-F_C$ AND OTHER CRYSTALLOGRAPHIC MAPS

Since model building and fitting into cryo-EM maps is a relatively new field, techniques and recommendations are still being established. It is often the case that researchers trained in crystallography apply their skills to cryo-EM, and it is natural for them to use tools available in crystallography. However, because of differences between crystallography and cryo-EM, tests and studies must be carried out to establish good practice. Some recent works have used such tools as $2F_o - F_c$ and FEMs for model building, and sometimes for illustration of the quality of maps and model (Khatter, Myasnikov, Natchiar, & Klaholz, 2015), potentially resulting in a misleading impression regarding the quality of the final map and atomic model.

There are several maps that are used in crystallographic model refinement and rebuilding, the most popular of which are the so-called $2F_o - F_c$ maps, in their various incarnations. It must be emphasized that Fourier coefficients (so-called structure factors) in crystallographic observations lack phases, which must be recovered from somewhere. These phases are replaced by model phases during refinement and model building, causing the resultant maps to more resemble the model rather than what is actually in the crystal. Main (1979) showed that if one calculates only $2F_o\exp(i\varphi_c)$ maps, then atoms absent in calculations (ie, the unmodeled parts of the molecule) would be present in the maps, in the best cases with half height. If one uses $(2F_o - F_c)\exp(i\varphi_c)$, then the peaks corresponding to the absent atoms will increase, and, in favorable cases where the atomic model is sufficiently complete and accurate, these peaks will be comparable to those corresponding atoms present in the model. In general, these maps will never be model bias free. Read (1986) showed that these coefficients can be extended to account for inaccuracies in atomic position, and, to a certain degree, experimental phase information. Now, usage of $2mF_o - DF_c$ maps is standard in crystallographic calculations. One of the main advantages, and perhaps the main difference between crystallographic and EM maps, is that EM maps do have experimental phase information. Moreover, the phases and amplitudes of Fourier coefficients cannot be considered as independent measurements. If one uses $2mF_o - DF_c$ maps without experimental phase information, then model bias is unavoidable.

Moreover, many procedures used to improve interpretability of maps for model building might not be suitable in cryo-EM maps. The recently popularized FEM (Afonine et al., 2015) is one such example. The basic idea behind FEM is to modify the map in order to make the histograms of different regions of the map become more similar to each other. This is a standard procedure in image processing, where the images are pure experimental data (Gonzalez & Woods, 2008). However, since $2F_o - F_c$ maps are necessarily model biased, such a procedure would enhance the observed features as well as features due to the bias toward model errors and thus they might create the impression of improved maps. Using $2F_o - F_c$ and related maps should in general be avoided in cryo-EM map interpretation. Even when phases are included with the assumption that they are perfect, ie, the figures of merit for the experimental phases are 1 (Pannu, Murshudov, Dodson, & Read, 1998), then the map coefficients will become $2F_o\exp(i\varphi_{\text{exp}}) - DF_c\exp(i\varphi_c)$. Subtracting $DF_c\exp(i\varphi_c)$ from the experimental maps can only add noise to the map, and does not add any new information. Thus, it is strongly recommended to avoid standard crystallographic procedures in model building with cryo-EM maps. Observed maps should always be used or consulted whenever new features are modeled; these maps are the last link between the data and the atomic models.

Fig. 3 demonstrates how if phases are ignored, then model bias in the $2F_o - F_c$ maps is unavoidable. As refinement is performed and the model changes its position, model bias is exhibited in different places, reflecting the model.

Another misuse of crystallographic tools on cryo-EM maps is using density modification for phase improvement (Hite et al., 2015). Again, since there is no phase recovery problem in cryo-EM, standard crystallographic density modification techniques should not be used for phase "improvement". They can result in wrong and misleading models. Maps in general can be improved in cryo-EM; however, new techniques must be developed that would use complex Fourier coefficients and their associated probability distribution. Information that may be added during "map improvement" should be analyzed very carefully using probabilistic approaches. Procedures are not ready yet for this purpose, and at the moment, the observed maps must always be consulted whenever new "improved" maps are generated. Illustration of modified maps should always be done for the parts of the map that have never seen atomic models.

8. FORM FACTORS

As stated earlier, despite there being many similarities between X-ray and electron scattering, there are also some differences between them. While

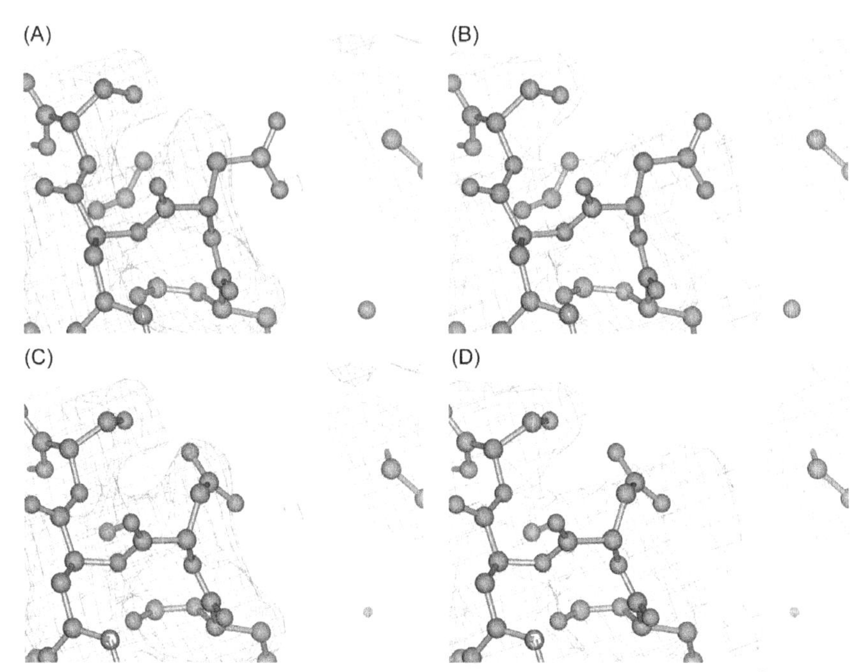

Fig. 3 Effect of $2F_o - F_c$ maps. (A) Original map (with blurring B-value 60 Å^{-2}) and the model 5a1a from the PDB, showing the region near residue 234 of chain A. (B) $2F_o - F_c$ map ignoring phases, corresponding to the same region as in subfigure (A). This map clearly shows bias toward the model used for phase calculation. (C) Same density as shown in subfigure (A) including the corresponding model after refinement. (D) Same density as shown in subfigure (B) including the corresponding model after refinement. These maps show that if experimental phases are ignored then model bias can be too strong to avoid.

X-ray photons "see" electron density, electrons interact with the charges in the molecule; unlike X-ray photons, electrons "see" electrostatic potential. Under the assumption that the first Born approximation with spherically symmetric potential is valid for electron diffraction (for details, see, for example, Spence, 2003), and that the electrostatic potential of a molecule can be written as the sum of atomic potentials, we can use the independent atom model to model the total electrostatic potential of a molecule. This essentially means that when electrons interact with the electrostatic potential they do not alter it significantly, and the effect of bonding electrons on molecular potential is negligible. Although these are very strong assumptions, in practice, they seem to be sufficiently valid to not affect the results of fitting into cryo-EM maps and refinement against X-ray or electron diffraction data. Nevertheless, to increase accuracy of atomic structure refinement, the atomic form factors need to be modeled for electron scattering. There have been several tabulations of the parameters of these form factors (Doyle & Turner, 1968; Peng, 1998; Peng, Ren, Dudarev, & Whelan, 1996) modeled as a sum of several Gaussians. However, they cannot be considered satisfactory in general for three main reasons: (1) form factors behave at high resolution as $1/|s|^2$, no Gaussian will behave like that; (2) charged atoms are tricky to model at lower resolution, if atoms were in a vacuum then their form factors would go to infinity when $|s|$ approaches 0, although in practice that never happens; and (3) at "low" resolution, which means ∞–3 Å, the effect of bonding electrons cannot be ignored, as they depend on the chemical environment of atoms.

Mott and Bethe (see Kirkland, 1998 and references therein), using the solution of the Poisson equation in Fourier space for isolated atoms and the fact that X-rays diffract from electron density while electrons diffract from electrostatic potential, derive the formula relating X-ray and electron form factors:

$$f_e(s) = \frac{me^2}{2h^2}\frac{Z_n - f_X(s)}{|s|^2} \tag{9}$$

where m is the mass of an electron, e is its charge, h is the Planck constant, and Z_n is the nuclear charge. Note that when $|s| = 0$ then $f_X(s) = Z_e$—the total charge of the electrons of the atom. If the atom is not neutral then the charge of the electron is different from that of the nucleus and therefore for charged atoms $f_e(s)$ when $|s|$ approaches to 0 would be infinity. If we assume screening a la Yukawa potential (Yukawa, 1935) for charged atoms in a solvent, then this behavior can be modified and the form factors would become:

$$f_e(s) = \frac{me^2}{2h^2}\frac{Z_n - f_X(s)}{|s|^2 + \lambda^2}$$

where λ is a screening parameter that may depend on the atom in the molecule and its environment. This function is finite when $s = 0$ and behaves like $1/|s|^2$ when $|s|$ becomes large.

Vainshtein (1964) gave a formula for Fourier coefficients of the electrostatic map using Mott Bethe formula:

$$\begin{aligned} F_e &= \frac{me^2}{2h^2}\frac{F_Z - F_X}{|s|^2} = \frac{me^2}{2h^2}\frac{F_{Z-X}}{|s|^2} \\ F_{Z-X} &= \sum \left(Z_j - f_{j,X}(s)\right) e^{-B_j|s|^2/2} \end{aligned} \tag{10}$$

where Z_j is the atomic number of the j-th atom, $f_{j,X}(|s|)$ is the X-ray form factor for this atom, B_j is its B-value, and x_j its position vector. F_Z are the Fourier coefficients of nuclear charges only, and F_X are those corresponding to X-ray diffraction. It is a curious fact that if we have Fourier coefficients for electron and X-ray diffraction, then Fourier coefficients corresponding to the positions of nuclei can be constructed, thus giving a more accurate map for atomic positions. This formula may be useful for accurate identification of hydrogen positions. Formula (10) has been implemented in the program REFMAC5 (Brown et al., 2015).

Formula (9) shows that tabulation of form factors for X-ray diffraction alone is sufficient, as they can be converted to electron scattering factors either directly using Eq. (9) or indirectly using Eq. (10). However, if there are charged atoms (which is always the case for macromolecules) in the molecule, then this formula will give infinity when $|s|$ approaches to 0. The reason why electrons diffract from crystals and cryo-EM produce finite values near to $s=0$ is that charged molecules are surrounded by solvent, which is a dielectric continuum, and there is a screening of these charges. It means that, in order to refine atomic models accurately against electron diffraction data and cryo-EM maps, we need to model the effect of solvent as accurately as possible. To account for the effect of solvent we need to solve the Poisson–Boltzmann equation (Fogolari, Brigo, & Molinari, 2002), which is a difficult problem, in general. However, under the drastically simplifying assumptions that the dielectric constant is the same everywhere (within the molecule and solvent) and that the Poisson Boltzmann equation can be linearized, then formula (10) with screening parameters becomes:

$$F_e = \frac{me^2}{2h^2} \frac{F_{Z-X}}{|s|^2 + k^2}$$

The parameter k is the solvent screening parameter, which must be refined if this formula is used.

Tabulation of very accurate form factors that would be useful for both X-ray and electron diffraction is a challenging problem. First of all, at low-resolution electron distribution depends on bonding atoms, meaning that for each chemical type there must be a different set of form factors. For example, sp^2 and sp^3 carbons should in general be treated differently. Second, slightest changes in the environment would change the distribution of the outer shell electrons, thus affecting the behavior of form factors at low resolution. Third, hydrogens pose a special problem: although for all heavy atoms it can be assumed that the position of the nucleus and the average position of electrons are the same, this is no longer true for hydrogens. The electrons of hydrogens are close to the atom that they are bonded to, with the positions of the protons being around 0.1 Å further away. This problem can be dealt with if we use two-center form factors for hydrogen atoms to account for the differences between electron and proton positions. This would improve behavior of hydrogens slightly, although it would not solve the problem completely. Electrons in general are not spherical, and the positions of hydrogen nuclei and electrons depend on their environments; sometimes hydrogen bonding can change the relative position of the proton significantly.

9. CONCLUSIONS

Single-particle Cryo-EM is a very powerful technique and has the potential to revolutionize structural biology. Now, the study of three-dimensional structures of large macromolecular complexes in different conformational states is realistic. However, high-resolution single-particle cryo-EM, especially refinement of atomic models against maps, is a relatively young field and is still establishing itself. As with all young fields, there is a strong desire to push the boundaries of the technique beyond what is currently possible. Unfortunately, not all techniques and methods are fully robust yet, and claims made about high resolution, high-quality maps can be based on very thin theoretical foundations and weak data, and are therefore sometimes wrong.

Maps used for visualization purposes are often sharpened to highlight detailed features in order to help model building and refinement. However, oversharpening amplifies noise and series termination effects, thus obscuring map features. Various resolution estimators either implicitly or explicitly rely on the overall *B*-values of maps, and oversharpening can give the impression of higher resolution. It is recommended that multiple maps with a range of sharpening parameters should be used in model building, and resolution should be estimated using unsharpened and unfiltered maps. Plotting average $\langle|F_o|\rangle$ vs resolution (as shown in Fig. 2A) can show the seriousness of oversharpening and might suggest reasonable values for sharpening parameters. In general, at the limit of resolution the average $\langle|F_o|\rangle$ should approach 0; if there is a sudden jump at the end of the resolution range, then series termination ripples are likely to make map interpretation very hard.

Estimation of resolution is another problem that needs to be considered very carefully. At the moment, a popular and robust tool for resolution estimation is the gold standard FSC with high-resolution noise substitution correction (Chen et al., 2013). If two half-maps used for FSC calculation are not independent, then the resolution will be overestimated, resulting in noisy maps, leading to the model being fitted into the noise. In this review, a simple technique for noise level estimation—the noise reconstruction method—is suggested. If this is applied, then exactly the same parameters must be used for image and noise reconstruction; there should be no hidden parameters that would depend on the particles and/or noise. It is often claimed that FSC between observed and model maps should be used to estimate the "true" resolution. Since the models before refinement do not reflect fully the information in the data, the FSC between model and map before refinement will be lower than "true" FSC. After refinement, overfitting of the model parameters into the noise is unavoidable, and such FSC will in general be overestimated. FSC between model and map can reflect true resolution, signal, and noise level if and only if the "true" structure of the molecule is known exactly independently of the current experiment; in which case there would be no point of doing the experiment.

Although there are many common features between fitting into cryo-EM maps and crystallographic refinement, the statistical properties of the data generated by these experiments are very different. It should be remembered that crystallography is concerned with phase recovery, and there are many tools to do that. These tools are not needed in cryo-EM model refinement, as the phases are available. The use of such crystallographic tools as $2F_o - F_c$ maps, FEM maps, and density modification for cryo-EM maps should be avoided; the use of such techniques might lead to wrong models.

Current methods for atomic structure refinement in cryo-EM use only the observed Fourier coefficients; they do not use any information about the variance of the noise in the data. Therefore, current techniques may produce overfitted models. Estimation of variance of the observed Fourier coefficients and their use in atomic structure refinement will reduce overfitting, thus producing better models.

It should be remembered that the purpose of cryo-EM is to study the structures of biologically relevant macromolecules. Such indicators as FSC and resolution are just curves or numbers; if reliably estimated, they can say something about the level of structural details that can be extracted from the data. However, these indicators are just numbers. For example, resolution is just the radius of a sphere in Fourier space within which reliably observed data are located; it says how much data there are in that sphere. It gives no information about the reliability of the data. Using effective resolution d_{eff}, instead of nominal resolution defined by FSC, may give a better idea about the information content of the map:

$$d_{\text{eff}} = d_{\text{nom}} \left(\frac{\sum M_i \text{FSC}_i}{M} \right)^{1/3} \tag{11}$$

where FSC_i is the Fourier shell correlation for i-th resolution bin, and d_{nom} is the nominal resolution, which can be taken as equal to the resolution where $\text{FSC}=0$, M_i is the number of Fourier coefficients in the i-th resolution shell, M is the total number of Fourier coefficients used in the calculations. In the formula (11), one of FSC_{full}, FSC_{half}, or FSC_{true} can be used. They will each give different numbers. However, if they are used consistently, then the qualities of different maps can be compared. This also means that all data out to complete noise can be used in map calculation and interpretation.

Since the field of single-particle cryo-EM is moving very fast, it is inevitable that at least some of the structures produced and deposited in the EMDB will have suboptimal quality; models are either incomplete, or, in some places, wrong. Moreover, the field is young, and computational technology at the moment is lagging behind the experimental techniques. Perhaps it is time to start an EMDB-redo project, similar to the PDB-redo (Joosten et al., 2012) initiative, to reinterpret and re-refine structures deposited to the EMDB. Within an EMDB-redo initiative, it would be possible to derive atomic models reflecting the current state of the technology. In order to be able to maximally exploit information contained in cryo-EM images in future, it is important to deposit as much as possible of the data. Single-particle cryo-EM experiments are different from crystallographic experiments. In crystallography, the integrated data (intensities of structure factors) contain almost as much information as raw images. In contrast, since processing raw data in cryo-EM is still under development, and it is not clear how much information, if any, is lost during reconstruction, depositing raw data might enable significantly improved map and model qualities in future, as and when new techniques become available. Finally, it can be expected that, in future, the refinement of atomic structures will be carried out against raw images. That, in theory, should increase the amount of information transferred from the data to the model. Moreover, if refinement against raw images is available, then many problems including crossvalidation and multiple conformations of molecules can be dealt with in a statistically robust way.

ACKNOWLEDGMENTS

This work was supported by the Medical Research Council (Grant number: MC US A025 0104). I would like to thank the members of the MRC Laboratory of Molecular Biology for inspirational and stimulating discussions that provoked this and many other works. I would also like to thank the CCP4 bulletin board community for never-ending and sometimes useful discussions that are an invaluable source of inspiration as well as problems that need to be solved. I would also like to thank Richard Henderson for his valuable suggestions. Robert Nicholls deserves special thanks for critical reading of the manuscript.

REFERENCES

Adams, P. D., Afonine, P. V., Bunkòczi, G., Chen, V. B., Davis, I. W., Echols, N., et al. (2010). *PHENIX*: A comprehensive python-based system for macromolecular structure solution. *Acta Crystallographica. Section D, 66*, 213–221.

Afonine, P. V., Moriarty, N. W., Mustyakimov, M., Sobolev, O. V., Terwilliger, T. C., Turk, D., et al. (2015). FEM: Feature-enhanced map. *Acta Crystallographica. Section D, 75*(3), 646–666.

Amunts, A., Brown, A., Bai, X.-C., Llàcer, J. L., Hussain, T., Emsley, P., et al. (2014). Structure of the yeast mitochondrial large ribosomal subunit. *Science, 348*, 1485–1489.

Bartesaghi, A., Merk, A., Banerjee, S., Matthies, D., Wu, X., Milne, J., et al. (2015). 2.2 A resolution cryo-EM structure of beta-galactosidase in complex with a cell-permeant inhibitor. *Science, 348*, 1147–1151.

Berman, H. M., Battostuz, T., Bhat, T. N., Bluhm, W. F., Bourne, P. E., Burkhardt, K., et al. (2002). The Protein Data Bank. *Acta Crystallographica. Section D, 58*, 899–907.

Bharat, T. A. M., Murshudov, G., Sachse, C., & Löwe, J. (2015). Structures of actin-like ParM filaments show architecture of plasmid-segregating spindles. *Nature, 523*, 106–110.

Brown, A., Long, F., Nicholls, R. A., Toots, J., Emsley, P., & Murshudov, G. (2015). Tools for macromolecular model building and refinement into electron cryo-microscopy reconstructions. *Acta Crystallographica. Section D, 71*(1), 136–153.

Brünger, A. T. (1992). Free R value: A novel statistical quantity for assessing the accuracy of crystal structures. *Nature, 355*, 472–475.

Chen, S., McMullans, G., Faruqi, A. R., Murshudov, G., Short, J. M., Scheres, S. H. W., et al. (2013). High-resolution noise substitution to measure overfitting and validate resolution in 3D structure determination by single particle electron cryomicroscopy. *Ultramicroscopy, 135*, 24–35.

Cheng, Y. (2015). Leading Edge Review: Single particle cryo-EM at crystallographic resolution. *Cell, 161*, 450–457.

Cheng, Y., Grigorieff, N., Penczek, P. A., & Walz, T. (2015). Leading Edge Primer: A primer to single-particle cryo-electron microscopy, commissioned primer. *Cell, 161*, 438–449.

Cowtan, M., & Main, P. (1998). Miscellaneous algorithms for density modification. *Acta Crystallographica. Section D, 54*, 487–493.

DiMiao, F., Zhang, J., Chiu, W., & Baker, D. (2013). Cryo-EM model validation using independent map reconstructions. *Protein Science, 22*, 865–868.

Doyle, P. A., & Turner, P. S. (1968). Realitivistic Hartree-Fock X-ray and electron scattering factors. *Acta Crystallographica. Section A, 24*, 390–397.

Eaton, M. L. (2007). *Multivariate statistics: A vector space approach*. Beachwood, OH: Institute of Mathematical Statistics.

Emsely, P., & Cowtan, K. D. (2004). Coot: Model-building tools for molecular graphics. *Acta Crystallographica. Section D, 60*, 2126–2132.

Emsley, P., Lohkamp, B., & Cowtan, K. D. (2010). Features and development in COOT. *Acta Crystallographica. Section D, 66*, 486–501.
Faruqi, A. R., & McMullan, G. (2011). Electronic detectors for electron microscopy. *Quarterly Reviews of Biophysics, 44*, 357–390.
Fernandez, I. S., Bai, X.-C., Hussain, T., Kelley, A. C., Lorsch, J. R., Ramakrishnan, V., et al. (2013). Molecular architecture of a eukaryotic translational initiation complex. *Science, 342*, 6160.
Fernandez, I. S., Bai, X. C., Murshudov, G., Scheres, S. H., & Ramakrishnan, V. (2014). Initiation of translation by cricket paralysis virus IRES requires its translocation in the ribosome. *Cell, 157*, 823–831.
Fogolari, F., Brigo, A., & Molinari, H. (2002). The Poisson–Boltzmann equation for biomolecular electrostatics: A tool for structural biology. *Journal of Molecular Recognition, 15*, 377–392.
Gonzalez, R. C., & Woods, R. E. (2008). *Digital image processing*. Upper Saddle River, NJ: Prentice Hall.
Henderson, R. (1995). The potential and limitations of neutrons, electrons and X-rays for atomic resolution microscopy of unstained biological molecules. *Quarterly Reviews of Biophysics, 18*(2), 171–193.
Hite, R. K., Yuan, Y., Li, Z., Hsuing, Y., Walz, T., & MacKinnon, R. (2015). Cryo-electron microscopy structure of the Slo2.2 Na(+)-activated K(+) channel. *Nature, 527*(12), 198–203.
Joosten, R. J., Joosten, K., Murshudov, G. N., & Perrakis, A. (2012). PDB_REDO: Constructive validation, more than just looking for errors. *Acta Crystallographica. Section D, 68*, 484–496.
Kay, S. M. (2010). *Fundamentals of statistical signal processing: Estimation theory*. Upper Saddle River, NJ: Pearson.
Khatter, H., Myasnikov, A. G., Natchiar, S. K., & Klaholz, B. P. (2015). Structure of the human 80S ribosome. *Nature, 520*, 640–645.
Kirkland, E. J. (1998). *Advanced computing in electron microscopy*. New York: Springer.
Kucukelbir, A., Sigworth, F. J., & Tagare, H. D. (2014). Quantifying the local resolution of cryo-EM density maps. *Nature Methods, 11*, 63–65.
Kühlbrandt, W. (2014). Cryo-EM enters a new era. *eLife, 3*, e03678.
Lawson, C. L., Baker, M. L., Best, C., Bi, C., Dougherty, M., Feng, P., et al. (2011). EMDataBank.org: Unified data resource for CryoEM. *Nucleic Acids Research, 39*, 456–464.
Li, X., Mooney, P., Zheng, S., Booth, C. R., Braunfield, M. B., Gubbens, S., et al. (2013). Electron counting and beam-induced motion correction enable near-atomic-resolution single-particle cryo-EM. *Nature Methods, 10*, 584–590.
Liao, M., Cao, E., Julius, D., & Cheng, Y. (2013). Structure of the TRPV1 ion channel determined by electron cryo-microscopy. *Nature, 504*, 107–112.
Luzzati, V. (1952). Traitement Statistique des Erreurs darts la Determination des Structures Cristallines. *Acta Crystallographica, 5*, 802–810.
Lyumkis, D., Brilot, A. F., Thebald, D. L., & Grigoreff, N. (2013). Likelihood-based classification of cryo-EM images using FREALIGN. *Journal of Structural Biology, 183*, 377–388.
Main, P. (1979). A theoretical comparison of the β, γ and $2F_o - F_c$ syntheses. *Acta Crystallographica. Section A, 35*(5), 779–785.
Murshudov, G. N., Skubak, P., Lebedev, A. A., Pannu, N. S., Steiner, R. A., Nicholls, R. A., et al. (2011). *REFMAC*5 for the refinement of macromolecular crystal structures. *Acta Crystallographica. Section D, 67*, 355–467.
Murshudov, G. N., Vagin, A. A., & Dodson, E. J. (1997). Refinement of macromolecular structures by maximum likelihood method. *Acta Crystallographica. Section D, 53*, 240–255.

Nicholls, R. A., Long, F., & Murshudov, G. N. (2012). Low-resolution refinement tools in REFMAC5. *Acta Crystallographica. Section D, 68*, 404–417.
O'Hagan, A. (1994). In *Kendall's advanced theory of statistics: , Vol. 2B. Bayesian inference*. London: A Hodder Arnold Publication.
Pannu, N. S., Murshudov, G. N., Dodson, E. J., & Read, R. J. (1998). Incorporation of prior phase information strengthens maximum-likelihood structure refinement. *Acta Crystallographica. Section D, 54*, 1285–1294.
Peng, L.-M. (1998). Electron scattering factors for ions and their parameterization. *Acta Crystallographica. Section D, 54*, 481–485.
Peng, L.-M., Ren, G., Dudarev, S. L., & Whelan, M. J. (1996). Robust parameterization of elastic absorptive electron atomic scattering factors. *Acta Crystallographica. Section A, 52*, 257–276.
Pettersen, E. F., Goddard, T. D., Huang, C. C., Couch, G. S., Greenblatt, D. M., Meng, E. C., et al. (2004). UCSF chimera—A visualization system for exploratory research and analysis. *Journal of Computational Chemistry, 25*, 1605–1612.
Printilie, G., & Chiu, W. (2012). Comparison of *Segger* and other methods for segmentation and rigid-body docking of molecular components in Cryo-EM density maps. *Biopolymers, 97*, 742–760.
Read, R. j. (1986). Improved Fourier coefficients for maps using phases from partial structures with errors. *Acta Crystallographica. Section A, 42*(2), 140–149.
Rosenthal, P. B., & Henderson, R. (2003). Optimal determination of particle orientation. *Journal of Molecular Biology, 333*, 721–745.
Scheres, S. H. W. (2012). RELION: Implementation of a Bayesian approach to cryo-EM structure determination. *Journal of Structural Biology, 180*(3), 519–530.
Scheres, S. H. W. (2014). Beam-induced motion correction for sub-megadalton cryo-EM particles. *eLife, 13*, e03665.
Scheres, S. H. W., & Chen, S. (2012). Prevention of overfitting in cryo-EM structure determination. *Nature Methods, 9*, 853–854.
Shröder, G. F., Brünger, A. T., & Levitt, M. (2010). Super-resolution biomolecular crystallography with low-resolution data. *Nature, 464*, 1218–1222.
Smith, M. R., & Rubinstein, J. L. (2014). Beyond blob-ology. *Science, 345*, 617–619.
Spence, J. H. (2003). *High-resolution electron microscopy*. Oxford: Oxford University Press.
Steiner, R., Lebedev, A., & Murshudov, G. N. (2003). Fisher's information matrix in maximum likelihood molecular refinement. *Acta Crystallographica. Section D, 9*, 2114–2124.
Stuart, A., Ord, K., & Arnold, S. (2009). *Kendall's advanced theory of statistics: , Vol. 2A. Classical inference*. New York: Wiley.
Terwilliger, T. C. (2003). Statistical density modification using local pattern matching. *Acta Crystallographica. Section D, 59*, 1688–1701.
Tikhonov, A. N., & Arsenin, V. Y. (1977). *Solution of Ill-posed problems*. Washington: Winston & Sons.
Topf, M., Lasker, K., Webb, B., Wolfson, H., Chiu, W., & Sali, M. (2008). Protein structure fitting and refinement guided by cryo-EM density. *Structure, 16*(2), 295–307.
Trabuco, L. G., Viia, E., Mitra, K., Frank, J., & Schulten, K. (2008). Flexible fitting of atomic structures into electron microscopy maps using molecular dynamics. *Structure, 16*, 673–683.
Vagin, A. A., Steiner, R. A., Lebedev, A. A., Potterton, L., McNicholas, S., Long, F., et al. (2004). REFMAC5 dictionary: Organization of prior chemical knowledge and guidelines for its use. *Acta Crystallographica. Section D, 60*, 2184–2195.
Vainshtein, B. K. (1964). *Structure analysis by electron diffraction*. Oxford: Pergamon Press.
Wilson, A. J. C. (1949). The probability distribution of X-ray intensities. *Acta Crystallographica, 2*, 318–321.

Winn, M. D., Ballard, C. C., Cowtan, K. D., Dodson, E. J., Emsley, P., Evans, P. R., et al. (2011). Overview of the *CCP4* suite and current developments. *Acta Crystallographica. Section D*, *67*(4), 235–242.

Wood, C., Burnley, T., Patwardhan, A., Scheres, S., Topf, M., Roseman, A., et al. (2015). Collaborative computational project for electron cryo-microscopy. *Acta Crystallographica. Section D*, *71*(1), 123–126.

Yukawa, H. (1935). On the interaction of elementary particles. *Proceedings of the Physico-Mathematical Society of Japan*, *17*, 48–56.

CHAPTER TWELVE

Cryo-EM Structure Determination Using Segmented Helical Image Reconstruction

S.A. Fromm, C. Sachse[1]
EMBL—European Molecular Biology Laboratory, Structural and Computational Biology Unit, Heidelberg, Germany
[1]Corresponding author: e-mail address: carsten.sachse@embl.de

Contents

Abstract

Treating helices as single-particle-like segments followed by helical image reconstruction has become the method of choice for high-resolution structure determination of well-ordered helical viruses as well as flexible filaments. In this review, we will illustrate how the combination of latest hardware developments with optimized image processing routines have led to a series of near-atomic resolution structures of helical assemblies. Originally, the treatment of helices as a sequence of segments followed by Fourier–Bessel reconstruction revealed the potential to determine near-atomic resolution structures from helical specimens. In the meantime, real-space image processing of helices in a stack of single particles was developed and enabled the structure determination of specimens that resisted classical Fourier helical reconstruction and also facilitated high-resolution structure determination. Despite the progress in real-space analysis, the combination of Fourier and real-space processing is still commonly used to better estimate the symmetry parameters as the imposition of the correct helical symmetry is essential for high-resolution structure determination. Recent hardware advancement by the introduction of direct electron detectors has significantly enhanced the image quality and together with improved image processing procedures has made segmented helical reconstruction a very productive cryo-EM structure determination method.

Methods in Enzymology, Volume 579
ISSN 0076-6879
http://dx.doi.org/10.1016/bs.mie.2016.05.034

1. INTRODUCTION

Structure determination by electron cryomicroscopy (cryo-EM) is producing a rapidly growing number of high-resolution structures. A series of recent technological breakthroughs in hardware and software has made three-dimensional (3-D) cryo-EM considerably more powerful. The fundamental discoveries leading to cryo-EM-based 3-D structure determination as we know it today date back to more than five decades of continuous scientific development. In fact, the first 3-D map computed from electron micrographs was of the helical tail of bacteriophage T4 and thus this study established the technique of helical reconstruction (DeRosier & Klug, 1968). In the early days the results were limited by the specimen preparation procedure of negative staining, which gave only a possibly distorted outline of the molecular shape. This problem was overcome by introducing vitrification of samples to observe them in a frozen-hydrated state, thus leading to the term cryo-EM (Adrian, Dubochet, Lepault, & McDowall, 1984). The first near-atomic resolution structure came from 2-D crystals of bacteriorhodopsin (Henderson et al., 1990). Unwin and colleagues adapted the analysis to helical membrane tubules of the acetylcholine receptor, which resulted in a series of high-resolution structures up to 4 Å (Miyazawa, Fujiyoshi, & Unwin, 2003; Unwin, 2005). Other successful examples of the Fourier helical processing approach included the 6.5 Å resolution Ca^{2+}-ATPase structure (Xu, Rice, He, & Stokes, 2002) and the near-atomic resolution structure of the bacterial flagellar filament (Yonekura, Maki-Yonekura, & Namba, 2003).

Traditionally periodic specimens were analyzed across many repeats to apply methods of Fourier decomposition efficiently, but this approach depends on a well-ordered arrangement of the molecules. It became clear, however, that long-range disorder of 2-D crystals limited resolution and computational methods were required to compensate for this by unbending the crystal lattice (Henderson, Baldwin, Downing, Lepault, & Zemlin, 1986). An alternative solution to deal with such long-range disorder in helical assemblies was to treat the helix as a sequence of small segments, as opposed to analyzing the whole helical entity (Beroukhim & Unwin, 1997). Following the development of image processing techniques for asymmetric single particles (Penczek, Grassucci, & Frank, 1994, Penczek, Radermacher, & Frank, 1992), a similar approach was used to study amyloid fibrils, where such segments were treated as single particles

(Jiménez et al., 1999). In this way, real-space analysis developed for single-particle reconstruction, with subsequent imposition of helical symmetry, led to the iterative helical real-space reconstruction (IHRSR) method (Egelman, 2000). This was particularly useful for studying flexible filaments with limited helical order such as Rad51 (Galkin et al., 2006). An extended single-particle-based helical approach with improved image processing protocols was applied to the well-ordered helical tobacco mosaic virus (TMV) resulting in a near-atomic resolution cryo-EM map (Sachse et al., 2007).

The most commonly studied helical specimens suitable for 3-D image reconstruction are built from proteins or protein–nucleic acid complexes with molecular weight (MW) ranges between 5 and 100 kDa. While in the case of single-particle cryo-EM, high-resolution structure determination is currently limited to a MW larger than 90 kDa (Bai et al., 2015; Merk et al., 2016), for helical assemblies no such MW limit of the individual unit exists. The reason is that small isolated single particles give rise to limited image contrast in vitreous ice, as opposed to the helical units that are packed regularly to form large macromolecular assemblies and thus produce high contrast when imaged by cryo-EM. Together with the fact that molecular conformations are significantly restrained in the helical lattice, this was one of the main reasons why helical structures could be determined to near-atomic resolution even before the age of direct electron detectors. In addition, the maps often possess a more homogeneous resolution distribution across the molecule (Fromm, Bharat, Jakobi, Hagen, & Sachse, 2015; Gutsche et al., 2015) when compared with isolated single-particle structures (Hoffmann et al., 2015; Voorhees, Fernandez, Scheres, & Hegde, 2014). Proteins forming helical assemblies cover a unique spectrum of macromolecules that cannot be easily studied by pure single-particle methods and often represent biologically relevant assembly states thus making helical image analysis an important part of the cryo-EM technique.

With the superior image quality obtainable with the advent of direct detector devices, these optimized image processing methods became significantly more powerful. The first structures determined at near-atomic resolution with data recorded on direct electron detectors were from single-particle samples of the ribosome and the 20S proteasome (Bai, Fernandez, McMullan, & Scheres, 2013; Li et al., 2013). Since then an avalanche of single-particle near-atomic resolution structures have been published. Similarly, a series of near-atomic resolution structures have become available for helical specimens using single-particle-based image processing workflows. Notably, maps were determined at 3.4 Å resolution from images of rigid

TMV (Fromm et al., 2015), at 4.3 Å of the nucleocapsid from measles virus (Gutsche et al., 2015), and at 4.0 Å of the hypothermophile SIRV2 (*Sulfolobus islandicus* rod-shaped virus 2) (DiMaio et al., 2015). In addition, more flexible helical structures of actin (von der Ecken et al., 2015) or actin-like ParM cytoskeletal filaments (Bharat, Murshudov, Sachse, & Löwe, 2015) and of MAVS filaments involved in signaling (Wu et al., 2014) were determined at 4.3, 3.7, and 3.6 Å resolution, respectively.

Helical macromolecular assemblies can be found in many contexts in biology where they often represent functionally critical assembly states. Depending on how close the helical units are positioned to the helix axis (Sachse, 2015), they can either give rise to filamentous rods like actin or tubular hollow cylinders like microtubules, viruses, and membrane tubules. The treatment of helices as single-particle-like segments has become the predominant way of determining structures from such assemblies. The real-space reconstruction approach made flexible helical specimens amenable to cryo-EM studies, which had previously resisted structure determination, and also produced a series of near-atomic resolution structures from well-ordered helical specimens. The foundations and principles of classical helical theory (Diaz, Rice, & Stokes, 2010; Moody, 2011; Stewart, 1988) as well as real-space helical reconstructions (Egelman, 2010; Sachse, 2015) have been discussed in-depth in numerous excellent reviews. Therefore, in this chapter, we will focus on describing how more recently the combination of hardware developments with optimized procedures have made segmented helical reconstruction a productive and powerful branch of cryo-EM structure determination.

2. ARCHITECTURE OF HELICAL ASSEMBLIES

A helical assembly can be described by geometric transformations that relate the units by a shift Δz along the helix axis and rotation $\Delta \varphi$ around the helix axis. In the physical helix, each packing unit interacts with identical neighbors to form the polymer. An equivalent way of describing helical symmetry is to use the pitch (P) of a generating helix and number of units per turn (N) required to revolve 360 degrees around the helical axis (Fig. 1A). These transformations are analogous to the x and y shifts required to relate the unit cells present in a 2-D crystal. Hence, a helix can also be understood as a 2-D lattice that is rolled up into a cylinder. Characteristically, when projection images of 2-D crystals are analyzed in the Fourier domain, they give rise to discrete diffraction spots. Similarly, when the

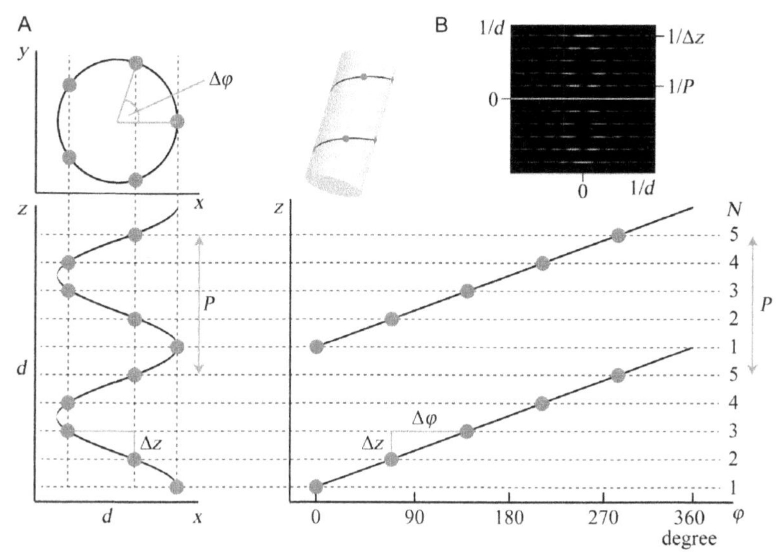

Fig. 1 Helical symmetry. (A) Helical symmetry can be unambiguously described by rise (Δz) and rotation ($\Delta\varphi$) per subunit (*red* (*gray* in the print version) *dots*). It is also useful to define the number N of subunits per turn ($N=2\pi/\Delta\varphi$) and the pitch P ($P=N\Delta z$) along the generating helix. *Left*: Projection of helical units from a helix *top* and *side view*. *Right*: Unrolled helical lattice with corresponding helical units from the left helix. (B) Fourier amplitude spectrum from a helical projection consists of stacked layer lines representing helical lattice lines. Layer lines are spaced apart by the reciprocal pitch distance $1/P$.

projection image of a helical crystal is converted into the Fourier domain, the amplitude spectrum is composed of a set of stacked layer lines (Fig. 1B). The characteristic position and shape of the set of layer lines represent the helical lattice and thereby encode the particular helical repeat of the assembly. Thus, an essential part of the structure determination procedure is to decompose the Fourier pattern in a process called indexing (see later).

3. SEGMENTED HELICAL IMAGE ANALYSIS

Helical specimens possess one critical advantage over other EM specimens in that a single helix already provides many different orientational views of the helical unit. Therefore, helical assemblies are ideally suited for reconstruction of a 3-D volume from a small number of images. This property was exploited by classical helical reconstruction methods in times of scarce computational resources. Helices were excised from digitized micrographs and further analyzed in Fourier space. Due to the helical organization, the Fourier amplitude spectra consist of a series of discrete stacked layer lines. Any further image processing such as correction of the contrast transfer function (CTF) and averaging could be reduced to these layer lines. Layer lines correspond mathematically to superimposed Bessel functions (Stewart, 1988) that represent periodic helical lattice lines in real space (Box 1). In order to maximize resolution and signal, long and straight particles were analyzed. As any deviation from an ideal helical arrangement resulted in attenuation of higher resolution amplitude and phase data, images were straightened or unbent by computational methods. Although successful for some well-ordered filaments and tubules (Jeng, Crowther, Stubbs, & Chiu, 1989; Toyoshima & Unwin, 1988), this procedure had limitations due to long-range deviations from ideal helices present in many helical specimens. In order to overcome this limitation, Unwin and colleagues extended the Fourier-based image processing procedure to break up helices into shorter segments in an effort to take advantage of the intact short-range helical order (Beroukhim & Unwin, 1997). This method generated several high-resolution image reconstructions up to 4 Å in resolution (Unwin, 2005; Yonekura et al., 2003).

BOX 1 Bessel functions and the Fourier transform of a helix

Fourier transforms of projections from 2-D crystals give rise to discrete reflections, whereas the transforms of helical specimens show layer lines. Discrete 2-D crystal reflections correspond to δ functions and helical layer lines are the result from the convolution of δ functions with a Bessel function due to the cylindrical shape of the helical lattice. The resulting functions are also called cylinder functions as in real space each Bessel function represents a wave revolving on the cylinder surface. Due to the angular periodicity on the cylinder, Bessel functions are limited to integer orders and the number of wave peaks corresponds to the Bessel order. A helical lattice is generated by superposition of many such helical waves.

Mathematically, Bessel functions $J_n(x)$ are reminiscent of oscillating cosine or sine functions that decay with increasing x. For Bessel orders $n = 1, 2, 3$, the primary maxima are located at $x = 1.8, 3.1, 4.2$, so as the Bessel order increases the primary peak moves away from the meridian (Fig. B.1A). In the Fourier transform, each Bessel function contribution depends on the helical radius r and reciprocal radius R.

$$J_n(2\pi^* R^* r)$$

BOX 1 Bessel functions and the Fourier transform of a helix—cont'd

Hence, the helical radius r of a cylindrical wave affects the position of the corresponding primary maximum, ie, smaller helical radii move the primary peak away from the meridian (Fig. B.1B). Each layer line thus represents the superposition of Bessel function contributions arising from features at different cylindrical radii in the structure. In Fourier transforms of helices excised from electron micrographs, interpretation is complicated as the amplitudes are modulated by the contrast transfer function of the microscope and high resolution layer lines fade due to contrast loss and disorder of the specimen.

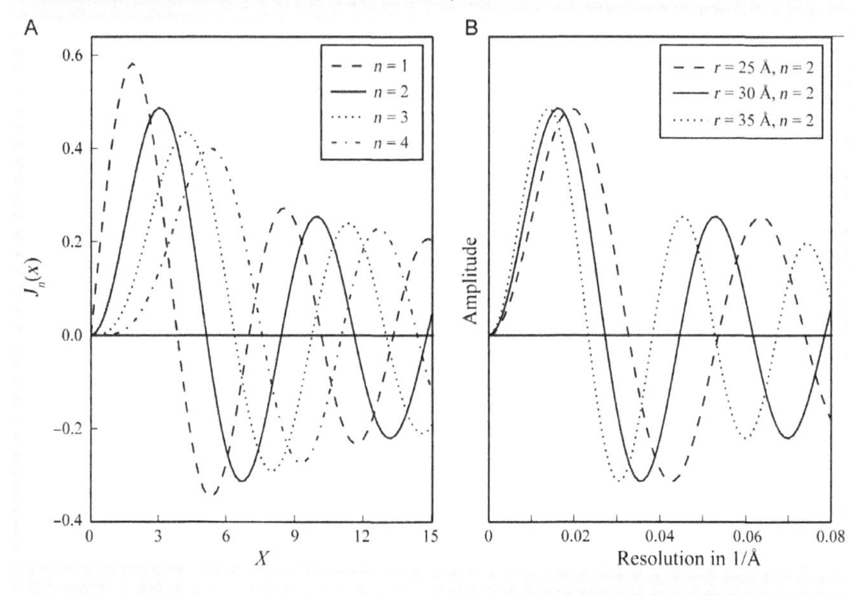

Different strategies have been used to extract segments along the helical path in steps from 10 to 100 s of Å or equal to the size of the segment (Fig. 2A). In principle, the smaller the step size the better helical bending can be compensated at the cost of increasing the data set size. As the step size is an important parameter to consider in order to make use of all helically related views, it is useful to choose it close to a multiple of the helical rise as described in the symmetrization step of the 3-D structure determination below (Sachse et al., 2007). For 2-D analysis, averaging the power spectra of in-plane rotated segments from multiple particles is an efficient way of inspecting helical order in Fourier space (Fig. 2B). While this approach

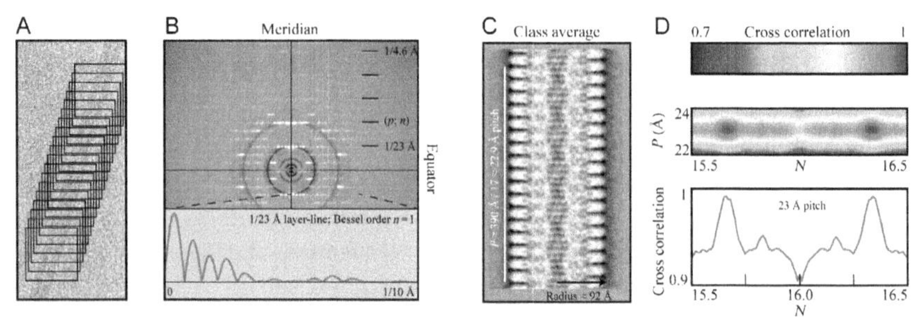

Fig. 2 2-D image analysis for segmented helical specimens. (A) Segmentation of image of TMV particle into overlapping segments. (B) Fourier analysis from sum of power spectra from overlapping segments of TMV. Layer lines can be indexed by assignment of position p and Bessel order n and thus helical symmetry can be derived. The uppermost highest resolution layer line corresponds to 1/4.6 Å. The *inset* shows the intensities of the 1/23 Å layer line with Bessel order $n=1$. (C) Real-space class averages can reveal gross morphological features such as helical outer radius r and pitch P of the generating helix on which all the subunits lie. (D) A grid of possible helical symmetry pairs can be assessed by comparing either class averages or power spectra from the experimental images with the reprojections of the 3-D model. *Top bar* shows the intensity scale for the crosscorrelation value, *middle bar* shows the crosscorrelation as a function of N, and *lower graph* shows trace through central bar. *Note*: number of units per turn $N=15.66$ and $N=16.34$ represent ambiguous solutions, while $P=23.0$ Å, $N=16.34$ is the correct symmetry for TMV. (See the color plate.)

enables fast analysis of many images into a merged data set, it loses the relevant phase information for further Fourier analysis. This information can be retrieved, however, by segment classification in real space and subsequent conversion to Fourier space. In comparison with the traditional Fourier analysis on rectangular boxes from individual particles, these class averages provide much better signal, thus enabling more detailed Fourier analysis.

In Fourier spectra it is useful to define the central horizontal line parallel to the layer lines as the equator and the central line perpendicular to this as the meridian (Fig. 2B). The high signal-to-noise views in class averages enable detailed analysis of principal helical features such as the outer helical radius (r) and pitch (P) (Fig. 2C). Furthermore, class averages can be Fourier transformed, and the resulting layer lines including amplitudes and phases can be analyzed. In order to determine the helical lattice in real space, the layer line pattern needs to be indexed. For this purpose, every layer line has to be assigned its position (p) in reciprocal Å and the Bessel order (n) in integer values. First, the layer line position (p) corresponds to the reciprocal distance from the equator. Second, the layer line order (n) can be derived from the distance of the first Bessel amplitude peak from the meridian and the helical radius (*r*) of the assembly (Box 1). In addition, the phase differences of 0 or 180 degrees between left and right side of the Fourier spectrum can be used to restrict the parity of Bessel order, ie, to clarify whether the Bessel order is even or odd, respectively. Once this position/order pair has been determined for all layer lines, the helical real-space lattice can be constructed thus representing the geometric transformations that specify the helical symmetry.

It should be noted, however, due to uncertainty in helical radius (*r*), unknown out-of-plane tilt and amplitude decay from long-range disorder and other imaging imperfections in the micrographs, the determination of Bessel order (*n*) can be problematic and give rise to ambiguous solutions of the helical lattice (Desfosses, Ciuffa, Gutsche, & Sachse, 2014; Egelman, 2010, 2014; Sachse, 2015). Alternatively, other approaches that use real-space information can be used to obtain symmetry estimates. First, the class averages can be used to reconstruct the initial 3-D models as a single helical view already contains the orientations of many helical units. Although not free of helical ambiguity, the maximum crosscorrelation of classes with reprojections of such 3-D initial models can also yield a set of helical parameters (Desfosses et al., 2014) (Fig. 2D). Second, subtomogram-based mapping of helical units on specimens works in tomograms (Bharat et al., 2012; Skruzny et al., 2015) thereby bypassing the helical ambiguity that arises from the 2-D helical projections. In most cases, the determined set of helical parameters present initial estimates that will need to be further refined (see later).

4. ITERATIVE HELICAL STRUCTURE REFINEMENT

Single-particle helical approaches make use of common projection matching routines (Penczek et al., 1994) by iterating through cycles of projection, alignment, and reconstruction until convergence with the specific addition of a helical symmetry imposition step. In this procedure, as in any single-particle processing scheme the critical step is to determine the orientation parameters including three Euler angles Φ, Θ, Ψ, and x/y position of the segment required for 3-D reconstruction (Fig. 3). Initially, after determining the CTF parameters from the micrographs, long particles are selected interactively and subsequently extracted by segmenting them into a set of overlapping squares. In contrast to single particles, elongated helices embedded in vitreous ice do not assume random orientations as their orientations vary only around their azimuth with little out-of-plane tilt. Although top

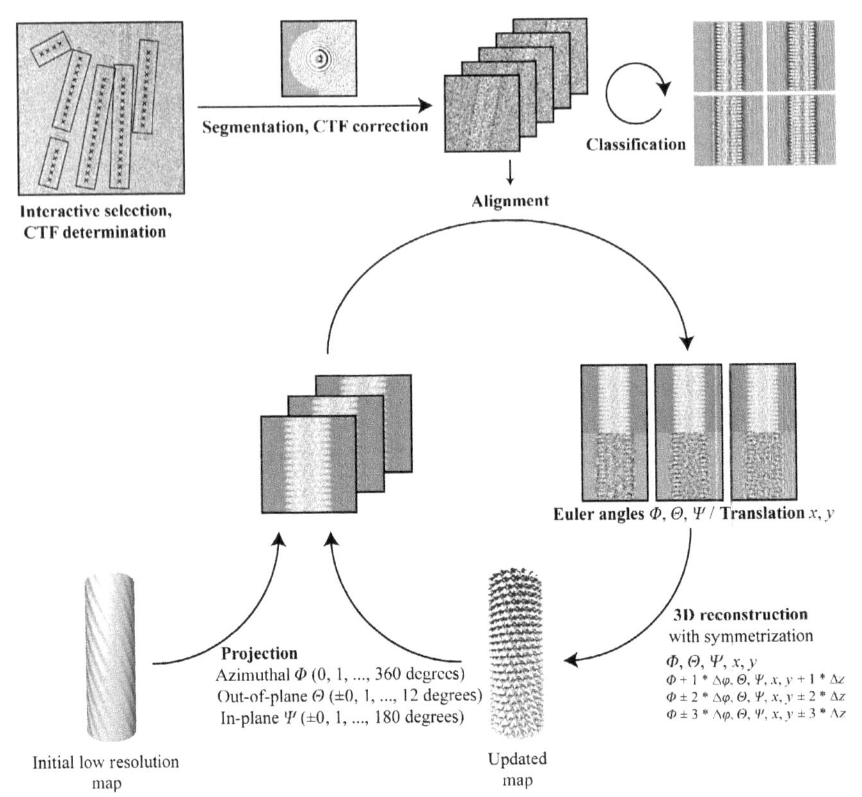

Fig. 3 3-D image processing workflow for single-particle-based helical reconstruction starting from raw micrographs finally leading to a 3-D EM density map. Segments are extracted and simultaneously convolved with the respective contrast transfer function. They are crosscorrelated with projections from the low-resolution model, starting with a cylinder, and thereby assigned a set of three Euler angles Φ, Θ, and Ψ and corresponding *x/y* translations. Subsequently, using these parameters, the segments are reconstructed including a symmetrization step that adds a series of symmetry related views into the 3-D volume. This projection matching cycle is iterated until convergence.

views of short helices occasionally exist and can be helpful for diagnostic purposes, they are not needed for 3-D reconstruction as azimuthal views suffice to generate isotropic volumes. The projections from a low-resolution starting model using the limited set of azimuthal and out-of-plane tilt angles are matched with the experimental images. For this purpose, the segments are convolved with the CTF in order to correct the phases and weight the amplitudes according to the expected signal imposed by the underfocus. Crosscorrelation-based matching assigns the three Euler angles Φ, Θ, Ψ, and x/y position for each segment required for 3-D reconstruction. To make use of all helically related views that are located between adjacent segments, duplicates of the aligned segments are generated and adjusted by multiples of the helical rise along the helix and by multiples of the helical rotation of the matched azimuthal angle. The set of helically related views is now included in the final 3-D reconstruction. Depending on the helical rise and the segmentation step size, this step can lead to an enormous increase in included views. For example, for TMV with a helical rise of 1.408 Å, a segmentation step of 70 Å is typically used, which results in the inclusion of an additional 50 views for each segment and is one of the reasons why ~4 Å resolution structures can be obtained with very small data sets from as few as 15 helical particles (Fromm et al., 2015). Obviously this is one of the most powerful features of helical reconstruction. It should be added that such a strong data multiplication bears the risk that imposition of incorrect helical symmetry will result in incorrect structures despite the fact that the refinement procedure appears stable and converges, thus making validation of the determined structure essential (see later). Subsequently, the reconstructed 3-D volume is deconvolved with the corresponding CTF average squared from all segments equivalent to a Wiener filter to correct for the weighted amplitudes (Böttcher et al., 1997; Grigorieff, 1998; Sachse et al., 2007). By analogy to single-particle refinement, the sequence of projection, alignment, and 3-D reconstruction including symmetrization is now iterated until the angular assignments converge, and the structure does not improve further in resolution.

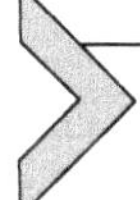

5. HIGH-RESOLUTION HELICAL IMAGE RECONSTRUCTIONS

Although treating helices as separate single particles has a series of advantages outlined earlier, the procedure also discards information on the continuity of helices. As the determination of single-particle orientation parameters is generally error-prone due to the low signal-to-noise ratio, the helical continuity can be used as a way to monitor accuracy of alignment. For example, many helical assemblies display polarity as helical units propagate in one direction. It is therefore impossible for adjacent segments to possess in-plane rotation angles differing by 180 degrees. Similarly, consecutive segments should possess closely related shifts perpendicular to the helix axis provided that the helix is intact. These dependencies of parameters within a helix can be used as exclusion criteria to remove poorly aligning segments from the data set. Either these segments are severely blurred by beam-induced motion, or they exhibit true distortions of the helical structure due to growth defects or mechanical rupture during plunge freezing. At the same time, monitoring the shift along the helix vs the azimuthal rotation provides additional information as to whether the helical symmetry is obeyed or has been correctly assigned. The critical factor in determining high-resolution structures for helical specimens as well as single particles remains the accurate determination of the five orientation parameters. In principle, the larger the molecular weight that is being projected by the electron beam, the more features giving rise to image contrast will be obtained yielding more accurate alignment estimates. For example, for the well-ordered specimen, TMV, segment sizes between 350 and 700 Å give rise to near-atomic resolution reconstructions. Nevertheless, large molecular assemblies tend to be more flexible and therefore lose long-range structural order. As a result, smaller segments are more appropriate in compensating deviations from ideal helices at the expense of the alignment accuracy. With the improvement of direct electron detectors underway, it can be expected that even smaller segments of more flexible specimens will have sufficient signal for the determination of high-resolution structures.

Another important factor for determining high-resolution cryo-EM structures from helical specimens is the accuracy of the helical symmetry determination. First, different helical symmetries might coexist within a single particle or across different particles within the same sample. In such cases structure refinement imposing a single symmetry will give rise to incorrect models. Therefore, it is essential to separate the populations of different symmetries through classification or multimodel refinement, ie, by offering sets of projections corresponding to multiple symmetries (Skruzny et al., 2015). Second, due to the strong imposition of helically related views, the helical parameters should be as close as possible to the true symmetry, which can be assessed by comparing the sum of power spectra from in-plane rotated segments with the sum of the power spectra from the 3-D reconstruction. In fact, this approach can be used to systematically optimize symmetry parameters in a grid search (Desfosses et al., 2014; Korkhov, Sachse, Short, & Tate, 2010; Low, Sachse, Amos, & Löwe, 2009). Although helical ambiguity cannot be resolved by this comparison (Desfosses et al., 2014), the consistency between experimental and model power spectrum is still an essential prerequisite for the correctness of the computed structure.

Various software implementations exist, which make use of segmented helical reconstruction. Apart from early developments based on classical helical reconstruction, which primarily worked in Fourier space (Toyoshima, 2000), most of the current solutions now make use of single-particle real-space image processing. Initially, the single-particle software suite SPIDER was used to process helical segments including helical volume symmetrization (Egelman, 2000; Jiménez et al., 1999). Further developments included CTF correction and improved 3-D reconstruction revealed the potential for high-resolution helical processing (Sachse et al., 2007; Sindelar & Downing, 2007). At the same time, the emergence of high-performance computing, bundling 1000s of CPUs into computer clusters, has made this approach significantly more powerful in the last decade as very large data sets with 100,000s of segments can be processed within a few days. One of the most commonly used implementations is the IHRSR procedure (Egelman, 2007). In the meantime, other dedicated software packages have been developed such as FREALIX, SPRING, and Helicon (Behrmann et al., 2012; Desfosses et al., 2014; Rohou & Grigorieff, 2014) all of which use helical restraints derived from the assembly architecture to monitor and improve orientation determination. More recently, maximum likelihood-based algorithms implemented in the package RELION (Scheres, 2012) have also been adopted for helical processing (Clemens, Ge, Lee, Horwitz, & Zhou, 2015). Currently, SPRING is the only single-particle-based software suite that includes tools to facilitate the critical indexing procedure and helical symmetry determination from segmented specimens. In the future, it would be helpful to develop automated or semiautomated procedures to overcome this bottleneck. Once the symmetry has been determined, a series of processing solutions exist, all of which have proven powerful, to generate high-resolution cryo-EM structures of helical specimens (Bharat et al., 2015; Clemens et al., 2015; Egelman, 2015; McCullough et al., 2015; von der Ecken et al., 2015).

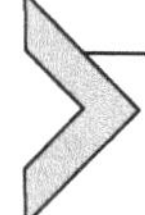

6. THE USE OF DIRECT ELECTRON DETECTORS FOR HELICAL STRUCTURES

The physical foundations of design and performance of direct detectors have been described and reviewed elsewhere (Faruqi & Henderson, 2007; McMullan, Chen, Henderson, & Faruqi, 2009) (see chapter "Direct Electron Detectors" by Henderson). In essence, due to the improved detective quantum efficiency of the new detectors in particular at higher spatial frequencies, the ability to capture signal has been boosted by factors of two in comparison with traditional EM detectors such as CCD camera and film (McMullan,

Faruqi, Clare, & Henderson, 2014). The first applications of these detectors to biological specimens led to impressive gains in resolution (Bai et al., 2013; Li et al., 2013) on previously studied single-particle structures such as the ribosome and the 20S proteasome. In order to assess directly the effect of different detectors on the raw image data for helical specimens, we compared TMV layer lines in averaged power spectra acquired on CCD camera, film, and direct electron detectors. The sum of ~2400 TMV power spectra segmented from a total length of 21.6 μm of viruses was averaged normal to the meridian (Fig. 4A). It is clear from examination of the layer line intensities that only for

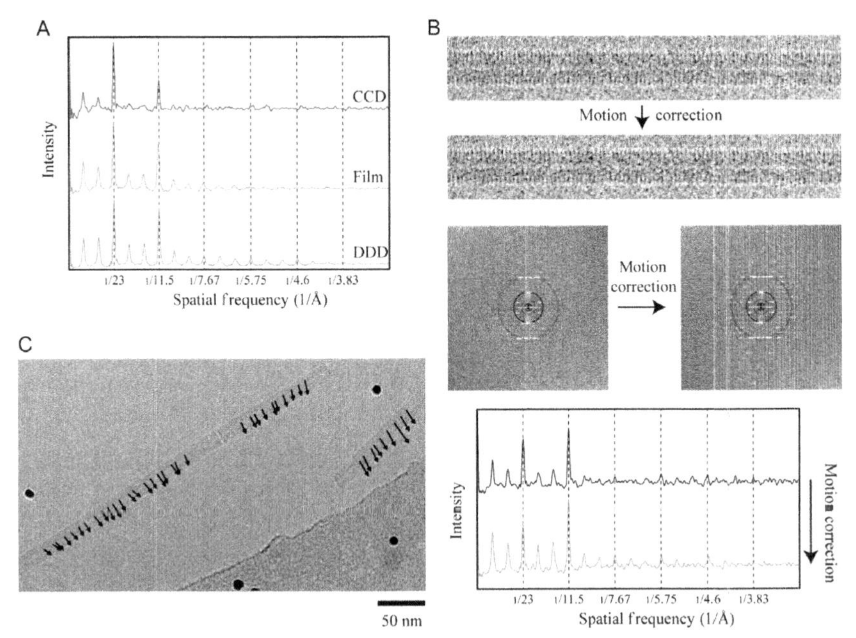

Fig. 4 Impact of direct electron detectors on image quality of helical specimens. (A) Comparison of summed power spectra calculated from TMV recorded with CCD (unpublished), film (Sachse et al., 2007), or direct electron detector (Fromm et al., 2015), respectively. In each case the trace represents the projection of the power spectrum onto the meridian, so the peaks show the relative strengths of the various layer lines. Power spectra sums were averaged from 2400 segments in all three cases. Note that high resolution features (<5 Å) are only preserved in the spectrum from data acquired with direct electron detectors. (B) Effect of micrograph-based motion correction on raw data. Improvement of data quality is visible in real space and in particular when comparing the layer line intensities beyond 1/11.5 Å before (*top*) and after (*bottom*) motion correction as evident from the power spectra and the collapsed power spectra. (C) Segment-based motion correction accounts for local movements during the exposure. *Arrows* indicate the shift of individual segments from the start to the end of the acquisition and are enlarged 30 × for visual purposes. Helical particles show correlated movements between helical segments.

data from the direct electron detector layer lines extend visibly up to 1/3.8 Å, whereas for film or CCD camera, the layer lines are limited to 1/6 or 1/11 Å, respectively. We also determined structures of TMV from images acquired in the same electron microscope with two commonly used direct electron detectors (Gatan K2 Summit and FEI Falcon II) in order to compare their performance (Fromm et al., 2015). We found that for very large structures such as TMV equally small data sets from either camera gave rise to ~4 Å reconstructions of comparable quality.

Due to the fast readout, the technology of direct electron detectors has opened up the possibility to record the classical low-dose exposure in a series of frames. By collecting movies of ice-embedded icosahedral viruses the imaging process could be studied in more detail (Brilot et al., 2012; Campbell et al., 2012). Grigorieff and coworkers showed that computational frame alignment could compensate for beam-induced movement and stage drift. Based on improved image detail (Brilot et al., 2012), recovery of high-resolution Thon rings (Li et al., 2013) or layer line intensities of periodic specimens like TMV (Fig. 4B), alignment of micrograph frames significantly boosts the transfer of high-resolution information (see chapter "Processing of Cryo-EM Movie Data" by Rubinstein). In addition to micrograph-based correction, per particle correction of large icosahedral viruses or segment correction of TMV has been shown to further improve the quality of the cryo-EM map (Campbell et al., 2012; Fromm et al., 2015). For icosahedral viruses, it was shown that the particles from different parts of the micrograph can move differently (Brilot et al., 2012). Elongated and rigid helical specimens such as TMV, segmented from a single virus spanning large parts of the micrograph, tend to move in correspondence (Fig. 4C) (Fromm et al., 2015). Here, the continuity of the helical particle imposes a strong physical restraint. Thus, movements can either be directly traced for large viruses or adjacent segments from helical specimens can be computationally averaged to move in correspondence. By contrast, smaller particles such as 20S proteasomes are too small to be individually traced by intraframe alignment, and therefore subregions of the micrograph have been aligned (Li et al., 2013) or the tracks from neighboring particles have been computationally restrained to move in patches (Scheres, 2014). The series of frames also enabled systematic investigation of the effect of electron dose and radiation damage on the structure by limiting the frames included in 3-D reconstruction to minimize radiation damage (Li et al., 2013). Examination of such high-resolution maps (Allegretti, Mills, McMullan, Kühlbrandt, & Vonck, 2014; Bartesaghi, Matthies, Banerjee, Merk, & Subramaniam, 2014) or reconstructions from individual frames (Fromm et al., 2015) revealed that increasing dose degrades the EM density of negatively charged side chains first in analogy to radical-mediated oxidation observed in X-ray crystallography experiments (Weik et al., 2000).

7. VALIDATION OF HELICAL STRUCTURES

The fact that single-particle-based helical reconstruction requires very few images to reconstruct high-resolution cryo-EM structures stems mainly from the efficient imposition of helical symmetry. While this makes helical reconstruction very powerful, if the helical symmetry parameters are incorrectly determined it also poses a significant risk in computing incorrect structures due to the strong reference bias (Stewart & Grigorieff, 2004) from comparing reprojections of symmetrized volumes with noisy helical segments. Therefore, convergence of refinement parameters must not be considered sufficient to assess the correctness of the structure. Even the consistency between power spectra obtained from the experimental segments and reprojections of the 3-D model is not sufficient to resolve the issue of helical ambiguity (Desfosses et al., 2014). As a larger number of atomic resolution maps are determined, structural features of a cryo-EM map can serve as evidence of a correct structure without the need of further experiments. For such high-resolution maps common features include the 5.3 Å α-helical pitch, the 4.8 Å β-strand distance and large side-chain densities emanating from the peptide backbone, which can be discerned from the maps. At subnanometer resolution, secondary structures of α-helices are visible as folds of tubular densities, and β-sheets are discernible and may be compared with reference structures. At lower resolution, additional independent data should be considered to validate the helical symmetry parameters. Subunit mapping in electron tomograms can serve as a direct tool to read out helical architecture while avoiding the pitfalls of helical ambiguity from 2-D helical projections (Bharat et al., 2012; Skruzny et al., 2015). Thus far, applications to large tubular assemblies have been reported, whereas determination of helical symmetry parameters from thin filamentous assemblies remains to be demonstrated. An alternative approach is the determination of mass per length of a helical assembly by scanning transmission electron microscopy (Wall & Hainfeld, 1986) as it restrains the number of units that make up a complete helical turn (Ciuffa et al., 2015). Due to specific hardware requirements, currently only Brookhaven National Laboratory can perform these experiments. An alternative approach to directly measure the mass-related scattering contribution of helices in unstained dehydrated specimens has been proposed by tilted beam EM (Chen, Thurber, Shewmaker, Wickner, & Tycko, 2009). A similar restraint can be also derived if occasional top views of class averages are available to directly reveal the number of units required for a turn. Due to the intricacies of Fourier or real-space analysis of helical assemblies, the bottleneck for rapid structure determination still remains the correct assignment of helical symmetry, which can often be only achieved by a supervised and systematic trial and error procedure.

8. CONCLUSION

Due to technological advances, the progress in structure determination of helical specimens follows the same trend as pure single-particle cryo-EM. Before the introduction of direct electron detectors, a few very well-ordered helical specimens were studied at near-atomic resolution (Ge & Zhou, 2011; Sachse et al., 2007; Unwin, 2005; Yonekura et al., 2003). Many tubules and filaments exhibiting structural flexibility and deviation from ideal helical periodicity, however, did not reach this resolution (Fujii, Iwane, Yanagida, & Namba, 2010; Low et al., 2009). Since the introduction of direct electron detectors, the number of helical structures with a resolution better than 5 Å ($n=36$ EMDB entries 2014/15) has increased at a similar rate as pure single-particle specimens ($n=182$ EMDB entries 2014/15). Segmented helical reconstruction has become the method of choice to deal with helical structures regardless of whether they are more flexible filaments or well-ordered helical viruses (Fig. 5) (Agirrezabala et al., 2015; Bharat et al., 2015; Kudryashev et al., 2015; von der Ecken et al., 2015). In particular for flexible filaments biochemical stability and conformational homogeneity is still critical to obtain high-resolution reconstructions. For example, actin-like ParM filaments could be resolved to 4.3 Å resolution but only in the presence of a transition state analog AMPPNP, whereas other nucleotides such as ATP and ADP limited the resolution to 7.5 or 11 Å due to conformational flexibility (Bharat et al., 2015). Similarly, the structure of F-actin could be resolved to 3.7 Å, whereas the much larger tropomyosin complex was limited to 6.5 Å resolution (von der Ecken et al., 2015). The structure of the type VI secretion sheath recently determined at 3.5 Å resolution exhibits high assembly stability due to the tight protein interaction surface between VipA and VipB (Kudryashev et al., 2015). Helical viruses such as TMV, SIRV2, barley stripe mosaic virus, and pepino mosaic virus, which enwrap genomic

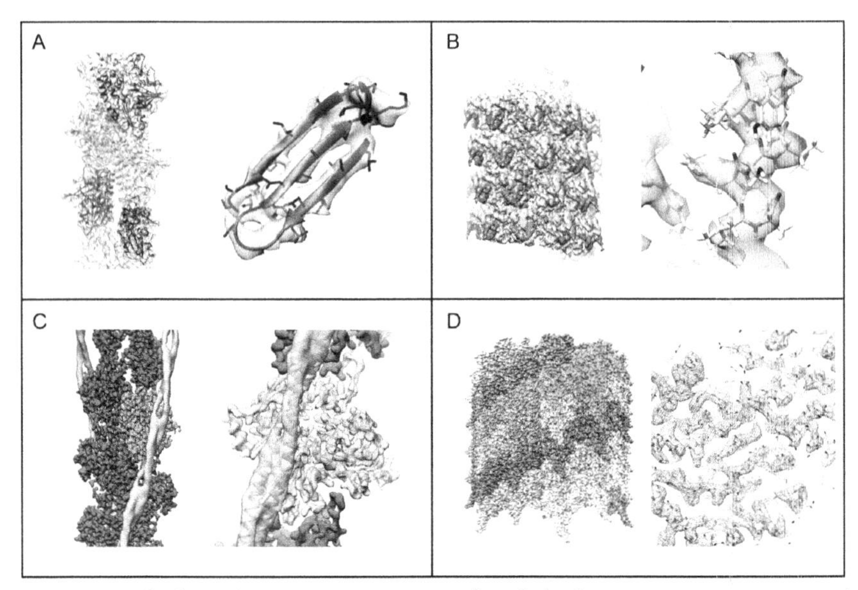

Fig. 5 Recent high-resolution cryo-EM structures from helical specimens using segmented helical reconstruction. (A) EMD-2850: 4.3 Å structure of the actin-like ParM protein bound to AMPPNP (Bharat et al., 2015). (B) EMD-3236: 3.9 Å structure of the Pepino Mosaic Virus (Agirrezabala et al., 2015). (C) EMD-6124: 3.7 Å structure of the F-actin-tropomyosin complex (von der Ecken et al., 2015). (D) EMD-2699: 3.5 Å structure of the VipA/VipB contractile sheath of the type VI secretion system (Kudryashev et al., 2015). (See the color plate.)

DNA, give rise to very stable helical assemblies as one of their functions is to protect the genome from the environment, which has enabled structure determination at resolutions between 3.3 and 4.5 Å (Agirrezabala et al., 2015; Clare et al., 2015; DiMaio et al., 2015; Fromm et al., 2015). The combination of the classical Fourier-based helical analysis with the versatility of the single-particle approach together with the improved detectors has made segmented helical image reconstruction a very powerful structural biology method that will continue to provide unique structural insights into these fundamental biological assemblies.

ACKNOWLEDGMENTS

We are grateful to Abul Tarafder and Martin Beck for critical reading and helpful suggestions on the manuscript. S.A.F. is supported by the EMBL International PhD Programme. We acknowledge image acquisition of discussed TMV data by Jamie Riches, James Chen, and Wim Hagen from CCD camera, film, and direct electron detectors, respectively. We thank Tanmay Bharat, Mikel Valle, Julian van der Ecken, and Misha Kudryashev for providing the images in Fig. 5.

REFERENCES

Adrian, M., Dubochet, J., Lepault, J., & McDowall, A. (1984). Cryo-electron microscopy of viruses. *Nature*, *308*, 32–36.

Agirrezabala, X., Méndez-López, E., Lasso, G., Sánchez-Pina, M. A., Aranda, M., & Valle, M. (2015). The near-atomic cryoEM structure of a flexible filamentous plant virus shows homology of its coat protein with nucleoproteins of animal viruses. *eLife*, *4*, e11795.

Allegretti, M., Mills, D. J., McMullan, G., Kühlbrandt, W., & Vonck, J. (2014). Atomic model of the F420-reducing [NiFe] hydrogenase by electron cryo-microscopy using a direct electron detector. *eLife*, *3*, e01963.

Bai, X.-C., Fernandez, I. S., McMullan, G., & Scheres, S. H. W. (2013). Ribosome structures to near-atomic resolution from thirty thousand cryo-EM particles. *eLife*, *2*, e00461.

Bai, X.-C., Yan, C., Yang, G., Lu, P., Ma, D., Sun, L., et al. (2015). An atomic structure of human γ-secretase. *Nature*, *525*, 212–217.

Bartesaghi, A., Matthies, D., Banerjee, S., Merk, A., & Subramaniam, S. (2014). Structure of β-galactosidase at 3.2-Å resolution obtained by cryo-electron microscopy. *Proceedings of the National Academy of Sciences of the United States of America*, *111*, 11709–11714.

Behrmann, E., Tao, G., Stokes, D. L., Egelman, E. H., Raunser, S., & Penczek, P. A. (2012). Real-space processing of helical filaments in SPARX. *Journal of Structural Biology*, *177*, 302–313.

Beroukhim, R., & Unwin, N. (1997). Distortion correction of tubular crystals: Improvements in the acetylcholine receptor structure. *Ultramicroscopy*, *70*, 57–81.

Bharat, T. A. M., Davey, N. E., Ulbrich, P., Riches, J. D., de Marco, A., Rumlova, M., et al. (2012). Structure of the immature retroviral capsid at 8 Å resolution by cryo-electron microscopy. *Nature*, *487*, 385–389.

Bharat, T. A. M., Murshudov, G. N., Sachse, C., & Löwe, J. (2015). Structures of actin-like ParM filaments show architecture of plasmid-segregating spindles. *Nature*, *523*, 106–110.

Böttcher, B., Wynne, S. A., & Crowther, R. A. (1997). Determination of the fold of the core protein of hepatitis B virus by electron cryomicroscopy. *Nature*, *386*, 88–91.

Brilot, A. F., Chen, J. Z., Cheng, A., Pan, J., Harrison, S. C., Potter, C. S., et al. (2012). Beam-induced motion of vitrified specimen on holey carbon film. *Journal of Structural Biology*, *177*, 630–637.

Campbell, M. G., Cheng, A., Brilot, A. F., Moeller, A., Lyumkis, D., Veesler, D., et al. (2012). Movies of ice-embedded particles enhance resolution in electron cryo-microscopy. *Structure (London, England: 1993)*, *20*, 1823–1828.

Chen, B., Thurber, K. R., Shewmaker, F., Wickner, R. B., & Tycko, R. (2009). Measurement of amyloid fibril mass-per-length by tilted-beam transmission electron microscopy. *Proceedings of the National Academy of Sciences of the United States of America*, *106*, 14339–14344.

Ciuffa, R., Lamark, T., Tarafder, A. K., Guesdon, A., Rybina, S., Hagen, W. J. H., et al. (2015). The selective autophagy receptor p62 forms a flexible filamentous helical scaffold. *Cell Reports*, *11*, 748–758.

Clare, D. K., Pechnikova, E. V., Skurat, E. V., Makarov, V. V., Sokolova, O. S., Solovyev, A. G., et al. (2015). Novel inter-subunit contacts in barley stripe mosaic virus revealed by cryo-electron microscopy. *Structure (London, England: 1993)*, *23*, 1815–1826.

Clemens, D. L., Ge, P., Lee, B.-Y., Horwitz, M. A., & Zhou, Z. H. (2015). Atomic structure of T6SS reveals interlaced array essential to function. *Cell*, *160*, 940–951.

DeRosier, D., & Klug, A. (1968). Reconstruction of three dimensional structures from electron micrographs. *Nature*, *217*, 130–134.

Desfosses, A., Ciuffa, R., Gutsche, I., & Sachse, C. (2014). SPRING—An image processing package for single-particle based helical reconstruction from electron cryomicrographs. *Journal of Structural Biology*, *185*, 15–26.

Diaz, R., Rice, W. J., & Stokes, D. L. (2010). Fourier-Bessel reconstruction of helical assemblies. *Methods in Enzymology, 482*, 131–165.
DiMaio, F., Yu, X., Rensen, E., Krupovic, M., Prangishvili, D., & Egelman, E. H. (2015). Virology. A virus that infects a hyperthermophile encapsidates A-form DNA. *Science (New York, NY), 348*, 914–917.
Egelman, E. H. (2000). A robust algorithm for the reconstruction of helical filaments using single-particle methods. *Ultramicroscopy, 85*, 225–234.
Egelman, E. H. (2007). The iterative helical real space reconstruction method: Surmounting the problems posed by real polymers. *Journal of Structural Biology, 157*, 83–94.
Egelman, E. H. (2010). Reconstruction of helical filaments and tubes. *Methods in Enzymology, 482*, 167–183.
Egelman, E. H. (2014). Ambiguities in helical reconstruction. *eLife, 3*, e04969.
Egelman, E. H. (2015). Three-dimensional reconstruction of helical polymers. *Archives of Biochemistry and Biophysics, 581*, 54–58.
Faruqi, A. R., & Henderson, R. (2007). Electronic detectors for electron microscopy. *Current Opinion in Structural Biology, 17*, 549–555.
Fromm, S. A., Bharat, T. A. M., Jakobi, A. J., Hagen, W. J. H., & Sachse, C. (2015). Seeing tobacco mosaic virus through direct electron detectors. *Journal of Structural Biology, 189*, 87–97.
Fujii, T., Iwane, A. H., Yanagida, T., & Namba, K. (2010). Direct visualization of secondary structures of F-actin by electron cryomicroscopy. *Nature, 467*, 724–728.
Galkin, V. E., Wu, Y., Zhang, X.-P., Qian, X., He, Y., Yu, X., et al. (2006). The Rad51/RadA N-terminal domain activates nucleoprotein filament ATPase activity. *Structure/Folding and Design, 14*, 983–992.
Ge, P., & Zhou, Z. H. (2011). Hydrogen-bonding networks and RNA bases revealed by cryo electron microscopy suggest a triggering mechanism for calcium switches. *Proceedings of the National Academy of Sciences of the United States of America, 108*, 9637–9642.
Grigorieff, N. (1998). Three-dimensional structure of bovine NADH:ubiquinone oxidoreductase (complex I) at 22 A in ice. *Journal of Molecular Biology, 277*, 1033–1046.
Gutsche, I., Desfosses, A., Effantin, G., Ling, W. L., Haupt, M., Ruigrok, R. W. H., et al. (2015). Structural virology. Near-atomic cryo-EM structure of the helical measles virus nucleocapsid. *Science (New York, NY), 348*, 704–707.
Henderson, R., Baldwin, J. M., Ceska, T. A., Zemlin, F., Beckmann, E., & Downing, K. H. (1990). Model for the structure of bacteriorhodopsin based on high-resolution electron cryo-microscopy. *Journal of Molecular Biology, 213*, 899–929.
Henderson, R., Baldwin, J. M., Downing, K. H., Lepault, J., & Zemlin, F. (1986). Structure of purple membrane from halobacterium halobium: Recording, measurement and evaluation of electron-micrographs at 3.5 Å resolution. *Ultramicroscopy, 19*, 147–178.
Hoffmann, N. A., Jakobi, A. J., Moreno-Morcillo, M., Glatt, S., Kosinski, J., Hagen, W. J. H., et al. (2015). Molecular structures of unbound and transcribing RNA polymerase III. *Nature, 528*, 231–236.
Jeng, T., Crowther, R., Stubbs, G., & Chiu, W. (1989). Visualization of alpha-helices in tobacco mosaic virus by cryo-electron microscopy. *Journal of Molecular Biology, 205*, 251–257.
Jiménez, J. L., Guijarro, J. I., Orlova, E., Zurdo, J., Dobson, C. M., Sunde, M., et al. (1999). Cryo-electron microscopy structure of an SH3 amyloid fibril and model of the molecular packing. *The EMBO Journal, 18*, 815–821.
Korkhov, V. M., Sachse, C., Short, J. M., & Tate, C. G. (2010). Three-dimensional structure of TspO by electron cryomicroscopy of helical crystals. *Structure (London, England: 1993), 18*, 677–687.

Kudryashev, M., Wang, R. Y.-R., Brackmann, M., Scherer, S., Maier, T., Baker, D., et al. (2015). Structure of the type VI secretion system contractile sheath. *Cell, 160*, 952–962.

Li, X., Mooney, P., Zheng, S., Booth, C. R., Braunfeld, M. B., Gubbens, S., et al. (2013). Electron counting and beam-induced motion correction enable near-atomic-resolution single-particle cryo-EM. *Nature Methods, 10*, 584–590.

Low, H. H., Sachse, C., Amos, L. A., & Löwe, J. (2009). Structure of a bacterial dynamin-like protein lipid tube provides a mechanism for assembly and membrane curving. *Cell, 139*, 1342–1352.

McCullough, J., Clippinger, A. K., Talledge, N., Skowyra, M. L., Saunders, M. G., Naismith, T. V., et al. (2015). Structure and membrane remodeling activity of ESCRT-III helical polymers. *Science (New York, NY), 350*, 1548–1551.

McMullan, G., Chen, S., Henderson, R., & Faruqi, A. R. (2009). Detective quantum efficiency of electron area detectors in electron microscopy. *Ultramicroscopy, 109*, 1126–1143.

McMullan, G., Faruqi, A. R., Clare, D., & Henderson, R. (2014). Comparison of optimal performance at 300 keV of three direct electron detectors for use in low dose electron microscopy. *Ultramicroscopy, 147*, 156–163.

Merk, A., Barthesaghi, A., Banerjee, S., Falconieri, V., Rao, P., Davis, M. I., et al. (2016). Breaking cryo-EM resolution barriers to facilitate drug discovery. *Cell*, http://dx.doi.org/10.1016/j.cell.2016.05.040.

Miyazawa, A., Fujiyoshi, Y., & Unwin, N. (2003). Structure and gating mechanism of the acetylcholine receptor pore. *Nature, 423*, 949–955.

Moody, M. F. (2011). *Structural biology using electrons and X-rays*. San Diego, CA: Academic Press.

Penczek, P. A., Grassucci, R. A., & Frank, J. (1994). The ribosome at improved resolution: New techniques for merging and orientation refinement in 3D cryo-electron microscopy of biological particles. *Ultramicroscopy, 53*, 251–270.

Penczek, P., Radermacher, M., & Frank, J. (1992). Three-dimensional reconstruction of single particles embedded in ice. *Ultramicroscopy, 40*, 33–53.

Rohou, A., & Grigorieff, N. (2014). Frealix: Model-based refinement of helical filament structures from electron micrographs. *Journal of Structural Biology, 186*, 234–244.

Sachse, C. (2015). Single-particle based helical reconstruction—How to make the most of real and Fourier space. *AIMS Biophysics, 2*, 219–244.

Sachse, C., Chen, J. Z., Coureux, P.-D., Stroupe, M. E., Fändrich, M., & Grigorieff, N. (2007). High-resolution electron microscopy of helical specimens: A fresh look at tobacco mosaic virus. *Journal of Molecular Biology, 371*, 812–835.

Scheres, S. H. W. (2012). RELION: Implementation of a Bayesian approach to cryo-EM structure determination. *Journal of Structural Biology, 180*, 519–530.

Scheres, S. H. (2014). Beam-induced motion correction for sub-megadalton cryo-EM particles. *eLife, 3*, e03665.

Sindelar, C. V., & Downing, K. H. (2007). The beginning of kinesin's force-generating cycle visualized at 9-A resolution. *The Journal of Cell Biology, 177*, 377–385.

Skruzny, M., Desfosses, A., Prinz, S., Dodonova, S. O., Gieras, A., Uetrecht, C., et al. (2015). An organized co-assembly of clathrin adaptors is essential for endocytosis. *Developmental Cell, 33*, 150–162.

Stewart, M. (1988). Computer image processing of electron micrographs of biological structures with helical symmetry. *Journal of Electron Microscopy Technique, 9*, 325–358.

Stewart, A., & Grigorieff, N. (2004). Noise bias in the refinement of structures derived from single particles. *Ultramicroscopy, 102*, 67–84.

Toyoshima, C. (2000). Structure determination of tubular crystals of membrane proteins. I. Indexing of diffraction patterns. *Ultramicroscopy, 84*, 1–14.

Toyoshima, C., & Unwin, N. (1988). Ion channel of acetylcholine receptor reconstructed from images of postsynaptic membranes. *Nature*, *336*, 247–250.
Unwin, N. (2005). Refined structure of the nicotinic acetylcholine receptor at 4A resolution. *Journal of Molecular Biology*, *346*, 967–989.
von der Ecken, J., Müller, M., Lehman, W., Manstein, D. J., Penczek, P. A., & Raunser, S. (2015). Structure of the F-actin-tropomyosin complex. *Nature*, *519*, 114–117.
Voorhees, R. M., Fernandez, I. S., Scheres, S. H. W., & Hegde, R. S. (2014). Structure of the Mammalian ribosome-sec61 complex to 3.4 Å resolution. *Cell*, *157*, 1632–1643.
Wall, J. S., & Hainfeld, J. F. (1986). Mass mapping with the scanning transmission electron microscope. *Annual Review of Biophysics and Biophysical Chemistry*, *15*, 355–376.
Weik, M., Ravelli, R. B., Kryger, G., McSweeney, S., Raves, M. L., Harel, M., et al. (2000). Specific chemical and structural damage to proteins produced by synchrotron radiation. *Proceedings of the National Academy of Sciences of the United States of America*, *97*, 623–628.
Wu, B., Peisley, A., Tetrault, D., Li, Z., Egelman, E. H., Magor, K. E., et al. (2014). Molecular imprinting as a signal-activation mechanism of the viral RNA sensor RIG-I. *Molecular Cell*, *55*, 511–523.
Xu, C., Rice, W. J., He, W., & Stokes, D. L. (2002). A structural model for the catalytic cycle of Ca(2+)-ATPase. *Journal of Molecular Biology*, *316*, 201–211.
Yonekura, K., Maki-Yonekura, S., & Namba, K. (2003). Complete atomic model of the bacterial flagellar filament by electron cryomicroscopy. *Nature*, *424*, 643–650.

Cryo-Electron Tomography and Subtomogram Averaging

W. Wan, J.A.G. Briggs[1]
Structural and Computational Biology Unit, European Molecular Biology Laboratory, Heidelberg, Germany
[1]Corresponding author: e-mail address: john.briggs@embl.de

Contents

Methods in Enzymology, Volume 579
ISSN 0076-6879
http://dx.doi.org/10.1016/bs.mie.2016.04.014

Abstract

Cryo-electron tomography (cryo-ET) allows 3D volumes to be reconstructed from a set of 2D projection images of a tilted biological sample. It allows densities to be resolved in 3D that would otherwise overlap in 2D projection images. Cryo-ET can be applied to resolve structural features in complex native environments, such as within the cell. Analogous to single-particle reconstruction in cryo-electron microscopy, structures present in multiple copies within tomograms can be extracted, aligned, and averaged, thus increasing the signal-to-noise ratio and resolution. This reconstruction approach, termed subtomogram averaging, can be used to determine protein structures in situ. It can also be applied to facilitate more conventional 2D image analysis approaches. In this chapter, we provide an introduction to cryo-ET and subtomogram averaging. We describe the overall workflow, including tomographic data collection, preprocessing, tomogram reconstruction, subtomogram alignment and averaging, classification, and postprocessing. We consider theoretical issues and practical considerations for each step in the workflow, along with descriptions of recent methodological advances and remaining limitations.

1. INTRODUCTION

A growing number of publications using single-particle reconstruction approaches illustrate the increasing power of cryo-electron microscopy (cryo-EM) to resolve high-resolution structures (Bartesaghi et al., 2015; Fischer et al., 2015; Grant & Grigorieff, 2015). Single-particle approaches generally require the molecules of interest to be isolated and purified from their native environments. For many biological molecules including membrane-associated complexes, large cellular ultrastructures, and pleomorphic viruses—their relevant structural forms may only exist in their native environments. Even in cases where purification is technically feasible, removing biological molecules from their native environments may eliminate interesting structural features including transient binding partners, environment-specific conformations, and other contextual information. There is therefore a need for reconstruction methods that can be applied "in situ," that is, within complex, native environments.

Single-particle and related reconstruction approaches can often not be applied in situ because molecules "above and below" the particles of interest are superimposed in the cryo-EM projection image; this overlapping information confounds image analysis. In these cases, the method of choice is cryo-electron tomography (cryo-ET) combined with subtomogram averaging, which is able to determine high-resolution structures of macromolecules in complex, heterogeneous environments. For recent reviews of cryo-ET and subtomogram averaging the reader is referred to Briggs (2013), Förster, Han, and Beck (2010), Förster and Hegerl (2007), Lučić, Rigort, and Baumeister (2013), Schmid (2011), and Yahav, Maimon, Grossman, Dahan, and Medalia (2011).

In cryo-ET, the sample is rotated through a set of defined tilts, and at each tilt a projection image is collected. From this set of projection images, ie, a tilt-series, a 3D reconstruction of the field of view can be generated (a tomogram). Confounding densities that would otherwise overlap with the object of interest in individual projection images are separated out in the tomogram. The objects of interest can then be extracted in 3D, free of overlapping densities, and further processed using subtomogram averaging.

Subtomogram averaging is analogous to single-particle reconstruction, but with the key distinction that the particles are represented by three-dimensional volumes rather than two-dimensional projections. Like single-particle reconstruction, subtomogram averaging works by iteratively aligning and averaging images, or in this case volumes, of a large number of copies of the particle, thereby increasing the signal-to-noise ratio (SNR) and the resolution (Fig. 1). Subtomogram averaging is less widely applied than single-particle reconstruction. At the time of writing, 11% of entries in the EMDataBank are from subtomogram averaging, compared to 77% from single-particle reconstruction. This reflects a number of limiting factors including the difficulty in collecting a large number of tomograms

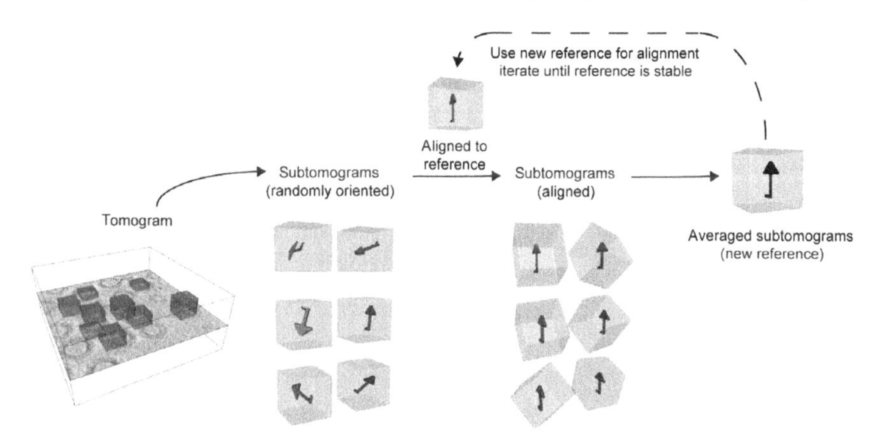

Fig. 1 Overview of subtomogram averaging workflow. Subtomograms are cubes extracted from the full tomogram; each subtomogram contains a randomly oriented copy of the molecule of interest. Subtomograms are aligned to the reference and a new reference is generated from the aligned particles. This process is iterated until the alignment converges to a stable reference. *Reproduced from Briggs, J. A. G. (2013). Structural biology in situ—The potential of subtomogram averaging.* Current Opinion in Structural Biology, 23, *261–267.*

suitable for high-resolution processing and the increased computational costs required for processing large 3D datasets. As such, where overlapping densities are not a limiting problem, single-particle reconstruction is the preferred method. We believe that the ability of subtomogram averaging to determine structures in situ will lead to it being even more widely applied in the future.

In this chapter, we seek to outline the general principles of subtomogram averaging and their distinctions with respect to single-particle approaches. We cover the complete workflow including tomographic data collection, raw data preprocessing, tomogram reconstruction, and subtomogram averaging (Fig. 2). It is still "early days" in subtomogram averaging and the

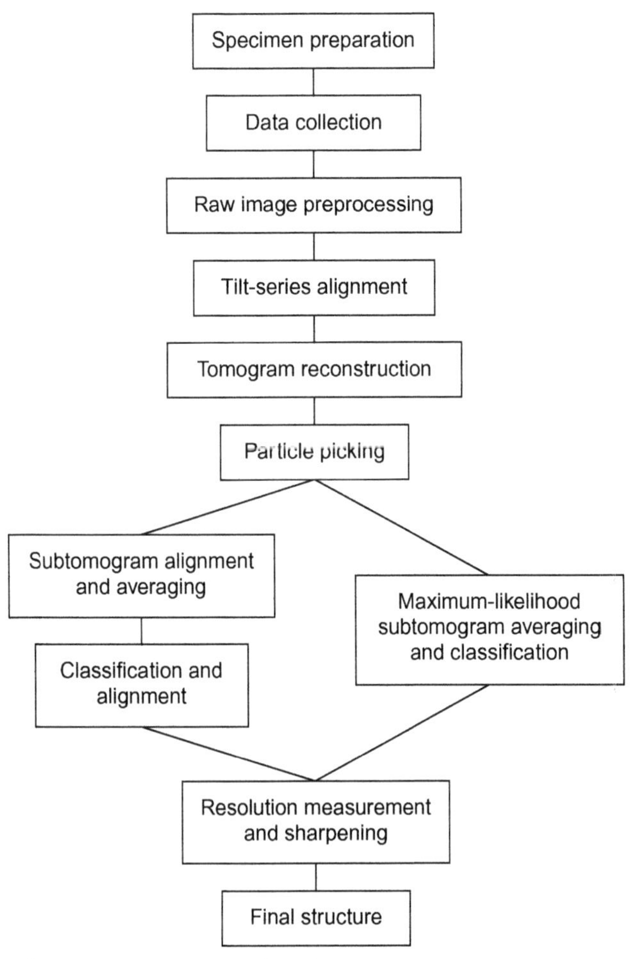

Fig. 2 Cryo-electron tomography and subtomogram averaging workflow. For details, see text.

method is still evolving. As such, rather than providing a step-by-step protocol, this chapter will discuss important aspects and considerations at each stage.

2. TOMOGRAPHIC DATA COLLECTION

2.1 General Principles

Cryo-electron tomograms are 3D reconstructions calculated from a series of cryo-EM images collected at defined tilts. In order to calculate the reconstructions, the relative orientations of the images within these tilt-series must be precisely determined; this is most commonly performed with the use of embedded fiducial markers. The aligned tilt-series are then computationally reconstructed into a 3D volume. The central slice theorem shows that the Fourier transform of each projection image corresponds to a planar slice through the Fourier transform of the 3D volume; tomographic reconstruction from a tilt-series is a means of filling 3D Fourier space with a series of planar slices oriented about the tilt axis (De Rosier & Klug, 1968). Practical factors such as the thickness of the sample and the sample holder geometry limit the maximum tilt angles that can be collected, leading to a "missing wedge" of information in Fourier space (Fig. 3) (Schmid & Booth, 2008). The conjugate effect in real space is a deformation along the direction of missing information. As such, tomograms are an incomplete 3D representation of the specimen.

While a tilt-series consists of a set of 2D cryo-EM images, tilt-series data collection nevertheless differs in a number of ways from 2D data collection for single-particle cryo-EM. Tilting the stage results in large movements of the specimen. Specimen movement parallel to the beam results in changes in the height of the specimen and thus changes in focus, while movements perpendicular to the beam result in shifts in the field of view. Tilting the specimen also introduces increased stage drift, increased specimen motion at higher tilt angles, increased specimen charging, and introduces defocus ramps due to height differences in a tilted specimen. Some of these problems can be addressed during tilt-series acquisition, while others can be dealt with during preprocessing of the raw data. Adequately addressing these considerations leads to longer data acquisition times; this is a necessary cost because quickly collected, poor data cannot yield high-resolution structures. Through the use of automated acquisition pipelines, large datasets can be efficiently collected while fulfilling all the stringent requirements outlined earlier.

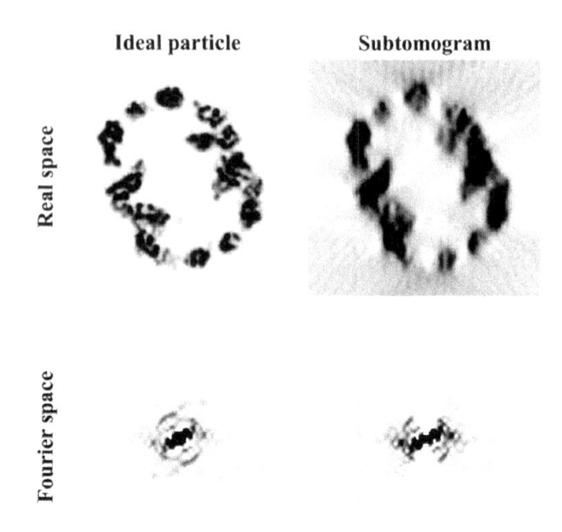

Fig. 3 Effects of the missing wedge on subtomograms. *Left column* is a simulated density map of an eightfold symmetric thermosome calculated from a crystal structure (PDB ID: 3j1b) using EMAN. To simulate a subtomogram affected by a missing wedge (*right column*), the thermosome density was projected from −60 to 60 degree in 3 degree increments and reconstructed by WBP using IMOD. The tilt axis is normal to the plane of the page, and the microscope column axis is oriented from top to bottom of the page. *Top row* shows a slice of the density maps in real space while *bottom row* shows a slice through Fourier space. Real-space and Fourier space slices of the ideal particle and subtomogram are each shown with the same *gray values*. In Fourier space, the subtomogram shows a missing wedge. This leads to smearing of the real-space density map.

The electron dose that a particular biological specimen can sustain is limited. In 2D data collection, this dose can be applied in one exposure, but for tilt-series acquisition this dose must be fractionated over 30–120 images. As such, the dose and therefore the SNR for each image within a tilt stack is much lower than that of a typical 2D image. A further loss in SNR results from the fact that tilting the specimen effectively increases its thickness.

2.2 Tomogram Acquisition

A useful tomogram acquisition pipeline should allow for the systematic identification of areas of interest as well as automated tilt-series collection. Multiscale mapping—iteratively mapping portions of the grid at different magnifications—is an efficient method for determining regions of interest (see chapter "Strategies for Automated CryoEM Data Collection Using Direct Detectors" by Cheng et al.). The initial maps are taken of the entire grid to determine grid squares that contain adequate ice. Grid squares of interest can then be selected and mapped at higher magnification to identify regions where a tilt-series could be collected. The best positions are then selected and stored for subsequent tilt-series collection. For example, we typically use a magnification of 175 × for full grid mapping and image-selected grid squares at 2250 ×, to find areas with suitable ice and objects of interest—a different magnification may be appropriate depending on the specimen and microscope configuration. Areas surrounded by thicker ice or contamination should be avoided, as these may come into the field of view during specimen tilting. It should also be noted that mapping results in preexposure of areas prior to data collection. As such, mapping should be performed with the minimal amount of electron exposure required to identify areas of interest. In an extension to this multiscale mapping approach, correlative light and electron microscopy can also be applied to identify areas of interest based on a fluorescent signal (Zhang, 2013).

Since tilting of the specimen leads to displacements and stage drift, it is necessary to perform tracking and autofocusing, and to wait for the stage to stabilize, before each image is collected. Automation of these steps is essential for efficient collection of high-quality data (Dierksen, Typke, Hegerl, Koster, & Baumeister, 1992; Koster, Van den Bos, & van der Mast, 1987). Tracking corrects for stage movement perpendicular to the electron beam, thus keeping the field of view constant in each tilt image. Autofocus corrects for stage movement parallel to the electron beam, allowing for reproducible defocus across a tilt-series. In order to accurately focus, autofocus should be performed on an area as close as the imaging area as possible along the tilt axis. This minimizes the defocus difference between the focus area and the imaged specimen, but cannot completely account for focal differences. Sufficient time between tilting and image acquisition is required in order to minimize mechanical drift. An arbitrary "wait time" can be set between tilting and acquisition to allow stage settling. This time may prove to be insufficient or excessive so it is preferable to use automated drift determination to wait until drift is below a given threshold (Hagen, Wan, & Briggs, 2016).

When collecting cryo-electron tomograms for subsequent subtomogram averaging, the data collection strategy must be designed with the singular goal of maintaining high-resolution information. This means that appropriate thresholds for focus and drift must be set. Setting a lower threshold tolerance for stage drift minimizes image blurring due to mechanical motion. Tracking, autofocus, and drift determination can be run iteratively for more precise and efficient determination of each parameter, as tracking and autofocus are performed while the stage is settling.

The pattern in which the stage is tilted during tilt-series acquisition is called a tilt scheme; tilt schemes determine how information is distributed across a tilt-series. For subtomogram averaging the tilt scheme must also be selected with the goal of maintaining high-resolution information. Tilt schemes essentially break down to three components: the total angular range covered, the angular increments between tilt images, and the order in which each tilt image is collected. The angular range is generally limited by the specimen geometry and stage mechanics—often it is ±70 degree. In general it is best to collect as wide a range as possible to minimize distortion caused by the missing wedge (Fig. 3). The angular increments used can be linear, ie, a constant value throughout the tilt-series, or nonlinear, where the angular increments are not equally spaced. Nonlinear tilt schemes have been used for well-defined specimens to strategically collect frequencies of interest (Saxton, Baumeister, & Hahn, 1984). However, we consider linear tilt increments to be sufficient for subtomogram averaging, as the ultimate goal is to fill Fourier space through alignment and averaging of multiple subtomograms.

There are two key factors to consider when selecting a tilt scheme. The first is that beam-induced damage of biological samples is cumulative, and that high-resolution information is degraded first (see chapter "Specimen Behavior in the Electron Beam" by Glaeser). This means that highest resolution information is only present in the early images in the tilt-series. The second is that at high tilts, the sample is thicker in the direction of the electron beam, leading to increased noise due to multiple scattering. The simplest scheme is the unidirectional tilt scheme, which sweeps from one angular extreme to the other (Fig. 4). Despite their simplicity,

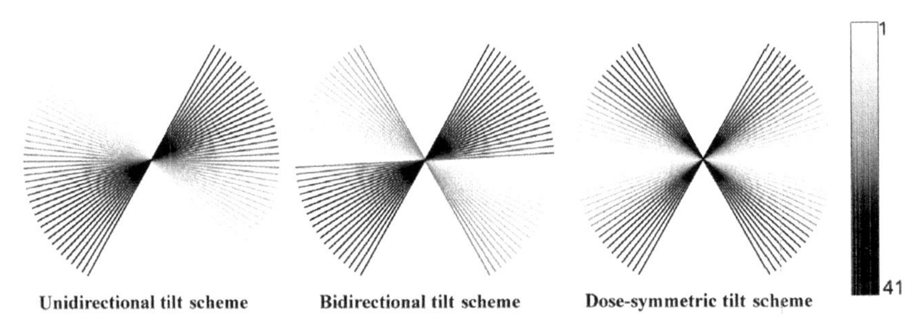

Fig. 4 Schematic showing the order in which tilts are collected in unidirectional, bidirectional, and "dual-walkup" tilt schemes. Tilts are shown from −60 to +60 degrees in 3 degree increments for a total of 41 tilts. *Gray values/colors* correspond to the collection order of each tilt according to the *color* map shown on the *right*. When tilts are collected with constant exposure times, tilt order is directly related to accumulated electron dose on each image. The unidirectional tilt scheme shows a linear sweep from one angular extreme to the other. The bidirectional tilt scheme shows the discontinuity when the tilt-increment direction is changed. The dual-walkup tilt scheme shows near-symmetric accumulated electron dose. (See the color plate.)

unidirectional tilt schemes are poor for preserving high-resolution information as the initial exposures, which contain the high-resolution information prior to beam-induced degradation, are concentrated on the high-tilt images where the sample appears thick. In the bidirectional tilt scheme (Fig. 4), image acquisition begins at zero-tilt and moves toward the first angular extreme, it then returns to zero and moves toward the other angular extreme to collect the second half of the tilt range. In this approach, earlier exposures are collected when the specimen is thinnest, better preserving high-resolution information. However, during the course of tilt-series collection, beam-induced deformations accumulate in the sample; the low-tilt images collected at the start of each half of the tilt-series are of samples with very different accumulated exposure, and there may be differences between the images. This can potentially introduce artifacts into the tomogram, and in some cases, prevent alignment of the tilt-series altogether. Hybrid approaches can give better results, for example, collecting from −20 to +60 degrees, thus sweeping through the central tilt angles before returning to −20 degrees and collecting the other half of the series (Pfeffer et al., 2015).

An optimal tilt scheme for preserving high-resolution information is dose symmetric: it begins at zero-tilt, followed by collecting incrementally larger tilts while oscillating the tilt direction; ie, 0, 3, −3, −6, 6 degree, ... (Hagen et al., 2016; Fig. 4). This scheme ensures that the highest resolution information is collected when the specimen is thinnest, provides a nearly symmetric distribution of information in Fourier space, and distributes beam-induced deformation evenly across the tilt-series.

There are currently a number of commercial and noncommercial software packages for automated tomographic data collection. In order to meet the requirements described earlier, a suitable package should allow for automated multiscale mapping, autofocusing, tracking, and a flexible scripting interface to allow for implementation of alternative tilt schemes. Several packages are available that fulfill some or all of the requirements outlined earlier (Carragher et al., 2000; Mastronarde, 2005; Zheng et al., 2007; Ziese et al., 2002; see also chapter "Strategies for Automated CryoEM Data Collection Using Direct Detectors" by Cheng et al.).

3. TILT-SERIES PROCESSING AND TOMOGRAM RECONSTRUCTION

3.1 Preprocessing of Raw Images

Raw images from tilt-series can be processed prior to tomogram reconstruction. For data collected in movie mode on a direct electron detector, an important initial processing step is frame alignment to correct for beam-induced motion (Brilot et al., 2012). A number of published algorithms for single-particle methods can be used to align frames (Campbell et al., 2012; Grant & Grigorieff, 2015; Li et al., 2013; see also chapter "Processing of CryoEM Movie Data" by Ripstein and Rubinstein), but since tilt-series frames generally have lower electron dose than single-particle frames, algorithms that use moving averages may prove more successful. The presence of high-contrast gold fiducials in the sample may facilitate frame alignment despite the lower dose.

Prior to tomogram reconstruction the individual tilt images can be "exposure filtered" to account for the effects of beam-induced information loss (Schur et al., 2016). Exposure filters essentially behave as dose-dependent low-pass filters, in which each image is filtered according to the accumulated electron dose; exposure filtering can be applied in a way that is analogous to that developed for single particles (Grant & Grigorieff, 2015). Exposure filtering results in reduced noise and higher contrast in the reconstructed tomograms, resulting in faster convergence of alignment during subtomogram averaging. However, exposure filtering also dampens Thon rings, so defocus determination should be performed on nonexposure-filtered tilt-series.

3.2 Defocus Determination of Tilt Images

Defocus determination for tilt images is typically more difficult than for standard single-particle images. When compared to single-particle images, images from a tilt-series have lower SNR, resulting in weaker Thon rings, making defocus determination by contrast transfer function (CTF) fitting more difficult. The low SNR results from a combination of the low dose applied to each tilt image; specimen tilt, which effectively increases sample thickness; and the fact that a typical cryo-ET specimen is thicker than a single-particle specimen (up to 200–500 nm). Low SNR is particularly challenging for data collected on CCD cameras. Another complication in CTF fitting is the mixing of defoci within a tilt image. Specimen tilt leads to differences in height across the image, resulting in a defocus gradient perpendicular to the tilt axis. Furthermore, within a thick sample objects at multiple heights within the ice layer are at different defoci, but are superimposed in the image. Despite these challenges, it is possible to determine defoci for tilt-series with sufficient accuracy for high-resolution structure determination (Schur, Hagen, de Marco, & Briggs, 2013).

Inaccuracy in defocus determination, along with the subsequent inaccuracy in CTF correction, can limit the achievable resolution of a dataset in the form of an envelope function (Jensen, 2001). The effect of defocus error has been simulated for individual tomograms and large tomographic datasets (Schur et al., 2013; Zanetti, Riches, Fuller, & Briggs, 2009). Defocus errors of 100 nm result in 50% signal transfer at ~8 Å, indicating that defocus errors within this range are acceptable for subnanometer structure determination. However, it should be noted that this defocus error envelope is one of many that limit information transfer, and as such does not act as a hard resolution limit. To a certain extent, defocus error can be overcome by increasing dataset size.

Low SNR and the resulting poor Thon ring signal make it difficult to perform reliable CTF fitting. Thon rings can be enhanced, as in single-particle images, by periodogram averaging or downsampling of the full image Fourier transform, as well as by radial averaging (Fernández, Li, & Crowther, 2006; Fernández, Sanjurjo, & Carazo, 1997; Rohou & Grigorieff, 2015). In cases where these enhancement methods still provide insufficient signal to measure the defocus, power spectra from multiple images can be averaged in order to enhance SNR (Schur et al., 2013). The CTF measured by averaging power spectra from images with varying defocus does not oscillate in phase with the theoretical CTF of the average defocus. The phase errors increase with defocus variation and resolution (Schur et al., 2013; Zanetti et al., 2009); it is therefore advisable in this case to use a data collection scheme with a low-tolerated defocus range.

In a tilted image there is a defocus gradient across the image perpendicular to the tilt axis (Fernández et al., 2006; Xiong, Morphew, Schwartz, Hoenger, & Mastronarde, 2009). For tilted images, it is only necessary to determine the mean defocus, or equivalently, the defocus at the tilt axis. The defocus gradient can then be accounted for because the tilt angle is known. Because of the defocus ramp, the Thon rings in the power spectrum of a tilted image reflect a number of different CTFs averaged together. Exactly as described earlier for averaging images of different defocus, this average CTF goes out of phase with the CTF of the mean defocus with increasing resolution (Zanetti et al., 2009). In addition, this averaging also produces an envelope function that dampens Thon ring signal. A number of algorithms have been developed to determine the mean defocus of a tilted image; these are generally based on periodogram averaging of strips that run parallel to the tilt axis. Within each strip, there is minimal variation in defocus. This approach provides a number of defocus measurements perpendicular to the tilt axis (Fernández et al., 2006; Mindell & Grigorieff, 2003; Xiong et al., 2009), which can then be used to calculate the mean defocus, the defocus gradient, or both. Since these strips incorporate only limited image area, it can be challenging to obtain accurate defocus measurements from the low SNR images from tilt-series.

3.3 CTF Correction of Tilt Images

CTF correction of tilt images is more complicated than for standard single-particle images due to the defocus gradient across the image. The simplest approach is to perform CTF correction on strips parallel to the tilt axis, each according to the local defocus. In this stripwise approach, the images in the tilt-series must first be rotated and shifted to align the tilt axis with an image axis (Fernández et al., 2006; Xiong et al., 2009). Overlapping strips are then extracted parallel to the tilt axis and CTF-corrected using a local defocus value; thin central strips, which have minimal defocus differences, are then extracted from the large corrected strips and combined to form a new corrected image (Fernández et al., 2006). Another approach is to also extract strips for CTF correction, but rather using only a central strip in the corrected image, the entire strip is used and interpolation is applied to areas of the corrected strips that overlap, smoothing the transitions between the corrected strips (Xiong et al., 2009). It is possible to avoid rotation of the tilt images by instead applying CTF correction to square tiles extracted in a gridwise pattern, but this has substantially greater computational cost (Zanetti et al., 2009).

A more complex approach to CTF correction of tilt images is to perform global correction; that is, CTF correction of a full image rather than piecewise correction across an image. The advantage to such an approach is that it avoids artifacts related to the size of the extracted areas and the interpolation involved in recombining locally corrected areas. Two optimized methods have been described: one for CTF correction per tilt image and one for 3D CTF correction during tomogram reconstruction (Voortman, Franken, van Vliet, & Rieger, 2012; Voortman, Stallinga, Schoenmakers, van Vliet, & Rieger, 2011). The per tilt-image correction is implemented as a processing filter similar to stripwise CTF correction, but with a CTF model that allows for CTF correction of the entire image. The 3D CTF correction takes into account defocus variation within the thickness of the specimen (Jensen & Kornberg, 2000; Voortman et al., 2012, 2011). While the

3D CTF-correction implementation described here is designed to be optimally efficient, it is still extremely computationally expensive, and therefore currently rarely applied during tomogram reconstruction. It may become more widely used as higher resolutions are targeted in thicker samples.

3.4 Local Defocus Determination and CTF Correction

In thick specimens there can be significant defocus differences between individual particles depending on whether they are at the top or bottom of the ice layer. As discussed earlier, one solution is to use 3D CTF-correction methods. An alternative solution is to determine the local defocus of each particle and perform local CTF correction.

To determine local defocus for each particle of interest it is first necessary to determine its 3D position in a reconstructed tomogram. The mean defocus must also be determined using the approaches described in the previous section. When the particles of interest are the only objects in the field of view, the mean defocus plane can be approximated as the center of mass of the particles; this approximation becomes less accurate when grid-support carbon or other Thon ring-producing objects are in the field of view. The height offset of each particle from the mean defocus plane can be calculated from its 3D position, and when combined with the distance of each particle from the tilt axis and the tilt angle for each image, the local defocus of each particle in each tilt image can be determined.

After determination of local defoci, local CTF correction can be performed; currently there are two described methods, each with their own benefits and problems. The first method is to extract the 2D projections of each particle from each tilt image, perform CTF correction, and then directly reconstruct a subtomogram containing each particle (without reconstructing the full tomogram; Chen & Förster, 2014; Zhang & Ren, 2012). This approach allows accurate CTF correction, as long as the projection that is corrected is large enough to contain the full point-spread function. There are, however, inaccuracies associated with directly reconstructing a subtomogram which is only part of the sample volume (Turoňová, Marsalek, & Slusallek, 2016), and doing so makes use of the subtomogram central section approximation (see Section 6).

The second method is to perform CTF correction on the Fourier slices of a subtomogram that has been extracted from a non-CTF-corrected tomogram (Bharat, Russo, Löwe, Passmore, & Scheres, 2015). This allows the CTF to be refined during reconstruction (see section 5). A possible problem with such an approach is that it cannot account for convolution of data from different Fourier planes that take place during subtomogram extraction—it implicitly makes use of the subtomogram central section approximation. An additional possible problem with this approach is that the size of each subtomogram must be large enough to contain the full point-spread function prior to CTF correction.

Despite the described problems, both of the earlier methods have been successfully applied (Bharat et al., 2015; Chen & Förster, 2014; Zhang & Ren, 2012). Nevertheless, further development of local CTF correction methods is required before they become a generally applicable part of the subtomogram workflow.

3.5 Tilt-Series Alignment

The relative orientations of all images within each tilt-series are only approximately known from the parameters of the tomographic data collection, and they must be more precisely calculated before tomogram reconstruction (Amat et al., 2010). This process, often called tilt-series or tomogram alignment, generally includes determining shifts between each image, refining the orientation of the tilt axis, refining the tilt angles, and compensating for image skewing or deformations. These alignment parameters are most commonly determined using high-contrast gold fiducials embedded within the ice throughout the field of view.

The positions of the fiducials first need to be measured. The most common fiducial markers used are gold beads, as they appear as highly contrasted discs in projection images. Their high contrast allows them to be identified by automatic picking algorithms. With good prior knowledge of the rotation and tilt angles, fiducials can generally be automatically tracked through a tilt-series. Manual intervention in fiducial tracking is often used to correct for clustered or overlapping fiducials, or for individual fiducials that may move during data collection. In some cases human judgment may be the most efficient way to identify the most well-behaved fiducial markers in order to yield an optimal alignment. Well-behaved fiducial markers may also be determined through statistical means; such approaches exhibit varying levels of success owing to the complexity of the fiducial models. Fully automated alignment algorithms that require almost no user intervention have recently become available (Han, Wang, Liu, Sun, & Zhang, 2015).

Tilt-series can also be aligned without the use of fiducial markers, for example, using "patch tracking." Patch tracking relies on calculating the cross correlation (CC) between image patches throughout the tilt-series; these values are then used to align the tilt stack (Amat et al., 2010; Brandt, 2005; Castaño-Díez, Scheffer, Al-Amoudi, & Frangakis, 2010). These approaches are often less accurate than fiducial-based alignment, but can be applied to samples such as focused ion beam scanning electron microscopy (FIB-SEM) lamellae, where it can be very difficult to prepare samples containing fiducial markers.

The positions of the fiducials within the tilt images are used to generate and refine models that describe the relationship of the tilt images to the 3D coordinate space in which the tomogram will be reconstructed. The refinement of the different parameters can be interrelated, and depending on the number of fiducials present the parameters may be underdetermined. For this reason the number of parameters being modeled should be minimized. Shifts must always be determined, as the field of view is never perfectly recentered after tilting. The angle of the tilt axis in each image is a combination of stage geometry, magnification, and beam settings; proper tuning of the microscope can effectively make this a constant at each magnification and remove the need for refinement during tilt-series alignment. Stage tilt angle is purely a mechanical property whose accuracy and precision can be determined prior to data collection—refinement of tilt angles is not always required during tilt-series alignment. The effects of beam-induced specimen deformations are minimized when the data are collected at high magnifications with correspondingly small fields of view—this is often the case for subtomogram averaging. Compensating for deformation is best avoided unless obviously necessary. In optimal cases, and with sufficient a priori knowledge about the microscope and stage, it may only be necessary to determine image shifts for tomogram reconstruction.

3.6 Tomogram Reconstruction

After preprocessing and alignment, the tilt-series can be used to generate a 3D reconstruction—a tomogram. A number of different methods are available to reconstruct tomograms. The conceptually simplest approach is Fourier synthesis, based on the central slice theorem. The Fourier transform of each projection image is calculated and placed into a plane of a 3D Fourier space. Once all tilts are included, this is then inverse-transformed to yield a real-space 3D reconstruction (De Rosier & Klug, 1968). Backprojection methods are real-space analogs to Fourier synthesis approaches, without the need to perform large Fourier transforms (Bracewell & Riddle, 1967).

In backprojection, the densities from the tilt images are projected into a real-space volume. It is most commonly applied as weighted backprojection (WBP) which more accurately reproduces the density within the specimen by reweighting overrepresented frequencies (Rademacher, 2005).

Widely used iterative reconstruction methods include the Algebraic Reconstruction Technique (ART; Gordon, Bender, & Herman, 1970; Marabini, Herman, & Carazo, 1998), the Simultaneous Iterative Reconstruction Technique (SIRT; Gilbert, 1972), and the Simultaneous Algebraic Reconstruction Technique (SART; Andersen & Kak, 1984). In such methods, the information from the tilt images is not projected to form the tomogram, but instead sets of linear equations corresponding to rays projecting into the tomogram are iteratively solved by minimizing discrepancies between calculated projections of the reconstruction and the tilt images. These algorithms are generally terminated after a user-defined number of iterations. Iterative methods yield tomograms with better contrast and less noise; they are generally easier to directly interpret than tomograms reconstructed with WBP.

The higher contrast in tomograms generated by iterative methods reflects better, low-resolution signal which is advantageous for subtomogram alignment. However, higher resolution information that is below the noise level in individual tomograms may be lost when applying these approaches. In WBP, the high-resolution signal that is substantially below the noise level is preserved such that it can be amplified and restored during subtomogram averaging. WBP is currently the most commonly used approach for subtomogram averaging.

New iterative methods are being developed specifically for subtomogram averaging applications; the goal of such methods is to achieve the high contrast of iterative methods while maintaining high-resolution information. The Iterative Nonuniform fast Fourier transform-based Reconstruction (INFR) method is an iterative method that is able to retrieve meaningful low-frequency information in missing regions of Fourier space (Chen & Förster, 2014). Supersampling SART is an implementation of SART that aims to minimize aliasing and improve high-resolution data restoration (Kunz & Frangakis, 2014). The progressive stochastic reconstruction technique samples the tomographic space with spheres and performs reconstruction by moving the spheres with a Metropolis–Hastings strategy and minimizing the difference between the calculated projections and experimental data, which allows for progressively finer sampling by reducing sphere size and localized refinement in regions of interest (Turoňová, Marsalek, Davidovič, & Slusallek, 2015). Each of these algorithms has been shown to perform better than WBP with low-resolution (20–30 Å) subtomogram averaging test sets. However, these methods have not been tested using datasets known to achieve subnanometer resolutions; whether these approaches maintain high-resolution data as well as WBP remains to be seen.

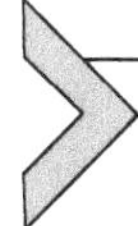

4. SUBTOMOGRAM AVERAGING

4.1 Particle Picking and Starting References

For subtomogram averaging of discrete single molecules or assemblies, particle picking is analogous to single-particle approaches. A number of packages allow for manual picking from tomograms with various options for tomogram visualization (Castaño-Díez, Kudryashev, Arheit, & Stahlberg, 2012; Galaz-Montoya, Flanagan, Schmid, & Ludtke, 2015). Filtering, contrast enhancement, or denoising of the tomograms may assist in manual picking (Hegerl & Frangakis, 2005). Automated particle picking generally relies on template matching approaches, where templates can be initial structures generated from manual picking, low-pass filtered structures that were previously determined, or simple shape phantoms (Böhm et al., 2000; Förster et al., 2010; Hrabe et al., 2012). For molecules attached to surfaces, such as glycoproteins, segmentation of the surfaces can reduce the search space for template matching (White et al., 2010).

Molecules that form densely packed 2D arrays can be efficiently picked by extracting subtomograms on a grid of points along the array surface. Examples of such arrays include helices or lattices assembled on membrane surfaces. Alignment of subtomograms extracted from an oversampled grid leads to clustering of the oversampled subunits; redundant subtomograms can then be removed. For regular surfaces such as spheres or tubes, the centers can be defined as points or splines, respectively. Using these center points and measured radii, regular lattices points can be calculated, for example (Briggs et al., 2009). Alternatively points can be extracted along a segmented surface, for example (Zanetti et al., 2013).

Because of the 3D nature of tomographic data, the orientation of the particle can often be roughly estimated prior to alignment based on the context of the particle within the tomogram. In the case of discrete molecules, visible structural features such as membrane tethers or characteristic shapes can allow rough orientations to be estimated. Such approaches have been implemented in some particle picking programs, where initial orientations can be manually assigned during particle picking. For particles within helical arrays or on membrane surfaces, angles normal to the surfaces can be calculated and used as starting orientations (Förster, Medalia, Zauberman, Baumeister, & Fass, 2005). This orientation information can greatly reduce the space that needs to be sampled during the subsequent angular search.

Just as with single-particle methods, the iterative nature of subtomogram averaging means that starting references have the potential to irreversibly bias the structure determination. Care must therefore be taken in deriving a starting reference. In practice, initial references can be generated by averaging particles according to predetermined starting orientations, or by generating appropriate phantom shapes. Small subsets of the data can be averaged and aligned to determine featured, low-resolution structures; these can then be used for alignment of the full dataset.

4.2 General Principles of Subtomogram Averaging

The core of the subtomogram averaging workflow is an iterative alignment and averaging procedure: a set of subtomograms are aligned with a reference so as to maximize a scoring function; the aligned subtomograms are then averaged to produce a structure which becomes the reference for the next round of alignment in an iterative process (Fig. 1). Alignments are generally performed by masking and rotating the reference structure into different orientations and comparing these with each subtomogram; spatial shifts and alignment scores are typically calculated by CC functions (Fig. 5; Förster & Hegerl, 2007). Alternate methods using fast rotational matching (FRM) have also been used (Bartesaghi et al., 2008; Chen & Förster, 2014; Xu, Beck, & Alber, 2012). At the end of each iteration, after optimal orientational parameters have been calculated, individual subtomograms are rotated and translated back to the reference frame and averaged. The average may require reweighting in Fourier space to account for anisotropic sampling caused by the missing wedge (Förster et al., 2005). This approach is analogous to many single-particle and helical reconstruction approaches, but in 3D. By going to 3D, both the memory and computation time required to process each particle increase significantly. In addition, the orientational search space expands; in 2D methods the search space consists of two shifts and three rotation axes, while in subtomogram averaging three shift directions and three rotations axes must be considered.

Each subtomogram has wedge of missing information in Fourier space as the consequence of 3D reconstruction from a limited range of tilt angles.

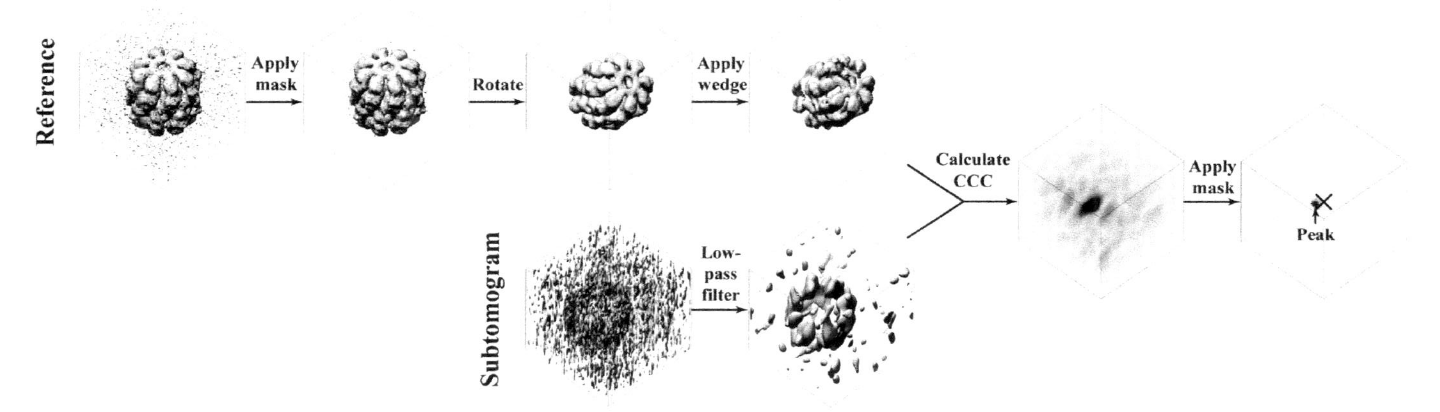

Fig. 5 Subtomogram alignment by angular search. In *gray* are the manipulations performed on the reference structure. In *purple* (*dark gray* in the print version) are the manipulations performed on each subtomogram. The referenced is masked and rotated into each orientation to be assessed and a wedge mask is applied. The subtomogram is filtered and compared to the reference by cross correlation. The CCC function is masked to limit the range of allowable shifts and the peak is identified. The peak position determines the translations; the rotation of the reference in which the highest CCC value is identified determines the rotation. Shown are simulated reference and subtomograms of eight-fold symmetric thermosomes. Density maps were calculated from a crystal structure (PDB ID: 3j1b) using EMAN; structures were low-pass filtered and Gaussian noise was added using MATLAB. The reference has a SNR of 0.75, while the subtomogram has a SNR of 0.1. In the masked CCC map (*rightmost panel*), the *X-mark* denotes the center of the *box* and the *arrow* indicates the CCC peak; the vector between the two points indicates the determined translation.

This missing wedge causes deformations in real space and can lead to erroneous subtomogram alignments (Figs. 3 and 5) (Förster et al., 2005; Schmid & Booth, 2008). This problem is especially severe when the reference volume also contains anisotropic information, for example, when the objects of interest have a preferred orientation within the sample. For alignment of an isotropic reference with a subtomogram, the missing wedge can be compensated for by applying an equivalent binary wedge-shaped mask to the reference in Fourier space (Bartesaghi et al., 2008; Förster et al., 2005; Frangakis et al., 2002; Schmid & Booth, 2008)—effectively the subtomogram is always compared with a reference that has a missing wedge in the same orientation. Such wedge-masked CC functions are generally called constrained cross-correlation (CCC) functions. For the alignment of two missing wedge-affected volumes, it is essential to compensate for each missing wedge to avoid alignment of the missing wedges regardless of the particle orientation. In this case wedge masks can be applied to both volumes so that only the common regions of Fourier space are considered during alignment (Förster, Pruggnaller, Seybert, & Frangakis, 2008).

To save computational resources, subtomogram alignments can be first performed with binned data; as the maximum resolution for each binning is reached, the data are unbinned and the alignments are further refined. Furthermore, as with single-particle methods, data should be low-pass filtered in order to remove high-resolution noise and to prevent overalignment. Especially during early stages of alignment, convergence is generally faster if only strong low-resolution information is included. The data should be filtered at a resolution below the measured resolution at each iteration. One approach is to limit the resolution used for alignment to the resolution where the Fourier shell correlation (FSC) reaches at 0.5 (Hrabe et al., 2012), though a more cautious filter may be needed to avoid correlations in Fourier space beyond the currently reliable resolution. Real-space masking can be used to remove regions of the volume that do not contain parts of the object of interest; this helps to remove noise and aids in normalization of the molecular density. Masks can be simple shapes such as spheres or cylinders or be designed to fit the molecules. As in single-particle reconstruction, mask edges should be mollified—this is particularly important for structures where molecular densities go to the volume edges.

4.3 Alignment by Angular Search

Subtomogram averaging is most commonly performed using an angular-search procedure to align the reference structure to each subtomogram

(Fig. 5) (Castaño-Díez et al., 2012; Förster et al., 2005; Hrabe et al., 2012; Nicastro et al., 2006). This proceeds by real-space masking the reference structure, rotating the masked reference, masking the missing wedge in Fourier space, and calculating the 3D CC between this volume and the subtomogram. The rotational search space is described by three Euler angles. There are various Euler angle conventions, for example: the first angle is an in-plane (azimuthal) rotation, followed by an out-of-plane (polar) rotation, and a final in-plane rotation. The first two angles can be envisaged as describing the subtomogram orientation as a position on the surface of a sphere; the third describes the subtomogram orientation in the plane of the sphere surface. The CC function is most efficiently calculated in Fourier space, which also allows for masking of the missing wedge (Frangakis et al., 2002). The position and value at each point in the resultant 3D CC volume describes the similarity of the two volumes with a given translation—the peak becomes the CCC score and the peak position provides the shift. Shifts can be limited to a certain ranges or directions by applying spherical or nonspherical binary masks to the CC volume. After optimal rotations and shifts have been determined for each subtomogram, they are applied to the subtomograms, which are then summed to generate the final average. In order to account for anisotropic sampling in Fourier space, wedge masks are rotated and summed as with the subtomograms, resulting in a Fourier space weighting mask. The Fourier-weighted subtomogram sum is then divided by the number of subtomograms to yield the final average.

In order to successfully identify the correct alignment for a subtomogram, the CC peak at the true angle and rotation must be higher than all other peaks in the search space. The chances of this being the case can be increased by masking and filtering the data in real and Fourier space to remove noise. It can be further increased by limiting the search space as much as possible. A priori information can be used to restrict the alignment search space in a number of ways. For discrete particles where rough orientations are known, the range of the angular search can be limited. For densely packed arrays, knowledge of the rough size of each subunit allows for estimation of the packing density, which limits the azimuthal and polar search space for each discrete subunit. Knowledge of symmetry can reduce the required search space. For special cases such as helices, all Euler angles can be restricted based on the geometries required for helical packing. In addition to angular restrictions, shifts can also be restricted using a priori information. Manual and automatic picking of discrete particles can generally be assumed to have a certain level of spatial precision; an estimate of this can be used to restrict shifts by applying masks to the CC volume. For oversampled, densely packed surfaces, shifts can be restricted by the sampling parameters, or by a priori knowledge of subunit sizes and packing distances. The more the search space can be limited by a priori information, the faster the search, the faster the convergence, and the greater the likelihood of avoiding local minima.

4.4 Alignment with Fast Rotational Matching

Subtomogram alignment is computationally expensive. Resource intensive processes include Fourier transforms and rotations of 3D volumes; rotations are generally the most costly as linear interpolation is required. In order to decrease computational cost, FRM-based approaches determine alignment parameters without the need to explicitly rotate volumes to perform angular searches (Kovacs & Wriggers, 2002). This is accomplished by dividing the alignment into two parts: the rotational search and the translational search. The rotational search is performed by the calculation of a 3D CC function between the reference and the subtomogram using spherical Fourier transforms; the peak of the spherical cross-correlation (SCC) map indicates the rotation that yields maximum similarity. The reference can be rotated by these angles and the Cartesian space CC function between the rotated reference and the subtomogram determines the optimal translation. An underlying problem with this approach is that calculation of the rotation by SCC requires a correctly centered particle while particle centering via CCC requires a correctly rotated particle. One solution is to use only the Fourier amplitudes in calculating the SCC, which yields a shift-invariant SCC (Bartesaghi et al., 2008; Xu et al., 2012). Another approach is to first calculate a shift-invariant SCC to obtain an initial rough rotation, and then to refine the alignment by iterating between translational and rotational searches (Chen, Pfeffer, Hrabe, Schuller, & Förster, 2013).

FRM-based methods are much faster than angular search-based methods for global alignment. Since each round of FRM alignment is a global alignment, they are also less prone to getting stuck in local alignment minima. Where fine angular-search increments are required, and reliable a priori information exists that allows the angular-search space to be limited, FRM-based approaches may be slower than standard angular-search approaches. The speed of FRM-based approaches depends on a number of parameters including maximal frequency and the bandwidth of spherical harmonics; more spherical harmonics terms are required to represent higher resolution features, resulting in increased computational cost with higher resolution alignments. A given estimate is that for a maximum frequency of 20 Å, one round of FRM is approximately as fast as calculating 108 angular comparisons (Chen et al., 2013). For high-resolution subtomogram averaging, combinations of FRM and angular search may be appropriate.

4.5 Postprocessing: Assessing and Reweighting the Final Structure

The final subtomogram averages can be assessed using statistical measures such as FSCs and spectral SNRs as in single-particle reconstruction. For continuous structures where structural density is present to the box edges, care must be taken to sufficiently mollify mask edges prior to FSC calculation, so as to not introduce artificial edges. It may be possible to further validate the final alignment parameters by visualizing the orientation and position of the aligned subtomograms in 3D—for example, membrane proteins should be distributed on a membrane surface, while the subunits of a helix should be distributed in a helical pattern.

As for single-particle reconstruction, subtomogram averages can be "sharpened" by appropriately reweighting high-resolution information. In practice the B factors are often higher than for single-particle maps due to factors including the more heterogeneous nature of the samples and overrepresentation of low-resolution terms in high-tilt images. Depending on the reconstruction pipeline, it may also be appropriate to apply other reweighting steps, for example, to upweight very low frequencies or to correct for an uneven defocus distribution.

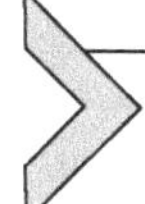

5. CLASSIFICATION IN SUBTOMOGRAM AVERAGING

5.1 General Principles

Classification allows for identification of false positives included in the dataset during the picking steps, and more importantly allows for the separation of heterogeneous assemblies or conformations. While classification is routine in single-particle approaches, it is more challenging in subtomogram averaging due to difficulties in implementation, computational cost, and the small size of typical datasets. Subtomogram classification methods broadly fall into two categories: classification performed after subtomogram alignment and classification performed during subtomogram alignment.

Classification after subtomogram alignment requires sufficient homogeneity to get an alignment in the first place. Classification is then based on assessing differences between individual subtomograms. Classification during alignment is generally performed by aligning each subtomogram against multiple references and assigning classes based on the similarities between each subtomogram and each reference.

5.2 Postalignment Classification

The simplest form of classification is to set a threshold value for the CCC between the aligned subtomogram and the reference; all subtomograms below the threshold are removed from the dataset. This can be useful when the majority of subtomograms contain the properly aligned structure of interest; in such cases CCC thresholding can be used to remove misaligned subtomograms or those containing other structures. Where there is a high degree of structural heterogeneity or misalignment, such approaches may fail because the CCC is calculated using a reference that is itself averaged from heterogeneous particles.

A more sophisticated approach to postalignment classification is to compare the subtomograms in the dataset and cluster the dataset into a number of different classes. This enables the separation of heterogeneous structures, unlike CCC thresholding, which only removes subtomograms dissimilar to the reference. In single-particle methods, classification is often performed in two stages: first the features in each image are reduced to a small number of attributes; these attributes are then used to cluster the images into a set of classes (Frank, 2006). In order to determine the features in each image, the aligned images are compared pixel by pixel, generally using correspondence analysis or principle component analysis (PCA). This results in the image information being reduced to a number of eigenvectors and eigenvalues; eigenvectors with high eigenvalues are then used for clustering, generally by *k*-means or hierarchical clustering. In subtomogram averaging, voxel-by-voxel comparisons are not generally performed directly on subtomograms, as the missing wedge becomes the main classification feature. An exception to this is in cases where the molecules have a preferred orientation with respect to the missing wedge, making the Fourier space sampling and real-space deformations roughly equivalent for each subtomogram (Taylor et al., 1999). In all other cases, clustering methods in subtomogram averaging must be performed on information that describes the differences in each subtomogram, but is unaffected by the missing wedge. These types of information can be obtained by modifying the subtomogram intensities to compensate for the missing wedge, or by calculating similarity measures that take into account or are not affected by the missing wedge.

Modifying subtomogram intensities to compensate for the missing wedge allows for voxel-by-voxel comparisons, analogous to single-particle methods. One approach is to calculate wedge-masked differences (WMDs) between each subtomogram and the averaged reference (Heumann, Hoenger, & Mastronarde, 2011). For calculation of WMDs, the subtomogram is first rotated into the same orientation as the reference. The reference is wedge-masked with a wedge in the same orientation as that in the rotated subtomogram, thereby reproducing the missing wedge-related distortions in the reference volume. The difference between the two volumes is then calculated. Classification is then performed by PCA on the WMDs, followed by *k*-means clustering of eigenvectors. If regions of expected variability are known a priori, classification can be improved by masking around these regions prior to PCA. Because eigenvolumes of the WMDs are determined, they can be used to visually assess variable regions of the structure, facilitating iterative clustering and masking.

Similarity-based clustering and classification is commonly performed using correlation matrices consisting of pairwise CCCs between all subtomograms in the dataset. CCCs are calculated after rotating each subtomogram into the reference orientation; as described earlier, CCCs between two missing wedge-affected volumes compare only the common parts of Fourier space. Clustering is then performed on the correlation matrix using PCA and *k*-means clustering, or hierarchical clustering (Castaño-Díez et al., 2012; Förster et al., 2008). Hierarchical clustering has the advantage that no explicit decision regarding the number of classes is required, though it does require choosing an arbitrary cutoff value to define classes. A priori knowledge about the regions of known variability can be used to aid classification by masking around these regions prior to correlation matrix calculation. A weakness of correlation matrix-based approaches is that the common area of Fourier space between two missing wedge-affected subtomograms that is considered during the CCC can be small, and information describing the variability may not be included. Correlation matrix-based approaches are generally lower dimensional problems than voxel comparisons, but have the additional computational cost of calculating CCCs.

In what can be considered a hybrid approach, areas of structural variance can first be identified through voxel comparisons, this information can be used to weight the subtomograms prior to calculation of a similarity score which is then used for classification (Xu et al., 2012). In this approach, areas of variability are determined by taking each voxel of an aligned subtomogram and calculating its covariance with respect to each neighboring voxel and with the same voxel from each subtomogram in the full dataset. The relationship between a voxel and its neighbors describes the continuity of a possible structural feature while its relationship with the entire dataset describes whether or not it is an area with variable structure. Variable areas that may contain useful classification features are then used to generate weighting masks before calculating a pairwise CCC matrix between each subtomogram. Classes are then determined by performing hierarchal clustering on this distance matrix.

These postalignment classification methods are effective at separating heterogeneous populations, but they have so far only been used on relatively small datasets. For large high-resolution subtomogram averaging datasets, the computational requirements for PCA and hierarchical clustering are daunting. Unlike subtomogram alignment, these clustering methods are not simply parallelizable problems; to our knowledge no parallelized PCA or hierarchical clustering methods have yet been implemented for subtomogram averaging.

5.3 Classification During Alignment

Classification during alignment is most commonly performed through multireference alignment. Multireference-based approaches require an estimate of the number of classes, which determines the number of references. Each subtomogram is assigned to a class by aligning it against all references and identifying the reference to which it is most similar. At the end of each iteration, the members of each class are averaged to generate a new set of references. Through iterative alignment and averaging the differences are amplified and the references are refined. In its simplest form, known references can be used to separate a heterogeneous dataset. Such approaches may be useful for classifying different molecules in visual proteomics experiments or identifying a subset of the data with a difference in subunit composition. For cases where there is no a priori knowledge of the structure, the generation of initial references is more difficult. One approach is to randomly split and average the dataset into a number of starting references. If the goal is to find sparsely populated classes, the number of references generated may need to be substantially higher than the number of classes required. After the references converge, the redundant classes are found using a correlation matrix and hierarchical clustering as described earlier. In this case, the information in each reference is isotropically distributed and has a higher SNR than individual subtomograms, making the clustering results more reliable than clustering between subtomograms (Hrabe et al., 2012). A simulated annealing approach can be added to the classification step to allow for jumping between classes, reducing the risk of entering local minima based on the starting references. With each new iteration, the annealing temperature is lowered, allowing for settling of the classes.

For particularly well-aligned datasets, the orientational search can also be omitted, effectively making multireference-based classification a postalignment approach. The advantage of such a postalignment multireference approach is that each of the references used has isotropically distributed information and a high SNR, making scoring by CCC more reliable.

Multireference-based classification can be used to sort according to local differences within similar structures by masking around known areas of variation (such as the binding sites of cofactors) before calculation of the CCC. If the dataset has already been partially aligned, variance maps may be used to determine areas of interest at each iteration. Masks for focused classification can be generated automatically; automatic mask generation during multireference alignment and classification can allow for targeting of the variable regions of the structure without a priori knowledge. One approach is to calculate the variation between each pair of references at every iteration in order to generate a set of pairwise focused masks (Chen, Pfeffer, Fernández, Sorzano, & Förster, 2014). Each subtomogram is scored against each reference pair using the corresponding focused masks and assigned to the highest scoring class. If there are more than two classes, the results of each pairwise comparison count as a vote; the subtomogram is assigned to the class with the largest number of votes.

The postalignment classification procedures described in the previous section and the "during-alignment" approaches described in this section can be combined and implemented within workflows that iterate between classification and multireference subtomogram alignment.

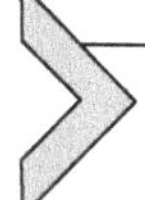

6. MAXIMUM-LIKELIHOOD APPROACHES FOR SUBTOMOGRAM ALIGNMENT AND CLASSIFICATION

6.1 General Principles

Maximum-likelihood (ML) approaches optimize a target function that describes the likelihood of observing the experimental data from the reference structure (Sigworth, Doerschuk, Carazo, & Scheres, 2010). This is achieved using expectation–maximization algorithms. First the probability of observing each possible orientation of each subtomogram, given the reference, is determined, while making use of prior knowledge about probability distributions. A new reference is then generated as a probability-weighted average of each subtomogram in each orientation. This process is iterated with finer samplings of orientational space until the model converges—the likelihood is maximized. ML approaches are distinct from the "maximum cross-correlation" approaches described earlier, in which structures are determined by finding the rotational and translational parameters that maximize the CCC between each subtomogram with the reference. Maximum CC methods seek to find the "correct" orientation for each subtomogram and each subtomogram contributes to the next reference in only one orientation. In contrast ML methods model the probability of each subtomogram in all orientations, and each subtomogram contributes to the next reference in all orientations, weighted by the determined probability. True ML approaches require evenly spaced complete sampling of parameter space to prevent orientational bias; ie, all possible angles and shifts must be sampled with evenly spaced steps. Due to this exhaustive sampling of the parameter space, ML approaches should provide the optimum solution for the given starting reference with minimal user input, but this exhaustive sampling results in substantially more computational cost than maximum CC approaches. Heuristic approaches can be used to limit the search space to more probable orientations; this can save computational costs but may reduce the likelihood of finding the true optimum solution. It is not necessarily the case that ML approaches provide inherently better results than maximum CC approaches.

Classification can be performed using ML approaches by increasing the number of reference structures (Scheres, 2010). This is a relatively simple addition to the statistical model as it only results in an additional parameter for which the probabilities are calculated. While this is a multireference approach, it is distinct from those normally found in maximum CC methods, as it does not require an explicit decision as to which class each subtomogram belongs to.

6.2 Current ML Implementations

Several ML approaches to subtomogram averaging have been described. While 2D ML approaches can be in real or Fourier space, ML approaches to subtomogram averaging work in Fourier space because of the requirement for missing wedge compensation. Subtomogram ML as implemented in the Xmipp package uses a relatively conservative approach to dealing with the missing wedge; the missing data are substituted with the previous reference, tempering the rate at which the new reference diverges from the old references (Scheres, Melero, Valle, & Carazo, 2009). Updating the reference is less sensitive to changes in undersampled regions of Fourier space, reducing problems with reference bias where the particles have a preferred orientation relative to the wedge. In MLTOMO, data in the missing region are simply not considered by applying a wedge mask to the reference structure (Stölken et al., 2011). To account for preferred orientations, a "compound wedge" is calculated by summing the missing wedges of all the oriented subtomograms, effectively producing a weighting mask describing sampling from the dataset. The compound wedge is applied to the subtomograms, thus giving the subtomogram and the reference the same wedge function prior to the calculation of probabilities.

The RELION approach for subtomogram averaging is distinct from other ML and maximum CC approaches in that it incorporates a CTF model into the ML optimization protocol (Bharat et al., 2015). As with the CTF model in single-particle RELION, the SNR is a parameter that is explicitly estimated in each iteration and used to calculate the structure using a Wiener filter leading to improved image restoration. Missing wedge compensation is combined into this CTF correction model—the missing regions are treated similarly to nodes in the CTF. The input data for RELION subtomogram averaging are non-CTF-corrected subtomograms. The CTF is corrected by Wiener filtering each corresponding Fourier slice (see Section 2). RELION has been successfully applied to determine subnanometer resolution structures of test samples (Bharat et al., 2015).

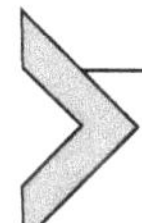

7. COMBINATIONS OF SUBTOMOGRAM AVERAGING AND 2D APPROACHES

7.1 Using 3D Data to Aid 2D Reconstructions

2D and 3D alignment and averaging approaches each have their own strengths and weaknesses; in certain situations, combining the two approaches can be an effective approach to structure determination. As noted earlier, subtomogram averaging has more complicated data collection and increased computational cost—it is generally not a good idea to use subtomogram averaging in situations where single-particle approaches are sufficient. However, subtomogram averaging has a number of features that can make it a useful tool to generate starting models for subsequent single-particle reconstruction. First, the initial tomographic reconstruction results in unbiased 3D reconstructions, as fiducial markers are used to drive the tilt-series alignment. Second, where averaging is required to generate a starting model, the higher dimensionality of subtomogram data may lead to more robust reference-free generation of starting models as compared to generating initial structures from projection images. Third, initial structures from tomographic data have absolute handedness, thus eliminating an ambiguity inherent to single-particle approaches.

For helical reconstruction, subtomogram averaging has been shown to useful in determining helical parameters (Bharat, Davey, et al., 2012; Bharat, Noda, et al., 2012; Bharat et al., 2011; Skruzny et al., 2015). Subtomogram alignment provides the coordinates of each subunit in 3D space, which can be used to directly determine average helical parameters, largely supplanting more difficult methods such as mass-per-length measurements from scanning transmission electron microscopy or helical indexing from X-ray diffraction. As with single-particle methods, subtomogram alignment also provides the absolute handedness of helical assemblies, supplanting methods such as metal shadowing, scanning electron microscopy, and atomic force microscopy. In situations where the helical parameters are variable, collection of a high-dose image followed by a tilt-series allows the local helical parameters to be determined by subtomogram averaging, and the high-resolution reconstruction to be generated from the 2D image (Bharat, Davey, et al., 2012; Bharat, Noda, et al., 2012).

7.2 Improving Subtomogram Averages with 2D Alignment

Alignment of the initial tomographic tilt-series is error prone. Beam-induced deformation may not be optimally determined using the fiducials present and may be difficult to compensate for. The tomogram may be optimally aligned in the center, but where the field of view is large, it may be poorly aligned toward the edges. Such distortions result in the particle projections in each tilt image not back projecting perfectly into the same volume. This effect can be seen by using the positions of aligned subtomograms to extract their particle projections from the tilt images; in distorted images, the particles will not appear centered in each projection. A number of approaches have been proposed to optimize tilt-series alignment after subtomogram alignment, guided by 2D alignment against the final reference generated by subtomogram averaging.

One method proposes iteratively reconstructing the subtomogram and using the subtomogram to refine the translations of the tilt-image particle projections (Zhang & Ren, 2012). In such an approach, the tilt angles are known, making the refinement problem two-dimensional. In some cases this may locally compensate for distortions across the tomogram and results in higher resolution subtomograms. A similar approach for 2D alignment has been applied to a high-resolution reconstruction approach called constrained single-particle tomography (CSPT; Bartesaghi, Lecumberry, Sapiro, & Subramaniam, 2012). This procedure begins with subtomogram averaging to determine a starting model and initial orientations. Based on this information the particle projections, their respective angles, and their individual defocus values in the original tilt-series are calculated. The particle projections are then CTF corrected and reconstructed into a reference structure, which is then used for refinement of projection rotations and translations using projection matching. The projection orientations for each particle are constrained to each other, enforcing the knowledge that they are views of the same particle from the same tilt-series. The new orientations are used to refine the tilt axis and tilt angle of the tilt-series, effectively transforming the particle projections into fiducial markers. The projection orientations and tomographic tilts are then iteratively refined until convergence. CSPT has been shown capable of determining subnanometer resolution structures.

Both of these approaches depend on the ability to perform projection matching between the particle projections in each tilt image and the subtomogram average. Successful projection matching depends on the ability to find a peak corresponding to the true orientation and translation within CC maps. Using the alignment data and constraints determined by subtomogram averaging limits the search space and increases the probability of success. Projection matching methods may not always work when overlapping densities are projected into the same volume, as these densities confound attempts to align individual subunits. These approaches may therefore not be applicable to many problems, such as complexes within cells, where subtomogram averaging is required—they have so far only been successfully applied where single-particle methods would suffice.

Performing 2D alignments of the projections of particles in the tilt-series against the subtomogram average is possible for a single-particle sample because in that case what we call the subtomogram central section approximation is valid. The subtomogram central section approximation is the approximation that the Fourier transform of the part of the raw tilt image that projects into the subtomogram is the same as the corresponding Fourier slice of the subtomogram. For single-particle samples this is a reasonable approximation. In most samples where subtomogram averaging is required the approximation is poor because objects above and below the object of interest in the direction of view contribute to the raw tilt image but have been removed from the subtomogram by boxing this region out of the full tomogram. Subtomogram extraction is a convolution in Fourier space that causes information from each angled slice in the Fourier transform of the tomogram to contribute to multiple slices in the Fourier transform of the subtomogram. The subtomogram central section approximation is also being used when subtomograms are generated directly from the tilt data without reconstructing the surrounding volume, or when CTF correction is applied to Fourier slices from the subtomogram. Caution is needed when using single-particle samples to test subtomogram averaging methods that make use of the subtomogram central section approximation.

Despite this caveat there are multiple samples where using a combination of subtomogram averaging and 2D approaches is likely to be effective.

8. SOFTWARE FOR SUBTOMOGRAM AVERAGING

A number of subtomogram averaging packages are available. The differences between packages include the programming languages they are written in, the formats in which they handle subtomogram and alignment data, and the parts of the overall processing workflow they perform. The TOM and AV3 packages are written in MATLAB, and together form a complete package for tomogram reconstruction and subtomogram averaging (Förster et al., 2005; Nickell et al., 2005). PyTOM is a similar package implemented in Python (Hrabe et al., 2012), but with an additional user interface and additional methods for subtomogram classification (Chen et al., 2014; Hrabe et al., 2012), FRM (Chen et al., 2013), and INFR tomogram reconstruction (Chen & Förster, 2014). IMOD and PEET in conjunction also provide an integrated workflow for tomographic reconstruction and subtomogram averaging (Kremer, Mastronarde, & McIntosh, 1996; Nicastro et al., 2006). Dynamo is a subtomogram averaging package developed in MATLAB that allows for particle picking, subtomogram averaging, and subtomogram classification (Castaño-Díez et al., 2012). MLTOMO is a ML-based subtomogram averaging workflow written in MATLAB (Stölken et al., 2011). A number of single-particle reconstruction packages have been expanded to include subtomogram averaging workflows. EMAN has been expanded to include subtomogram averaging methods which can be used along with its single-particle tools (Galaz-Montoya et al., 2015). The Xmipp package contains a number of programs for ML-based subtomogram averaging (Scheres et al., 2009). ML-based subtomogram averaging has also been implemented in the RELION package (Bharat et al., 2015).

9. CONCLUSIONS

Subtomogram averaging is a structure determination method that uniquely allows for studying macromolecules in their native environments. While the highest resolutions of published subtomogram averaging structures are behind those of single-particle structures, most of them have been determined in situ, that is, the macromolecular complexes have not been purified out of their native environments. There is no theoretical basis restricting the attainable resolution of subtomogram averaging, and higher resolution structures will be determined in the future.

Even without trying to perform high-resolution structure determination, subtomogram averaging can provide useful information toward solving 2D averaging problems. The 3D nature of tomographic data may make the reference-free generation of 3D models by subtomogram averaging more straightforward than with 2D methods. The 3D nature of tomographic data also provides information on absolute hand absent in projection images. For helical reconstruction, subtomogram averaging provides a way of determining helical parameters directly from the data that will be used for 3D reconstruction. As such, subtomogram averaging should not be overlooked as useful complementary method.

Despite recent advances in the field of cryo-EM, many technical limitations still remain; this is particularly true in cryo-ET and subtomogram averaging, which bring their own set of additional challenges. Specimen movement, poor SNR, and challenging defocus determination and CTF correction are inherent issues in cryo-EM; low electron dose, increased stage movement, and specimen tilt exacerbate these problems in cryo-ET. A problem that is particularly limiting in cryo-ET is data collection throughput. It is time consuming to collect large numbers of high-quality tilt-series. The higher dimensionality and subsequently increased computing requirements, as well as the missing wedge problem, are issues that are unavoidable, making subtomogram averaging inherently more complex than single-particle methods.

The cryo-EM and ET fields are making progress toward solving the problems outlined earlier. In the future we expect that the growing user base for subtomogram averaging will lead to further, cryo-ET specific, progress. Recent developments in highly stable specimen supports have been shown to provide some improvement in subtomogram averaging (Bharat et al., 2015; Russo & Passmore, 2014). Recent developments in phase plates have also shown substantial improvements in contrast (Danev, Buijsse, Khoshouei, Plitzko, & Baumeister, 2014; Danev, Kanamaru, Marko, & Nagayama, 2010; Malac, Beleggia, Kawasaki, Li, & Egerton, 2012); contrast enhancement with phase plates may expand the range of problems that are tractable with cryo-ET and subtomogram averaging by allowing for averaging of thicker specimens and smaller molecules (Asano et al., 2015; Fukuda, Laugks, Lučić, Baumeister, & Danev, 2015; Murata et al., 2010). Automated microscope alignment and increasingly powerful automated tomogram collection software will likely make high-quality cryo-ET data collection more accessible. The continuing development of methods for automated tilt-series alignment and tomogram reconstruction algorithms will likely eliminate some of the current workflow bottlenecks, better preserve high-resolution data, and improve signal to noise. With the use of current and developing methods, subtomogram averaging will continue to advance, providing an increasingly powerful and unique method for high-resolution structure determination in situ.

ACKNOWLEDGMENTS

We would like to thank Wim J.H. Hagen and Florian K.M. Schur for helpful discussions regarding the manuscript. W.W. was supported by an EMBO long-term fellowship ALTF 748-2014.

REFERENCES

Amat, F., Castaño-Diez, D., Lawrence, A., Moussavi, F., Winkler, H., & Horowitz, M. (2010). Alignment of cryo-electron tomography datasets. In G. J. Jensen (Ed.), *Methods in enzymology: Vol. 482* (pp. 343–367). New York, NY: Academic Press. chapter 13.

Andersen, A. H., & Kak, A. C. (1984). Simultaneous algebraic reconstruction technique (SART): A superior implementation of the art algorithm. *Ultrasonic Imaging*, *6*, 81–94.

Asano, S., Fukuda, Y., Beck, F., Aufderheide, A., Förster, F., Danev, R., et al. (2015). A molecular census of 26S proteasomes in intact neurons. *Science*, *347*, 439–442.

Bartesaghi, A., Lecumberry, F., Sapiro, G., & Subramaniam, S. (2012). Protein secondary structure determination by constrained single-particle cryo-electron tomography. *Structure*, *20*, 2003–2013.

Bartesaghi, A., Merk, A., Banerjee, S., Matthies, D., Wu, X., Milne, J. L. S., et al. (2015). 2.2 Å Resolution cryo-EM structure of β-galactosidase in complex with a cell-permeant inhibitor. *Science*, *348*, 1147–1151.

Bartesaghi, A., Sprechmann, P., Liu, J., Randall, G., Sapiro, G., & Subramaniam, S. (2008). Classification and 3D averaging with missing wedge correction in biological electron tomography. *Journal of Structural Biology, 162*, 436–450.

Bharat, T. A. M., Davey, N. E., Ulbrich, P., Riches, J. D., de Marco, A., Rumlova, M., et al. (2012). Structure of the immature retroviral capsid at 8 Å resolution by cryo-electron microscopy. *Nature, 487*, 385–389.

Bharat, T. A. M., Noda, T., Riches, J. D., Kraehling, V., Kolesnikova, L., Becker, S., et al. (2012). Structural dissection of Ebola virus and its assembly determinants using cryo-electron tomography. *Proceedings of the National Academy of Sciences of the United States of America, 109*, 4275–4280.

Bharat, T. A. M., Riches, J. D., Kolesnikova, L., Welsch, S., Krähling, V., Davey, N., et al. (2011). Cryo-electron tomography of Marburg virus particles and their morphogenesis within infected cells. *PLoS Biology, 9*, e1001196.

Bharat, T. A. M., Russo, C. J., Löwe, J., Passmore, L. A., & Scheres, S. H. W. (2015). Advances in single-particle electron cryomicroscopy structure determination applied to sub-tomogram averaging. *Structure, 23*, 1743–1753.

Böhm, J., Frangakis, A. S., Hegerl, R., Nickell, S., Typke, D., & Baumeister, W. (2000). Toward detecting and identifying macromolecules in a cellular context: Template matching applied to electron tomograms. *Proceedings of the National Academy of Sciences of the United States of America, 97*, 14245–14250.

Bracewell, R. N., & Riddle, A. (1967). Inversion of fan-beam scans in radio astronomy. *The Astrophysical Journal, 150*, 427.

Brandt, S. S. (2005). Markerless alignment in electron tomography. In J. Frank (Ed.), *Electron tomography* (pp. 187–215). New York, NY: Springer.

Briggs, J. A. G. (2013). Structural biology in situ—The potential of subtomogram averaging. *Current Opinion in Structural Biology, 23*, 261–267.

Briggs, J. A. G., Riches, J. D., Glass, B., Bartonova, V., Zanetti, G., & Kräusslich, H.-G. (2009). Structure and assembly of immature HIV. *Proceedings of the National Academy of Sciences of the United States of America, 106*, 11090–11095.

Brilot, A. F., Chen, J. Z., Cheng, A., Pan, J., Harrison, S. C., Potter, C. S., et al. (2012). Beam-induced motion of vitrified specimen on holey carbon film. *Journal of Structural Biology, 177*, 630–637.

Campbell, M. G., Cheng, A., Brilot, A. F., Moeller, A., Lyumkis, D., Veesler, D., et al. (2012). Movies of ice-embedded particles enhance resolution in electron cryo-microscopy. *Structure, 20*, 1823–1828.

Carragher, B., Kisseberth, N., Kriegman, D., Milligan, R. A., Potter, C. S., Pulokas, J., et al. (2000). Leginon: An automated system for acquisition of images from vitreous ice specimens. *Journal of Structural Biology, 132*, 33–45.

Castaño-Díez, D., Kudryashev, M., Arheit, M., & Stahlberg, H. (2012). Dynamo: A flexible, user-friendly development tool for subtomogram averaging of cryo-EM data in high-performance computing environments. *Journal of Structural Biology, 178*, 139–151.

Castaño-Díez, D., Scheffer, M., Al-Amoudi, A., & Frangakis, A. S. (2010). Alignator: A GPU powered software package for robust fiducial-less alignment of cryo tilt-series. *Journal of Structural Biology, 170*, 117–126.

Chen, Y., & Förster, F. (2014). Iterative reconstruction of cryo-electron tomograms using nonuniform fast Fourier transforms. *Journal of Structural Biology, 185*, 309–316.

Chen, Y., Pfeffer, S., Fernández, J. J., Sorzano, C. O. S., & Förster, F. (2014). Autofocused 3D classification of cryoelectron subtomograms. *Structure, 22*, 1528–1537.

Chen, Y., Pfeffer, S., Hrabe, T., Schuller, J. M., & Förster, F. (2013). Fast and accurate reference-free alignment of subtomograms. *Journal of Structural Biology, 182*, 235–245.

Danev, R., Buijsse, B., Khoshouei, M., Plitzko, J. M., & Baumeister, W. (2014). Volta potential phase plate for in-focus phase contrast transmission electron microscopy.

Proceedings of the National Academy of Sciences of the United States of America, *111*, 15635–15640.

Danev, R., Kanamaru, S., Marko, M., & Nagayama, K. (2010). Zernike phase contrast cryo-electron tomography. *Journal of Structural Biology*, *171*, 174–181.

De Rosier, D. J., & Klug, A. (1968). Reconstruction of three dimensional structures from electron micrographs. *Nature*, *217*, 130–134.

Dierksen, K., Typke, D., Hegerl, R., Koster, A. J., & Baumeister, W. (1992). Towards automatic electron tomography. *Ultramicroscopy*, *40*, 71–87.

Fernández, J. J., Li, S., & Crowther, R. A. (2006). CTF determination and correction in electron cryotomography. *Ultramicroscopy*, *106*, 587–596.

Fernández, J.-J., Sanjurjo, J., & Carazo, J.-M. (1997). A spectral estimation approach to contrast transfer function detection in electron microscopy. *Ultramicroscopy*, *68*, 267–295.

Fischer, N., Neumann, P., Konevega, A. L., Bock, L. V., Ficner, R., Rodnina, M. V., et al. (2015). Structure of the E. coli ribosome-EF-Tu complex at <3 A resolution by Cs-corrected cryo-EM. *Nature*, *520*, 567–570.

Förster, F., Han, B.-G., & Beck, M. (2010). Visual proteomics. In G. J. Jensen (Ed.), *Methods in enzymology: Vol. 483* (pp. 215–243). New York, NY: Academic Press. chapter 11.

Förster, F., & Hegerl, R. (2007). Structure determination in situ by averaging of tomograms. In J. R. McIntosh (Ed.), *Methods in cell biology: Vol. 79* (pp. 741–767). New York, NY: Academic Press.

Förster, F., Medalia, O., Zauberman, N., Baumeister, W., & Fass, D. (2005). Retrovirus envelope protein complex structure in situ studied by cryo-electron tomography. *Proceedings of the National Academy of Sciences of the United States of America*, *102*, 4729–4734.

Förster, F., Pruggnaller, S., Seybert, A., & Frangakis, A. S. (2008). Classification of cryo-electron sub-tomograms using constrained correlation. *Journal of Structural Biology*, *161*, 276–286.

Frangakis, A. S., Böhm, J., Förster, F., Nickell, S., Nicastro, D., Typke, D., et al. (2002). Identification of macromolecular complexes in cryoelectron tomograms of phantom cells. *Proceedings of the National Academy of Sciences of the United States of America*, *99*, 14153–14158.

Frank, J. (2006). *Three-dimensional electron microscopy of macromolecular assemblies: Visualization of biological molecules in their native state*. New York, NY: Oxford University Press.

Fukuda, Y., Laugks, U., Lučić, V., Baumeister, W., & Danev, R. (2015). Electron cryotomography of vitrified cells with a Volta phase plate. *Journal of Structural Biology*, *190*, 143–154.

Galaz-Montoya, J. G., Flanagan, J., Schmid, M. F., & Ludtke, S. J. (2015). Single particle tomography in EMAN2. *Journal of Structural Biology*, *190*, 279–290.

Gilbert, P. (1972). Iterative methods for the three-dimensional reconstruction of an object from projections. *Journal of Theoretical Biology*, *36*, 105–117.

Gordon, R., Bender, R., & Herman, G. T. (1970). Algebraic reconstruction techniques (ART) for three-dimensional electron microscopy and X-ray photography. *Journal of Theoretical Biology*, *29*, 471–481.

Grant, T., & Grigorieff, N. (2015). Measuring the optimal exposure for single particle cryo-EM using a 2.6 Å reconstruction of rotavirus VP6. *eLife*, *4*, e06980.

Hagen, W.J.H., Wan, W., & Briggs, J.A.G. (2016). Implementation of a cryo-electron tomography tilt-scheme optimized for high resolution subtomogram averaging, submitted for publication.

Han, R., Wang, L., Liu, Z., Sun, F., & Zhang, F. (2015). A novel fully automatic scheme for fiducial marker-based alignment in electron tomography. *Journal of Structural Biology*, *192*, 403–417.

Hegerl, R., & Frangakis, A. S. (2005). Denoising of electron tomograms. In J. Frank (Ed.), *Electron tomography* (pp. 331–352). New York, NY: Springer.

Heumann, J. M., Hoenger, A., & Mastronarde, D. N. (2011). Clustering and variance maps for cryo-electron tomography using wedge-masked differences. *Journal of Structural Biology*, *175*, 288–299.

Hrabe, T., Chen, Y., Pfeffer, S., Kuhn Cuellar, L., Mangold, A.-V., & Förster, F. (2012). PyTom: A python-based toolbox for localization of macromolecules in cryo-electron tomograms and subtomogram analysis. *Journal of Structural Biology*, *178*, 177–188.

Jensen, G. J. (2001). Alignment error envelopes for single particle analysis. *Journal of Structural Biology*, *133*, 143–155.

Jensen, G. J., & Kornberg, R. D. (2000). Defocus-gradient corrected back-projection. *Ultramicroscopy*, *84*, 57–64.

Koster, A. J., Van den Bos, A., & van der Mast, K. D. (1987). An autofocus method for a TEM. *Ultramicroscopy*, *21*, 209–222.

Kovacs, J. A., & Wriggers, W. (2002). Fast rotational matching. *Acta Crystallographica Section D, Biological Crystallography*, *58*, 1282–1286.

Kremer, J. R., Mastronarde, D. N., & McIntosh, J. R. (1996). Computer visualization of three-dimensional image data using IMOD. *Journal of Structural Biology*, *116*, 71–76.

Kunz, M., & Frangakis, A. S. (2014). Super-sampling SART with ordered subsets. *Journal of Structural Biology*, *188*, 107–115.

Li, X., Mooney, P., Zheng, S., Booth, C., Braunfeld, M. B., Gubbens, S., et al. (2013). Electron counting and beam-induced motion correction enable near atomic resolution single particle cryoEM. *Nature Methods*, *10*, 584–590.

Lučić, V., Rigort, A., & Baumeister, W. (2013). Cryo-electron tomography: The challenge of doing structural biology in situ. *The Journal of Cell Biology*, *202*, 407–419.

Malac, M., Beleggia, M., Kawasaki, M., Li, P., & Egerton, R. F. (2012). Convenient contrast enhancement by a hole-free phase plate. *Ultramicroscopy*, *118*, 77–89.

Marabini, R., Herman, G. T., & Carazo, J. M. (1998). 3D reconstruction in electron microscopy using ART with smooth spherically symmetric volume elements (blobs). *Ultramicroscopy*, *72*, 53–65.

Mastronarde, D. N. (2005). Automated electron microscope tomography using robust prediction of specimen movements. *Journal of Structural Biology*, *152*, 36–51.

Mindell, J. A., & Grigorieff, N. (2003). Accurate determination of local defocus and specimen tilt in electron microscopy. *Journal of Structural Biology*, *142*, 334–347.

Murata, K., Liu, X., Danev, R., Jakana, J., Schmid, M. F., King, J., et al. (2010). Zernike phase contrast cryo-electron microscopy and tomography for structure determination at nanometer and subnanometer resolutions. *Structure*, *18*, 903–912.

Nicastro, D., Schwartz, C., Pierson, J., Gaudette, R., Porter, M. E., & McIntosh, J. R. (2006). The molecular architecture of axonemes revealed by cryoelectron tomography. *Science*, *313*, 944–948.

Nickell, S., Förster, F., Linaroudis, A., Net, W. D., Beck, F., Hegerl, R., et al. (2005). TOM software toolbox: Acquisition and analysis for electron tomography. *Journal of Structural Biology*, *149*, 227–234.

Pfeffer, S., Burbaum, L., Unverdorben, P., Pech, M., Chen, Y., Zimmermann, R., et al. (2015). Structure of the native Sec61 protein-conducting channel. *Nature Communications*, *6*, 8403.

Rademacher, M. (2005). Weighted back-projection methods. In J. Frank (Ed.), *Electron tomography* (pp. 245–273). New York, NY: Springer.

Rohou, A., & Grigorieff, N. (2015). CTFFIND4: Fast and accurate defocus estimation from electron micrographs. *Journal of Structural Biology*, *192*, 216–221.

Russo, C. J., & Passmore, L. A. (2014). Ultrastable gold substrates for electron cryomicroscopy. *Science*, *346*, 1377–1380.

Saxton, W. O., Baumeister, W., & Hahn, M. (1984). Three-dimensional reconstruction of imperfect two-dimensional crystals. *Ultramicroscopy*, *13*, 57–70.

Scheres, S. H. W. (2010). Classification of structural heterogeneity by maximum-likelihood methods. In G. J. Jensen (Ed.), *Methods in enzymology: Vol. 482* (pp. 295–320). New York, NY: Academic Press. chapter 11.
Scheres, S. H. W., Melero, R., Valle, M., & Carazo, J.-M. (2009). Averaging of electron subtomograms and random conical tilt reconstructions through likelihood optimization. *Structure, 17*, 1563–1572.
Schmid, M. F. (2011). Single-particle electron cryotomography (cryoET). In J. L. Steven & B. V. V. Prasad (Eds.), *Advances in protein chemistry and structural biology: Vol. 82* (pp. 37–65). New York, NY: Academic Press. chapter 2.
Schmid, M. F., & Booth, C. R. (2008). Methods for aligning and for averaging 3D volumes with missing data. *Journal of Structural Biology, 161*, 243–248.
Schur, F.K.M., Obr, M., Hagen, W.J.H., Wan, W., Jakobi, A.J., Kirkpatrick, J.M., et al. (2016). An atomic model of the HIV-1 capsid-SP1 layer reveals structures regulating assembly and maturation, submitted for publication.
Schur, F. K. M., Hagen, W. J. H., de Marco, A., & Briggs, J. A. G. (2013). Determination of protein structure at 8.5 Å resolution using cryo-electron tomography and sub-tomogram averaging. *Journal of Structural Biology, 184*, 394–400.
Sigworth, F. J., Doerschuk, P. C., Carazo, J.-M., & Scheres, S. H. W. (2010). An introduction to maximum-likelihood methods in cryo-EM. In G. J. Jensen (Ed.), *Methods in enzymology: Vol. 482* (pp. 263–294). New York, NY: Academic Press. chapter 10.
Skruzny, M., Desfosses, A., Prinz, S., Dodonova, S. O., Gieras, A., Uetrecht, C., et al. (2015). An organized co-assembly of clathrin adaptors is essential for endocytosis. *Developmental Cell, 33*, 150–162.
Stölken, M., Beck, F., Haller, T., Hegerl, R., Gutsche, I., Carazo, J.-M., et al. (2011). Maximum likelihood based classification of electron tomographic data. *Journal of Structural Biology, 173*, 77–85.
Taylor, K. A., Schmitz, H., Reedy, M. C., Goldman, Y. E., Franzini-Armstrong, C., Sasaki, H., et al. (1999). Tomographic 3D reconstruction of quick-frozen, Ca^{2+}-activated contracting insect flight muscle. *Cell, 99*, 421–431.
Turoňová, B., Marsalek, L., Davidovič, T., & Slusallek, P. (2015). Progressive stochastic reconstruction technique (PSRT) for cryo electron tomography. *Journal of Structural Biology, 189*, 195–206.
Turoňová, B., Marsalek, L., & Slusallek, P. (2016). On geometric artifacts in cryo electron tomography. *Ultramicroscopy, 163*, 48–61.
Voortman, L. M., Franken, E. M., van Vliet, L. J., & Rieger, B. (2012). Fast, spatially varying CTF correction in TEM. *Ultramicroscopy, 118*, 26–34.
Voortman, L. M., Stallinga, S., Schoenmakers, R. H. M., van Vliet, L. J., & Rieger, B. (2011). A fast algorithm for computing and correcting the CTF for tilted, thick specimens in TEM. *Ultramicroscopy, 111*, 1029–1036.
White, T. A., Bartesaghi, A., Borgnia, M. J., Meyerson, J. R., de la Cruz, M. J. V., Bess, J. W., et al. (2010). Molecular architectures of trimeric SIV and HIV-1 envelope glycoproteins on intact viruses: Strain-dependent variation in quaternary structure. *PLoS Pathogens, 6*, e1001249.
Xiong, Q., Morphew, M. K., Schwartz, C. L., Hoenger, A. H., & Mastronarde, D. N. (2009). CTF determination and correction for low dose tomographic tilt series. *Journal of Structural Biology, 168*, 378–387.
Xu, M., Beck, M., & Alber, F. (2012). High-throughput subtomogram alignment and classification by Fourier space constrained fast volumetric matching. *Journal of Structural Biology, 178*, 152–164.
Yahav, T., Maimon, T., Grossman, E., Dahan, I., & Medalia, O. (2011). Cryo-electron tomography: Gaining insight into cellular processes by structural approaches. *Current Opinion in Structural Biology, 21*, 670–677.

Zanetti, G., Prinz, S., Daum, S., Meister, A., Schekman, R., Bacia, K., et al. (2013). The structure of the COPII transport-vesicle coat assembled on membranes. *eLife*, *2*, e00951.

Zanetti, G., Riches, J. D., Fuller, S. D., & Briggs, J. A. G. (2009). Contrast transfer function correction applied to cryo-electron tomography and sub-tomogram averaging. *Journal of Structural Biology*, *168*, 305–312.

Zhang, P. (2013). Correlative cryo-electron tomography and optical microscopy of cells. *Current Opinion in Structural Biology*, *23*, 763–770.

Zhang, L., & Ren, G. (2012). IPET and FETR: Experimental approach for studying molecular structure dynamics by cryo-electron tomography of a single-molecule structure. *PloS One*, *7*, e30249.

Zheng, S. Q., Keszthelyi, B., Branlund, E., Lyle, J. M., Braunfeld, M. B., Sedat, J. W., et al. (2007). UCSF tomography: An integrated software suite for real-time electron microscopic tomographic data collection, alignment, and reconstruction. *Journal of Structural Biology*, *157*, 138–147.

Ziese, U., Janssen, A. H., Murk, J. L., Geerts, W. J. C., Van der Krift, T., Verkleij, A. J., et al. (2002). Automated high-throughput electron tomography by pre-calibration of image shifts. *Journal of Microscopy*, *205*, 187–200.

CHAPTER FOURTEEN

High-Resolution Macromolecular Structure Determination by MicroED, a cryo-EM Method

J.A. Rodriguez*, T. Gonen†,1
*UCLA-DOE Institute, University of California, Los Angeles CA, United States
†Janelia Research Campus, Howard Hughes Medical Institute, Ashburn VA, United States
[1]Corresponding author: e-mail address: gonent@janelia.hhmi.org

Contents

Abstract

Microelectron diffraction (MicroED) is a new cryo-electron microscopy (cryo-EM) method capable of determining macromolecular structures at atomic resolution from vanishingly small 3D crystals. MicroED promises to solve atomic resolution structures from even the tiniest of crystals, less than a few hundred nanometers thick. MicroED complements frontier advances in crystallography and represents part of the rebirth of cryo-EM that is making macromolecular structure determination more accessible

Methods in Enzymology, Volume 579
ISSN 0076-6879
http://dx.doi.org/10.1016/bs.mie.2016.04.017

for all. Here we review the concept and practice of MicroED, for both the electron microscopist and crystallographer. Where other reviews have addressed specific details of the technique (Hattne et al., 2015; Shi et al., 2016; Shi, Nannenga, Iadanza, & Gonen, 2013), we aim to provide context and highlight important features that should be considered when performing a MicroED experiment.

1. INTRODUCTION

Over the past century, X-ray crystallography has revealed the structures of thousands of macromolecules. Today, the frontier of structural biology is rapidly being expanded by cryo-electron microscopy (cryo-EM) thanks to a new generation of electron microscopes and detectors, and the emergence of new techniques (Cheng, Grigorieff, Penczek, & Walz, 2015; Shi, Nannenga, Iadanza, & Gonen, 2013). Four cryo-EM methods benefit from these improvements: tomography, single particle, 2D electron crystallography, and microelectron diffraction (MicroED). MicroED is a cryo-EM method that determines atomic resolution structures from three-dimensional protein crystals only hundreds of nanometers in thickness and in doing so promises to provide a new approach to structure determination of macromolecular structures (Shi et al., 2013). MicroED leverages infrastructure created by X-ray crystallography for structure determination while exploiting the strong interaction between electrons and the crystal (Henderson, 1995; Nannenga & Gonen, 2014). This strong interaction arises from the intrinsic nature of electrons as charged particles. Electrons have a higher scattering cross-section and are sensitive to charge. Magnetic lenses can be used to focus electrons and produce high-resolution images or diffraction and, on a practical note, electron microscopes are more accessible and less expensive than high-flux X-ray sources. New electron sources can even generate ultrafast pulses that probe matter at femtosecond timescales (Weathersby et al., 2015). Empowered by these new methods, electron microscopists can now determine the structures of protein molecules at high resolutions from images (Cheng et al., 2015) or diffraction (Rodriguez et al., 2015), and produce atomic resolution structures from crystals one protein layer thick (Gonen et al., 2005). In what follows we present a brief overview of electron diffraction and review the tools and techniques used by MicroED.

2. BACKGROUND

2.1 Origins of Electron Diffraction

The wave nature of the electron was first confirmed by diffraction of electrons from a crystal of nickel in 1927. The demonstration by Davisson and Germer using low-energy electrons (Davisson & Germer, 1927a, 1927b) linked their scattering properties to the de Broglie wavelength. The experiment was echoed shortly thereafter by Thomson and Reid who used higher energy electrons to observe diffraction from a thin celluloid film (Thomson & Reid, 1927). These first experiments highlight unique properties of electrons compared to X-rays: electrons benefit from intrinsically higher scattering and are easier to manipulate than X-rays, but are also limited in ways that X-rays are not. The inelastic mean free path of electrons is much shorter than for X-rays (Bethe, Rose, & Smith, 1938). Accordingly, samples analyzed by electrons must be smaller than those analyzed by X-rays. This limits electron diffraction (particularly using low-energy electrons) to the analysis of surface atomic layers in a thick crystal (Held, 2012). Higher energy electrons can penetrate thin inorganic nanoparticles and submicron thick organic structures. Using high-energy electron diffraction, structures can be obtained from micron-sized organic crystals that are thin and radiation resistant (Dorset & Hauptman, 1976). However, structure determination by electron diffraction from macromolecular crystals that are sensitive to radiation has required the development of specialized methods in cryo-EM (Shi et al., 2013).

2.2 Transmission Electron Microscopy and Diffraction

The first transmission electron micrograph was captured from magnified images of mesh grids in 1931 through the use of magnetic lenses. For their continuous improvement of electron microscope designs and progress toward the modern microscope, Ernst Ruska, Gerd Binnig, and Heinrich Rohrer would go on to win the Nobel prize in physics in 1986 (Robinson, 1986). Technical improvements and the invention of new devices for electron manipulation during the age of film detectors would build toward the success of modern microscopes. New lens configurations (Cowley, 1969), spatial filters (Danev & Nagayama, 2001), energy discriminators (Henkelman & Ottensmeyer, 1974), and aberration correctors

(Haider et al., 1998) would improve image quality and reduce artifacts (Batson, Dellby, & Krivanek, 2002). Automation and digitization have increased microscope stability and reproducibility of image acquisition (Koster, Chen, Sedat, & Agard, 1992). For biological specimens, cryogenic techniques (Adrian, Dubochet, Lepault, & McDowall, 1984; Christensen, 1971; Cowley, 1964; Taylor & Glaeser, 1976), detector improvements (Cheng et al., 2015; Ruskin, Yu, & Grigorieff, 2013), and new data collection and analysis methods (Shi et al., 2013) have driven the modern revolution in cryo-EM.

The transmission electron microscope (TEM) has made high-resolution imaging and diffraction accessible, but concerns about absorption and multiple-scattering phenomena have curbed the widespread determination of atomic resolution structures obtained using electron diffraction measurements. Electron diffraction even from thin inorganic crystals is influenced by nonkinematical scattering that perturbs its intensities, limiting determination of their structure by crystallographic means (Stern & Taub, 1970). Despite these challenges, in 1976 direct phasing from electron diffraction data was accomplished on an organic compound (Dorset & Hauptman, 1976), demonstrating the potential of electron crystallography at atomic resolution. Meanwhile, in the 1960s, Aaron Klug and colleagues pioneered the use of quantitative electron microscopy for structural studies of biomolecules at the MRC Laboratory of Molecular Biology in Cambridge (Crowther, DeRosier, & Klug, 1970; DeRosier & Klug, 1968). Their efforts forged the path to computerized image processing of TEM images and the 3D reconstruction of macromolecules by electron microscopy. These milestones set the stage for the use of electron diffraction in high-resolution structure determination from ordered protein assemblies, an achievement realized by Henderson and Unwin with their structure of bacteriorhodopsin (Henderson & Unwin, 1975). This was an effort nearly 20 years in the making that culminated in an atomic model of the protein in the early 1990s from 2D crystals (Henderson et al., 1990). Since these initial demonstrations, 2D electron crystallography has produced a handful of high-resolution structures of membrane proteins.

2.3 Electron Diffraction of Protein Assemblies

The study of ordered protein assemblies by electron microscopy dates back to the structure of T4 bacteriophage tails (DeRosier & Klug, 1968). Reconstruction of these assemblies relied on the helical nature of the tails. Similarly,

2D arrays of simple molecules with regular packing could be studied. Since an electron beam can diffract from even a single layer of molecules, thin crystals became an ideal target for electron diffraction. Among the first of these assemblies to be investigated was of a light-driven proton pump in *Halobacterium salinarum*, a transmembrane protein named bacteriorhodopsin or bR for short. This channel constitutes the protein fraction of a 2D crystal known as purple membrane, which exists naturally as part of the membrane of *H. salinarum*. By the mid-1970s Henderson and Unwin had achieved a 7 Å model of bR from a combination of electron diffraction patterns and electron micrographs (Henderson & Unwin, 1975). Over the next decades, their structure would improve to reach atomic resolution (Henderson et al., 1990). While the photoreaction complex would be the first membrane protein structure uncovered by crystallography using X-ray methods (Deisenhofer, Epp, Miki, Huber, & Michel, 1985), the structure of bR would open the door to 2D electron crystallography of a number of other membrane proteins. Structures of the light-harvesting chlorophyll a/b-protein complex (Kühlbrandt, Wang, & Fujiyoshi, 1994), of PhoE porin (Jap, Downing, & Walian, 1990), of aquaporin (Gonen, Sliz, Kistler, Cheng, & Walz, 2004; Hiroaki et al., 2006; Walz et al., 1997), and several others have been deciphered by 2D electron crystallography. Of these, the structure of the lens-specific water pore, aquaporin-0, shows the highest resolution at 1.9 Å (Gonen et al., 2005). The atomic model built for this structure includes coordinates for surrounding lipids and shows details of channel–lipid interactions. It is important to note that for a 2D crystal the phases of the diffraction spots can be recovered from Fourier transforms of images of the crystal, so a 3D map can be directly calculated.

2.4 Electron Diffraction of 3D Protein Crystals: MicroED

The success of 2D electron crystallography is limited by the need to produce well-ordered 2D crystals from proteins (Walz & Grigorieff, 1998). While collection of data is well established (Gonen, 2013), growing 2D crystals remains a challenge. Meanwhile cryo-EM has expanded to include a new method developed by Tamir Gonen and colleagues that marries approaches from electron and X-ray crystallography (Shi et al., 2013). The method, named MicroED, is short for 3D microelectron diffraction (Nannenga & Gonen, 2014). In MicroED, diffraction patterns are collected from submicron thick 3D crystals using a focused low-dose electron beam under cryogenic temperatures. Sampling a crystal at different orientations reveals its

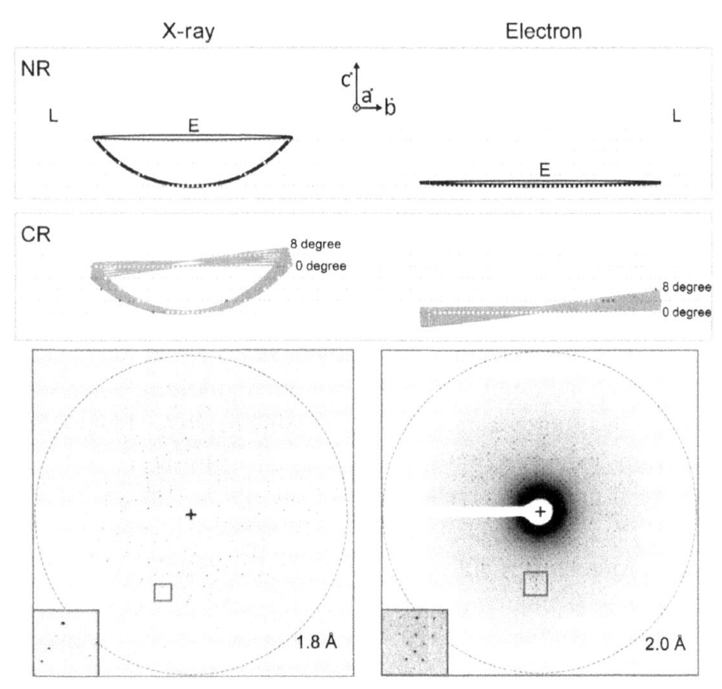

Fig. 1 Comparison between X-ray and electron crystallography. *Left* and *right* panels show schematic representations of two experimental geometries, X-ray (*left*) and electron (*right*) diffraction; distances and images are not to scale. The Ewald sphere (*E*), the *black shell outline* in both, displays significant curvature in X-ray diffraction; its curvature is hardly visible in the case of electron diffraction. The upper panel demonstrates a case where the crystal is kept fixed (*NR*, no rotation) and reflections are measured (*yellow, light gray* in the print version) as they intersect the Ewald sphere. The reciprocal lattice (*L*) is shown in light gray. Reciprocal lattice vectors are represented by a*, b*, and c*. The *middle panel* demonstrates the case where the crystal undergoes a uniform unidirectional rotation of 8 degree (*CR*, continuous rotation). As before, when reflections intersect the Ewald sphere they appear colored, first *yellow* (*light gray* in the print version), *blue* (*gray* in the print version), *green* (*gray* in the print version), *orange* (*light gray* in the print version), and *magenta* (*dark gray* in the print version). The *bottom panel* shows representative diffraction micrographs collected by X-ray and electron diffraction from 3D protein crystals. Insets show magnified areas of interest; the *circles denote resolution*.

reciprocal lattice and allows indexing of its structure factors (Fig. 1). The appeal of MicroED comes in part from its requirement for a simple reconfiguration of readily accessible, commercially available electron microscopy equipment (Table 1 and Fig. 2). MicroED allows macromolecular structure determination from submicron thick protein crystals whose structures might otherwise remain inaccessible. The method was first

Table 1 Technical Specifications of a MicroED Experiment

Property	**Range**
Energy	80–300 keV
Wavelength	0.0418–0.0197 Å
Camera length	200–6000 mm
Tilt range	±70 degree
Energy spread	0.2–3.0 eV

Ranges for energy, wavelength, camera length, tilt range, and energy spread are provided. Typical values for MicroED experiments using a standard TEM instrument are shown.

demonstrated on the well-known structure of lysozyme (Shi et al., 2013). Lysozyme crystals were preserved in a cryogenic state on a standard microscopy grid and diffracted to high resolution. Initially, diffraction patterns were indexed and integrated using custom software (Iadanza & Gonen, 2014). A set of low-dose diffraction patterns acquired at discrete angular increments resulted in a nearly complete data set that was phased by molecular replacement to produce a structure of lysozyme at 2.9 Å resolution (Shi et al., 2013). Shortly after that first demonstration, a continuous rotation approach to data collection (Fig. 1) produced better data with fewer artifacts from partial reflections and diminished multiple-scattering effects. The usefulness of the continuous rotation method was demonstrated on crystals of both lysozyme and catalase (Nannenga, Shi, Hattne, Reyes, & Gonen, 2014; Nannenga, Shi, Leslie, & Gonen, 2014) (Fig. 3). Meanwhile, a group led by Koji Yonekura in Japan worked to solve the structures of catalase and calcium ATPase from 3D crystals using a variation of the MicroED method (Yonekura, Kato, Ogasawara, Tomita, & Toyoshima, 2015). Added to this list is a recent high-resolution structure of proteinase K, determined by MicroED to 1.75 Å resolution (Hattne, Shi, de la Cruz, Reyes, & Gonen, 2016). In a collaborative effort, the groups of Tamir Gonen and David Eisenberg used MicroED to obtain the structure of two amyloid segments of 10 and 11 amino acids from the protein alpha-synuclein which forms pathological deposits in Parkinson's disease. The segments formed crystals a mere 200 by 200 by 1500 nanometers in size, only about an order of magnitude larger than an amyloid fibril. These are the first new structures determined by this method, and at 1.4 Å, the highest resolution structures determined by cryo-EM to date, and allowed determination of the positions of some protons (Rodriguez et al., 2015). In MicroED, as in X-ray

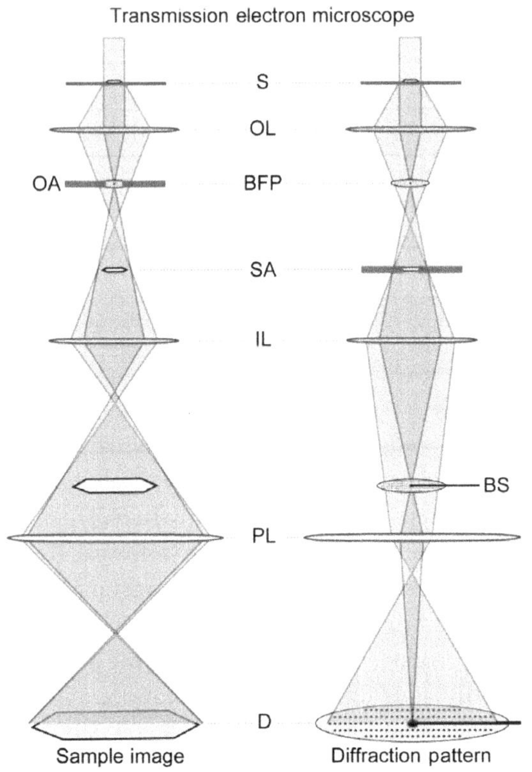

Fig. 2 Comparison of configurations of a transmission electron microscope for imaging and MicroED. A simplified ray diagram demonstrates the principle of each; components and distances are not to scale. A nearly parallel electron beam is shown impinging on a crystal sample (*S*), imaged by an objective lens (*OL*). The *left panel* shows the microscope configuration for obtaining an image of the sample in question; the *right* shows how to configure the microscope to obtain a diffraction pattern. The back focal plane (*BFP*) is shown in each. In imaging mode, a spatial filter, an objective aperture (*OA*) is placed at this plane to provide contrast. The selected area aperture (*SA*) is used to restrict diffraction to a specific region using an aperture as a spatial filter. An intermediate lens system follows (*IL*) which focuses on either the conjugate image plane (*left panel*) or back focal plane (*right panel*) of the objective lens to produce a magnified image or diffraction pattern. In diffraction mode, a beam stop (*BS*) can be placed within or after the projection lens system to obstruct measurement of the focused electron beam. A projection lens system (*PL*) renders the final image or diffraction pattern onto a detector (*D*).

diffraction, only the amplitudes of the diffracted beams are measured but the phases cannot be directly recovered from images in the way they are for 2D crystals, so structures must be solved by indirect means such as molecular replacement.

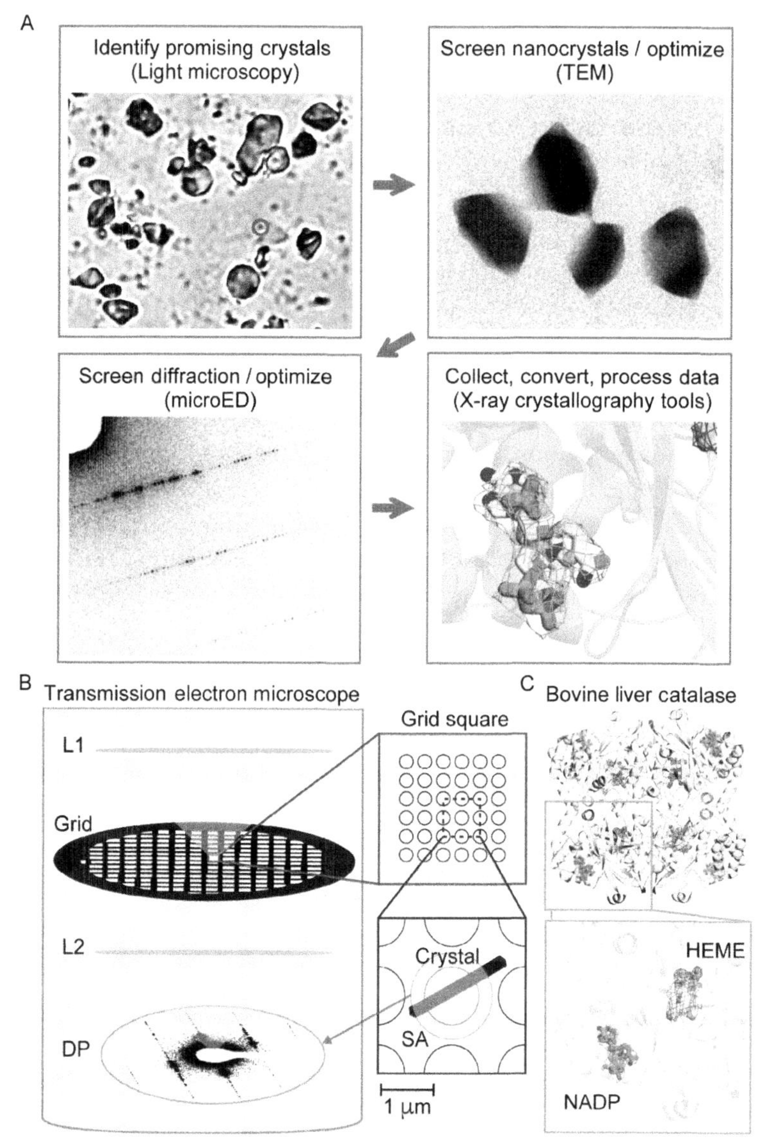

Fig. 3 Pipeline for structure determination by MicroED. (A) A simplified four step process is shown to summarize the process for macromolecular structure determination by MicroED. First, promising protein mixtures are screened by light microscopy for the presence of crystals then by electron microscopy for adequately sized crystals (*red circle* shows a 5 μm diameter). If crystals are present, they can immediately be screened for diffraction on the electron microscope. Diffraction can be optimized until high-resolution diffraction tilt series can be obtained for a given protein. These tilt series images can be converted to SMV format and processed using standard crystallography (*Continued*)

3. SAMPLE PREPARATION

3.1 Approaches to Growth and Screening of Crystals for MicroED

A variety of techniques has been developed for growing protein crystals. At least two of these techniques are used routinely in MicroED experiments: the vapor diffusion technique and the batch approach to crystallization. In vapor diffusion protocols, a drop is prepared that contains both the macromolecule of interest and mother liquor solution, a cocktail of precipitants and ions. This drop is held in a closed environment in the presence of a reservoir solution. Over time, equilibration drives the protein concentration in the drop toward saturation and crystals form. In the batch experiment, a protein sits in mother liquor solution until crystals appear. Both of these preparations are screened visually for crystal formation using optical microscopes (Fig. 3). This screening process presents inherent limitations. If the crystals formed are smaller than the wavelength of visible light, resolving individual crystals by eye becomes difficult. This means submicron thick crystals ideal for MicroED might be missed, and since these crystals may not grow any larger, an opportunity to determine their structure is lost.

To overcome the bottleneck present in screening for tiny crystals, new high-throughput methods are being engineered. One possible solution is

Fig. 3—Cont'd software packages. Structure determination then follows the procedures established for X-ray crystallography to phase by molecular replacement. (B) Process by which diffraction is obtained from a 3D protein crystal using a transmission electron microscope. An electron beam is condensed (*L1*) and illuminates a protein crystal that sits vitrified on a holey carbon film (*inset, top*) within a meshed electron microscopy grid. A representation of the magnified *grid square* is shown to demonstrate the size of the illuminated area relative to the crystal, and the selected area (*inset, bottom*) from which a diffraction pattern (*DP*) is obtained at the back focal plane of a second lens or lens system (*L2*). Projection lenses (not shown, see Fig. 2) image this pattern onto an electronic detector, which digitally records this pattern. The grid containing the crystal can be rotated in a discrete or continuous fashion to produce patterns at various orientations from a single or multiple crystals. These patterns can be processed and phased to determine the structure of the molecule that formed the crystal. (C) In this example, the protein catalase was solved by MicroED to a resolution of 3.2 Å using the method of continuous rotation. A ribbon diagram of the protein is shown along with density for two ligands, a HEME group (*red mesh, orange sticks*), and an NADP molecule (*blue mesh, purple sticks*). These molecules were omitted from an initial model search during the molecular replacement protocol; their density was recovered during the refinement process (Nannenga, Shi, Hattne, et al., 2014). Renderings were generated using PDB ID 3J7B. (See the color plate.)

provided by second order nonlinear imaging of chiral crystals (SONICC), an optical technique that can be used to coarsely visualize crystals in solution. However, SONICC cannot universally detect protein crystals in solution, is limited by crystal size, is not useful for certain crystal symmetries, and thus cannot generally assess the diffracting quality of crystals (Haupert & Simpson, 2011). Instead, a more reliable approach to screening is by preparation of EM grids from a suspension of crystals and direct visualization and diffraction on an electron microscope (Fig. 3). Even the simple preparation of negative-stained crystal samples can provide fast insight into crystal quality. This makes evident the need for an electron microscope capable of routine and rapid screening of MicroED samples, to achieve high-throughput evaluation of potential crystals. Alternatively, X-ray-based screening can be conducted either by obtaining powder-like diffraction from a crystal slurry using large, weak X-ray probes or proof of single-crystal diffraction from single crystals using microfocus X-ray sources.

3.2 Cryo-Preservation of Nanocrystals for MicroED

Provided a crystal is found suitable for MicroED by the aforementioned screening methods, the crystal must then be cryo-preserved for subsequent diffraction. Cryo-sample preparation has been extensively reviewed for single particle and cellular studies (Glaeser, 2008; Grassucci, Taylor, & Frank, 2007; Iancu et al., 2007). Here we briefly introduce the concept of cryo-sample preparation of crystals, which draws strongly from preestablished methods. To summarize the procedure, we begin by placing a small volume of crystals suspended in mother liquor onto an EM grid. For ease of blotting, Quantifoil grids are used; these contain a thin carbon film (5–30 nm thick) with regularly sized and spaced holes, on a standard copper support mesh (typically 200–400 mesh). In these grids, a typical mesh square measures 40–80 μm on a side. The concentration of crystals can be optimized such that no more than a few microcrystals end up on any given grid square. Once a droplet with crystals is placed on a grid, freezing can proceed using manual instruments or robots that wick or blot excess mother liquor within a few seconds, then plunge the grid at high speed into a bath of liquid ethane or other cryogen (Dobro, Melanson, Jensen, & McDowall, 2010). Frozen grids can be transferred to liquid nitrogen for temporary or long-term storage and ultimately loaded onto a cryo-holder under liquid nitrogen. Although light microscopes can visualize grids under a liquid nitrogen environment to screen for quality of freezing (Lepper, Merkel, Sartori, Cyrklaff, & Frischknecht, 2010), their utility is limited for imaging crystals whose size is comparable to or smaller than the wavelength of light.

4. INSTRUMENTATION AND DATA COLLECTION

4.1 Electron Source Specifications

While electron diffraction can be performed on almost any electron microscope, high-resolution diffraction is most likely to be obtained using modern microscopes equipped with field emission electron guns (FEGs). The electron gun in a microscope ultimately dictates the quality of the beam used for diffraction. In field emission guns the ejection of electrons is induced by an electrostatic field from a narrow tip. Compared to filament-based sources, FEGs produce electron beams that are brighter, stable, and more coherent. Diffraction is subject to the effects of both temporal and spatial coherence; these have been extensively reviewed elsewhere (Morishita, Yamasaki, & Tanaka, 2013; Zuo et al., 2004). We present a brief summary of temporal and lateral coherence and their relevance to 3D protein crystals. Temporal coherence is dictated by the chromatic or energy spread of the electron beam. Both the energy of an electron beam and its intrinsic spread are determined by the configuration of its source. Modern field emission guns can operate at a range of energies, with stable configurations typically ranging from 80 to 300 keV. The energy spread for field emission sources is narrower than for thermal emitters (eg, filaments), and can be as low as a fraction of an eV, or less than 0.0005% for a 200 keV beam (Table 1). A simple test for overall coherence is diffraction from slits or a small aperture. Provided proper beam collimation is achieved, a sufficiently coherent electron beam can be obtained (Morishita et al., 2013).

4.2 Electron Optics Specifications

The design of lens configurations varies between electron microscopes, even for a given manufacturer. Rather than present a particular configuration, we assume a general design in which the primary beam is condensed by a two or three lens system onto a sample, then imaged by an objective lens and focused onto a detector by a projector lens system as summarized in Fig. 2. The condenser lens system includes one or more apertures that serve to collimate the incident beam and improve its coherence. Apertures are typically positioned at the back focal plane of a corresponding lens and spatially restrict electrons that cross that plane. The sample stage is sandwiched closely between a condenser and objective lens pair (Fig. 2). Following the objective lens are two apertures, one positioned at its back focal plane and a second at its conjugate image plane. The aperture at its back focal plane is termed the objective aperture (Fig. 2). It acts as a spatial frequency filter in the Fourier domain, akin to a digital low-pass filter. The second aperture is termed the selected area aperture (Fig. 2). This aperture limits the area to be imaged by the projector lens system. When this area is limited, the projector lens system can produce a diffraction pattern from the selected area alone (Fig. 3). This mode is termed selected area electron diffraction. As an electron diffraction technique, MicroED is performed by measuring either a full diffraction image from a given field of view, or a selected area diffraction image from a limited field of view.

Several systematic errors can occur in the beam path of an electron microscope that degrade electron micrographs. These must be avoided in a diffraction experiment. No microscope is perfect, so deviations from an ideal microscope have been the subject of many studies and have inspired the creation of corrective optics for electron lenses (Batson et al., 2002; Haider et al., 1998). The most relevant to diffraction-based techniques, including MicroED are astigmatism in the diffraction lens (the first projection lens) and beam collimation. Because electron diffraction patterns are generated by the objective and projector lens systems in a microscope, imperfections in these lenses limits the ability to properly focus the diffracted reflections onto the detector and can substantially interfere with the measured intensities in diffraction micrographs. To evaluate these errors, a test sample can be measured; catalase crystals, evaporated gold, and graphene are common test samples. Given the well-known structure of these samples, the effect of astigmatisms, aberrations, and misalignments in the electron optics can be assessed. The specifications for all protein crystals whose structures have been determined by MicroED so far, and which have been used for test purposes are detailed in Table 2. The structures of four prion peptides have also been solved at atomic resolution (Table 2). The structures of these segments from the yeast prion protein Sup35 were determined using ab initio methods from data collected by MicroED (Sawaya et al., in preparation).

A focused electron beam is prevented from impinging directly onto the detector by a beam stop positioned after the objective lens. In principle a Faraday device can be used as a beam stop to measure the transmitted and focused electron beam (Grubb, 1971). Beam stops can also be designed to minimize the obstruction of low-resolution information. Ultimately the lowest resolution sampling achievable is dictated by the effective camera length of a particular diffraction setting. The camera length is determined by the projection lens system and governs the resolution range and sampling of the reciprocal lattice. For example, crystals with large unit cells produce closely spaced reflections. The camera length must be optimized so that closely spaced reflections

Table 2 Structures Determined by MicroED

Sample	PDB ID	EMDB ID	Resolution (Å)	Space Group	Cell Dimensions a, b, c (Å) | α, β, γ (degree)
Lysozyme	3J4G	EMD-2945	2.9	P 43 21 2	77, 77, 37 | 90, 90, 90
Lysozyme	3J6K	EMD-6313	2.5	P 43 21 2	76, 76, 37 | 90, 90, 90
Catalase	3J7B	EMD-6314	3.2	P 21 21 21	68, 172, 182 | 90, 90, 90
Catalase	3J7U	–	3.2	P 21 21 21	69, 174, 206 | 90, 90, 90
Ca-ATPase	3J7T	–	3.4	C 1 2 1	166, 64, 147 | 90, 98, 90
Proteinase K	5I9S	EMD-8077	1.8	P 43 21 2	67, 67, 102 | 90, 90, 90
NACore	4RIL	EMD-3028	1.4	C 1 2 1	70.8, 4.8, 16.8 | 90, 106, 90
PreNAC	4ZNN	EMD-3001	1.4	P 1 21 1	17.9, 4.7, 33 | 90, 94, 90
Zn-NNQQNY	5K2E	EMD-8196	1.0	P 21	21.5, 4.9, 23.9 | 90, 104, 90
Cd-NNQQNY	5K2F	EMD-8197	1.0	P 21	22.1, 4.9, 23.5 | 90, 104.3, 90
GNNQQNY1	5K2G	EMD-8198	1.1	P 21	22.9, 4.9, 24.2 | 90, 107.8, 90
GNNQQNY2	5K2H	EMD-8199	1.05	P 21 21 21	23.2, 4.9, 40.5 | 90, 90, 90

can be accurately sampled as individual spots on a diffraction image. Measurement of reflections less than 50 Å in resolution proves a challenge on most conventional microscopes, so MicroED is not suitable for low-resolution studies.

4.3 Energy Filtering

As the thickness of a specimen grows, the ratio of inelastically to elastically scattered electrons can become significant. To alleviate this effect, a number of tools have been developed that can filter out inelastically scattered electrons and reduce their impact on measured diffraction patterns. This is because inelastically scattered electrons emerge from a sample with an energy that differs from that of their elastically scattered counterparts and the primary beam (Henkelman & Ottensmeyer, 1974; Zanchi, Sevely, & Jouffrey, 1977). Several types of energy filters are currently manufactured; all spatially separate the scattered electrons based on their respective energy. Some energy filters are located in column, and make up part of the optical train, while others are postcolumn filters that are placed just upstream of and couple to the detector. The energy spread of the primary beam is determined by the properties of its source, while electrons that lose energy during their interactions with a crystal will have energies that differ beyond that intrinsic spread. Nonkinematic scattering of electrons can be further reduced by operating a microscope at higher energies, but under these conditions other damage mechanisms become prominent (Thomas, 1970; Zanchi et al., 1977).

4.4 Low-Dose Data Collection

To obtain diffraction micrographs from a single crystal of biological material at multiple orientations without destroying the crystal, a strategy must be employed to limit the dose given to the sample during the experiment. In MicroED, loss of high-resolution information can occur with doses over $9e^{-}/Å^{2}$ (Shi et al., 2013) in still diffraction but in continuous rotation with doses as low as $5e^{-}/Å^{2}$ (Nannenga, Shi, Hattne, et al., 2014; Nannenga, Shi, Leslie, et al., 2014). A very low-dose data collection strategy must be employed to overcome this limitation. The most efficient of these strategies involves a crystal being dosed in a discrete series of micrographs collected at various angles. However, this series of "still" diffraction patterns measures only partial reflections and is therefore not the most accurate way to recover true reflection intensities (Fig. 1). To integrate over full reflections, a unidirectional rotation of the crystal can be performed during continuous exposure to the electron beam (Nannenga, Shi, Leslie, et al., 2014). In this mode, the detector can be operated with line-by-line readout, termed rolling shutter integration. While better sampling full reflections, this type of readout suffers from higher detector noise and allows the crystal to be continuously dosed during data collection. Alternatively, as has been demonstrated with dose-insensitive inorganic specimens, precession can be used to collect still diffraction images that suffer less from partiality (Oleynikov, Hovmöller, & Zou, 2007). In such a case, instead of rotating the sample, the beam is tilted (Table 1). This efficiently exposes the crystal only during meaningful data acquisition intervals. In this scheme, the detector readout time can be longer and therefore images with lower readout noise can be acquired.

The dose of electrons can be tuned on a microscope by condensing or spatially restricting the electron beam. One way to achieve this is by limiting the spot size of the beam on the specimen as a result of a strong crossover produced by the first condenser lens. Condenser apertures can further reduce the size of the illuminating beam. While commercial field emission guns can achieve electron currents of 10^{8} $e^{-}/Å^{2}/s$, the flux experienced by a crystal in a MicroED experiment is typically limited to 0.01 $e^{-}/Å^{2}/s$ or lower (Shi et al., 2013).

Detailed protocols for setting up the microscope for MicroED, data collection and analysis have been published recently (Shi et al. 2016; Hattne et al. 2015).

4.5 Detector Specifications

A recent revolution in electron detectors has produced devices capable of rapidly counting electrons. However, a typical electron microscope might be equipped with any number of detectors, ranging from film systems to the newest electron detectors. Rather than cover such a broad spectrum of options, we focus on the important features required of a detector to collect high-quality diffraction micrographs in a MicroED experiment. Because most MicroED data reported to date have been recorded on a CMOS camera manufactured by the TVIPS corporation (TVIPS TemCam model F416) (Hattne et al., 2015), we highlight features present in this system that make MicroED possible. This model contains a square sensor with 4000 pixels on each side, where each square pixel is approximately 15 μm on a side. The total sensor size is thus about 60 mm on a side. For the purposes of noise reduction and memory efficiency, the images are compressed on camera using a 2-by-2 binning operation. The lowest measurable resolution is dictated by beam stop geometry and position in the electron microscope. Ultimately, a sensor size and camera length must be carefully chosen to match the desired resolution range for a given micrograph. CMOS sensors are preferred over slower CCDs for high-speed data collection due to their faster write times and since CMOS detectors do not suffer from blooming artifacts as do CCDs. A scintillator optimized for sensitivity is also beneficial. However, devices and/or settings with less sensitivity and slower read out times can still be used in precession electron diffraction. These considerations change when other types of electron detectors are used. Evaluation of electron diffraction measurements using new devices is ongoing (Nederlof, van Genderen, Li, & Abrahams, 2013). Across the different types of direct electron detectors, dynamic range must be considered to avoid detector damage and maintain accurate counts for all measured reflections.

5. PROCESSING OF MicroED DATA

5.1 Conversion and Processing of Diffraction Images

The types of detectors used in MicroED data collection are not currently suited to save measured data in typical crystallographic formats. While software has been written to specifically process still diffraction images collected by MicroED (Iadanza & Gonen, 2014), conversion tools have also been developed to port data collected on TVIPS CMOS detectors to the SMV crystallographic data format (Hattne et al., 2015) and are available for download at https://www.janelia.org/sites/default/files/tvips-tools-0.0.0.tgz. The benefit of the latter is that most software toolboxes for crystallographic data processing can recognize the SMV format, making processing of MicroED data as facile and rapid as is processing of data collected at modern X-ray sources. Several factors must be considered when interpreting MicroED data using standard crystallographic analysis pipelines. We review these briefly since many have been detailed elsewhere (Hattne et al., 2015).

First we consider differences in the curvature of the Ewald sphere between X-rays and electron diffraction data. With X-rays, the curvature of the sphere is relevant and often obvious in diffraction images by the appearance of loons (Fig. 1). In contrast, because of the miniscule wavelength of the electrons used for MicroED, at 200 KeV, where the wavelength is approximately 0.025 Å, the curvature of the Ewald sphere is negligible (Fig. 1). For example, at this energy, a 1 Å reflection would experience only a mere fraction of a degree of curvature on the Ewald sphere. By comparison, to achieve similar resolution using monochromatic 12 KeV X-rays that curvature grows to tens of degrees. This difference has practical implications. One is that in diffraction images collected by MicroED information in the direction normal to the sphere is limited. This is a major reason for why indexing from a single-diffraction image is not possible (Hattne et al., 2015; Shi et al., 2013) unless one has a priori knowledge of the unit cell dimensions (Jiang, Georgieva, Zandbergen, & Abrahams, 2009). Indexing without a priori knowledge of the unit cell can be achieved in practice, as has been demonstrated in MicroED with data that covers a phi range of ~30 degree (Nannenga, Shi, Leslie, et al., 2014). A benefit is that particular orientations of a crystal can lead to the Ewald sphere intercepting an entire zone of the reciprocal lattice, simultaneously revealing many Bragg reflections. Another important consequence is the need for accurate knowledge of experimental geometry during data collection, including beam position, camera length, and tilt ranges (Table 1). When partial reflections are measured, this problem is confounded and can lead to further inaccuracies in assignment of intensities to particular reflections.

Second we consider crystal mosaicity, a concept common to macromolecular X-ray crystallography. A model for mosaicity within a macroscopic protein crystal illuminated by a perfectly parallel X-ray beam presents the crystal as an ensemble of lattice blocks with small deviations from the global orientation of the ensemble. The size of these blocks and their relative deviations from the ensemble orientation can be modeled based on the reflections that appear in a diffraction image (Leslie, 2006; Leslie & Powell, 2007). This model makes simultaneous refinement of mosaicity, experimental geometry, and beam divergence difficult (Leslie, 2006). As crystals shrink to the size of a single mosaic block, the overall lattice should appear ideal. In MicroED, diffracted crystals are only a few hundred nanometers thin, and no larger than a few microns in length and width. Such small crystals should in principle contain fewer mosaic blocks and bear greater influence from the persistence length of the lattice (Dorset, 1980; Nederlof, Li, van Heel, & Abrahams, 2013; Subramanian, Basu, Liu, Zuo, & Spence, 2015). However, if after refinement of experimental parameters errors remain unaccounted for, these may become absorbed by mosaicity and limit its representation of lattice disorder (Hattne et al., 2015).

Lastly, we consider the integration of intensities and its sources of error. As with macromolecular X-ray crystallography, in MicroED, identification of reflections is followed by integration and background correction. The accuracy of integration depends on proper fitting of spot profiles (Kabsch, 2010; Leslie, 1999, 2006) (Fig. 4). Since in MicroED most crystals are small enough to be fully bathed by the electron beam, the shapes of reflections are dictated primarily by the shape transform of the crystal, its unit cell count, and lattice order (Robinson, Vartanyants, Williams, Pfeifer, & Pitney, 2001). The value of background pixels in regions that surround Bragg reflections have also been investigated, with new corrections introduced for truncated values that appear when using equipment not sensitive to low-intensity measurements (Hattne et al., 2015). Several software packages for crystallographic data reduction can accurately estimate reflection profiles during integration (Kabsch, 2010; Leslie & Powell, 2007). Local background near the integrated reflections must also be accounted for. In MicroED this is in principle affected by detector noise, stray light sources in the column, and other sources of incoherent scattering including inelastically scattered electrons. Despite these potential sources of error, MicroED images can show high-quality diffraction even at atomic resolution (Fig. 4).

Multiple-scattering phenomena have been predicted to influence and perhaps overwhelm structure determination by electron diffraction from 3D crystals (Diaz-Avalos et al., 2003; Glaeser & Ceska, 1989; Grigorieff & Henderson, 1996; Spence, 2013; Subramanian et al., 2015). However, recent structures obtained from crystals hundreds of nanometers thick (Table 2) suggest that the theory does not agree with experiment and a more comprehensive theory may be required to account for data obtained by MicroED.

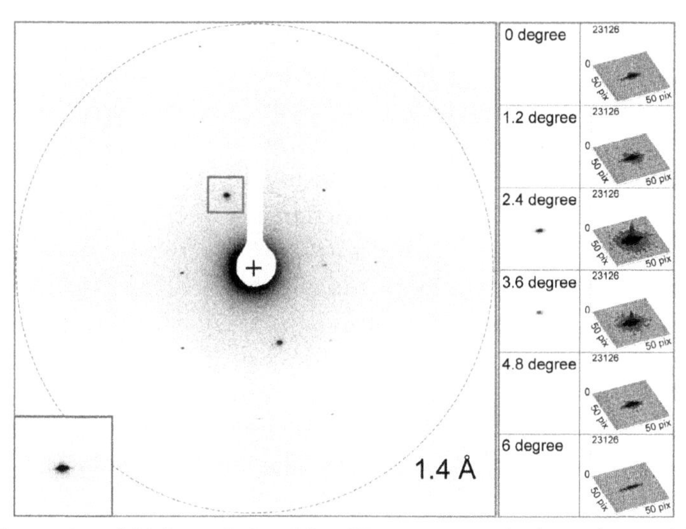

Fig. 4 Properties of high-resolution MicroED patterns. A high-resolution diffraction micrograph (*left*) obtained by MicroED from a crystal of an amyloid peptide (Rodriguez et al., 2015) is shown. An inset and corresponding *red* (*gray* in the print version) *box* highlight a single Bragg reflection that crosses the Ewald sphere at this crystal orientation. The resolution at the edge of the detector is marked. This same reflection is tracked over an angular range of 6 degree (*right*). Profiles of the spot are shown as well as three-dimensional plots of the intensities for the pixels that comprise the spot within a 50-by-50-pixel area. Intensity is shown on the *z*-axis and ranges from 0 to 23126 counts. Each image represents a 1.2 degree wedge through reciprocal space. The start of the wedge is marked on the upper left hand corner of each image; the first of these begins at 0 degree.

Improvements to the theory will require better treatment of crystalline order, eg, lattice bending and disorder, a more comprehensive evaluation of zones, other than major zone axes, an account of rotation or precession during measurement, and a better treatment of experimental sources of noise and error.

5.2 Structure Determination and Refinement

The accurate determination of high-resolution structures in MicroED relies on the same standards for data quality required by X-ray crystallography. Metrics that are optimized to ensure success in structure determination include errors in integration and merging, completeness, and redundancy. The values found in MicroED data for these metrics are generally within the realm encountered in an X-ray crystallography experiment (Hattne et al., 2015; Nannenga, Shi, Hattne, et al., 2014; Nannenga, Shi, Leslie, et al., 2014; Rodriguez et al., 2015; Shi et al., 2013). When crystal symmetry is low and crystals lie randomly on an EM grid, diffraction data from multiple crystals can be merged to increase completeness. Continuous rotation data collection improves the accuracy of estimates for reflection intensities and therefore reduces the errors observed in data reduction (Nannenga, Shi, Leslie, et al., 2014; Rodriguez et al., 2015).

For MicroED data that meet these criteria, molecular replacement using known structures as probes provides accurate structure solutions, even for previously unknown structures (Rodriguez et al., 2015). Several crystallographic structure determination software packages have been used to solve macromolecular structures from MicroED data; these use electron form factors for structure determination and refinement (Adams et al., 2010; Murshudov et al., 2011). While all structures determined by MicroED have thus far required known atomic models as probes, the accuracy of these solutions is strengthened by the recovery of ligands and side chains during refinement that were omitted from the probe structure (Fig. 3) (Nannenga, Shi, Hattne, et al., 2014; Shi et al., 2013). Errors in the final structures have been comparable to those observed for macromolecular structures determined by X-ray crystallographic methods.

6. SUMMARY AND OUTLOOK

MicroED is a new cryo-EM method. It joins the present revolution in electron microscopy that drives renewed interest in fundamental properties of electrons and their interactions with matter, improved software algorithms for data collection and reduction, and new and more efficient instruments and detectors. These improvements are sure to drive the method beyond its current limits. With further improvements, questions arise: What are the size limits, large and small, for crystals in MicroED? Will very large unit cell dimensions be limiting? How much crystalline disorder can be tolerated? Can dynamics be probed? What general phasing methods are applicable to MicroED? How much damage is inflicted on crystals during MicroED and can structures be obtained with minimal to no visible damage? Can the structure of any well-ordered submicron thick crystal be determined by MicroED? The answer to these questions will herald a new age for the marriage of macromolecular crystallography and cryo-EM.

ACKNOWLEDGMENTS

We thank HHMI for support. J.A.R. was supported by the Giannini Foundation.

REFERENCES

Adams, P. D., Afonine, P. V., Bunkóczi, G., Chen, V. B., Davis, I. W., Echols, N., et al. (2010). PHENIX: A comprehensive Python-based system for macromolecular structure solution. *Acta Crystallographica. Section D, Biological Crystallography, 66*, 213–221.

Adrian, M., Dubochet, J., Lepault, J., & McDowall, A. W. (1984). Cryo-electron microscopy of viruses. *Nature, 308*, 32–36.

Batson, P. E., Dellby, N., & Krivanek, O. L. (2002). Sub-ångstrom resolution using aberration corrected electron optics. *Nature, 418*, 617–620.

Bethe, H. A., Rose, M. E., & Smith, L. P. (1938). The multiple scattering of electrons. *Proceedings of the American Philosophical Society, 78*, 573–585.

Cheng, Y., Grigorieff, N., Penczek, P. A., & Walz, T. (2015). A primer to single-particle cryo-electron microscopy. *Cell, 161*, 438–449.

Christensen, A. K. (1971). Frozen thin sections of fresh tissue for electron microscopy, with a description of pancreas and liver. *The Journal of Cell Biology, 51*, 772–804.

Cowley, C. W. (1964). Cryobiology as viewed by the engineer. *Cryobiology, 51*, 40–43.

Cowley, J. M. (1969). Image contrast in a transmission scanning electron microscope. *Applied Physics Letters, 15*, 58–59.

Crowther, R. A., DeRosier, D. J., & Klug, A. (1970). The reconstruction of a three-dimensional structure from projections and its application to electron microscopy. *Proceedings of the Royal Society A: Mathematical, Physical & Engineering Sciences, 317*, 319–340.

Danev, R., & Nagayama, K. (2001). Transmission electron microscopy with Zernike phase plate. *Ultramicroscopy, 88*, 243–252.

Davisson, C., & Germer, L. H. (1927a). Diffraction of electrons by a crystal of nickel. *Physics Review, 30*, 705–740.

Davisson, C., & Germer, L. H. (1927b). The scattering of electrons by a single crystal of nickel. *Nature, 119*, 558–560.

Deisenhofer, J., Epp, O., Miki, K., Huber, R., & Michel, H. (1985). Structure of the protein subunits in the photosynthetic reaction centre of Rhodopseudomonas viridis at 3 Å resolution. *Nature, 318*, 618–624.

DeRosier, D. J., & Klug, A. (1968). Reconstruction of three dimensional structures from electron micrographs. *Nature, 217*, 130–134.

Diaz-Avalos, R., Long, C., Fontano, E., Balbirnie, M., Grothe, R., Eisenberg, D., et al. (2003). Cross-beta order and diversity in nanocrystals of an amyloid-forming peptide. *Journal of Molecular Biology, 330*, 1165–1175.

Dobro, M. J., Melanson, L. A., Jensen, G. J., & McDowall, A. W. (2010). Chapter three—Plunge freezing for electron cryomicroscopy. In G. J. Jensen (Ed.), *Methods in enzymology* (pp. 63–82): Academic Press.

Dorset, D. L. (1980). Electron diffraction intensities from bent molecular organic crystals. *Acta Crystallographica. Section A, 36*, 592–600.

Dorset, D. L., & Hauptman, H. A. (1976). Direct phase determination for quasi-kinematical electron diffraction intensity data from organic microcrystals. *Ultramicroscopy, 1*, 195–201.

Glaeser, R. M. (2008). Cryo-electron microscopy of biological nanostructures. *Physics Today, 61*, 48–54.

Glaeser, R. M., & Ceska, T. A. (1989). High-voltage electron diffraction from bacteriorhodopsin (purple membrane) is measurably dynamical. *Acta Crystallographica. Section A, 45*, 620–628.

Gonen, T. (2013). The collection of high resolution electron diffraction data. In I. Schmidt-Krey & Y. Cheng (Eds.), *Electron crystallography of soluble and membrane proteins* (pp. 153–169): Humana Press.

Gonen, T., Cheng, Y., Sliz, P., Hiroaki, Y., Fujiyoshi, Y., Harrison, S. C., et al. (2005). Lipid–protein interactions in double-layered two-dimensional AQP0 crystals. *Nature, 438*, 633–638.

Gonen, T., Sliz, P., Kistler, J., Cheng, Y., & Walz, T. (2004). Aquaporin-0 membrane junctions reveal the structure of a closed water pore. *Nature, 429*, 193–197.
Grassucci, R. A., Taylor, D. J., & Frank, J. (2007). Preparation of macromolecular complexes for cryo-electron microscopy. *Nature Protocols, 2*, 3239–3246.
Grigorieff, N., & Henderson, R. (1996). Comparison of calculated and observed dynamical diffraction from purple membrane: Implications. *Ultramicroscopy, 65*, 101–107.
Grubb, D. T. (1971). The calibration of beam measurement devices in various electron microscopes, using an efficient Faraday cup. *Journal of Physics E, 4*, 222.
Haider, M., Rose, H., Uhlemann, S., Schwan, E., Kabius, B., & Urban, K. (1998). A spherical-aberration-corrected 200 kV transmission electron microscope. *Ultramicroscopy, 75*, 53–60.
Hattne, J., Reyes, F. E., Nannenga, B. L., Shi, D., de la Cruz, M. J., Leslie, A. G. W., et al. (2015). MicroED data collection and processing. *Acta Crystallographica. Section A, Foundations and Advances, 71*, 353–360.
Hattne, J., Shi, D., de la Cruz, J., Reyes, F. E., & Gonen, T. (2016). Modeling truncated intensities of faint reflections in MicroED images. *Journal of Applied Crystallography, 49*. http://dx.doi.org/10.1107/S1600576716007196.
Haupert, L. M., & Simpson, G. J. (2011). Screening of protein crystallization trials by second order nonlinear optical imaging of chiral crystals (SONICC). *Methods (San Diego Calif.), 55*, 379–386.
Held, G. (2012). Low-energy electron diffraction: Crystallography of surfaces and interfaces. In R. Schäfer & P. C. Schmidt (Eds.), *Methods in physical chemistry* (pp. 625–642). Weinheim: Wiley-VCH Verlag GmbH & Co. KGaA.
Henderson, R. (1995). The potential and limitations of neutrons, electrons and X-rays for atomic resolution microscopy of unstained biological molecules. *Quarterly Reviews of Biophysics, 28*, 171–193.
Henderson, R., Baldwin, J. M., Ceska, T. A., Zemlin, F., Beckmann, E., & Downing, K. H. (1990). Model for the structure of bacteriorhodopsin based on high-resolution electron cryo-microscopy. *Journal of Molecular Biology, 213*, 899–929.
Henderson, R., & Unwin, P. N. T. (1975). Three-dimensional model of purple membrane obtained by electron microscopy. *Nature, 257*, 28–32.
Henkelman, R. M., & Ottensmeyer, F. P. (1974). An energy filter for biological electron microscopy. *Journal of Microscopy, 102*, 79–94.
Hiroaki, Y., Tani, K., Kamegawa, A., Gyobu, N., Nishikawa, K., Suzuki, H., et al. (2006). Implications of the aquaporin-4 structure on array formation and cell adhesion. *Journal of Molecular Biology, 355*, 628–639.
Iadanza, M. G., & Gonen, T. (2014). A suite of software for processing MicroED data of extremely small protein crystals. *Journal of Applied Crystallography, 47*, 1140–1145.
Iancu, C. V., Tivol, W. F., Schooler, J. B., Dias, D. P., Henderson, G. P., Murphy, G. E., et al. (2007). Electron cryotomography sample preparation using the Vitrobot. *Nature Protocols, 1*, 2813–2819.
Jap, B. K., Downing, K. H., & Walian, P. J. (1990). Structure of PhoE porin in projection at 3.5 Å resolution. *Journal of Structural Biology, 103*, 57–63.
Jiang, L., Georgieva, D., Zandbergen, H. W., & Abrahams, J. P. (2009). Unit-cell determination from randomly oriented electron-diffraction patterns. *Acta Crystallographica. Section D, Biological Crystallography, 65*, 625–632.
Kabsch, W. (2010). XDS. *Acta Crystallographica. Section D, Biological Crystallography, 66*, 125–132.
Koster, A. J., Chen, H., Sedat, J. W., & Agard, D. A. (1992). Automated microscopy for electron tomography. *Ultramicroscopy, 46*, 207–227.
Kühlbrandt, W., Wang, D. N., & Fujiyoshi, Y. (1994). Atomic model of plant light-harvesting complex by electron crystallography. *Nature, 367*, 614–621.

Lepper, S., Merkel, M., Sartori, A., Cyrklaff, M., & Frischknecht, F. (2010). Rapid quantification of the effects of blotting for correlation of light and cryo-light microscopy images. *Journal of Microscopy*, *238*, 21–26.
Leslie, A. G. W. (1999). Integration of macromolecular diffraction data. *Acta Crystallographica. Section D, Biological Crystallography*, *55*, 1696–1702.
Leslie, A. G. W. (2006). The integration of macromolecular diffraction data. *Acta Crystallographica. Section D, Biological Crystallography*, *62*, 48–57.
Leslie, A. G. W., & Powell, H. R. (2007). Processing diffraction data with mosflm. In R. J. Read & J. L. Sussman (Eds.), *Evolving methods for macromolecular crystallography* (pp. 41–51). The Netherlands: Springer.
Morishita, S., Yamasaki, J., & Tanaka, N. (2013). Measurement of spatial coherence of electron beams by using a small selected-area aperture. *Ultramicroscopy*, *129*, 10–17.
Murshudov, G. N., Skubák, P., Lebedev, A. A., Pannu, N. S., Steiner, R. A., Nicholls, R. A., et al. (2011). *REFMAC* 5 for the refinement of macromolecular crystal structures. *Acta Crystallographica. Section D, Biological Crystallography*, *67*, 355–367.
Nannenga, B. L., & Gonen, T. (2014). Protein structure determination by MicroED. *Current Opinion in Structural Biology*, *27*, 24–31.
Nannenga, B. L., Shi, D., Hattne, J., Reyes, F. E., & Gonen, T. (2014b). Structure of catalase determined by MicroED. *eLife*, *3*, e03600.
Nannenga, B. L., Shi, D., Leslie, A. G. W., & Gonen, T. (2014a). High-resolution structure determination by continuous-rotation data collection in MicroED. *Nature Methods*, *11*, 927–930.
Nederlof, I., Li, Y. W., van Heel, M., & Abrahams, J. P. (2013). Imaging protein three-dimensional nanocrystals with cryo-EM. *Acta Crystallographica. Section D, Biological Crystallography*, *69*, 852–859.
Nederlof, I., van Genderen, E., Li, Y.-W., & Abrahams, J. P. (2013). A Medipix quantum area detector allows rotation electron diffraction data collection from submicrometre three-dimensional protein crystals. *Acta Crystallographica. Section D, Biological Crystallography*, *69*, 1223–1230.
Oleynikov, P., Hovmöller, S., & Zou, X. D. (2007). Precession electron diffraction: Observed and calculated intensities. *Ultramicroscopy*, *107*, 523–533.
Robinson, A. L. (1986). Electron microscope inventors share Nobel physics prize: Ernst Ruska built the first electron microscope in 1931; Gerd Binnig and Heinrich Rohrer developed the scanning tunneling microscope 50 years later. *Science*, *234*, 821–822.
Robinson, I. K., Vartanyants, I. A., Williams, G. J., Pfeifer, M. A., & Pitney, J. A. (2001). Reconstruction of the shapes of gold nanocrystals using coherent X-Ray diffraction. *Physical Review Letters*, *87*, 195505.
Rodriguez, J. A., Ivanova, M. I., Sawaya, M. R., Cascio, D., Reyes, F. E., Shi, D., et al. (2015). Structure of the toxic core of α-synuclein from invisible crystals. *Nature*, *525*, 486–490.
Ruskin, R. S., Yu, Z., & Grigorieff, N. (2013). Quantitative characterization of electron detectors for transmission electron microscopy. *Journal of Structural Biology*, *184*, 385–393.
Shi, D., Nannenga, B. L., de la Cruz, J. M., Liu, J., Sawtelle, S., Calero, G., et al. (2016). The collection of MicroED data for macromolecular crystallography. *Nature Protocols*, *11*, 895–904.
Shi, D., Nannenga, B. L., Iadanza, M. G., & Gonen, T. (2013). Three-dimensional electron crystallography of protein microcrystals. *eLife*, *2*, e01345.
Spence, J. C. H. (2013). *High-resolution electron microscopy*. Oxford: OUP.
Stern, R. M., & Taub, H. (1970). An introduction to the dynamical scattering of electrons by crystals. *C R C Critical Reviews in Solid State Sciences*, *1*, 221–302.
Subramanian, G., Basu, S., Liu, H., Zuo, J.-M., & Spence, J. C. H. (2015). Solving protein nanocrystals by cryo-EM diffraction: Multiple scattering artifacts. *Ultramicroscopy*, *148*, 87–93.

Taylor, K. A., & Glaeser, R. M. (1976). Electron microscopy of frozen hydrated biological specimens. *Journal of Ultrastructure Research*, *55*, 448–456.
Thomas, L. E. (1970). The diffraction-dependence of electron damage in a high voltage electron microscope. *Radiation Effects*, *5*, 183–194.
Thomson, G. P., & Reid, A. (1927). Diffraction of cathode rays by a thin film. *Nature*, *119*, 890.
Walz, T., & Grigorieff, N. (1998). Electron crystallography of two-dimensional crystals of membrane proteins. *Journal of Structural Biology*, *121*, 142–161.
Walz, T., Hirai, T., Murata, K., Heymann, J. B., Mitsuoka, K., Fujiyoshi, Y., et al. (1997). The three-dimensional structure of aquaporin-1. *Nature*, *387*, 624–627.
Weathersby, S. P., Brown, G., Centurion, M., Chase, T. F., Coffee, R., Corbett, J., et al. (2015). Mega-electron-volt ultrafast electron diffraction at SLAC National Accelerator Laboratory. *The Review of Scientific Instruments*, *86*, 073702.
Yonekura, K., Kato, K., Ogasawara, M., Tomita, M., & Toyoshima, C. (2015). Electron crystallography of ultrathin 3D protein crystals: Atomic model with charges. *Proceedings of the National Academy of Sciences*, *112*, 3368–3373.
Zanchi, G., Sevely, J., & Jouffrey, B. (1977). An energy filter for high voltage electron microscopy. *Journal of Electron Spectroscopy and Related Phenomena*, *2*, 95–104.
Zuo, J. M., Gao, M., Tao, J., Li, B. Q., Twesten, R., & Petrov, I. (2004). Coherent nano-area electron diffraction. *Microscopy Research and Technique*, *64*, 347–355.

CHAPTER FIFTEEN

Databases and Archiving for CryoEM

A. Patwardhan*[,1], C.L. Lawson[†,1]
*Protein Data Bank in Europe, European Molecular Biology Laboratory, European Bioinformatics Institute, Wellcome Genome Campus, Hinxton, United Kingdom
[†]Research Collaboratory for Structural Bioinformatics, Center for Integrative Proteomics Research, Rutgers, The State University of New Jersey, Piscataway, NJ, United States
[1]Corresponding authors: e-mail address: ardan@ebi.ac.uk; cathy.lawson@rutgers.edu

Contents

Abstract

CryoEM in structural biology is currently served by three public archives—EMDB for 3DEM reconstructions, PDB for models built from 3DEM reconstructions, and EMPIAR for the raw 2D image data used to obtain the 3DEM reconstructions. These archives play a vital role for both the structural community and the wider biological community in making the data accessible so that results may be reused, reassessed, and integrated with other structural and bioinformatics resources. The important role of the archives is underpinned by the fact that many journals mandate the deposition of data to PDB and EMDB on publication. The field is currently undergoing transformative changes where on the one hand high-resolution structures are becoming a routine occurrence

Methods in Enzymology, Volume 579
ISSN 0076-6879
http://dx.doi.org/10.1016/bs.mie.2016.04.015

while on the other hand electron tomography is enabling the study of macromolecules in the cellular context. Concomitantly the archives are evolving to best serve their stakeholder communities.

In this chapter, we describe the current state of the archives, resources available for depositing, accessing, searching, visualizing and validating data, on-going community-wide initiatives and opportunities, and challenges for the future.

1. INTRODUCTION

In recent years cryo-electron microscopy (cryoEM) and electron tomography (cryoET) have become indispensable tools for molecular and cellular structural biology. In the past they were commonly used to complement the more established techniques of X-ray crystallography and nuclear magnetic resonance (NMR) spectroscopy. Single-particle EM enables the study of large macromolecular assemblies and complexes in a close-to-native environment without the need for generating large amounts of purified material, forming crystals, or isotopic labeling. Single-particle cryoEM even forgoes the requirement for extreme sample homogeneity, either compositional or conformational, since multiple states can be computationally separated into different 3D classes. CryoET is the method of choice when studying pleomorphic structures such as the HIV virus or structures in the cellular context.

Traditionally, EM-based techniques have yielded 3D structures with limited resolution, preventing the direct unambiguous interpretation of the data in terms of biological entities. Electron diffraction and imaging have been used successfully on helical and 2D crystalline arrays to overcome this hurdle and obtain structures to atomic resolution, eg, the $\alpha\beta$ tubulin dimer (Nogales, Wolf, & Downing, 1998), but researchers using diffraction methods face the traditional challenge of obtaining well-ordered crystals. These issues have prevented wider use but electron crystallography has found a niche in structure determination of membrane proteins, eg, aquaporin at 1.9 Å resolution (Gonen et al., 2005). More typically, interpretation of lower resolution 3D maps has been aided by fitting of atomic coordinate models derived from other experiments, by comparing maps of related structures, or by segmenting the structure using other biochemical information or prior knowledge. For cryoET the problem of limited resolution has been even more severe owing to intrinsic limitations of the technique such as missing wedges, radiation damage from imaging the same specimen area multiple times, and specimen tilting. However, the resolution may be adequate for the purpose of the experiment, for example, to examine the distribution or organization of a complex and molecular assembly in the cell (Brandt, Carlson, Hartl, Baumeister, & Grunewald, 2010). In cases where there is ambiguity, other methods can be used to robustly identify targets, for example, correlative light microscopy with fluorescent tagging (Kukulski, Schorb, Kaksonen, & Briggs, 2012). Subtomogram averaging (see chapter "Cryo-Electron Tomography and Subtomogram Averaging" by Briggs)—a technique similar to single-particle methodology but involving the averaging and classification of 3D subvolumes, can be used to improve resolution and overcome tomographic artifacts. Using classification techniques, subtomogram averaging enables visualization of structural variability in a cellular context.

In the past few years there have been major technological advances in the field, including the introduction of the direct electron detector, that have enabled the determination of single-particle structures to atomic resolution, and cryoET has also benefitted from the improved resolution. At the same time there has been an increased emphasis on combining different structural techniques to build up a holistic understanding of the biological problem at hand. Here electron tomography and correlative light microscopy has been vital in providing the cellular context to the macromolecular world (Zeev-Ben-Mordehai, Hagen, & Grunewald, 2014). Other notable developments include that for the first time there are phase plate technologies sufficiently robust for routine adoption (Danev & Baumeister, 2016) and that 3D electron diffraction has been successfully used to determine structures to 1.4 Å resolution (Rodriguez et al., 2015).

The structural biology community was one of the first to recognize the value of providing public open access to data from X-ray crystallography with the inception of the Protein Data Bank (PDB) as an archive for atomic coordinate models in 1971 (Berman, Kleywegt, Nakamura, & Markley, 2012). Open access to data provides a means to independent validation, reuse, and integration of structural information. The PDB has served as a source of data for methods development and teaching, and for driving the field forward. It has also been a focal point for community-wide efforts on many issues including standardization and validation that have benefitted the field. Today the PDB archive comprises over 120,000 structures, including over 1000 structures determined using EM-based techniques (Fig. 1, green bars). Deposition of experimentally derived atomic coordinate model structures to PDB is mandatory upon publication for most relevant journals.

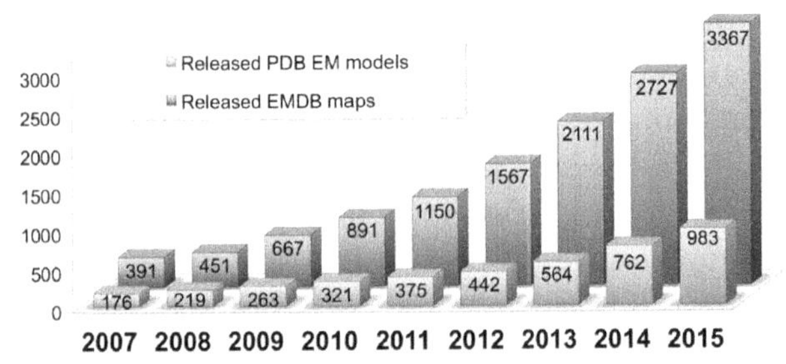

Fig. 1 Released 3DEM entries in EMDB and PDB, cumulative by year, since the start of the EMDataBank Project in 2007.

The PDB is managed by the members of the Worldwide Protein Databank (wwPDB; http://wwpdb.org; Berman, Henrick, & Nakamura, 2003): the Research Collaboratory for Structural Bioinformatics PDB (RCSB PDB), the Protein Data Bank in Europe (PDBe) at the European Bioinformatics Institute (EMBL-EBI), the Protein Data Bank Japan (PDBj) at the Institute for Protein Research in Osaka University, and the Biological Magnetic Resonance Bank (BMRB) at the University of Wisconsin-Madison.

In the same vein, in the late 1990s and early 2000s there was a growing realization by EM researchers of the need for a similar resource for EM-derived structures. At that time, most EM structures were not solved to a resolution where an atomic coordinate structure could be built from the 3D EM volume so it was critical for the volume itself to be stored. The Electron Microscopy Data Bank (EMDB) was set up in 2002 at EMBL-EBI as an archive for 3DEM reconstructions (Tagari, Newman, Chagoyen, Carazo, & Henrick, 2002). It now comprises over 3400 structures (Fig. 1, purple bars) from a variety of EM techniques including single-particle, electron tomography, subtomogram averaging, and 2D and 3D electron diffraction (Fig. 2). Current trends in the field reflect directly on depositions to EMDB. Fig. 3 shows how the number of structures deposited at better than 4 Å resolution has increased dramatically in the past few years and Fig. 4 highlights the importance of the direct electron detector in advancing the field. Map volume deposition rates for published 3DEM structures have been gradually increasing as the potential of 3DEM methods is recognized. Many journals have implemented policies requiring experimental data to be deposited for EM-based studies. Nowadays many EM experiments involve coordinated depositions of the 3DEM volume to EMDB and fitted or built atomic coordinates to PDB. Another trend is toward hybrid experiments where constraints from several different methods are combined to obtain a structure. Notable examples include the nuclear pore complex

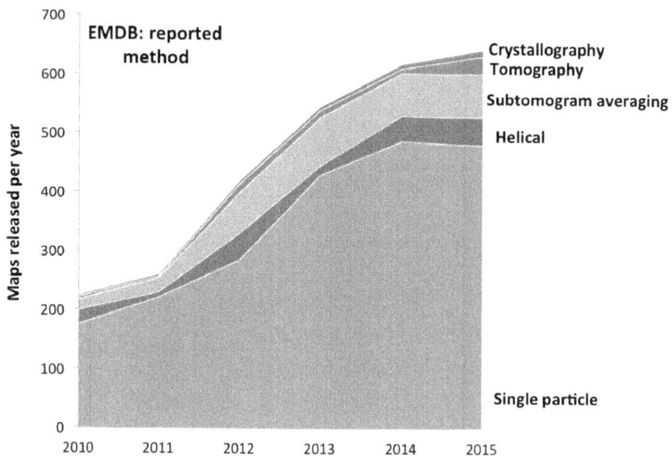

Fig. 2 Trend in reported 3DEM method for EMDB entries released between 2010 and 2015, showing annual releases. (See the color plate.)

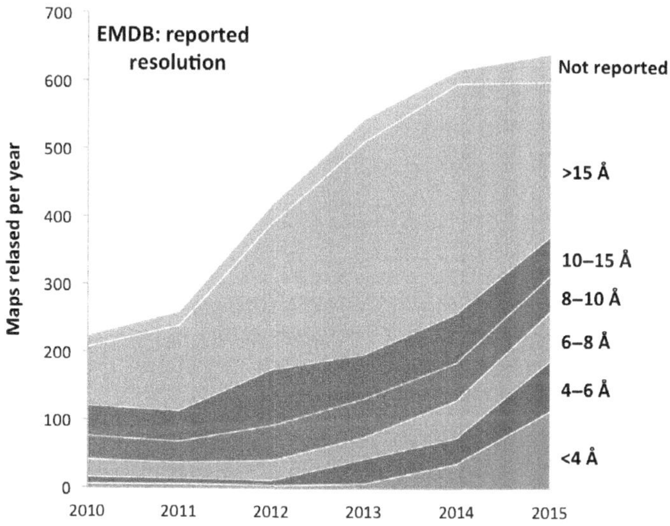

Fig. 3 Trend in reported resolution for EMDB entries released between 2010 and 2015, showing annual releases. (See the color plate.)

(Alber et al., 2007) and amyloid fibrils (Fitzpatrick et al., 2013). The current data archives do not fully support the range of possible hybrid experimental data; the challenges are discussed in more detail by Sali et al. (2015).

The EMDB archives the final 3D reconstructions (map volumes) from EM experiments. There have been growing calls from the EM community

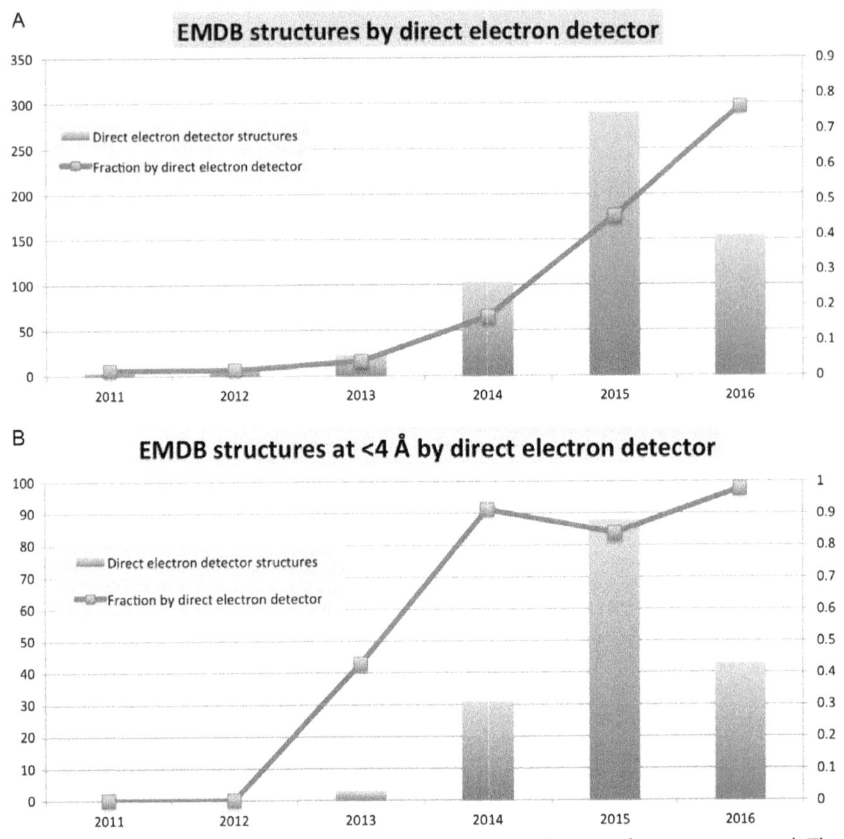

Fig. 4 Trends in released EMDB entries where a direct electron detector was used. The numbers for 2016 are for the period up to Mar. 3, 2016. (A) The column chart shows the total number of entries where a direct electron detector was used (axis to the *left*) and the line chart shows the fraction of all entries (axis to the *right*). (B) These charts are similar to (A) except that the resolution is restricted to 4 Å or better.

for the public archiving of the raw EM image data, both to serve as benchmarks (Henderson et al., 2012) and to allow others to perform a full validation of the experimental results (Glaeser, 2013; Henderson, 2013). The raw data are often orders of magnitude larger in size than the final 3D reconstructions and the EMDB infrastructure is not able to cope with the storage or transfer of these large datasets. In 2014, PDBe created EMPIAR (Iudin, Korir, Salavert-Torres, Kleywegt, & Patwardhan, 2016), a dedicated archive for raw EM image data designed to handle large data set transfers from the outset. EMPIAR now comprises over 45 datasets averaging 700 GB in size with six datasets over a TB in size. In its short existence, EMPIAR has already been cited over 16 times and EMPIAR data are downloaded at an average rate of over 10 TB per month, underlining the important role it is playing for the EM community. EMPIAR data are used for a range of purposes including validation, methods development, testing, and training. Based on input from the community, PDBe is also working on extending EMPIAR to support related imaging modalities including 3D scanning electron microscopy, soft X-ray tomography, and correlative light and electron microscopy.

2. RESOURCES

The EMDataBank website (http://emdatabank.org) provides a unifying portal to resources relating to 3DEM map and model data deposited to EMDB and PDB. The EMDataBank project is a joint effort among PDBe, RCSB PDB, and the National Center for Macromolecular Imaging (NCMI) at Baylor College of Medicine (Lawson et al., 2011, 2016). Resources for EMDB and EMPIAR from PDBe may also be accessed via the links http://pdbe.org/emdb and http://pdbe.org/empiar, respectively.

2.1 Searching, Browsing, and Visualizing Data

Two search services are currently available through EMDataBank. EMSEARCH (Lawson et al., 2011), hosted at RCSB, facilitates simple searches of EMDB based on author name, title, entry ID, sample name, citation abstract content, aggregation type, resolution, and/or release date. Advanced EMDB search (http://pdbe.org/emsearch; Gutmanas et al., 2014), hosted at PDBe, provides additional capabilities such as searches by sample molecular weight, taxonomy, reconstruction software package, microscope model, and parameters, and has the ability to filter and further refine initial search results. Both search sites provide results listings with links to individual entry pages, where one can view overview information and access links for visualization tools and file downloads. The EMDB archive can also be investigated using the EMStats statistics service (http://pdbe.org/emstats; Gutmanas et al., 2014).

Three types of web-based visualization are available for 3DEM structures. First, a Java-applet-based volume viewer permits 3D visualization of maps and their associated PDB coordinate models (Lagerstedt et al., 2013; Lawson et al., 2011). Second, a volume slicer is available for inspecting 2D slices of EMDB entries from three orthogonal orientations (Salavert-Torres et al., 2016). Third, visual analysis pages (Gutmanas et al., 2014;

Lagerstedt et al., 2013; Patwardhan et al., 2012) facilitate analysis and validation of maps, tomograms and models by providing static images of map orthogonal projections and central slices as well as graphs of Fourier-Shell correlation (FSC) curves, map density distribution, rotationally averaged power spectrum, volume estimation vs contour level, and model atom inclusion at the recommended contour level. These visualization tools have been designed to help nonexperts and experts alike to gain insight into the content and assess the quality of 3DEM structures in EMDB and PDB without the need to install specialized software or to download large amounts of data from the structural data archives.

Maps and associated model files may also be downloaded for local analysis via links on individual entry pages. EMDB maps can be viewed along with associated models using locally installed software such as UCSF Chimera (https://www.cgl.ucsf.edu/chimera/; Pettersen et al., 2004), Pymol (https://www.pymol.org), VMD (http://www.ks.uiuc.edu/Research/vmd/; Humphrey, Dalke, & Schulten, 1996), and Coot (http://www2.mrc-lmb.cam.ac.uk/personal/pemsley/coot/; Emsley, Lohkamp, Scott, & Cowtan, 2010), enabling investigation with an extensive set of tools.

2.2 Validation

2.2.1 Visual Analysis Pages

The functionality of the visual analysis pages has been described above and they provide a basic simple first check of the entry based on the information available. They are available from links of the form: http://pdbe.org/emd-####/analysis, eg, http://pdbe.org/emd-2852/analysis for EMDB entry EMD-2852.

2.2.2 Stand-Alone Validation Servers

Two validation servers have been developed at PDBe for eventual integration into the 3DEM validation pipeline (see Section 2.2.3).

FSC (Harauz & van Heel, 1986) is the most commonly reported method for estimating the resolution of single-particle maps. However, the estimated resolution depends critically on the threshold criteria used, and several different conventions are followed. In order to simplify deposition of FSC curve data to EMDB, a web service for calculating FSC curves has been developed, community-tested, and placed into production (http://pdbe.org/fsc). A user can upload two independent maps, receive the calculated

FSC curve in a standardized format for deposition into EMDB, and view and download a plot of the curve. Several reconstruction packages also produce FSC files suitable for direct upload to EMDB including EMAN2 (http://blake.bcm.edu/emanwiki/EMAN2; Tang et al., 2007), RELION (http://www2.mrc-lmb.cam.ac.uk/relion/index.php/Main_Page; Scheres, 2012), and Bsoft (http://lsbr.niams.nih.gov/bsoft/; Heymann & Belnap, 2007). More than 120 map entries in EMDB now include deposited FSC curves.

Tilt-pair analysis (Henderson et al., 2011; Rosenthal & Henderson, 2003) is a useful method for validating the hand and overall shape of a map, particularly for lower resolution maps in which secondary structure features are absent. A tilt-pair validation server developed by the Rosenthal group (Wasilewski & Rosenthal, 2014) has made the method generally accessible. This server has now been migrated to PDBe and is available for public use (http://pdbe.org/tiltpair).

2.2.3 Validation Pipeline

An initial EM validation report has been developed for use in the wwPDB Deposition and Annotation system (see Section 3.1). The format closely follows the validation reports produced by wwPDB for structures from X-ray crystallography (Read et al., 2011) and NMR (Montelione et al., 2013), and is based on the same underlying validation software pipeline (Gore, Velankar, & Kleywegt, 2012). We have initially focused on providing map-independent assessments of model quality. Model assessments include standard geometry (bonds, angles, and torsion angles), close contacts, protein and nucleic acid backbone geometry, and ligand geometry. A slider graphic compares the quality of the given structure, for key indicators, to all EM structures in the PDB archive, as well as all structures in the PDB archive.

Recognizing that nearly one-quarter of all 3DEM models in the PDB contain polymers represented as atom traces (Cα-atom only for protein chains; P-atom only for nucleic acid chains), we are actively investigating new assessments for trace atom model geometry. Consecutive Cα–Cα distances are reported as outliers if they fall outside of $\pm 3\sigma$ limits for *cis* and *trans* peptide distributions; consecutive P–P distances are reported as outliers if they are shorter than 4.4 Å or longer than 8.0 Å.

The EM validation report also provides a table of basic information about the map, eg, the reconstruction method, reported resolution, resolution method, imposed symmetry, number of images used, microscope, imaging parameters, and detector. Future report versions will, with guidance from the EM Validation Task Force (Henderson et al., 2012 and see later), add validation components for the map as well as for the fit of the model to the map, as the relevant methods and metrics evolve and become accepted community standards. We will encourage depositors to include the validation report when submitting manuscripts for review and encourage journal editors and reviewers to request these reports.

2.3 Programmatic Access

A REST API providing easy access to EMDB metadata and PDB data in JSON format is available from http://pdbe.org/api. A web service based on the SOAP protocol for accessing EMDB metadata is available from http://emdatabank.org/webservice.html.

3. DEPOSITION AND ANNOTATION

3.1 EMDB and PDB

The wwPDB partners and the EMDataBank project recently launched a new Deposition and Annotation system that supports structures determined using 3DEM, NMR, and X-ray, neutron, and electron crystallography. New entries can be submitted at http://deposit.wwpdb.org/deposition/. Depositors will be able to complete map-only (EMDB) and combined map + model (EMDB + PDB) submissions, providing information tailored to the particular 3DEM method selected (single particle, helical, subtomogram average, tomography, or electron crystallography). The new system produces an EM-specific validation report and features a revised and expanded metadata dictionary for 3DEM experiments (Patwardhan et al., 2012). For example, hierarchical sample description is implemented in way that can be tied to map segmentations (Patwardhan et al., 2014), and extensions for each 3DEM method have been added, following community-based recommendations (Henderson et al., 2012; Patwardhan et al., 2014). The new system also supports a rich set of auxiliary data including half maps used for validation, masks, and FSC curves. Legacy systems for EM deposition (EMDEP for EMDB deposition and EM-ADIT (RCSB) or AUTODEP (PDBe)) will be kept running for a transitional phase, particularly to allow on-going depositions to complete. The shutdown of these systems will be announced well ahead of time on relevant forums in order to give the EM community time to prepare for the changes.

3.2 EMPIAR

A web-based deposition system for EMPIAR can be accessed via the "Deposition" tab at the top of the EMPIAR home page (pdbe.org/empiar). This is a user-based system that allows users to manage multiple depositions and to share access to depositions among multiple users. For EM experiments that require mandatory deposition of the final 3D reconstruction to EMDB, we require that an EMDB deposition is associated with the EMPIAR deposition. It is however also possible to deposit data relating to 3D SEM and soft X-ray tomography. For these imaging modalities the requirement is that the data are associated with a journal publication. As with the EMDB depositions, once a deposition is submitted, the entry is curated and released as per the instructions of the depositor. In contrast to EMDB, EMPIAR entries do not follow a weekly release process, but are released when they are ready. It should be noted though that this process is not instantaneous and multi terabyte datasets may take days to release from when the instruction is received.

4. RECENT COMMUNITY-WIDE INITIATIVES

The structural archives serve a greater role than as mere data repositories. Reuse of data makes apparent issues related to data and metadata formats, data storage and transfer, integration of data with other forms of structural data and other bioinformatics data, and data validation. The organizations, and partners involved in the running of the EMDB, PDB, and EMPIAR archives play key roles as facilitators in helping bring about consensus and agreement on a range of issues to the wider benefit of the structural community. Here we provide an overview of some of the key initiatives and workshops that have helped move the field forward in recent years.

4.1 EM Validation Task Force

Assessment of structural data crucially requires community-accepted validation criteria (Montelione et al., 2013; Read et al., 2011). However, methods for validation of 3DEM structures are still in early development and are applied inconsistently (Glaeser, 2013; Henderson, 2013; Subramaniam, 2013; van Heel, 2013). EMDataBank has been working with the 3DEM community to establish data validation methods that can be used in the structure determination process, to define key indicators of a well-determined structure that should accompany every structure deposition, and to implement appropriate validation procedures into a 3DEM validation pipeline.

In 2010 the EM Validation Task Force (EM VTF) was established. The international group of ~30 3DEM experts explored how to assess maps, models, and other data that are deposited into the EM Data Bank (EMDB) and PDB public data archives. Overall the need for more research and development of validation criteria for maps and map-derived models was strongly articulated (Henderson et al., 2012). *For deposited maps*, the EM VTF recognized a critical need to develop standards for assessing map resolution and accuracy, and recommended reporting of map resolution in accordance with visible features, deposition of annotations specific to each map type, and validation of map symmetry. *For deposited map-derived models*, EM VTF recommendations included establishment of criteria for assessing models both with and without regard to the fit to the map, creation of community-wide benchmarks for modeling methods, sequence annotation of all map components, and capability to archive coarse-grained representations of models. *Additional recommendations* included establishing deposition guidelines for publication of 3DEM structures in journals, and expanding the role of EMDataBank to work together with the 3DEM community to provide unified access to 3DEM structures and to facilitate development of validation and data standards. The EM VTF also recommended providing full FSC curves with deposited maps, indicating whether or not maps used for FSC calculation are fully independent, establishing benchmark datasets for maps and models, and providing multiple types of assessments for models. Participants at the 2011 "Data management challenges in 3D electron microscopy" workshop (Patwardhan et al., 2012) reiterated the EM VTF's advice and provided further recommendations regarding development of data models and validation-related services.

EMDataBank is following up on these recommendations with research efforts, community-wide challenges, and validation pipeline development.

4.2 "Data Management Challenges in 3D Electron Microscopy" Workshop

The aim of this expert workshop held in Dec. 2011 was to discuss the growing challenges of storing, sharing, transferring, analyzing, viewing, validating, and annotating 3DEM data. The outcomes of the meeting included an acknowledgement for the need to set up an archive for raw image data relating to EMDB entries as vital for validation, laying the foundations for the initial ideas that would eventually result in the establishment of EMPIAR

(Patwardhan et al., 2012). Overall the participants endorsed the vital role that EMDataBank and its partners could play in improving reporting standards for validation in 3DEM, and providing validation tools for 3DEM data. The participants also felt that it would be desirable to have community-wide efforts on the subject of format standardization and harmonization. Subsequent to this meeting there has been an initiative to try and clarify and standardize the definition of the MRC map format (Cheng et al., 2015) and to develop an EM exchange format that could be used to represent EM-related metadata such as particle coordinates and coordinate transformations (Marabini et al., 2016). Meeting participants also recognized a problem with the deposition rates for electron tomography lagging behind that of single-particle EM. It was agreed that there were several reasons for this, partly having to do with the cellular community being more closely related to the cellular imaging community than to the structural community where the concept of deposition had been long since established. Also the EMDB data model did not adequately capture metadata for cellular EM, nor was integration adequately promoted as the support for segmentation data was rudimentary. In order to improve deposition of electron tomography EMDataBank highlighted the issue on the 3DEM bulletin board and at meetings and managed to garner agreement on the mandatory deposition of subtomogram averages, and a "strong recommendation" for the deposition of at least one representative tomogram from every electron tomography publication.

4.3 "A 3D Cellular Context for the Macromolecular World" Workshop

This expert workshop (Patwardhan et al., 2014) expanded on the discussions on archiving in cellular EM initiated at the "Data management challenges in 3D electron microscopy" workshop. A key question discussed was whether archiving needed to be expanded to support other imaging modalities to adequately provide a cellular context, and how to integrate structural data at different scales of imaging. There was a further endorsement for the development of an archive for raw image data with an eye to extend this to be able to easily accommodate new modalities of cellular imaging data. The participants recognized that the scope for this could be very wide and that it would make sense to focus on a few but important modalities, and the ones suggested were 3D SEM, soft X-ray tomography, and correlative light and electron microscopy. In terms of data integration the participants recognized the need to capture segmentation data in EMDB and to ensure that segmentations were annotated with biological terms from relevant biological ontologies and bioinformatics databases, and that tools and formats were created that facilitated this annotation and subsequent deposition. In 2014 PDBe received funding from the MRC/BBSRC for the very purpose of working on these outcomes from the meeting—the development of EMPIAR, the development of integrated web-based visualization of molecular and cellular structural data, and the development of a web-based tool to facilitate the biological annotation of segmentations. A slight digression but nevertheless of importance for the field was a discussion on the need for archiving movies relating to EMDB and PDB entries usually included in the supplementary materials for journal publications. The motivation was the lack of consistency in annotation and presentation between journals and questions about long-term sustainability—there was anecdotal evidence that many movies included as supplementary information were no longer available after some period of time. More recently PDBe (work by Vladislav Lysenkov) has developed a prototype movie archive, which subject to support from the EM community and publishers and additional funding will be further developed into a full fledged movie archive for EMDB and PDB entries.

4.4 wwPDB Hybrid/Integrative Methods Task Force Workshop

This workshop was convened to bring together experts in the field to discuss steps to enable better support for the archiving of hybrid methods experiments in the public domain (Sali et al., 2015). There are several structural biological problems that cannot be tackled by a single structural technique alone and require the close integration of information from both established structural techniques and other sources such small-angle X-ray scattering (SAXS), fluorescence microscopy, and mass spectrometry. The information that can currently be deposited to public databases is often insufficient to provide a holistic view of hybrid methods experiments.

The participants agreed that all relevant experimental data should be archived, but it would be up to the expert communities to drive decisions on what should be archived and how. A flexible model representation needed to be developed to accommodate models at different scales—as integrative modeling would often yield coarse-grained nonatomistic models (showing, for example, the positioning of domains rather than domain detail) and to accommodate multistate models and model ensembles. Additionally methods for estimating the uncertainty needed to be established.

Finally, due to the wide array of methods, user communities and funding sources involved a single archive model was deemed unrealistic and instead the approach would be to create a federated system of model and data archives.

4.5 EMDataBank Map and Model Validation Challenges

EMDataBank is sponsoring two new community challenges in 2015–2016 to raise awareness of the need for structure validation as a routine part of 3DEM studies and publications. Additional goals are to develop suitable sets of benchmark data, establish best practices, evolve criteria for validation, and compare and contrast different 3DEM methodologies. The new challenges have been formulated by committees composed of 3DEM community members, with benchmark targets of varying size and complexity selected from recently deposited 3DEM structures based on current state-of-the-art detectors and processing methods, in the resolution range 2.2–4.5 Å. The new challenges follow in the positive spirit of previous community-based challenge activities for particle picking (Zhu et al., 2004), modeling (Ludtke, Lawson, Kleywegt, Berman, & Chiu, 2012), and CTF correction (Ludtke et al., 2012; Marabini et al., 2015; Zhu et al., 2004). We anticipate that the community-developed benchmarks will prove useful for methods evaluation, even beyond these challenge activities (Editorial, 2015).

For the map challenge, participants have been asked to create and submit reconstructions using supplied image data. The map challenge data, which totals 12 TB for seven benchmark targets and includes raw movie frame images, have been provided by the original authors of the selected targets, and are stored in EMPIAR.

For the model challenge, participants have been asked to create and submit atomic coordinate models using supplied maps. Following a key recommendation of cryoEM specialists and modeling software developers at a planning workshop organized in Jun. 2015, each benchmark target is an unfiltered, unsharpened map. Half maps used for FSC curve calculation are also available for participants to try out various refinement and validation strategies. Maps for the eight targets have been provided by the original authors of the target structures and are stored as supplemental files associated with EMDB entries.

Each challenge has challenger submission and results assessment phases. Follow-up discussions via participant workshops are planned, as well as dissemination of results in journal special issues. Anyone from the scientific community is welcome to participate as a challenger and/or assessor. Both challenges are hosted at http://challenges.emdatabank.org/.

5. CHALLENGES AND OPPORTUNITIES

5.1 Rise of Multiuser Facilities, CCP-EM, and Prospects for Data Harvesting

The rising costs of purchasing state-of-the-art cryoEM microscopy systems and maintaining the supporting infrastructure are putting them beyond the reach of many individual institutions. A growing trend is therefore for a more coordinated approach often involving regional or national collaborations between multiple institutions to set up multiuser facilities similar to the beamlines at synchrotron facilities. Examples include Necen in the Netherlands (http://www.necen.nl/), the electron Bio-Imaging Centre (eBIC; Saibil, Grunewald, & Stuart, 2015) at the Diamond Light Source in the UK, the National Resource for Automated Molecular Microscopy in New York (NRAMM; http://nramm.nysbc.org/), and the National Center for Macromolecular Imaging in Houston (NCMI; http://ncmi.bcm.edu/ncmi/). A number of issues need to be considered to maximize the efficiency and throughput of these centers. For instance, the availability of lower end microscopes that can be used for screening and fine-tuning to maximize the chance of getting high quality datasets when time is finally allocated. Automation of the imaging session using software such as EPU (https://www.fei.com/software/epu/), Leginon (Suloway et al., 2005), and SerialEM (Mastronarde, 2005) is essential to maximize the amount of data that can be collected. Data management and image processing can also pose major challenges. However, the development of coherent software pipelines also gives rise to the opportunity for harvesting data directly to EMDB and EMPIAR, which could greatly facilitate the deposition process. In the context of the EMDataBank project we have previously demonstrated the feasibility of harvesting a partially populated metadata XML file following the EMDB XML schema and populating relevant form fields.

The Collaborative Computational Project for Electron cryo-Microscopy (CCP-EM; Wood et al., 2015) was established following the model in the United Kingdom for Collaborative Computational Projects (eg, CCP4 for macromolecular crystallography) for providing long-term support to scientific areas that require significant computation. The aims are to build a coherent cryoEM community, supporting users of cryoEM software, and supporting developers in producing and distributing software. Funding for the CCP-EM project was recently renewed with an additional emphasis on helping develop software pipelines and infrastructure for eBIC. In this context the prospect of direct harvesting of data from eBIC to EMPIAR and EMDB will be considered with PDBe, and any developments are likely to benefit other multiuser facilities as well.

5.2 New Imaging Modalities

With EMPIAR now accepting data from 3D SEM and soft X-ray tomography experiments it may be asked whether public archiving can and should be extended to other imaging modalities, including fluorescence microscopy, to accommodate for the rapidly changing landscape of cellular structural biology? The approach to archiving followed by EMDB and EMPIAR remains fairly traditional with centralized archiving, well-structured data, disciplined practices, and transparent processes. The strengths of such a system are its robustness, high availability of data, and ease of access. The high coherency and consistency in describing the data simplifies reuse and integration. However, with centralized archiving costs can increase prohibitively for the archive provider if data volumes expand too quickly. Furthermore in fields such as super-resolution microscopy where there are a multitude of variations in how experiments are conducted, it may be difficult to abstract a coherent set of metadata beyond a very basal level. On the opposite end of the spectrum is the prospect of distributed archiving, and unstructured data. From our perspective there are no "right" solutions and the relative merits of these solutions need to be considered on a case-by-case basis and may change over time. On an even more fundamental level a question that needs to be posed is what purpose public archiving of this data will serve? On the molecular end of structural biology it has largely been the community itself that has driven the need for public access of data for the purposes of reproducibility, validation, reuse of data, and data integration. Even here some compromises have had to be struck, for example, it has been deemed impractical to archive raw X-ray imaging data in PDB and instead where possible links are maintained to the local sites where these data are stored. Similarly these considerations need to be made very carefully for cellular imaging data in order to arrive at solutions that are viable and sustainable.

ACKNOWLEDGMENTS

We thank the many current and past colleagues who have made significant contributions to the development of data archiving for 3DEM methods. EMDataBank Unified Data Resource is funded by National Institutes of Health GM079429 to Baylor College of Medicine (Wah Chiu, PI), Rutgers University (Helen Berman, co-PI), and EMBL-EBI

(Gerard Kleywegt, co-PI). Work on EMDB and EMPIAR at EMBL-EBI is also supported by the UK Medical Research Council with cofunding from the UK Biotechnology and Biological Sciences Research Council (BBSRC; grant MR/L007835), the BBSRC (grant BB/M018423/1), the Wellcome Trust (grants 088944 and 104948), and the European Commission Framework 7 Programme (grant 284209).

REFERENCES

Alber, F., Dokudovskaya, S., Veenhoff, L. M., Zhang, W., Kipper, J., Devos, D., et al. (2007). The molecular architecture of the nuclear pore complex. *Nature*, *450*, 695–701.

Berman, H., Henrick, K., & Nakamura, H. (2003). Announcing the worldwide Protein Data Bank. *Nature Structural Biology*, *10*, 980.

Berman, H. M., Kleywegt, G. J., Nakamura, H., & Markley, J. L. (2012). The Protein Data Bank at 40: Reflecting on the past to prepare for the future. *Structure*, *20*, 391–396.

Brandt, F., Carlson, L. A., Hartl, F. U., Baumeister, W., & Grunewald, K. (2010). The three-dimensional organization of polyribosomes in intact human cells. *Molecular Cell*, *39*, 560–569.

Cheng, A., Henderson, R., Mastronarde, D., Ludtke, S. J., Schoenmakers, R. H., Short, J., et al. (2015). MRC2014: Extensions to the MRC format header for electron cryo-microscopy and tomography. *Journal of Structural Biology*, *192*, 146–150.

Danev, R., & Baumeister, W. (2016). Cryo-EM single particle analysis with the Volta phase plate. *eLife*, *5*.

Editorial. (2015). The difficulty of a fair comparison. *Nature Methods*, *12*, 273.

Emsley, P., Lohkamp, B., Scott, W. G., & Cowtan, K. (2010). Features and development of Coot. *Acta Crystallographica, Section D: Biological Crystallography*, *66*, 486–501.

Fitzpatrick, A. W., Debelouchina, G. T., Bayro, M. J., Clare, D. K., Caporini, M. A., Bajaj, V. S., et al. (2013). Atomic structure and hierarchical assembly of a cross-beta amyloid fibril. *Proceedings of the National Academy of Sciences of the United States of America*, *110*, 5468–5473.

Glaeser, R. M. (2013). Replication and validation of cryo-EM structures. *Journal of Structural Biology*, *184*, 379–380.

Gonen, T., Cheng, Y., Sliz, P., Hiroaki, Y., Fujiyoshi, Y., Harrison, S. C., et al. (2005). Lipid-protein interactions in double-layered two-dimensional AQP0 crystals. *Nature*, *438*, 633–638.

Gore, S., Velankar, S., & Kleywegt, G. J. (2012). Implementing an X-ray validation pipeline for the Protein Data Bank. *Acta Crystallographica, Section D: Biological Crystallography*, *68*, 478–483.

Gutmanas, A., Alhroub, Y., Battle, G. M., Berrisford, J. M., Bochet, E., Conroy, M. J., et al. (2014). PDBe: Protein Data Bank in Europe. *Nucleic Acids Research*, *42*, D285–D291.

Harauz, G., & van Heel, M. (1986). Exact filters for general geometry three dimensional reconstruction. *Optik*, *73*, 146–156.

Henderson, R. (2013). Avoiding the pitfalls of single particle cryo-electron microscopy: Einstein from noise. *Proceedings of the National Academy of Sciences of the United States of America*, *110*, 18037–18041.

Henderson, R., Chen, S., Chen, J. Z., Grigorieff, N., Passmore, L. A., Ciccarelli, L., et al. (2011). Tilt-pair analysis of images from a range of different specimens in single-particle electron cryomicroscopy. *Journal of Molecular Biology*, *413*, 1028–1046.

Henderson, R., Sali, A., Baker, M. L., Carragher, B., Devkota, B., Downing, K. H., et al. (2012). Outcome of the first electron microscopy validation task force meeting. *Structure*, *20*, 205–214.

Heymann, J. B., & Belnap, D. M. (2007). Bsoft: Image processing and molecular modeling for electron microscopy. *Journal of Structural Biology*, *157*, 3–18.

Humphrey, W., Dalke, A., & Schulten, K. (1996). VMD: Visual molecular dynamics. *Journal of Molecular Graphics*, *14*(33–38), 27–38.

Iudin, A., Korir, P. K., Salavert-Torres, J., Kleywegt, G. J., & Patwardhan, A. (2016). EMPIAR: A public archive for raw electron microscopy image data. *Nature Methods*, *13*, 387–388. http://dx.doi.org/10.1038/nmeth.3806.

Kukulski, W., Schorb, M., Kaksonen, M., & Briggs, J. A. (2012). Plasma membrane reshaping during endocytosis is revealed by time-resolved electron tomography. *Cell*, *150*, 508–520.

Lagerstedt, I., Moore, W. J., Patwardhan, A., Sanz-Garcia, E., Best, C., Swedlow, J. R., et al. (2013). Web-based visualisation and analysis of 3D electron-microscopy data from EMDB and PDB. *Journal of Structural Biology*, *184*, 173–181.

Lawson, C. L., Baker, M. L., Best, C., Bi, C., Dougherty, M., Feng, P., et al. (2011). EMDataBank.org: Unified data resource for CryoEM. *Nucleic Acids Research*, *39*, D456–D464.

Lawson, C. L., Patwardhan, A., Baker, M. L., Hryc, C., Garcia, E. S., Hudson, B. P., et al. (2016). EMDataBank unified data resource for 3DEM. *Nucleic Acids Research*, *44*, D396–D403.

Ludtke, S. J., Lawson, C. L., Kleywegt, G. J., Berman, H., & Chiu, W. (2012). The 2010 cryo-EM modeling challenge. *Biopolymers*, *97*, 651–654.

Marabini, R., Carragher, B., Chen, S., Chen, J., Cheng, A., Downing, K. H., et al. (2015). CTF challenge: Result summary. *Journal of Structural Biology*, *190*, 348–359.

Marabini, R., Ludtke, S. J., Murray, S. C., Chiu, W., de la Rosa-Trevin, J. M., Patwardhan, A., et al. (2016). The Electron Microscopy eXchange (EMX) initiative. *Journal of Structural Biology*, *194*, 156–163.

Mastronarde, D. N. (2005). Automated electron microscope tomography using robust prediction of specimen movements. *Journal of Structural Biology*, *152*, 36–51.

Montelione, G. T., Nilges, M., Bax, A., Guntert, P., Herrmann, T., Richardson, J. S., et al. (2013). Recommendations of the wwPDB NMR Validation Task Force. *Structure*, *21*, 1563–1570.

Nogales, E., Wolf, S. G., & Downing, K. H. (1998). Structure of the alpha beta tubulin dimer by electron crystallography. *Nature*, *391*, 199–203.

Patwardhan, A., Ashton, A., Brandt, R., Butcher, S., Carzaniga, R., Chiu, W., et al. (2014). A 3D cellular context for the macromolecular world. *Nature Structural & Molecular Biology*, *21*, 841–845.

Patwardhan, A., Carazo, J. M., Carragher, B., Henderson, R., Heymann, J. B., Hill, E., et al. (2012). Data management challenges in three-dimensional EM. *Nature Structural & Molecular Biology*, *19*, 1203–1207.

Pettersen, E. F., Goddard, T. D., Huang, C. C., Couch, G. S., Greenblatt, D. M., Meng, E. C., et al. (2004). UCSF Chimera—A visualization system for exploratory research and analysis. *Journal of Computational Chemistry*, *25*, 1605–1612.

Read, R. J., Adams, P. D., Arendall, W. B., 3rd, Brunger, A. T., Emsley, P., Joosten, R. P., et al. (2011). A new generation of crystallographic validation tools for the protein data bank. *Structure*, *19*, 1395–1412.

Rodriguez, J. A., Ivanova, M. I., Sawaya, M. R., Cascio, D., Reyes, F. E., Shi, D., et al. (2015). Structure of the toxic core of alpha-synuclein from invisible crystals. *Nature*, *525*, 486–490.

Rosenthal, P. B., & Henderson, R. (2003). Optimal determination of particle orientation, absolute hand, and contrast loss in single-particle electron cryomicroscopy. *Journal of Molecular Biology*, *333*, 721–745.

Saibil, H. R., Grunewald, K., & Stuart, D. I. (2015). A national facility for biological cryo-electron microscopy. *Acta Crystallographica, Section D: Biological Crystallography, 71*, 127–135.

Salavert-Torres, J., Iudin, A., Lagerstedt, I., Sanz-Garcia, E., Kleywegt, G. J., & Patwardhan, A. (2016). Web-based volume slicer for 3D electron-microscopy data from EMDB. *Journal of Structural Biology, 194*, 164–170.

Sali, A., Berman, H. M., Schwede, T., Trewhella, J., Kleywegt, G., Burley, S. K., et al. (2015). Outcome of the first wwPDB Hybrid/Integrative Methods Task Force Workshop. *Structure, 23*, 1156–1167.

Scheres, S. H. (2012). RELION: Implementation of a Bayesian approach to cryo-EM structure determination. *Journal of Structural Biology, 180*, 519–530.

Subramaniam, S. (2013). Structure of trimeric HIV-1 envelope glycoproteins. *Proceedings of the National Academy of Sciences of the United States of America, 110*, E4172–E4174.

Suloway, C., Pulokas, J., Fellmann, D., Cheng, A., Guerra, F., Quispe, J., et al. (2005). Automated molecular microscopy: The new Leginon system. *Journal of Structural Biology, 151*, 41–60.

Tagari, M., Newman, R., Chagoyen, M., Carazo, J. M., & Henrick, K. (2002). New electron microscopy database and deposition system. *Trends in Biochemical Sciences, 27*, 589.

Tang, G., Peng, L., Baldwin, P. R., Mann, D. S., Jiang, W., Rees, I., et al. (2007). EMAN2: An extensible image processing suite for electron microscopy. *Journal of Structural Biology, 157*, 38–46.

van Heel, M. (2013). Finding trimeric HIV-1 envelope glycoproteins in random noise. *Proceedings of the National Academy of Sciences of the United States of America, 110*, E4175–E4177.

Wasilewski, S., & Rosenthal, P. B. (2014). Web server for tilt-pair validation of single particle maps from electron cryomicroscopy. *Journal of Structural Biology, 186*, 122–131.

Wood, C., Burnley, T., Patwardhan, A., Scheres, S., Topf, M., Roseman, A., et al. (2015). Collaborative computational project for electron cryo-microscopy. *Acta Crystallographica, Section D: Biological Crystallography, 71*, 123–126.

Zeev-Ben-Mordehai, T., Hagen, C., & Grunewald, K. (2014). A cool hybrid approach to the herpesvirus 'life' cycle. *Current Opinion in Virology, 5C*, 42–49.

Zhu, Y., Carragher, B., Glaeser, R. M., Fellmann, D., Bajaj, C., Bern, M., et al. (2004). Automatic particle selection: Results of a comparative study. *Journal of Structural Biology, 145*, 3–14.

AUTHOR INDEX

Note: Page numbers followed by "*f*" indicate figures, and "*t*" indicate tables.

D

H

I

J

K

L

M

N

O

P

T

U

V

Z

SUBJECT INDEX

Note: Page numbers followed by "*f*" indicate figures, "*t*" indicate tables, and "*b*" indicate boxes.

A

N

S

T

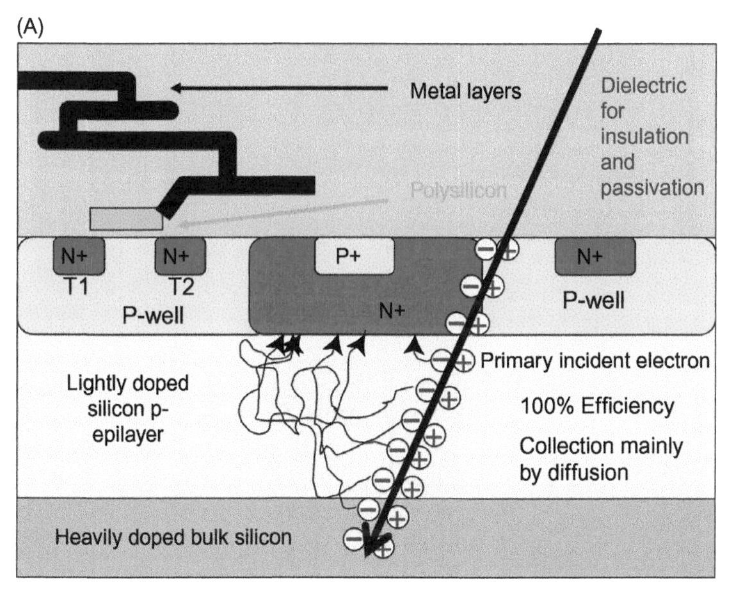

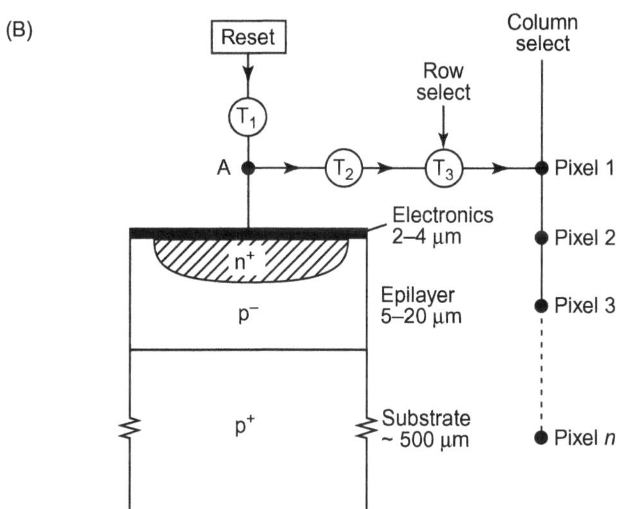

G. McMullan *et al.*, Fig. 1 (A) Schematic diagram viewed parallel to the sensor surface, of a single pixel in a typical CMOS sensor. The incident electron is represented by the *black arrow* from *top right* to *bottom left*. The + and − symbols indicate the electron–hole pairs that are created by the transient electric field as the high-energy electron passes. The mini-electrons are collected by the N-well whose potential drops during the exposure. (B) Schematic of readout for a single pixel, showing the 3T logic and its relationship to the N+ diode/capacitor. *Panel (A): Adapted from Turchetta, R. (2003). CMOS monolithic active pixel sensors (MAPS) for scientific applications. In:* 9th Workshop on electronics for LHC experiments Amsterdam, 2003 *(pp. 1–6). http://lhc-electronics-workshop.web.cern.ch/lhc-electronics-workshop/2003/PLENARYT/TURCHETT.PDF. Panel (B): Reproduced from Faruqi, A. R. (2007). Direct electron detectors for electron microscopy.* Advances in Imaging and Electron Physics, 145, *55–93.*

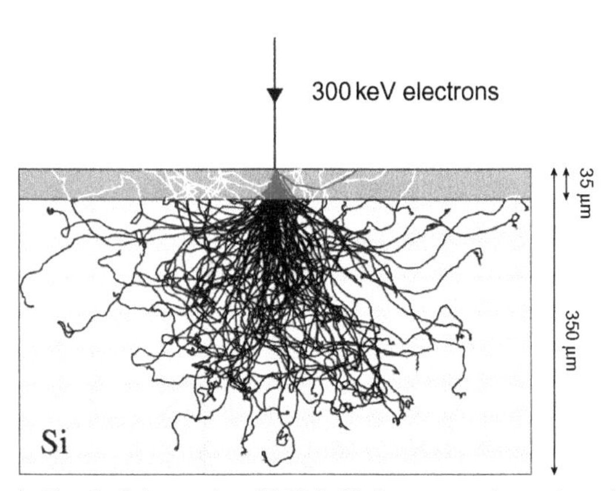

G. McMullan *et al.*, Fig. 3 Schematic of 300 keV electron trajectories, showing a Monte Carlo simulation of 300 keV electron tracks in silicon. After backthinning to 35 μm, only those parts of the electron tracks highlighted in *red* would contribute to the recorded signal. Before backthinning, the additional *white tracks* would contribute a low-resolution component to the signal together with contributions to the noise at all spatial frequencies. The overall thickness of the silicon in the figure is 350 μm with the 35 μm layer that remains after backthinning shown in *gray*. *Reproduced from McMullan, G., Faruqi, A. R., Henderson, R., Guerrini, N., Turchetta, R., Jacobs, A., & van Hoften, G. (2009). Experimental observation of the improvement in MTF from backthinning a CMOS direct electron detector.* Ultramicroscopy, 109*(9), 1144–1147.*

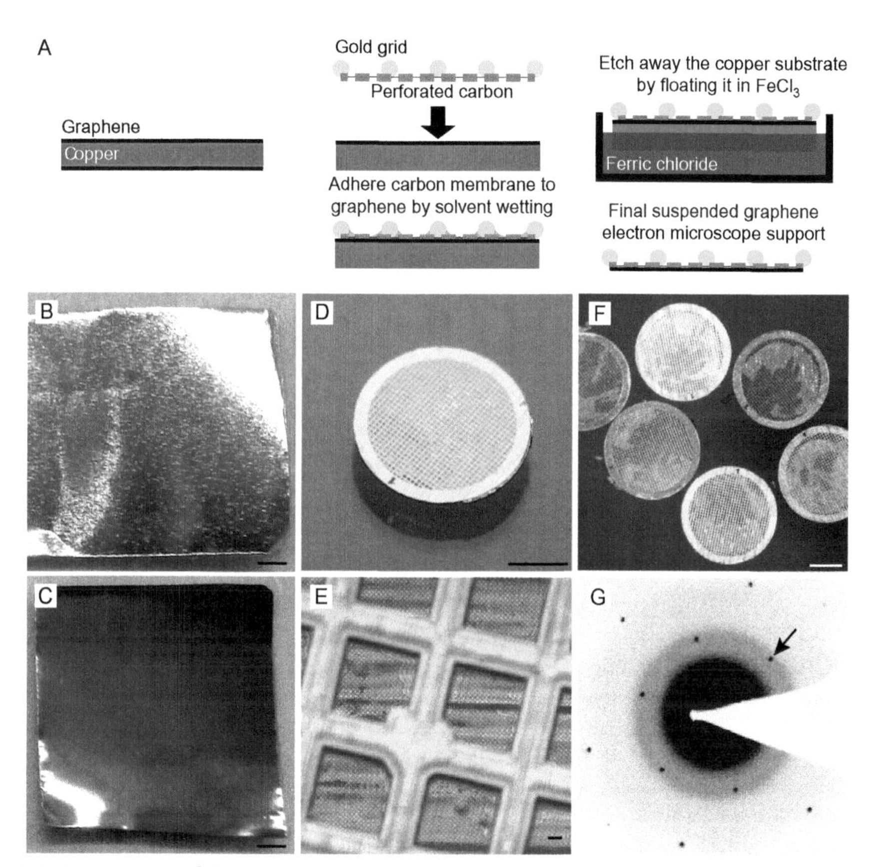

L.A. Passmore and C.J. Russo, Fig. 7 *Graphene transfer onto supports with carbon foils.* The process is diagrammed in panel A. Panels B and C show copper heated to 150°C for 10 min in air, where B is fully covered in graphene so does not oxidize while C has no graphene and turns color due to oxidization, scale bars are 3 mm. This simple test is used to map the location of the graphene on the foil. Panels D and E show the grid–graphene–copper sandwich, scale bars are 1 mm and 10 μm, respectively. Panel F is the sandwich floating in the etchant, where the partially etched grains of copper are visible (scale 1 mm). Panel G is an electron diffraction pattern of suspended graphene with ice, where the arrow points to the 2.1 Å reflection from the graphene lattice.

Ted Pella easyGlow (c. 2015)

Edwards S150B (c.1995)

Edwards 12E6 (c.1962)

L.A. Passmore and C.J. Russo, Fig. 8 *Plasmas generated by glow discharge*. Residual air plasma generation by three different instruments is shown. All are effective in increasing hydrophilicity of support surfaces but have varying degrees of reproducibility and can damage the supports. Note that the plasmas in glow discharge apparatuses are often nonuniform which can vary the exposure dose significantly, even in a single batch. Scale bars are 20 mm.

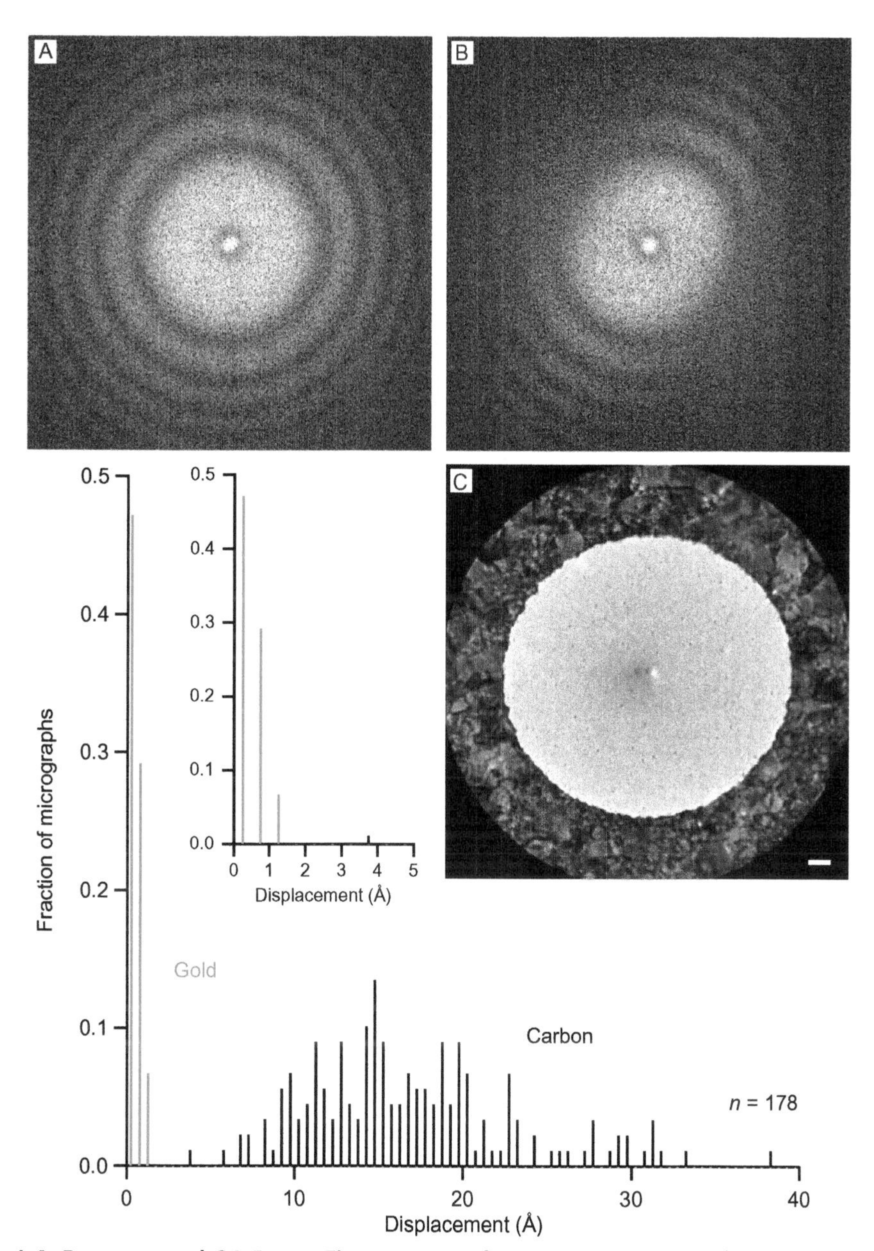

L.A. Passmore and C.J. Russo, Fig. 11 *Testing for specimen movement during or after data collection*. Large amounts of specimen movement or stage drift can be detected during data acquisition using real-time fast Fourier transforms (FFTs) of the collected micrographs. FFTs of specimens are shown with no stage drift (A) and 10 Å/s temperature induced stage drift (B). Micrograph C shows the recommended, symmetric illumination of a frozen specimen suspended across a hole in an all-gold support foil. Histogram is the in-plane movement statistics for 1 s micrographs (16 $e^{-}/Å^{2}$) on all-gold supports vs amorphous carbon on gold (Quantifoil) under the same symmetric illumination conditions shown in C. Inset is enlargement of histogram near origin.

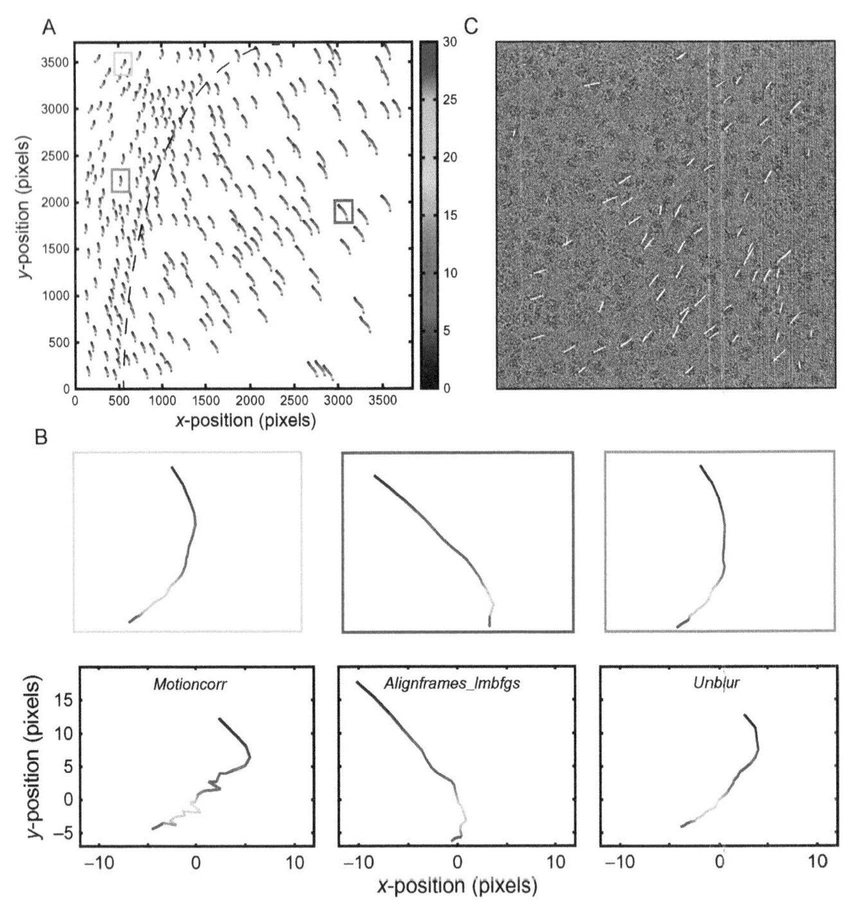

Z.A. Ripstein and J.L. Rubinstein, Fig. 2 (A) A plot of individual particle trajectories from *alignparts_lmbfgs* applied to unaligned movie frames. Each line in the plot indicates the trajectory of a single particle from frame 1 (*black*) to frame 30 (*blue*), exaggerated by a factor of 5. The broken line indicates the approximate edge of the carbon support film, with particles to the left of the line lying on the carbon support and particles to the right being in unsupported ice. (B) Inspection of individual particle trajectories from three regions of the micrograph shows smoothed particle trajectories that differ across the micrograph (three upper panels, *color* coded such that they correspond to the three small boxes in panel A). The whole-frame alignment programs *Motioncorr, alignframes_lmbfgs*, and *Unblur* all find somewhat different solutions to the alignment problem that produce similar power spectra from the average of aligned frames, each of which approximately matches the individual particle trajectories in different regions of the micrograph (lower panels). (C) Linear particle trajectories calculated using *Relion* particle polishing applied to whole-frame aligned movies. Particle trajectories start at the *green* dot and move to the *red* dot. Trajectories are from a different movie than used for parts A and B and are exaggerated by a factor of 50. *Source: Reproduced with permission from reference Scheres, S. H. (2014). Beam-induced motion correction for sub-megadalton cryo-EM particles.* Elife, 3, *e03665.*

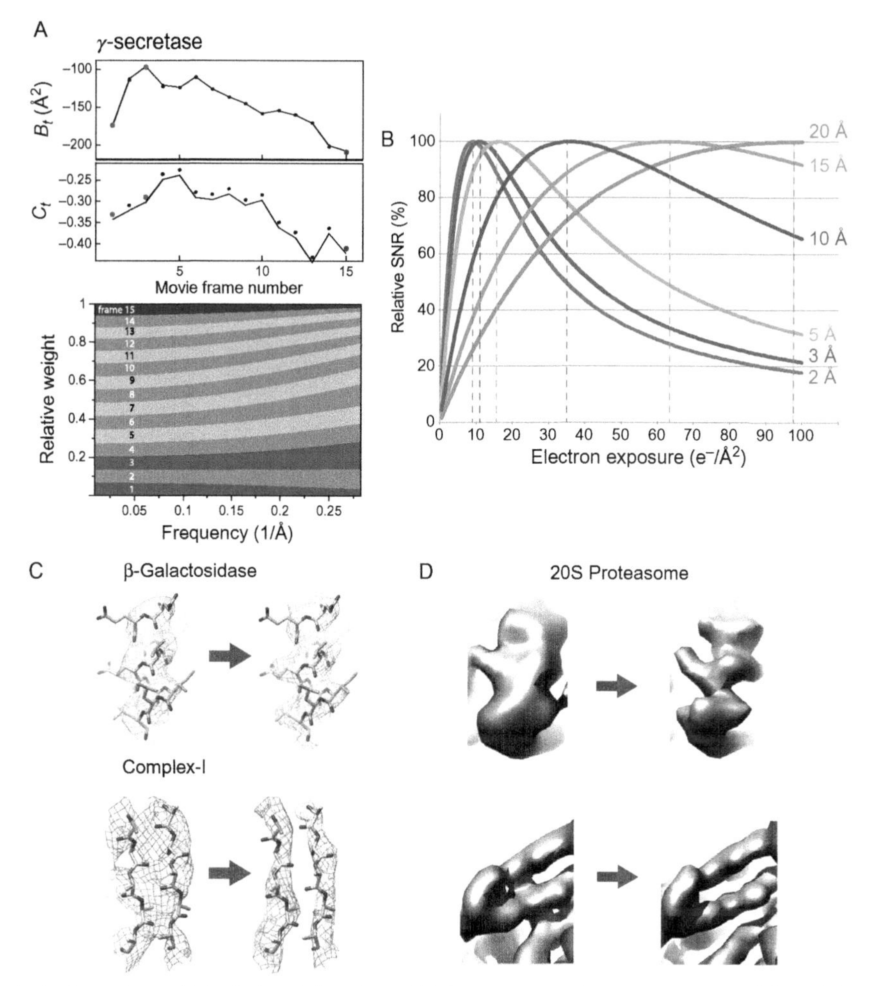

Z.A. Ripstein and J.L. Rubinstein, Fig. 3 (A) An example of measurements of B_t and C_t from *Relion* particle polishing. Lower panel shows the resulting weights of frames used. (B) Optimal exposure weighting curve used in *alignparts_lmbfgs* and *Unblur*. (C) Examples of map improvement upon motion correction and frame weighting with *Relion* particle polishing (Scheres, 2014). (D) Examples of map improvement upon motion correction and exposure weighting in *alignparts_lmbfgs. Source: (A) and (C) Reproduced with permission from Scheres, S. H. (2014). Beam-induced motion correction for sub-megadalton cryo-EM particles.* Elife, 3, *e03665. (B) Figure adapted from Baker, L. A., Smith, E. A., Bueler, S. A., & Rubinstein, J. L. (2010). The resolution dependence of optimal exposures in liquid nitrogen temperature electron cryomicroscopy of catalase crystals.* Journal of Structural Biology, 169 *(3), 431437. doi: 10.1016/j.jsb.2009.11.014 using the critical exposures measured by Grant and Grigorieff (2015) for virus particles at 300 kV.*

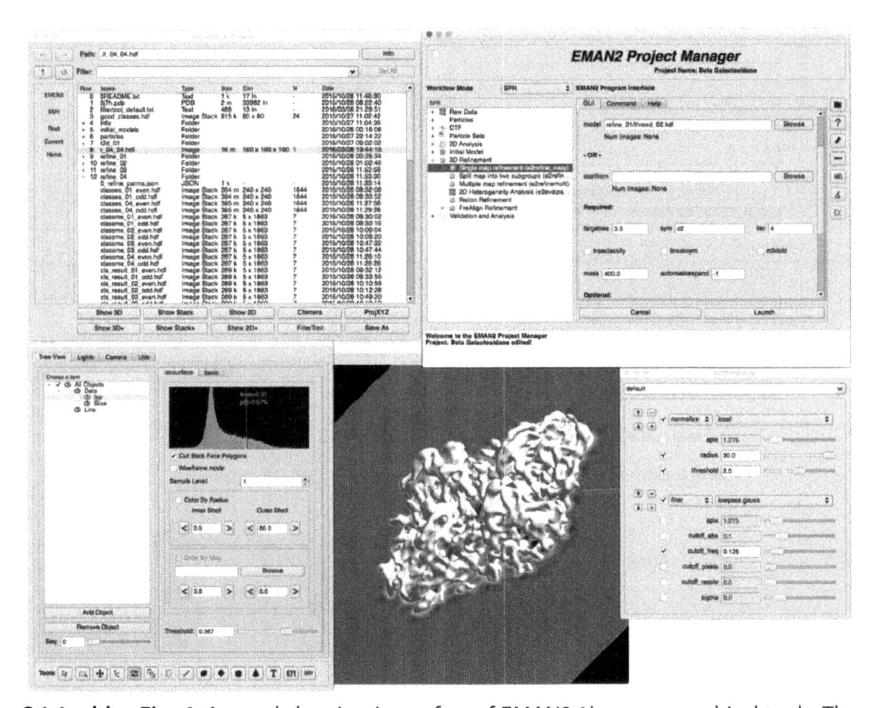

S.J. Ludtke, Fig. 4 A panel showing just a few of EMAN2.1's many graphical tools. The main project manager is shown in the *upper right*, with a portion of the dialog for a run of *e2refine_easy* shown. On the *upper left* is the file browser, note the metadata displayed in the columns of the browser, as well as the actions available at the *bottom* of the window specific to the selected file type. The *bottom* three windows show *e2filtertool* operating on a 3-D volume. The 3-D display in the *center* shows an isosurface with a slice through the *middle* and a 3-D *arrow* annotation. The control panel, which permits modifying this display, is shown on the *left*. On the *right* is the *e2filtertool* dialog, where a sequence of two image processing operations has been added: first, a "local normalization" operation, which helps compensate for low-resolution noise and ice gradients; second, a Gaussian low-pass filter set to 8 Å resolution. The parameters of these operations can be adjusted in real time with corresponding updates in the 3-D display. 2-D images and stacks can also be processed using this program. Any of over 200 image processing operations can be used from this interface, and command-line parameters are provided mimicking the final adjusted parameters.

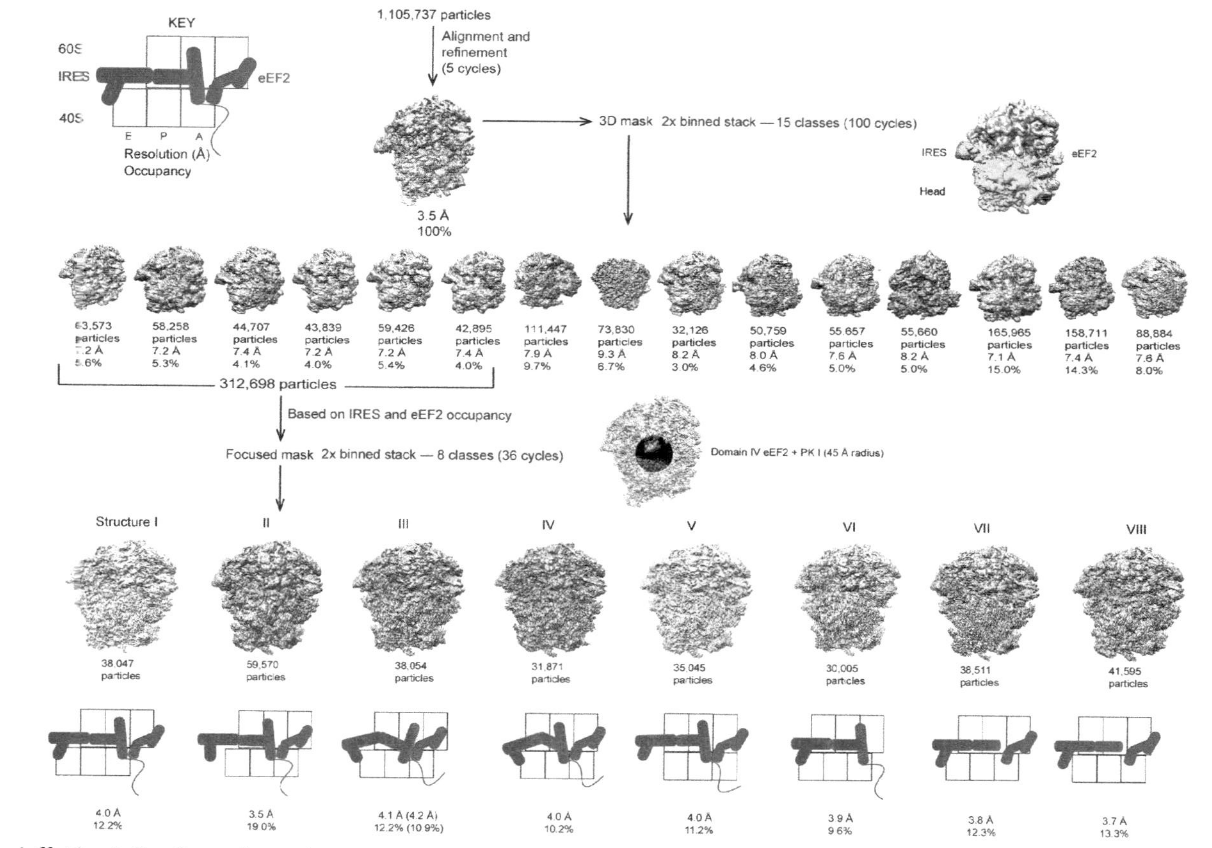

N. Grigorieff, Fig. 6 See figure legend on next page.

N. Grigorieff, Fig. 6 Example of a classification scheme using 2D and 3D masking. The dataset consisted of 80S ribosomes prepared with the Taura syndrome virus internal ribosome entry site (TSV IRES) and elongation factor 2 (eEF2) (Abeyrathne, Koh, Grant, Grigorieff, & Korostelev, 2016). The complete dataset of 1,105,737 images of 80S–IRES–eEF2 complex was initially aligned against a density map calculated from the atomic model of the nonrotated 80S ribosome bound with 2 tRNAs (PDB: 3J78, Svidritskiy, Brilot, Koh, Grigorieff, & Korostelev, 2014). This initial alignment was performed on data with a pixel size of 1.64 Å and limited to 20 Å resolution, resulting in a 3.5-Å resolution reconstruction. After 5 cycles of refinement the data were 2 × binned (by Fourier cropping using the `resample.exe` tool, new pixel size = 3.28 Å) and subjected to classification into 15 classes using a 3D mask that contained the IRES, eEF2, and head domain of the small subunit. Six of the resulting classes (312,698 particle images) contained density for the IRES and eEF2 and were further classified into eight classes. For this classification, a 2D mask was applied around the ribosomal A site to include IRES pseudoknot I and eEF2 domain IV. The figure shows this mask as a sphere which, when projected according to the orientation of a particle, results in a 2D mask correctly placed on the region of interest. In the case of the 80S–IRES–eEF2 complex, this focused classification resulted in the separation of different translocation states of the IRES, catalyzed by eEF2, as shown schematically below each reconstruction. The structures I–V containing clear density for the IRES and eEF2 are *highlighted in color*.

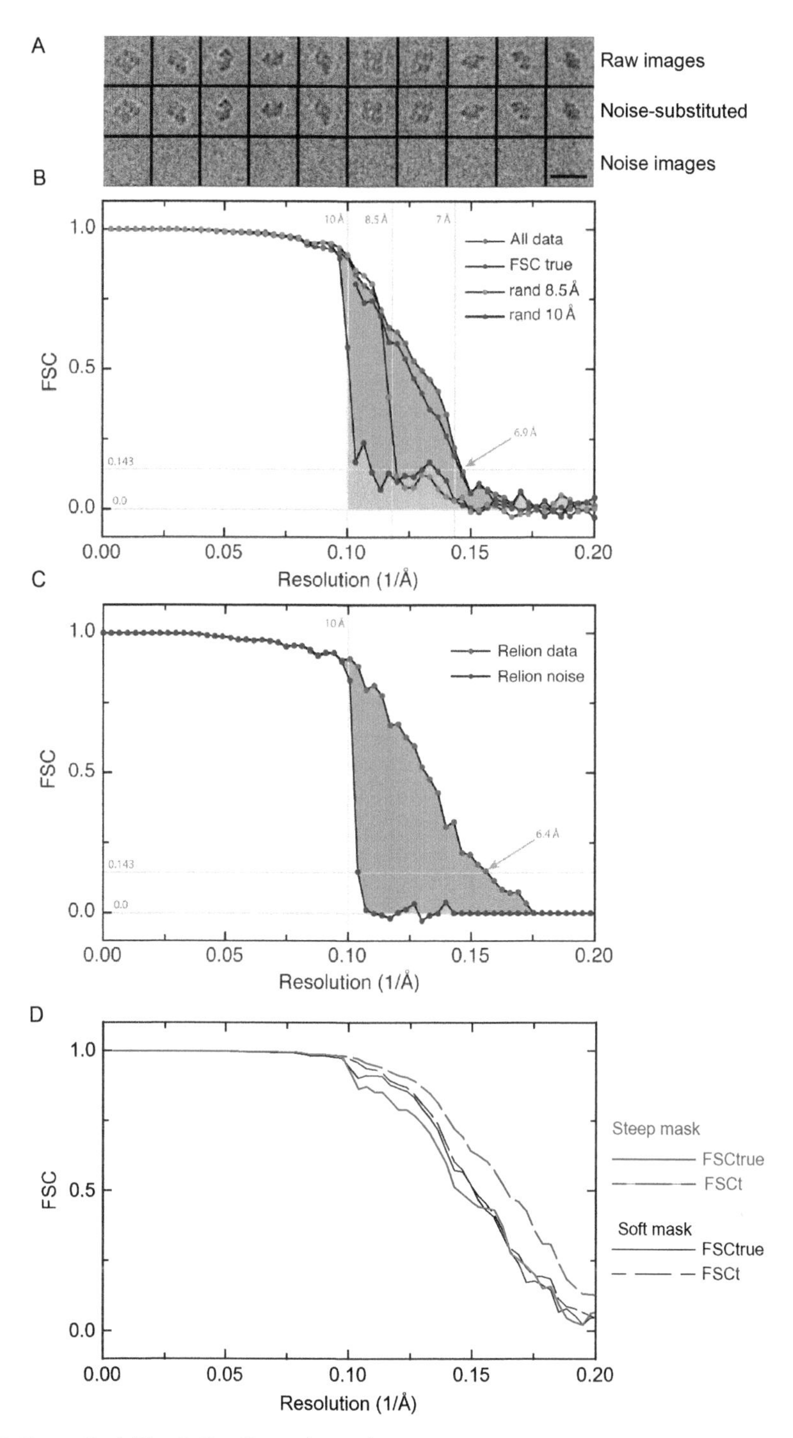

P.B. Rosenthal, Fig. 5 See figure legend on next page.

P.B. Rosenthal, Fig. 5 High-resolution noise-substitution test to validate resolution assessment by the Fourier shell correlation (FSC). (A) Single-particle images of β-galactosidase. *Top row*: raw images, *middle row*: images with phases randomized higher than 10 Å have a similar appearance, *bottom row*: noise images obtained by selecting regions of ice only from the same images where particle images are obtained. (B) Case where particle orientation refinement overweights high-resolution data leading to over-fitting. FSC is shown between half-maps for original image data (*red circles*) and image data with randomized phases beyond 10 Å (*blue circles*) or beyond 8.5 Å (*green circles*). Portion of FSC signal due to over-fitting is shown by the shaded *blue region* (FSCn), and the portion due to true structural signal is given by the shaded *pink region* or the curve FSCtrue (*purple circles*). (C) FSCt between half-maps refined using completely independent half-maps and FSC weighting (program *Relion*) for original image data (*red circles*) or image data with randomized phases beyond 10 Å (*blue circles*). FSCn (shaded *blue*) indicates that there is no over-fitting. (D) Comparison of FSCt and FSCn for cases where a sharp-edge mask or a soft-edge mask is used to calculate the FSC between half-maps. Exaggerated resolution results from a tight-edge mask but not from a soft-edge mask. *Panel (A) is from fig. 1 of, (B and C) are from fig. 3A and B of, and (D) shows replotted graphs from fig. 4A and C of Chen, S., McMullan, G., Faruqi, A. R., Murshudov, G. N., Short, J. M., Scheres, S. H., et al. (2013). High-resolution noise substitution to measure overfitting and validate resolution in 3D structure determination by single particle electron cryomicroscopy.* Ultramicroscopy, 135, *24–35.*

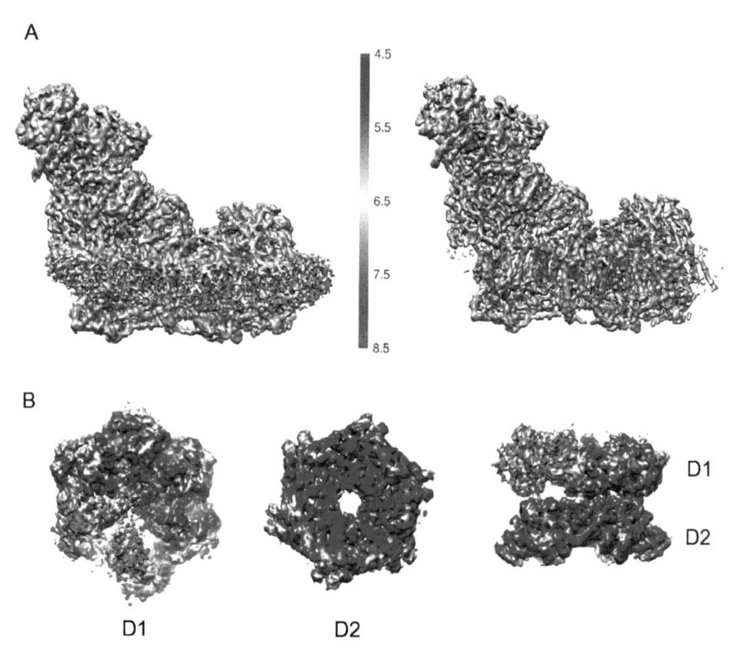

P.B. Rosenthal, Fig. 6 ResMap analysis of local resolution for two single-particle maps (A) integral membrane protein mammalian mitochondrial complex I (EMD-2676) shows that the detergent-phospholipid belt surrounding the transmembrane region on map (*left*) is at lower resolution than other parts of the map (*right*). (B) NSF (*N*-ethylmaleimide sensitive factor, EMD-6204) ATPase shows uniform resolution for subunits of the D2 ring, but variable resolution for two subunits in the D1 ring. *Color legend* indicates resolution in Å units.

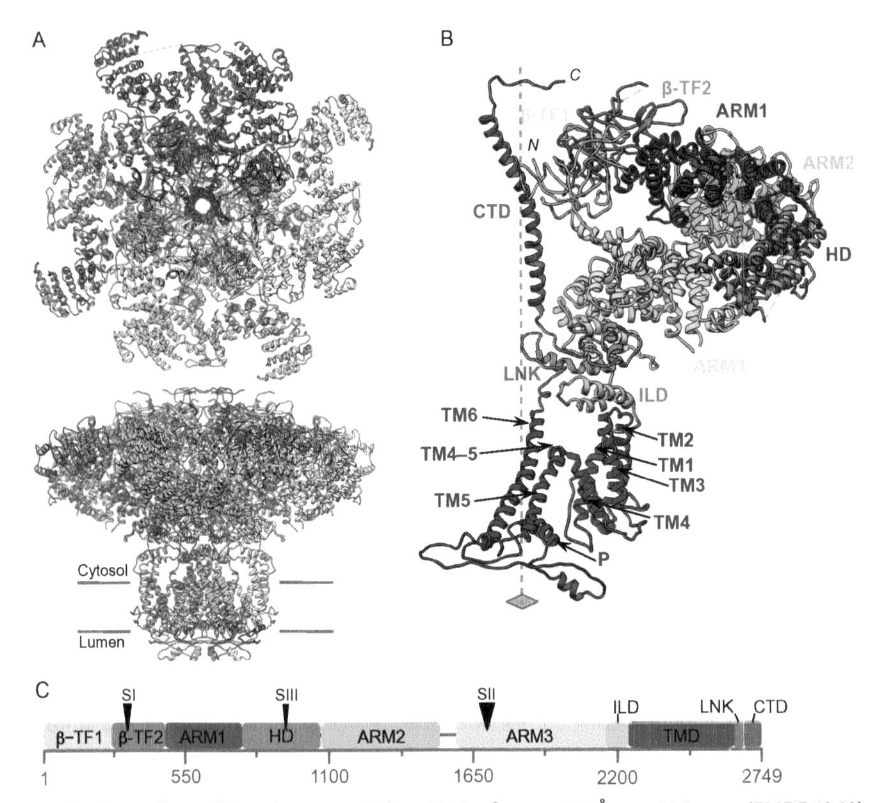

F. DiMaio and W. Chiu, Fig. 5 Modeling IP3R1 from a 4.7 Å cryoEM map (EMDB6369) (Fan et al., 2015). (A) The model (PDB 3JAV) was built using a variety of modeling protocols, shown from two views. The model is of the entire tetramer with 85% chain connectivity per chain, partly due to the presence of isoforms at the SI, SII, and SIII sites causing specimen heterogeneity and partly due to the limited map resolution. (B) The annotation of the 10 structural domains of a single IP3R1 subunit with 2700 amino acids. (C) A schematic of the corresponding domains in the linear sequence. *Reproduced from Fan, G., Baker, M. L., Wang, Z., Baker, M. R., Sinyagovskiy, P. A., Chiu, W., et al. (2015). Gating machinery of InsP3R channels revealed by electron cryomicroscopy. Nature, 527, 336–341 and provided by Matthew L. Baker.*

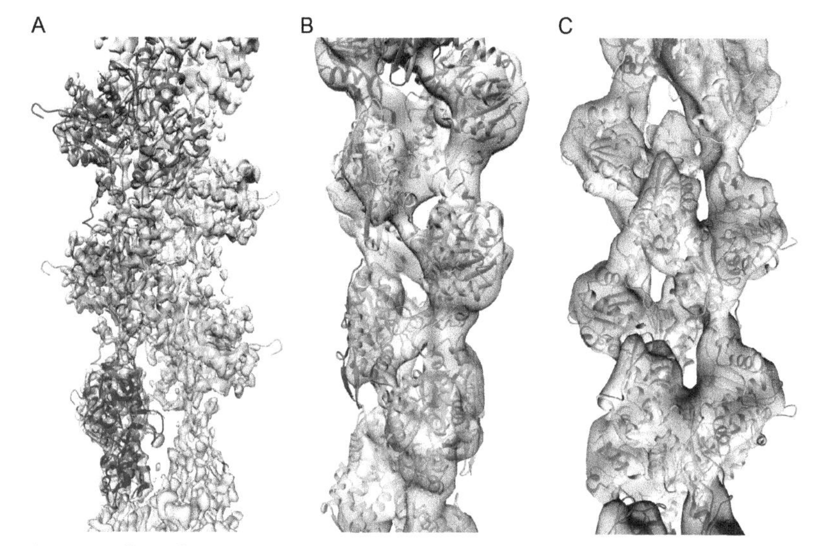

F. DiMaio and W. Chiu, Fig. 7 An example of model optimization using DireX to model distinct conformational states of F-actin from a 4.8 Å cryoEM map (Galkin, Orlova, Vos, Schroder, & Egelman, 2015). (A) A 4.8 Å resolution reconstruction of F-actin (EMDB6179) into which a model has been built and optimized (PDB 3J8I). (B and C) Two alternate, low-occupancy conformations of actin, titled T1 and T2, into which the initial model has been refined. Even though the data are of relatively low resolution, DireX attempts to maintain as many contacts as possible during refinement.

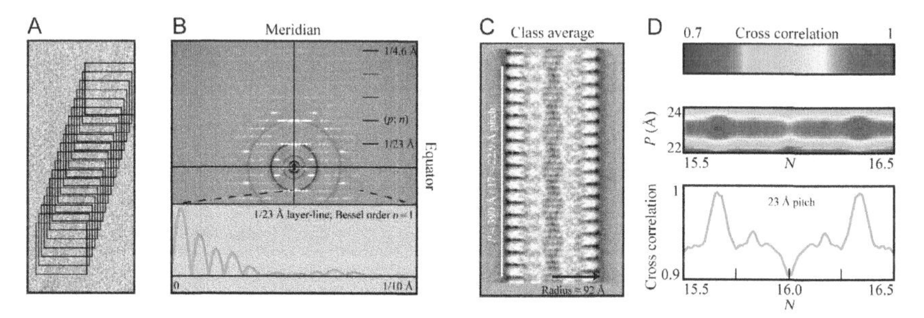

S.A. Fromm and C. Sachse, Fig. 2 2-D image analysis for segmented helical specimens. (A) Segmentation of image of TMV particle into overlapping segments. (B) Fourier analysis from sum of power spectra from overlapping segments of TMV. Layer lines can be indexed by assignment of position *p* and Bessel order *n* and thus helical symmetry can be derived. The uppermost highest resolution layer line corresponds to 1/4.6 Å. The *inset* shows the intensities of the 1/23 Å layer line with Bessel order $n = 1$. (C) Real-space class averages can reveal gross morphological features such as helical outer radius *r* and pitch *P* of the generating helix on which all the subunits lie. (D) A grid of possible helical symmetry pairs can be assessed by comparing either class averages or power spectra from the experimental images with the reprojections of the 3-D model. *Top bar* shows the intensity scale for the crosscorrelation value, *middle bar* shows the crosscorrelation as a function of *N*, and *lower graph* shows trace through central bar. *Note*: number of units per turn $N = 15.66$ and $N = 16.34$ represent ambiguous solutions, while $P = 23.0$ Å, $N = 16.34$ is the correct symmetry for TMV.

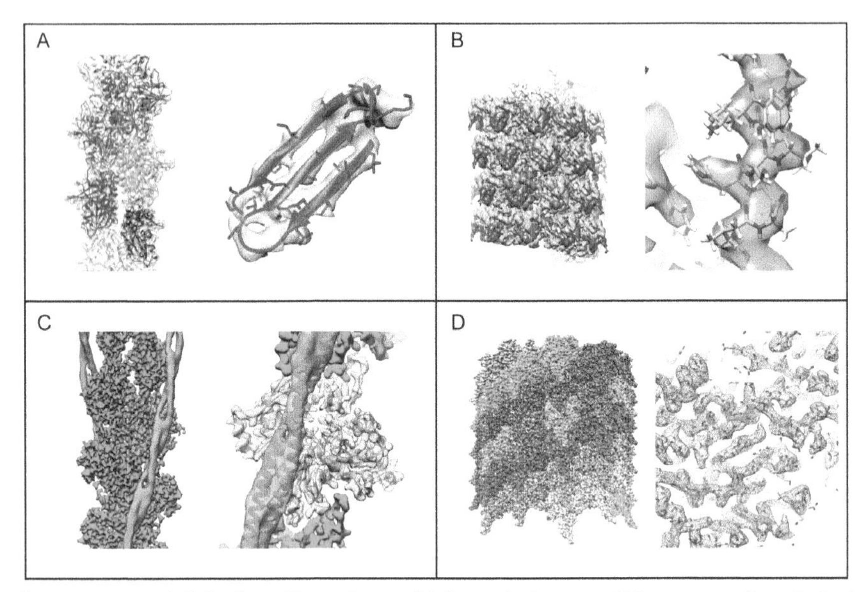

S.A. Fromm and C. Sachse, Fig. 5 Recent high-resolution cryo-EM structures from helical specimens using segmented helical reconstruction. (A) EMD-2850: 4.3 Å structure of the actin-like ParM protein bound to AMPPNP (Bharat et al., 2015). (B) EMD-3236: 3.9 Å structure of the Pepino Mosaic Virus (Agirrezabala et al., 2015). (C) EMD-6124: 3.7 Å structure of the F-actin-tropomyosin complex (von der Ecken et al., 2015). (D) EMD-2699: 3.5 Å structure of the VipA/VipB contractile sheath of the type VI secretion system (Kudryashev et al., 2015).

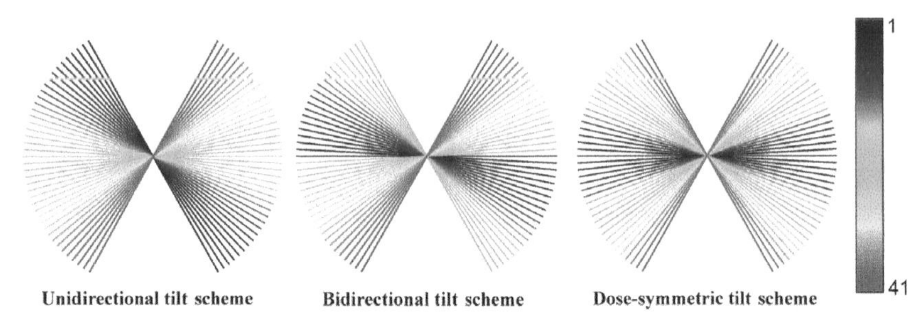

W. Wan and J.A.G. Briggs, Fig. 4 Schematic showing the order in which tilts are collected in unidirectional, bidirectional, and "dual-walkup" tilt schemes. Tilts are shown from −60 to +60 degrees in 3 degree increments for a total of 41 tilts. *Gray values/colors* correspond to the collection order of each tilt according to the *color* map shown on the *right*. When tilts are collected with constant exposure times, tilt order is directly related to accumulated electron dose on each image. The unidirectional tilt scheme shows a linear sweep from one angular extreme to the other. The bidirectional tilt scheme shows the discontinuity when the tilt-increment direction is changed. The dual-walkup tilt scheme shows near-symmetric accumulated electron dose.

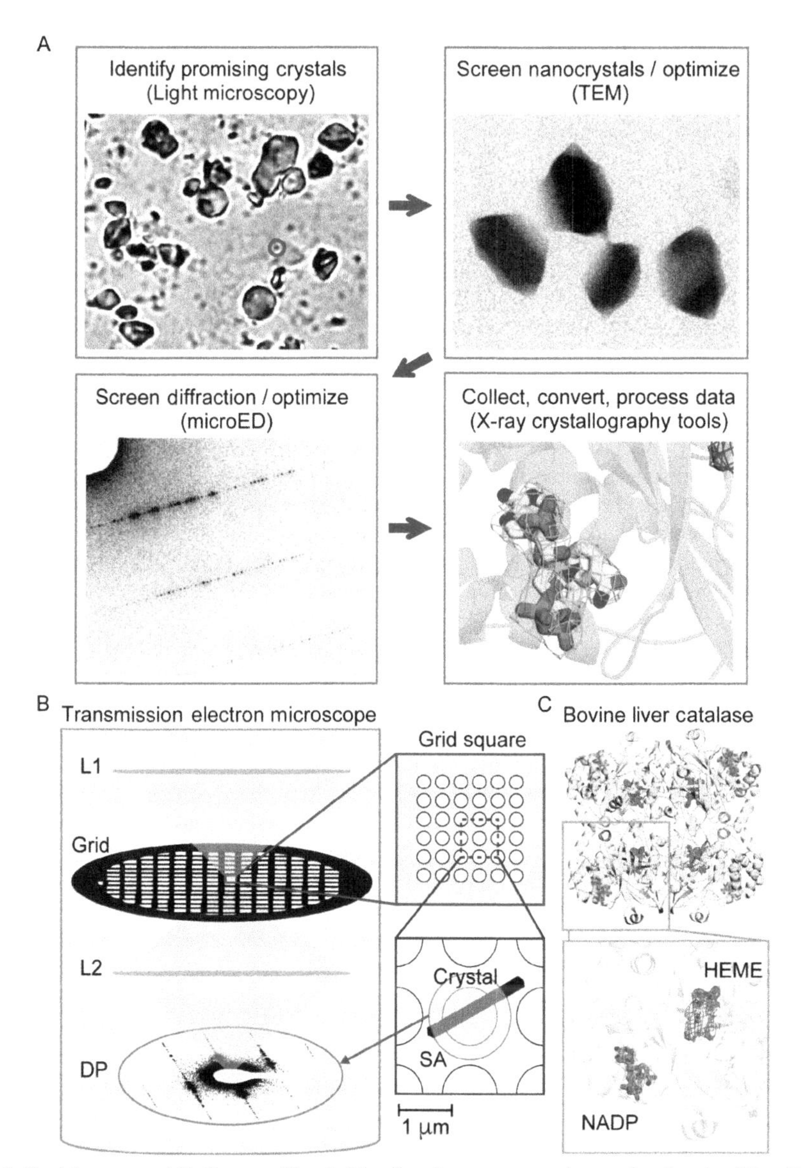

J.A. Rodriguez and T. Gonen, Fig. 3 Pipeline for structure determination by MicroED. (A) A simplified four step process is shown to summarize the process for macromolecular structure determination by MicroED. First, promising protein mixtures are screened by light microscopy for the presence of crystals then by electron microscopy for adequately sized crystals (*red circle* shows a 5 μm diameter). If crystals are present, they can immediately be screened for diffraction on the electron microscope. Diffraction can be optimized until high-resolution diffraction tilt series can be obtained for a given protein. These tilt series images can be converted to SMV format and processed using standard crystallography software packages. Structure determination then follows the

(*Continued*)

J.A. Rodriguez and T. Gonen, Fig. 3—Cont'd procedures established for X-ray crystallography to phase by molecular replacement. (B) Process by which diffraction is obtained from a 3D protein crystal using a transmission electron microscope. An electron beam is condensed (*L1*) and illuminates a protein crystal that sits vitrified on a holey carbon film (*inset, top*) within a meshed electron microscopy grid. A representation of the magnified *grid square* is shown to demonstrate the size of the illuminated area relative to the crystal, and the selected area (*inset, bottom*) from which a diffraction pattern (*DP*) is obtained at the back focal plane of a second lens or lens system (*L2*). Projection lenses (not shown, see Fig. 2) image this pattern onto an electronic detector, which digitally records this pattern. The grid containing the crystal can be rotated in a discrete or continuous fashion to produce patterns at various orientations from a single or multiple crystals. These patterns can be processed and phased to determine the structure of the molecule that formed the crystal. (C) In this example, the protein catalase was solved by MicroED to a resolution of 3.2 Å using the method of continuous rotation. A ribbon diagram of the protein is shown along with density for two ligands, a HEME group (*red mesh, orange sticks*), and an NADP molecule (*blue mesh, purple sticks*). These molecules were omitted from an initial model search during the molecular replacement protocol; their density was recovered during the refinement process (Nannenga, Shi, Hattne, et al., 2014). Renderings were generated using PDB ID 3J7B.

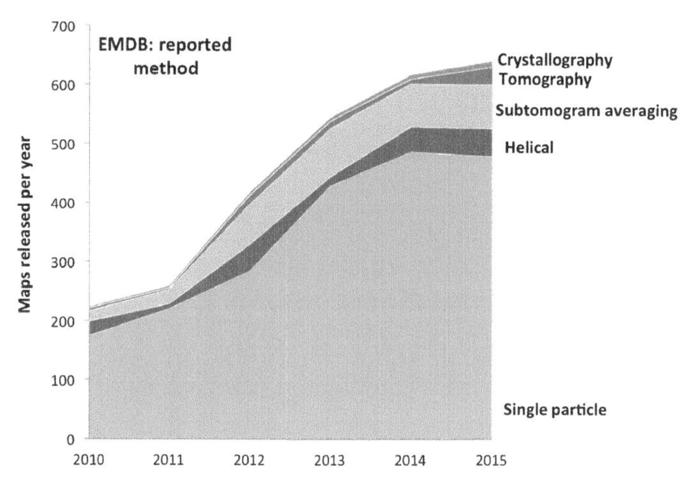

A. Patwardhan and C.L. Lawson, Fig. 2 Trend in reported 3DEM method for EMDB entries released between 2010 and 2015, showing annual releases.

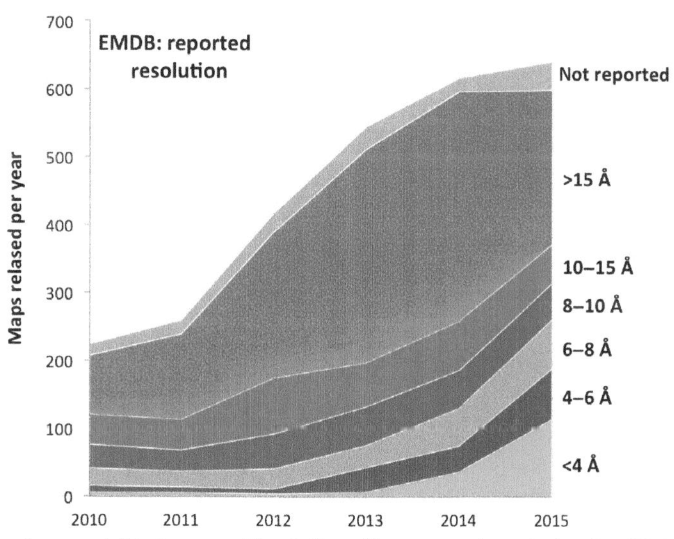

A. Patwardhan and C.L. Lawson, Fig. 3 Trend in reported resolution for EMDB entries released between 2010 and 2015, showing annual releases.

Edwards Brothers Malloy
Ann Arbor MI. USA
August 29, 2016